Mathematical and Computational Modeling of Phenomena Arising in Population Biology and Nonlinear Oscillations

CONTEMPORARY MATHEMATICS

793

Mathematical and Computational Modeling of Phenomena Arising in Population Biology and Nonlinear Oscillations

Abba B. Gumel
Editor

AMERICAN MATHEMATICAL SOCIETY

2020 *Mathematics Subject Classification.* Primary 34C15, 34D05, 34D20, 34D23, 37M20, 39A28, 39A30, 92B05.

For additional information and updates on this book, visit
www.ams.org/bookpages/conm-793

Library of Congress Cataloging-in-Publication Data

Names: Gumel, Abba B., 1966- editor. | Mickens, Ronald E., 1943- honoree.
Title: Mathematical and computational modeling of phenomena arising in population biology and nonlinear oscillations / Abba B. Gumel, editor.
Description: Providence, Rhode Island : American Mathematical Society, [2024] | Series: Contemporary mathematics, 0271-4132 ; volume 793 | Includes bibliographical references.
Identifiers: LCCN 2023045277 | ISBN 9781470471040 (paperback) | ISBN 9781470476069 (ebook)
Subjects: LCSH: Differential equations–Numerical solutions. | Finite differences. | Population biology–Mathematical models. | Nonlinear oscillations–Mathematical models. | Mickens, Ronald E., 1943- | AMS: Ordinary differential equations – Qualitative theory – Nonlinear oscillations, coupled oscillators. | Ordinary differential equations – Stability theory – Asymptotic properties. | Ordinary differential equations – Stability theory – Stability. | Ordinary differential equations – Stability theory – Global stability. | Dynamical systems and ergodic theory – Approximation methods and numerical treatment of dynamical systems – Computational methods for bifurcation problems. | Difference and functional equations – Difference equations – Bifurcation theory. | Difference and functional equations – Difference equations – Stability theory. | Biology and other natural sciences – Mathematical biology in general – General biology and biomathematics.
Classification: LCC QA372 .M386 2024 | DDC 515/.35–dc23/eng/20231130
LC record available at https://lccn.loc.gov/2023045277

♾ The paper used in this book is acid-free and falls within the guidelines
established to ensure permanence and durability.
Visit the AMS home page at `https://www.ams.org/`

10 9 8 7 6 5 4 3 2 1 29 28 27 26 25 24

To Professor Ronald E. Mickens on the occasion of his 80th birthday

Photo credit: Bryan A. Briones

Contents

Preface

This book contains a series of chapters, contributed in honor of the 80th birthday of Professor Ronald E. Mickens (Distinguished Callaway Professor of Physics Emeritus, Clark Atlanta University, Georgia, USA), that address some of the current challenges and advances on the study of the mathematical and computational modeling of real-life phenomena arising in population biology and nonlinear oscillation. Professor Mickens is renowned for his stellar contributions in nonlinear oscillations and computational mathematics, particularly the design and analysis of robust and dynamically-consistent *nonstandard finite difference* methods for discretizing continuous-time dynamical systems arising from the mathematical modeling of phenomena in the natural and engineering sciences. The papers contained in this volume are contributed by collaborators of, and researchers whose work was inspired by the works of, Professor Mickens. Some of the contributors spoke at the AMS Special Session on Advances in the Applications of Nonstandard Finite Difference Methods, January 6-9, 2021, Atlanta, Georgia.

The volume contains papers in three main areas, notably the design, analysis and applications of nonstandard finite difference discretization of continuous-time dynamical systems in the natural and engineering sciences, mathematical population biology (with emphasis on the design, analysis, parameterizations of mathematical models for gaining insight into the transmission dynamics and control of emerging and re-emerging infectious diseases, such as COVID-19 and malaria), and nonlinear oscillations. The volume is a collection of 16 papers, containing some new results and techniques. In particular, the papers highlight numerous new advances and challenges associated with the design of dynamically-consistent discrete-time models for approximating the solutions of differential equations (both ordinary and partial), based on Mickens's nonstandard finite-difference discretization framework. Some new higher-order techniques for designing nonstandard finite-difference methods for phenomena in the biological and chemical processes, as well as a generalized autonomous dynamical system, are proposed. Furthermore, the volume contains papers that focus on the design and analysis of robust models for gaining insight and understanding on the spread and mitigation of infectious diseases. In particular, the book contains chapters on modeling the spread and control of the 2019 novel Coronavirus pandemic (COVID-19, which represents the greatest public health problem human have faced since the 1918/1919 influenza pandemic), on the development of a novel genetic-epidemiology modeling framework for malaria disease (the most important parasitic disease of mankind) and a discrete-time epidemic model with Ricker growth. The book also contains papers on mathematical ecology, including works on the effects of temperature and prescribed fire on the dynamics of tick-borne diseases, discretization of models for microbial population with environmental stress, Monod kinetics with half saturation in a chemostat setting etc. Each of the papers is peer-reviewed by at least two anonymous reviewers.

The editor acknowledges the support of the American Mathematical Society. In particular, special thanks to Ms. Christine M. Thivierge (Editorial Assistant, AMS) and the Production and Technical Staff for the great work throughout this process. I am grateful for their careful reading, diligence and efficiency. I am immensely thankful to the authors who contributed their excellent chapters and to the many anonymous reviewers for their constructive and diligent reviews.

Abba B. Gumel

May 2023

Career of Professor Ronald E. Mickens

Professor Mickens graduated summa cum laude from Fisk University in 1964 with a B.A. degree in Physics and Mathematics. His graduation honors included election to Phi Beta Kappa. Doctoral studies in theoretical Physics were completed in 1968 at Vanderbilt University. For the next two years, as a National Science Foundation Postdoctoral Fellow at the Center for Theoretical Physics, the Massachusetts Institute of Technology, he did research on calculating and studying the scattering amplitudes for strongly interacting elementary particles at high energies. In 1970, he returned to Fisk University to teach and do research in Mathematical Physics.

Professor Mickens' major research activities from 1966-1977 were centered on the application of scattering theory to problems in both elementary particle physics and chemical reaction rates. He showed that the chemical equilibrium coefficient could be directly constructed from the general principles of scattering theory. He also applied these techniques to obtain new results for the low temperature behavior of recombination reaction rate functions. In addition, Mickens showed how to generalize the fundamental Pomeranchuk Theorem, which relates various physical properties of particle and anti-particle collisions at very high energies, to complex processes having two-and many-particle final states.

During 1977-1981, Professor Mickens began a study of second-order, ordinary differential equations which model nonlinear oscillatory dynamical systems. One of the results from this research was the publication of the book, Nonlinear Oscillations (Cambridge University Press, New York, 1981). Subsequent work in this area led to more than 50 publications related to the analysis and construction of uniformly valid perturbation and global methods to calculate analytical approximations to the solutions of such equations. A major tool used for these investigations was the rewriting of the original second-order ODEs to system form and then studying the behavior of the trajectories in the resulting two-dimensional phase-space.

In 1981, Professor Mickens was a Visiting Research Fellow at the Joint Institute for Laboratory Astrophysics located in Boulder, CO. There he began a study of difference equations and the instabilities that can arise in the numerical integrations of differential equations by the use of finite-difference methods. This work was the genesis of a research program, which allowed the construction of what is now called "nonstandard finite-difference schemes." Discrete models based on these techniques provide superior numerical solutions, as compared with many standard methods, and allow the direct incorporation of important physical properties of the dynamical systems such as positivity and boundedness. The general philosophy of these techniques, as well as their application to problems in the natural and engineering sciences are given in two books: R.E. Mickens: Finite Difference Models of Differential Equations (World Scientific, Singapore, 1994) and R.E. Mickens (editor): Applications of Nonstandard Finite Difference Schemes (World Scientific, Singapore, 2000).

One of the highlights of Professor Mickens' career was the publication of a book that provided a broad introduction to finite-difference equations. This book, R.E. Mickens: Difference Equations (Van Nostrand Reinhold, New York, 1987) was one of the first modern textbooks on the subject. The demand for this textbook, coupled with a general public interest in difference equations, iteration procedures and discrete modeling, led to a second greatly expanded edition, R.E. Mickens: Difference Equations, Theory and Applications (Van Nostrand Reinhold, New York, 1990). A significant feature of the latter book is its final chapter. In it, various difference equations arising from the modeling of phenomena in physics, chemistry, the biosciences, numerical integration and asymptotics are presented and discussed, along with references to the appropriate research literature. A third edition was published in 2015.

Throughout his career, Professor Mickens has received numerous fellowships, some of which are: Woodrow Wilson Fellowship (1964-1965); Danforth Foundation Fellowship (1965-1968); National Science Foundation Postdoctoral Fellowship (1968-1970); and joint Institute for Laboratory Astrophysics Visiting Research Fellowship (1981-1982). His awards include: selection as the Distinguished Fuller E. Callaway Professor at Atlanta University (now Clark Atlanta University) in 1985 to the present; the National Association of Mathematicians Distinguished Service Award (1996); election to Fellowship in the American Physical Society (1999); and service for a two-year term (2000-2002) as a Distinguished Lecturer of Sigma Xi, the Scientific Research Society of America. In 2018, Mickens was the recipient of the Blackwell-Tapia Prize in Mathematics. Since 1970, his research efforts have been continuously funded by grants from the Army Research Office, Department of Energy, NASA, National Institutes of Health, and National Science Foundation.

Professor Mickens has made more than 600 invited presentations at national and international scientific/mathematical conferences and at various colleges and universities. He has served as a member of the editorial board of several research journals; these include the Journal of Difference Equations and Applications, Journal of Sound and Vibration, and Journal on Applied Sciences and Computation. He has published more than 350 refereed research articles, written 12 books and edited seven volumes. Of interest to those with a philosophical leaning is the following volume that deals with the relationship between science and mathematics, R.E. Mickens (editor): Mathematics and Science (World Scientific, Singapore, 1990).

In the early 2000's, Professor Mickens extended the concept of "dynamic consistency" and used it to place limitations on the structure of mathematical models for physical systems and their discretizations. He demonstrated that models may not satisfy all the relevant fundamental laws and other requirements of physics, yet still be useful for predicting the properties of particular aspects of a system. He also has investigated periodic solutions associated with simple, closed, convex curves in the plane, with the origin an interior point of the curve, and showed that three periodic functions could be generated. This work is summarized in the book, R.E. Mickens: Generalized Trigonometric and Hyperbolic Functions (CRC Press, 2019).

Finally, Professor Mickens' current research involves (a) the construction of approximate and exact solutions to both linear and nonlinear difference and differential equations; (b) the study of the qualitative properties of solutions to strongly nonlinear, oscillatory, second-order differential equations; (c) the continuation of work on nonstandard finite difference schemes for the numerical integration of differential equations; and, (d) the gathering of relevant materials and the publication of scholarly articles on the history and sociology of science as it relates to African Americans. All of these exciting efforts are supported by a growing list of national and international collaborators who provide not only valued scientific input and stimulating questions/comments/critiques, but who are also both good colleagues and friends.

Selective List of Major Oral Interviews

(1) American Institute of Physics (2020 August 5 -7, 10, 11, and 13). Web Site:
`https://www.aip.org/history-programs/niels-bohr-library/oral-histories/47213`

(2) Atomic Heritage Foundation (July 30, 2018). Web Site:
`https://ahf.nuclearmuseum.org/voices/oral-histories/ronald-e-mickens-interview/`

(3) The HistoryMakers (December 11, 2006). Web Site:
`https://www.thehistorymakers.org/biography/ronald-mickens-41`

Toni L. Fanin,
Office of the President,
Clark Atlanta University,
Atlanta, GA 30314

Contemporary Mathematics
Volume **793**, 2024
https://doi.org/10.1090/conm/793/15904

A second-order nonstandard finite difference scheme and application to a model of biological and chemical processes

Roumen Anguelov and Jean M.-S. Lubuma

This paper is dedicated to Professor Ronald E. Mickens, Father of the Nonstandard Finite Difference Method, on the occasion of his 80th *birthday.*

ABSTRACT. We consider a dynamical system defined by a system of autonomous differential equations on a set $\Omega \subset \mathbb{R}^n$. By using Mickens' rule on the nonlocal approximation of nonlinear terms, we construct an implicit Nonstandard Finite Difference (NSFD) scheme which is an explicit time reversible scheme. Apart from being elementary stable, we show that the NSFD scheme is of second-order and domain-preserving, thereby solving a pending problem on the construction of higher-order nonstandard schemes without spurious solutions. It is shown that the new scheme applies directly for mass action-based models of biological and chemical processes.

1. Introduction

The general setting of this work is a dynamical system, on a compact set $\Omega \subset \mathbb{R}^n$, defined by the following system of autonomous ordinary differential equations:

$$\frac{dx}{dt} \equiv \dot{x} = f(x). \tag{1.1}$$

The solutions, $x(t)$, of the continuous system are approximated at the discrete times, $t_k = hk$, $h > 0$, $k = 0, 1, 2, \cdots$, by a sequence (x_k) obtained recursively through a numerical method of the following form:

$$x_{k+1} = F(h, x_k) \equiv F(h)(x_k). \tag{1.2}$$

For both the continuous and the discrete systems, it is assumed that the functions $f \equiv f(x) : \mathbb{R}^n \to \mathbb{R}^n$ and $F \equiv F(h, x) \equiv F(h)(x) : \mathbb{R} \times \mathbb{R}^n \to \mathbb{R}^n$ are as smooth as

---

2020 *Mathematics Subject Classification.* Primary 37C79, 65L99, 92D30; Secondary 34C45, 92E99.

Key words and phrases. Nonstandard finite difference method, elementary stability, order of convergence, time-reversible schemes, mass action principle.

R. Anguelov was partially supported by the DSI/NRF SARChI Chair on Mathematical Models and Methods in Bioengineering and Biosciences at the University of Pretoria. He acknowledges the hospitality of the University of the Witwatersrand where this work was finalized during his research visit. J. Lubuma is grateful to the NRF for financial support under the Competitive Programme for Rated Researchers (CPRR) (Grant No. 138013). He also acknowledges the support of the University of the Witwatersrand under the Science Faculty Sart-up Funds.

needed. Naturally, the consistency requirements,

$$F(0)(x) = x \text{ and } \frac{dF(0)}{dh}(x) = f(x), \tag{1.3}$$

are assumed to hold, and thus we suppose that F is at least continuously differentiable on h. Our aim is that the numerical method (1.2) be reliable in the three important directions below. The numerical method must be second-order convergent, elementary stable, and preserve the property of having the set Ω forward invariant with respect to the continuous system.

Classical higher-order numerical methods such as the linear multistep methods and the explicit Runge-Kutta methods for ordinary differential systems are important as far as the computational analysis, including the calculation of the rate of convergence, is concerned. However, when the order of convergence is higher or equal to two, these classical methods are in general not elementary stable in the sense of [**A4**] recalled in Theorem 2.1 below, as they exhibit spurious fixed-points, and they generally fail to preserve the qualitative properties of the system being approximated. A comment is made in Remark 2.4 below about implicit Runge-Kutta methods. The nonstandard finite difference (NSFD) approach initiated by Mickens more than three decades ago aims at addressing these shortcomings. Since the publication of his pioneering book [**M2**], the NSFD method has been extensively developed for differential equation models originating from problems in life science, engineering, chemistry, etc. Paraphrasing Mickens himself from his latest book [**M3**], "the NSFD method is becoming an established field whose the concept of dynamic consistency and associated techniques is increasingly and quickly accepted for the construction of numerical schemes for differential equations."

Despite the progress made, the construction of higher-order NSFD schemes that are dynamically consistent with the underlying features of the continuous differential equation models, particularly for those with transient dynamics, is a pending problem. Several authors have attempted to address this problem. These include the nonstandard versions of the classical θ-method, with $\theta = 1/2$, developed in [**A1, A2, A3, L**] where the positivity of the discrete solution is achieved by also using Mickens' rule on a nontrivial denominator function for the discrete derivative. In the same vein, we mention the recent work [**K**], valid for a scalar differential equation, where the second-order accuracy and elementary stability are achieved by a modified nonstandard θ method, $0 \leq \theta \leq 1$, in which the nontrivial denominator function varies at each iteration.

The purpose of this book chapter is to develop a theoretical and systematic framework to design numerical methods that are characterized as follows [**H**]:

Definition 1.1. A numerical scheme (1.2) is called symmetric or (time) reversible if

$$x_k = F(-h, x_{k+1}). \tag{1.4}$$

More generally, the scheme is said to be reversible for all x whenever $y = F(h, x)$ implies that $x = F(-h, y)$.

We construct a NSFD scheme that is time reversible and show that, in essence, this property makes our scheme of second-order accuracy and elementary stable (Section 2). The same property is used to establish a discrete analogue of the tangent condition for forward invariance of a convex set, which implies that the domain under consideration is forward invariant with respect to our NSFD scheme (Section 2). The case of mass action models of biological and chemical processes is considered in details (Section 3), with a vector-borne disease model [**M1**], as an illustrative example (Section 4). Concluding remarks, with possible extensions of this work, are given in Section 5.

2. Nonstandard finite difference method of order 2

It is well known that the discrete models of differential equations have a larger parameter space than the corresponding differential equations. Adding to this Mickens' Rule 1, which says that the orders of the discrete derivatives must be exactly equal to the orders of the corresponding derivatives of the differential equations, it is convenient for nonlocal approximations (i.e. Mickens' Rule 3) to view the right-hand side of the system (1.1) as a restriction, on the diagonal $z = y$, of a certain function of two variables $\varphi \equiv \varphi(y, z)$ such that $f(x) = \varphi(x, x)$ and hence the system takes the form

$$\dot{x} = \varphi(x, x), \quad x \in \Omega \subseteq \mathbb{R}^n. \tag{2.1}$$

We consider the numerical method,

$$\frac{x_{k+1} - x_k}{h} = \frac{1}{2}(\varphi(x_{k+1}, x_k) + \varphi(x_k, x_{k+1})), \tag{2.2}$$

which is a NSFD scheme [**A4**], on which only the nonlocal approximation of the right-hand side of (2.1) is used, the rule on the nontrivial denominator function of the derivatives being excluded. The NSFD scheme (2.2) is implicit. The existence and uniqueness of solution of the respective system of algebraic equation is an issue which requires attention on its own. To make this NSFD scheme a generalized dynamical system on Ω [**S**], we will assume here that

(2.3) (i) given $x_k \in \Omega$, equation (2.2) has a unique solution x_{k+1} in Ω;
(ii) different values of x_k result in different solutions for x_{k+1}.

Finding general conditions for (2.3) to hold is challenging, as it is the case for general equations of the form $G(x, y) = 0$ solved by using for instance the implicit function theorem, sophisticated fixed-point iterations, *etc.* However, in particular settings, as the ones considered in the sequel such conditions are relatively easy to formulate.

THEOREM 2.1. *Assume that* (2.3) *holds. Then:*

a) *The NSFD scheme* (2.2) *is time reversible;*
b) *The NSFD scheme* (2.2) *is elementary stable in the sense that, irrespective of the value of the step size h, (i) its fixed-points are precisely the equilibrium points of the system* (2.1), *and (ii) it replicates the linear stability of equilibrium points of* (2.1) *that are all assumed to be hyperbolic;*
c) *The NSFD scheme* (2.2) *is a second-order scheme for the equation* (2.1).

PROOF. The assumption (2.3) implies that the NSFD scheme (2.2) can be written in the explicit form (1.2) where the function F satisfies the equation

$$F(h, x) - x = \frac{h}{2}\left[\varphi(F(h, x), x) + \varphi(x, F(h, x))\right]. \tag{2.4}$$

a) If in the equation (2.2) we interchange x_k and x_{k+1} and replace h by $-h$, the equation remains the same. Therefore (1.4) holds, which means that the NSFD scheme is time-reversible.

b) It is clear that $\overline{x}$ is a fixed point of (2.2) if, and only if, $\varphi(\overline{x},\overline{x}) = 0$, *i.e.* $\overline{x}$ is an equilibrium point of (2.1). This proves the first part of item (b).

Regarding the second part (ii) of item (b), let us use the explicit form (1.2) of (2.2). From (2.4), the Jacobian matrix (in the x variable) of F satisfies

$$\begin{aligned}\frac{\partial F(h,x)}{\partial x} - I = \frac{h}{2}\Bigg(&\frac{\partial \varphi}{\partial y}(F(h,x),x)\frac{\partial F(h,x)}{\partial x} + \frac{\partial \varphi}{\partial z}(F(h,x),x)\\ &+\frac{\partial \varphi}{\partial y}(x,F(h,x)) + \frac{\partial \varphi}{\partial z}(x,F(h,x))\frac{\partial F(h,x)}{\partial x}\Bigg)\end{aligned} \tag{2.5}$$

To obtain the right hand side of (2.5), we used the chain rule, noting that for the function of two arguments $\varphi = \varphi(y,z)$, we set in (2.4) $y(x) = F(h,x)$ and $z(x) = x$ for $\varphi(F(h,x),x)$ and, vice versa, for the term $\varphi(x,F(h,x))$. Let $\overline{x}$ be an arbitrary fixed point of (2.2). At $x = \overline{x}$, using $F(h,\overline{x}) = \overline{x}$, equation (2.5) simplifies to

$$\left(I - \frac{h}{2}\left(\frac{\partial \varphi}{\partial y}(\overline{x},\overline{x}) + \frac{\partial \varphi}{\partial z}(\overline{x},\overline{x})\right)\right)\frac{\partial F(h,\overline{x})}{\partial x} = I + \frac{h}{2}\left(\frac{\partial \varphi}{\partial y}(\overline{x},\overline{x}) + \frac{\partial \varphi}{\partial z}(\overline{x},\overline{x})\right),$$

or, equivalently,

$$\left(I - \frac{h}{2}\frac{\partial f(\overline{x})}{\partial x}\right)\frac{\partial F(h,\overline{x})}{\partial x} = I + \frac{h}{2}\frac{\partial f(\overline{x})}{\partial x}. \tag{2.6}$$

Let $\lambda \equiv \Re\lambda + \imath\Im\lambda$ be an eigenvalue of $\dfrac{\partial f(\overline{x})}{\partial x}$ with associated left eigenvector v. Then multiplying both sides of (2.6) on the left by v we obtain

$$\left(1 - \frac{h}{2}\lambda\right) v\,\frac{\partial F(h,\overline{x})}{\partial x} = \left(1 + \frac{h}{2}\lambda\right) v$$

It follows from (2.3) that the matrix $\dfrac{\partial F(h,\overline{x})}{\partial x}$ is not singular. Then, due to the obvious impossibility for $1 - \frac{h}{2}\lambda$ and $1 + \frac{h}{2}\lambda$ to be both zero, neither of them is. Hence

$$v\,\frac{\partial F(h,\overline{x})}{\partial x} = \frac{1 + \frac{h}{2}\lambda}{1 - \frac{h}{2}\lambda}\,v.$$

Therefore, to every eigenvalue λ of $\dfrac{\partial f(\overline{x})}{\partial x}$ with associated left eigenvector v, there corresponds an eigenvalue $\mu = \dfrac{1 + \frac{h}{2}\lambda}{1 - \frac{h}{2}\lambda}$ of the matrix $\dfrac{\partial F(h,\overline{x})}{\partial x}$ with the same left eigenvector v. After some technical manipulation we obtain

$$\begin{aligned}|\mu|^2 &= \frac{(1 + \frac{h}{2}\Re(\lambda))^2 + \frac{h^2}{4}\Im^2(\lambda)}{(1 - \frac{h}{2}\Re(\lambda))^2 + \frac{h^2}{4}\Im^2(\lambda)}\\ &= \frac{1 + \frac{h^2}{4}|\lambda|^2 + h\Re(\lambda)}{1 + \frac{h^2}{4}|\lambda|^2 - h\Re(\lambda)}\end{aligned} \tag{2.7}$$

If follows from (2.7) that

$$\Re(\lambda) < 0 \iff |\mu| < 1,$$

which implies that a hyperbolic equilibrium point $\overline{x}$ of (2.1) is asymptotically stable if, and only if, it is asymptotically stable as a fixed-point of (2.2).

d) Let $x \in \Omega$ and let u denote the solution of (1.1) with $u(0) = x$. Denote

$$\xi = \frac{1}{2}(F(h)(x) + x).$$

By Taylor expansion about $h = 0$ and the consistency conditions (1.3), we have (2.8)

$$\xi = \frac{1}{2}\left(x + F(0,x) + h\frac{dF}{dh}(0,x) + O(h^2)\right) = x + \frac{h}{2}f(x) + O(h^2) = u\left(\frac{h}{2}\right) + O(h^2).$$

Then, for the local truncation error,

$$E(h) = u(h) - F(h,x) = u(h) - x - \frac{h}{2}\varphi(F(h)(x), x) - \frac{h}{2}\varphi(x, F(h)(x)),$$

of the NSFD scheme (1.2) & (2.4), Taylor expansions of u about $\frac{h}{2}$ and of φ about $(y,z) = (\xi, \xi)$ yield

$$\begin{aligned} E(h) &= h\dot{u}\left(\frac{h}{2}\right) - \frac{h}{2}\left(\varphi(\xi,\xi) + \frac{\partial\varphi(\xi,\xi)}{\partial y}(F(h)(x) - \xi) + \frac{\partial\varphi(\xi,\xi)}{\partial z}(x - \xi)\right) \\ &\quad - \frac{h}{2}\left(\varphi(\xi,\xi) + \frac{\partial\varphi(\xi,\xi)}{\partial y}(x - \xi) + \frac{\partial\varphi(\xi,\xi)}{\partial z}(F(h)(x) - \xi)\right) + O(h^3) \\ &= h\left(f\left(u\left(\frac{h}{2}\right)\right) - f(\xi)\right) + O(h^3). \end{aligned}$$

Then using (2.8), we obtain $E(h) = O(h^3)$ and second order accuracy follows from the standard theory for one-step numerical methods. □

REMARK 2.2. Since the scheme (2.2) is symmetric, the statement c) of Theorem 2.1 can be derived from the general theory of symmetric schemes [**H**].

By definition of a dynamical system, the set Ω is forward invariant with respect to the differential equation (1.1) or (2.1). Therefore, it is natural to demand that this set be invariant under the flow $F(h,.)$ for $h \in (0, \bar{h}]$ or that the NSFD scheme (1.2) be dynamically consistent with respect to this property. This is achieved in Theorem 2.3 below. Before stating the theorem, we make the following assumptions on the underlying set and its boundary: The compact set Ω with its interior, $\overset{\circ}{\Omega}$, and boundary, $\partial\Omega$, are such that

$$\Omega = closure(\overset{\circ}{\Omega}) = \overset{\circ}{\Omega} \cup \partial\Omega. \tag{2.9}$$

The set Ω is assumed to be convex. Finally, we assume that the boundary $\partial\Omega$ is piecewise smooth, namely it consists of a finite number of smooth parts, Γ_i, (i.e., $\partial\Omega = \cup_i^N \bar{\Gamma}_i$), the outer normal vector at $x \in \Gamma_i$ being denoted by ${}^i n(x)$. The announced qualitative result reads as follows:

THEOREM 2.3. *Let a scheme* (1.2) *be reversible for all* $h \in (0, \overline{h}]$ *and let the assumptions made above about* Ω *and its boundary hold. If for every* $h \in (0, \overline{h}]$ *and* $x \in \Gamma_i \subset \partial\Omega$, *we have*

$$\langle {}^i n(x), (F(-h,x) - x)\rangle \geq 0, \; i = 1, \cdots, N, \tag{2.10}$$

where $\langle .,. \rangle$ *denotes the inner product of two vectors, then the set* Ω *is invariant under* $F(h,\cdot)$ *for every* $h \in (0, \overline{h}]$.

Theorem 2.3 is proved in [**A5**] in the general setting of a set Ω, which is not piecewise smooth, using an extended concept of outer normal vector due to [**B2**] (see also [**W**]). The relation (2.10) is the discrete tangent condition proposed in [**A5**]. Roughly, it states that the vector $F(-h,x) - x$ points to the outside of the set Ω for all involved h. This compares nicely with the tangent condition by Bony [**B2**] for the continuous model (1.1) that states that the vector $f(x)$ is directed to the interior of Ω or is tangent to its boundary at the point $x \in \partial\Omega$, and formally reads as follows:

$$\langle n(x), f(x)\rangle \leq 0, \text{ for } x \in \partial\Omega. \tag{2.11}$$

REMARK 2.4. For the model (1.1), one can derive the function φ in (2.1) in many different ways. For example, if we choose $\varphi(x,y) = \frac{1}{2}(f(x) + f(y))$, then the numerical method (2.2) is exactly the trapezoidal rule, which is well known to be symmetric and second-order accurate. The power of the proposed approach is precisely in the freedom of choosing φ. More precisely, we can choose φ in such a way that it takes advantage of the specific form of the function f. In essence, this is the philosophy behind the use of nonlocal approximations in the design of NSFD schemes, a fact highlighted in [**A6**]. In the next section where this approach is applied to mass action models, we show that φ can be selected in a such a way that Theorem 2.3 permits to prove that the NSFD scheme (1.2) preserves the domain of the continuous model it approximates. We note that several classical numerical methods, including the trapezoidal rule, fail to replicate this essential property of the original model. More about the advantages of the nonstandard discretizations and the disadvantages of classical schemes such as the midpoint, the Heun and the explicit Runge-Kutta methods is well reported in the literature. See, for instance, [**A2, A3, A4, A6, M3**]. As a matter of fact, Heun method and explicit Runge-Kutta methods of order $p \geq 2$ are not elementary stable as they exhibit spurious/ghost solutions. The class of higher-order implicit Runge-Kutta methods which are reputed to be very stable and computationally effective include the *SDIRK* (Singulary Diagonally Implicit Runge-Kutta) and *ROW* (Rosenbrok-Wanner) methods [**C, R**]. Whether the domain of the continuous model is forward invariant with respect to the latter numerical methods requires further investigation.

3. Schemes for mass action models

Modeling of biological and chemical processes by applying the mass action principle for representing the interaction of the involved species typically results in a system of the form (2.1) in Ω, a convex and compact subset of $\mathbb{R}^n_+$, where the function φ is linear in both its arguments. Then φ can be represented as

$$\varphi(x,y) = P(x)y + A(x+y) + b = Q(y)x + A(x+y) + b \tag{3.1}$$

where P and Q are $n \times n$ matrix functions of x and y, respectively. The matrix A and the vector b are constant. In this particular setting the numerical method (2.2) can be written in the form

$$\frac{x_{k+1} - x_k}{h} = \frac{1}{2}(P(x_k)x_{k+1} + Q(x_k)x_{k+1}) + A(x_{k+1} + x_k) + b \tag{3.2}$$

or equivalently,

$$\left(I - \frac{h}{2}\left(P(x_k) + Q(x_k)\right) - hA\right) x_{k+1} = (I + hA)x_k + hb, \tag{3.3}$$

where I is the $n \times n$ identity matrix. Furthermore, and also expressing it reversibility, the equation (3.2) can be written in the form

$$\left(I + \frac{h}{2}\left(P(x_{k+1}) + Q(x_{k+1})\right) + hA\right) x_k = (I - hA)x_{k+1} - hb. \tag{3.4}$$

Considering that Ω is compact, there exists $\overline{h}$ such that the matrices

$$\left\{\begin{array}{l} I - \frac{h}{2}(P(x) + Q(x)) - hA \text{ and } I + \frac{h}{2}(P(x) + Q(x)) + hA \\ \text{are both (row or column) strictly diagonally dominant matrices} \\ \text{for all } x \in (0, \overline{h}] \text{ and } x \in \Omega. \end{array}\right. \tag{3.5}$$

Then, (3.3) can be written explicitly as

$$x_{k+1} = F(h, x_k) \tag{3.6}$$

where $F(h, x)$ is the solution of the linear system

$$\left(I - \frac{h}{2}(P(x) + Q(x)) - hA\right) F(h, x) = (I + hA)x + hb \tag{3.7}$$

or, equivalently

$$F(h, x) = \left(I - \frac{h}{2}(P(x) + Q(x)) - hA\right)^{-1} ((I + hA)x + hb). \tag{3.8}$$

Further, the map F can be represented in the usual form of one-step numerical methods as

$$F(h, x) = x + h\left(I - \frac{h}{2}(P(x) + Q(x)) - hA\right)^{-1} f(x). \tag{3.9}$$

Indeed, from (3.8) we have

$$\begin{aligned} F(h, x) &= x + \left(I - \frac{h}{2}(P(x) + Q(x)) - hA\right)^{-1} \\ &\quad \left((I + hA)x + hb - \left(I - \frac{h}{2}(P(x) + Q(x)) - hA\right) x\right) \\ &= x + \frac{h}{2}\left(I - \frac{h}{2}(P(x) + Q(x)) - hA\right)^{-1} \\ &\quad (P(x)x + 2Ax + b + Q(x)x + 2Ax + b) \\ &= x + \frac{h}{2}\left(I - \frac{h}{2}(P(x) + Q(x)) - hA\right)^{-1} (f(x) + f(x)) \\ &= x + h\left(I - \frac{h}{2}(P(x) + Q(x)) - hA\right)^{-1} f(x). \end{aligned}$$

The inverse map is given by

$$\begin{aligned} F^{-1}(h, x) = F(-h, x) &= \left(I + \frac{h}{2}(P(x) + Q(x)) + hA\right)^{-1} ((I - hA)x - hb) \\ &= x - h\left(I + \frac{h}{2}(P(x) + Q(x)) + hA\right)^{-1} f(x). \end{aligned} \tag{3.10}$$

THEOREM 3.1. *Let there exist $\bar{h} > 0$ such that* (3.5) *holds and Ω is an invariant set of the map $F(h, \cdot)$ for all $h \in (0, \bar{h}]$. Then the scheme* (3.2) *satisfies the properties a)–c) in Theorem* 2.1.

PROOF. The scheme (3.2) is constructed in the general form (2.2). Therefore, it is sufficient to show that Condition (2.3) holds. The solution is given in the explicit form (3.6). Therefore, it exists for every $h \in (0, \bar{h}]$ and it is unique. Further, it belongs to Ω, due to the assumption that Ω is invariant. □

4. Application to an epidemiological model

To be more explicit on the theoretical and systematic framework presented in the previous sections for the design of reliable NSFD schemes, we consider the following model for vector borne diseases with temporal immunity, as given in [**M1**]:

$$\frac{dS_v}{dt} = \Lambda_v - pS_vI - \mu_vS_v \tag{4.1}$$

$$\frac{dI_v}{dt} = pS_vI - \mu_vI_v \tag{4.2}$$

$$\frac{dS}{dt} = \Lambda - qSI_v - \mu S + \gamma R \tag{4.3}$$

$$\frac{dI}{dt} = qSI_v - (\mu + \alpha + \delta)I \tag{4.4}$$

$$\frac{dR}{dt} = \alpha I - (\mu + \gamma)R \tag{4.5}$$

where, with all the parameters being positive and bearing in mind the dimensions/units, we have the description below.

- S_v and I_v denote the susceptible and infective vectors, while S, I and R are the susceptible, infective and recovered (with temporary immunity) hosts, e.g. humans.
- Λ_v and Λ as well as μ_v and μ are the recruitment constants and the natural death rates for vectors and hosts, respectively.
- p and q are the probabilities of sufficient contact for transmission from host to vector and from vector to host, respectively. (Note that an additional parameter, a, in [**M1**] is absorbed here in the parameters p and q)
- The average period of infectivity of host is α^{-1}, while the average duration of immunity of host is γ^{-1}.
- In contrast to [**M1**], we added a disease-induced death rate, δ. By adding the equations of the hosts, we obtain the conservation law,

  $$\frac{dN}{dt} = \Lambda - \mu N - \delta I,$$

  which is an underdetermined differential equation or an underconstrained equation in the constraint counting theory. In order to study this conservation law, we couple the system (4.1)-(4.5) with the equation

  $$\frac{dD}{dt} = \delta I - \mu D. \tag{4.6}$$

  Similar to constrained optimization and optimal control problems where new variables such as the Lagrange multipliers and the co-state variables arise, the compartment D in (4.6) represents additional (*a priori*) information on the unknown variables that leads to the well-posed conservation law (4.9) below. Hence, the variable D accounts for the demographic effect of the disease or the burden of the disease (e.g. malaria in Africa) on the total host population, as explained in the comments related to Eq. (4.8).

Let $x = (S_v, I_v, S, I, R, D)^T$. Then, the system (4.1)–(4.6) can be written in the form (2.1) with φ given by (3.1), where

(4.7)

$$P(x) = \begin{pmatrix} -pI & 0 & 0 & 0 & 0 & 0 \\ pI & 0 & 0 & 0 & 0 & 0 \\ 0 & 0 & -qI_v & 0 & & 0 \\ 0 & 0 & qI_v & 0 & 0 & 0 \\ 0 & 0 & 0 & 0 & 0 & 0 \\ 0 & 0 & 0 & 0 & 0 & 0 \end{pmatrix}, \quad Q(x) = \begin{pmatrix} 0 & 0 & 0 & -pS_v & 0 & 0 \\ 0 & 0 & 0 & pS_v & 0 & 0 \\ 0 & -qS & 0 & 0 & 0 & 0 \\ 0 & qS & 0 & 0 & 0 & 0 \\ 0 & 0 & 0 & 0 & 0 & 0 \\ 0 & 0 & 0 & 0 & 0 & 0 \end{pmatrix},$$

$$A = \frac{1}{2}\begin{pmatrix} -\mu_v & 0 & 0 & 0 & 0 & 0 \\ 0 & -\mu_v & 0 & 0 & 0 & 0 \\ 0 & 0 & -\mu & 0 & \gamma & 0 \\ 0 & 0 & 0 & -(\mu+\alpha+\delta) & 0 & 0 \\ 0 & 0 & 0 & \alpha & -(\mu+\gamma) & 0 \\ 0 & 0 & 0 & \delta & 0 & -\mu \end{pmatrix}, \quad b = \begin{pmatrix} \Lambda_v \\ 0 \\ \Lambda \\ 0 \\ 0 \\ 0 \end{pmatrix}.$$

The set of equations (4.1)–(4.6) defines a dynamical system on the biologically feasible compact and convex region

$$\Omega = \left\{ x \in \mathbb{R}^6_+ : S_v + I_v \le \frac{\Lambda_v}{\mu_v},\ S + I + R + D \le \frac{\Lambda}{\mu} \right\}.$$

It is easy to see that for the populations, $N_v = S_v + I_v$, of vectors and, $N = S+I+R$, of hosts, we have

$$\lim_{t\to+\infty} N_v(t) = \frac{\Lambda_v}{\mu_v} \text{ and } \lim_{t\to+\infty} (N(t) + D(t)) = \frac{\Lambda}{\mu}, \tag{4.8}$$

which shows that growing populations of vectors or hosts may not increase beyond the upper bounds set in the inequalities defining Ω. Furthermore, the equation

$$\frac{d(N+D)}{dt} = \Lambda - \mu(N+D). \tag{4.9}$$

whose the equilibrium $\frac{\Lambda}{\mu}$ is globally asymptotically stable, clarifies the biological meaning of D, namely $D(t)$ is the difference between the size of a disease-free population and the size of population when the disease is present at time t. If $\overline{N}$ is an equilibrium of the host population, then the equilibrium $\overline{D}$ of the compartment D is precisely the difference between the disease-free equilibrium $\dfrac{\Lambda}{\mu}$ and $\overline{N}$. Specifically, when $\delta > 0$, then $\overline{D} > 0$ indicates an endemic equilibrium, while $\overline{D} = 0$ indicates eventual disease extinction.

REMARK 4.1. Clearly the compartments R and D can be decoupled from the system (4.1)-(4.6). Similarly, they can be decoupled from the numerical computations if the values of $R(t)$ and $D(t)$ are not specifically required. However, it is more convenient to carry out the theoretical analysis of both the model and its NSFD discretization for the full system, especially that the equation $S+I+R+D=\frac{\Lambda}{\mu}$ defines an invariant set.

The set Ω is forward invariant with respect to the system (4.1)–(4.5), owing to the tangent condition that holds for each one of the hyperplanes $\Gamma_1=\{S_v=0\}$, $\Gamma_2=\{I_v=0\}$, $\Gamma_3=\{S=0\}$, $\Gamma_4=\{I=0\}$, $\Gamma_5=\{R=0\}$, $\Gamma_6=\{D=0\}$, $\Gamma_7=\{S_v+I_v=\frac{\Lambda_v}{\mu_v}\}$, and $\Gamma_8=\{S+I+R+D=\frac{\Lambda}{\mu}\}$ that form its boundary $\partial\Omega$. Note that, in terms of Eq. (2.11) written for each hyperplane, the said-tangent condition reads:

$$\langle {}^i n(x), f(x)\rangle \leq 0, \text{ for } x\in\Gamma_i, \tag{4.10}$$

where $f(x)=\varphi(x,x)$ is the vector field defined by the right hand side of (4.1)–(4.6). The relation is straightforward for $1\leq i\leq 6$ because the outer normal vector in this case is ${}^i n=-e^i$ where $\{e^1,e^2,e^3,e^4,e^5,e^6\}$ is the canonical basis of the vector space $\mathbb{R}^6$. For the last two hyperplanes, the outer normal vectors are ${}^7 n=(1,1,0,0,0,0)^T$ and ${}^8 n=(0,0,1,1,1,1)^T$, so that the tangent condition on Γ_7 and Γ_8 is given by:

$$\langle {}^7 n, f(x)\rangle = \Lambda_v-\mu_v(S_v+I_v)=0,\ x\in\Gamma_7, \tag{4.11}$$

$$\langle {}^8 n, f(x)\rangle = \Lambda-\mu(S+I+R+D)=0,\ x\in\Gamma_8. \tag{4.12}$$

We consider the numerical method (3.6) for the system (4.1)–(4.6) on Ω. Our aim is to apply Theorem 3.1. From the structure of the matrices P, Q and A, it follows that the matrices $I-\frac{h}{2}(P(x)+Q(x))-hA$ and $I+\frac{h}{2}(P(x)+Q(x))+hA$ are row strictly diagonally dominant if, and only if, the following set of inequalities holds:

$$\begin{aligned} h(pI+pS_v+\mu_v) &< 2,\\ h(qI_v+qS+\mu) &< 2,\\ h(qI_v+qS+\mu+\alpha+\delta) &< 2,\\ h(\mu+\alpha+\gamma) &< 2,\\ h(\mu+\delta) &< 2. \end{aligned} \tag{4.13}$$

Therefore, condition (3.5) holds for all $x\in\Omega$ and $h\in(0,\overline{h}]$, where

$$\overline{h}=\min\left\{\frac{1}{p\frac{\Lambda}{\mu}+p\frac{\Lambda_v}{\mu_v}+\mu_v},\ \frac{1}{q\frac{\Lambda_v}{\mu_v}+q\frac{\Lambda}{\mu}+\mu+\alpha+\delta},\ \frac{1}{\mu+\alpha+\gamma},\ \frac{1}{\mu+\gamma+\frac{\alpha\gamma}{\mu}}\right\}. \tag{4.14}$$

The reason for including the last fraction in the set in (4.14) is clarified in (4.21) below.

It remains to show the invariance of Ω under the flow $F(h,\cdot)$. We apply Theorem 2.3 and the involved discrete tangent condition (2.10) on each hyperplanes Γ_7, Γ_8, Γ_1 and Γ_3 of the boundary. Using (3.10), we have

$$\left(I+\frac{h}{2}(P(x)+Q(x))+hA\right)(F(-h,x)-x)=-hf(x). \tag{4.15}$$

It is easy to see from the structure of P, Q and A in (4.7) that

$$
\begin{aligned}
{}^7n^TP = {}^8n^TP = {}^7n^TQ = {}^8n^TQ = (0,0,0,0,0,0)^T,\\
{}^7n^TA = -\frac{1}{2}\mu_v{}^7n, \quad {}^8nA = -\frac{1}{2}\mu{}^8n.
\end{aligned} \tag{4.16}
$$

Putting together (4.15) and (4.16), we obtain

$$
\begin{aligned}
&\left(1-\frac{h}{2}\mu_v\right)\langle{}^7n,(F(-h,x)-x)\rangle\\
&\quad = {}^7n^T\left(I+\frac{h}{2}(P(x)+Q(x))+hA\right)(F(-h,x)-x)\\
&\quad = -h\langle{}^7n,f(x)\rangle \geq 0,
\end{aligned}
$$

in view also of (4.11). Using the fact that $h \leq \overline{h} < \frac{1}{\mu_v}$, and again (4.11), we have

$$
\langle{}^7n,(F(-h,x)-x)\rangle = -\frac{h}{1-\frac{h}{2}\mu_v}\langle{}^7n,f(x)\rangle \geq 0 \tag{4.17}
$$

for all x on the hyperplane Γ_7 and $h \in (0,\overline{h}]$. Similarly, from (4.15), (4.16) and (4.12), we obtain

$$
\langle{}^8n,(F(-h,x)-x)\rangle = -\frac{h}{1-\frac{h}{2}\mu}\langle{}^8n,f(x)\rangle \geq 0 \tag{4.18}
$$

for all x on the hyperplane Γ_8 and $h \in (0,\overline{h}]$.

On the boundary Γ_1, the coefficient matrix of the system in (4.15) simplifies, due to $S_v = 0$, and we have

$$
{}^1n^T\left(I+\frac{h}{2}(P(x)+Q(x))+hA\right) = \left(1-\frac{h}{2}(pI+\mu_v)\right){}^1n^T, \quad x \in \Gamma_1.
$$

Then, similar to Γ_7 and Γ_8, we obtain

$$
\langle{}^1n,(F(-h,x)-x)\rangle = -\frac{h}{1-\frac{h}{2}(pI+\mu_v)}\langle{}^1n,f(x)\rangle \geq 0 \tag{4.19}
$$

for all x on the hyperplane Γ_1 and $h \in (0,\overline{h}]$.

Next, we consider the hyperplane Γ_3. We need to show that on Γ_3 we have

$$
\langle{}^3n,(F(-h,x)-x)\rangle = -(F_3(-h,x)-S) \geq 0. \tag{4.20}
$$

For $S = 0$, the coefficient matrix in (4.15) has a block-triangular form, so that $F_3(-h,x)-S = F_3(-h,x)$ is obtained as a part of the solution of a 3×3 system. Using Cramer's rule, $F_3(-h,x)$ is non-positive if, and only if,

$$
d = \begin{vmatrix}
\Lambda+\gamma R & 0 & \frac{h}{2}\gamma\\
-(\mu+\alpha+\delta)I & 1-\frac{h}{2}(\mu+\alpha+\delta) & 0\\
\alpha I-(\mu+\gamma)R & \frac{h}{2}\alpha & 1-\frac{h}{2}(\mu+\gamma)
\end{vmatrix} \geq 0.
$$

Expanding the determinant and using the fact that $I \le \frac{\Lambda}{\mu}$ as well as the choice of $\overline{h}$ in (4.14), we have

$$\begin{aligned} d &= \Lambda\left(1-\frac{h}{2}\left(2\mu+\gamma+\delta+\alpha+\alpha\gamma\frac{I}{\Lambda}\right)\right)+\gamma R\left(1-\frac{h}{2}(\mu+\alpha+\gamma)\right)\\ &\quad+\frac{h^2}{4}\Lambda(\mu+\alpha+\delta)(\mu+\gamma)\\ &\ge \Lambda\left(1-\frac{h}{2}(\mu+\gamma+\delta)-\frac{h}{2}\left(\mu+\alpha+\frac{\alpha\gamma}{\mu}\right)\right)+\gamma R\left(1-\frac{h}{2}(\mu+\alpha+\gamma)\right)\\ &\ge 0 \end{aligned} \tag{4.21}$$

for all $h \in (0, \overline{h}]$. Hence, (4.20) holds.

Thus, we have established that the NSFD scheme (3.6) inherits from the system (4.1)–(4.5) the continuous tangent condition (4.10), (4.11) & (4.12) in the form of the discrete tangent condition (4.17), (4.18), (4.19) & (4.20) along the boundaries Γ_7, Γ_8, Γ_1 and Γ_3, respectively. Then, from Theorem 2.3, we obtain that if $x \in \Omega$ and $h \in (0, \overline{h}]$, then $F(h, x)$ is in the region bounded by Γ_1, Γ_3, Γ_7 and Γ_8.

Regarding the hyperplanes Γ_i, $i = 2, 4, 5, 6$, the situation is much easier because the sub-system of (3.7) for the variables I_v, I, R, D can be written in the form

$$\begin{pmatrix} 1+\frac{h}{2}\mu_v & -\frac{h}{2}pS_v & 0 & 0\\ -\frac{h}{2}qS & 1+\frac{h}{2}(\mu+\alpha+\delta) & 0 & 0\\ 0 & -\frac{h}{2}\alpha & 1+\frac{h}{2}(\mu+\gamma) & 0\\ 0 & -\frac{h}{2}\delta & 0 & 1+\frac{h}{2}\mu \end{pmatrix}\begin{pmatrix} F_2(h,x)\\ F_4(h,x)\\ F_5(h,x)\\ F_6(h,x)\end{pmatrix} = \begin{pmatrix} (1-\frac{h}{2}\mu_v)I_v+\frac{h}{2}pIF_1(h,x)\\ (1-\frac{h}{2}(\mu+\alpha+\delta))I+\frac{h}{2}qI_vF_3(h,x)\\ \frac{h}{2}\alpha I+(1-\frac{h}{2}(\mu+\gamma))R\\ \frac{h}{2}\delta I+(1-\frac{h}{2}\mu)D \end{pmatrix}. \tag{4.22}$$

For $h \in (0, \overline{h}]$, the coefficient matrix of the linear system (4.22) is an nonsingular M-matrix. Therefore, its inverse is nonnegative [**B1**]. Further, using the already established relations $F_1(h, x) \ge 0$ and $F_3(h, x) \ge 0$, the right hand side in (4.22) is also nonnegative. Hence, the solution $(F_2(h, x), F_4(h, x), F_5(h, x), F_6(h, x))^T$ is nonnegative.

Combining all the steps considered above, we have proved that $F(h, x) \in \Omega$ for all $x \in \Omega$ and $h \in (0, \overline{h}]$, that is Ω is forward invariant for all maps $F(h, \cdot)$, $h \in (0, \overline{h}]$. Consequently, the NSFD scheme satisfies the properties a)–c) in Theorem 2.1 and Theorem 3.1.

REMARK 4.2. Though our focus in this paper is not on Mickens' NSFD schemes with nontrivial denominator functions of the discrete derivative, the parameter $\overline{h}$ in (4.14) can in some sense be linked to that approach. To see this, let $J(\tilde{x})$ be the Jacobian matrix of the vector field of the the system (4.1)–(4.5), evaluated at the equilibrium points $\tilde{x}$. We set

$$K := \max_{\tilde{x}} \max_{1\le j\le 6} \sum_{i=1}^{6} |J_{ij}(\tilde{x})|.$$

Then, the said parameter satisfies the relation $K^{-1} \leq \bar{h}$. Then the following NSFD forward Euler scheme is elementary stable ([**A3, M2**]):

$$\frac{x_{k+1} - x_k}{(1 - e^{-r\Delta t})/r} = f(x_k)$$

where the parameter $r \geq K \geq (\bar{h})^{-1}$ captures the dynamics of the system. Moreover, the set Ω is forward invariant under all maps $F(h, \cdot)$, $h \in (0, K^{-1}]$ and the NSFD scheme (3.6) satisfies properties a)–c) in Theorem 2.1 and Theorem 3.1.

The power of the NSFD approach over classical schemes is well documented, as mentioned in Remark 2.4. A comparative study, including numerical simulations and the calculation of the convergence rate, were done in [**A2**] for the $MSEIR$ model, one of the general compartmental epidemic models. A similar study can easily be done for the model (4.1)-(4.6).

5. Conclusion

The literature on high-order numerical methods for dynamical systems defined by ordinary differential equations is rich. However, several of these classical schemes can exhibit spurious/ghost solutions or other elementary instability that do not correspond to the feature of the continuous equations. We overcome this difficulty by constructing a time-reversible nonstandard finite difference (NSFD) scheme, which is directly applicable to mass action-type models in biology. This NSFD scheme is second-order convergent and elementary stable. Moreover, using a discrete analog of the tangent condition, it is shown that the NSFD scheme preserves the property of the continuous model stating that its feasible region is forward invariant.

A possible extension of this study is to consider it for a dynamical system with nonhyperbolic equilibrium points, an approach investigated in [**A7**] for the stability analysis of one-dimensional models.

The decomposition (3.1) allows for polynomial nonlinearities in the system (1.1), as refected by $f(x) = \varphi(x, x)$ in (2.1). The question of considering for the function $\varphi(x, y)$ a decomposition that is not linear in the dependent variables is of interest for future research. The investigation and impementation of this general case is certainly challenging; it will require, among others solving a nonlinear system in each time step.

Acknowledgments

The two authors would like to thank Professor Abba B. Gumel, Editor, for his kind invitation to contribute this book chapter. They are grateful to anonymous reviewers for their constructive comments and suggestions.

References

[A1] Roumen Anguelov, Tsanou Berge, Michael Chapwanya, Jules Kamdem Djoko, Phumezile Kama, Jean M.-S. Lubuma, and Yibeltal Terefe, *Nonstandard finite difference method revisited and application to the Ebola virus disease transmission dynamics*, J. Difference Equ. Appl. **26** (2020), no. 6, 818–854, DOI 10.1080/10236198.2020.1792892. MR4141413

[A2] Roumen Anguelov, Yves Dumont, Jean M.-S. Lubuma, and Meir Shillor, *Dynamically consistent nonstandard finite difference schemes for epidemiological models*, J. Comput. Appl. Math. **255** (2014), 161–182, DOI 10.1016/j.cam.2013.04.042. MR3093413

[A3] Roumen Anguelov, Phumezile Kama, and Jean M.-S. Lubuma, *On non-standard finite difference models of reaction-diffusion equations*, J. Comput. Appl. Math. **175** (2005), no. 1, 11–29, DOI 10.1016/j.cam.2004.06.002. MR2105666

[A4] Roumen Anguelov and Jean M.-S. Lubuma, *Contributions to the mathematics of the nonstandard finite difference method and applications*, Numer. Methods Partial Differential Equations **17** (2001), no. 5, 518–543, DOI 10.1002/num.1025. MR1849163

[A5] Roumen Anguelov and Jean M.-S. Lubuma, *Forward invariant set preservation in discrete dynamical systems and numerical schemes for ODEs: application in biosciences*, Adv. Contin. Discrete Models, posted on 2023, Paper No. 38, 22, DOI 10.1186/s13662-023-03784-2. MR4644983

[A6] Roumen Anguelov and Jean M.-S. Lubuma, *Nonstandard finite difference method by nonlocal approximation*, Math. Comput. Simulation **61** (2003), no. 3-6, 465–475, DOI 10.1016/S0378-4754(02)00106-4. MODELLING 2001 (Pilsen). MR1984145

[A7] Roumen Anguelov, Jean M.-S. Lubuma, and Meir Shillor, *Topological dynamic consistency of non-standard finite difference schemes for dynamical systems*, J. Difference Equ. Appl. **17** (2011), no. 12, 1769–1791, DOI 10.1080/10236198.2010.488226. MR2854823

[B1] Abraham Berman and Robert J. Plemmons, *Nonnegative matrices in the mathematical sciences*, Classics in Applied Mathematics, vol. 9, Society for Industrial and Applied Mathematics (SIAM), Philadelphia, PA, 1994. Revised reprint of the 1979 original, DOI 10.1137/1.9781611971262. MR1298430

[B2] Jean-Michel Bony, *Principe du maximum, inégalite de Harnack et unicité du problème de Cauchy pour les opérateurs elliptiques dégénérés* (French, with English summary), Ann. Inst. Fourier (Grenoble) **19** (1969), no. fasc. 1, 277–304 xii. MR262881

[C] Frank Cameron, Mikko Palmroth, and Robert Piché, *Quasi stage order conditions for SDIRK methods*, Appl. Numer. Math. **42** (2002), no. 1-3, 61–75, DOI 10.1016/S0168-9274(01)00142-8. Ninth Seminar on Numerical Solution of Differential and Differential-Algebraic Equations (Halle, 2000). MR1921329

[H] Ernst Hairer, Christian Lubich, and Gerhard Wanner, *Geometric numerical integration: Structure-preserving algorithms for ordinary differential equations*, 2nd ed., Springer Series in Computational Mathematics, vol. 31, Springer-Verlag, Berlin, 2006. MR2221614

[K] Hristo V. Kojouharov, Souvik Roy, Madhu Gupta, Fawaz Alalhareth, and John M. Slezak, *A second-order modified nonstandard theta method for one-dimensional autonomous differential equations*, Appl. Math. Lett. **112** (2021), Paper No. 106775, 6, DOI 10.1016/j.aml.2020.106775. MR4154331

[L] Jean M.-S. Lubuma and Alet Roux, *An improved theta-method for systems of ordinary differential equations*, J. Difference Equ. Appl. **9** (2003), no. 11, 1023–1035, DOI 10.1080/1023619031000146904. MR2027165

[M1] Maia Martcheva, *An introduction to mathematical epidemiology*, Texts in Applied Mathematics, vol. 61, Springer, New York, 2015, DOI 10.1007/978-1-4899-7612-3. MR3409181

[M2] Ronald E. Mickens, *Nonstandard finite difference models of differential equations*, World Scientific Publishing Co., Inc., River Edge, NJ, 1994. MR1275372

[M3] Ronald E. Mickens, *Nonstandard finite difference schemes—methodology and applications*, World Scientific Publishing Co. Pte. Ltd., Hackensack, NJ, [2021] ©2021. Expanded second edition of [1275372]. MR4274594

[M4] Ronald E. Mickens, *Nonstandard finite difference methods*, Advances in the applications of nonstandard finite difference schemes, World Sci. Publ., Hackensack, NJ, 2005, pp. 1–9, DOI 10.1142/9789812703316_0001. MR2212196

[R] Joachim Rang, *An analysis of the Prothero-Robinson example for constructing new DIRK and ROW methods*, J. Comput. Appl. Math. **262** (2014), 105–114, DOI 10.1016/j.cam.2013.09.062. MR3162307

[S] Andrew M. Stuart and Antony R. Humphries, *Dynamical systems and numerical analysis*, Cambridge Monographs on Applied and Computational Mathematics, vol. 2, Cambridge University Press, Cambridge, 1996. MR1402909

[W] Wolfgang Walter, *Ordinary differential equations*, Graduate Texts in Mathematics, vol. 182, Springer-Verlag, New York, 1998. Translated from the sixth German (1996) edition by Russell Thompson; Readings in Mathematics, DOI 10.1007/978-1-4612-0601-9. MR1629775

Department of Mathematics and Applied Mathematics, University of Pretoria, Pretoria 0002, South Africa; and Institute of Mathematics and Informatics, Bulgarian Academy of Sciences, Sofia, Bulgaria

Email address: roumen.anguelov@up.ac.za

School of Computer Science and Applied Mathematics, University of the Witwatersrand, Johannesburg, Braamfontein 2000, South Africa

Email address: jean.lubuma@wits.ac.za (Corresponding author)

Contemporary Mathematics
Volume **793**, 2024
https://doi.org/10.1090/conm/793/15882

A generalized nonstandard finite difference method for a class of autonomous dynamical systems and its applications

Manh Tuan Hoang

This paper is dedicated in honor of Ronald E. Mickens' 80th birthday.

ABSTRACT. In this work, a class of continuous-time autonomous dynamical systems describing many important phenomena and processes arising in real-world applications is considered. We apply the nonstandard finite difference (NSFD) methodology proposed by Mickens to design a generalized NSFD method for the dynamical system models under consideration. This method is constructed based on a novel non-local approximation for the right-hand side functions of the dynamical systems. It is proved by rigorous mathematical analyses that the NSFD method is dynamically consistent with respect to positivity, asymptotic stability and three classes of conservation laws, including direct conservation, generalized conservation and sub-conservation laws. Furthermore, the NSFD method is easy to be implemented and can be applied to solve a broad range of mathematical models arising in real-life. Finally, a set of numerical experiments is performed to illustrate the theoretical findings and to show advantages of the proposed NSFD method.

1. Introduction

Many important problems in real-world situations can be mathematically modeled by autonomous dynamical systems of the general form:

$$\frac{dy(t)}{dt} = f(y(t)), \quad t \geq 0, \qquad y(0) = y_0 \in \mathbb{R}^n, \tag{1.1}$$

where y is a vector function of t with n components; f is a function of y and it is assumed to satisfy suitable conditions that guarantee solutions of the model (1.1) exist and are unique (see, for instance, [**4, 14, 32, 49, 54, 57, 84, 85**]). The solutions of the model (1.1) often possess several important properties, typically positivity, asymptotic stability, boundedness, periodicity and physical properties.

In this work, we consider the model (1.1) under the following two conditions:

- **(C1):** There exists a real number α such that

$$f(y) + \alpha y \geq 0 \quad \text{for all} \quad y \geq 0. \tag{1.2}$$

  Here the notation $\geq$ is meant entry-wise for vectors.
- **(C2):** The set of equilibria is finite and all the equilibria are hyperbolic.

2020 *Mathematics Subject Classification.* Primary 34A45, 65L05.
Key words and phrases. Mickens' methodology, nonstandard finite difference method, dynamic consistency, positivity, stability, conservation laws.

The condition **(C1)** implies that the set $\mathbb{R}^n_+ = \{y \in \mathbb{R}^n | y \geq 0\}$ is a positively invariant set of the model (1.1), that is $y(t) \geq 0$ whenever $y_0 \geq 0$ (see [**45, 84**]), meanwhile, the condition **(C2)** means that the local asymptotic stability (LAS) of all the equilibria can be determined by the Lyapunov's indirect method (see [**49, 52, 53, 85**]). Also, in many cases the model (1.1) can satisfy three classes of conservation laws, which include direct conservation, generalized conservation and sub-conservation laws [**65**]. It is easy to find a large number of mathematical models in real-world applications with the above-mentioned properties. Below, we list some notable models that satisfy the conditions **(C1)** and **(C2)** and/or the conservation laws

- mathematical models in biology, ecology and epidemiology [**4, 14, 32, 46, 47, 54**];
- models for interacting species [**14**];
- chemostat systems [**84**];
- epidemic models for infectious diseases [**39, 58**];
- metapopulation models [**5, 48**];
- single-species population models [**18**];
- predator-prey models with linear prey growth and Beddington-DeAngelis functional response [**27, 31**];
- a simple vaccination model with multiple endemic states [**51**];
- a vaccination model with non-linear incidence [**38**];
- a two-stage epidemic model with generalized non-linear incidence [**77**];
- a generalization of the Kermack-McKendrick epidemic model [**15**].

It should be emphasized that dynamically consistent nonstandard finite difference (NSFD) schemes for most of the above models have not been studied.

Our main objective is to construct a numerical method preserving the positivity, LAS and conservation laws of the model (1.1) for all the values of the step size. To achieve this goal, we will apply the *Mickens' methodology* [**59–63**] to propose a dynamically consistent NSFD method for the model (1.1). We recall that the concept of NSFD schemes was first introduced by Mickens to overcome usual numerical instabilities caused by standard finite difference (SFD) schemes as well as drawbacks of SFD schemes [**59–64**]. One of the main and outstanding advantages of NSFD schemes is that they have the ability to preserve essential mathematical properties of the corresponding differential equation models regardless of the values of the step size; furthermore, they are easy to be implemented. Therefore, NSFD schemes are suitable and effective to simulate dynamics of differential equation models over very long time periods. During the past several decades, NSFD schemes have been strongly developed by mathematicians and engineers and have become one of the most powerful methods for solving differential equations. Nowadays, NSFD schemes have been widely used for ordinary differential equations, partial differential equations, delay differential equations, fractional differential equations and integro-differential equations (see, for instance, [**1, 2, 6–10, 12, 16, 19, 20, 28–30, 33, 35, 36, 41, 50, 59–76, 78–83, 86–89**] and references therein). Here, we refer the readers to [**62, 78, 79**] for notable reviews of NSFD schemes.

It is well-known that the positivity and LAS are essential properties of most systems in biology, ecology and epidemiology (see, for example, [**4, 14, 32, 49, 54, 84, 85**]). For this reason, the construction of NSFD schemes preserving these properties is very important and has attracted the attention of many researchers. Below, we mention some notable results in this topic.

In an early work [**6**], Anguelov and Lubuma introduced an NSFD method preserving monotonic properties of solutions of differential equations and the LAS of hyperbolic equilibrium points. In 2005 and 2007, Dimitrov and Kojouharov in [**28, 29**] analyzed stability-preserving NSFD schemes based on the θ-methods and the second-order Runge-Kutta methods for general autonomous dynamical systems. Note that the NSFD schemes constructed in [**6, 28, 29**] are not positive. In [**30, 87**], dynamically consistent NSFD schemes for general productive-destructive systems were formulated. In 2015, Wood and Kojouharov in [**88**] designed an NSFD method preserving the positivity of solutions and the local behavior of dynamical systems near equilibria. In [**19**], Cresson and Pierret studied the construction of NSFD schemes for a general class of two-dimensional dynamical systems including many models in population dynamics. In 2020, Dang and Hoang constructed and analyzed high-order nonstandard Runge-Kutta methods for a class of autonomous dynamical systems possessing the positivity and LAS. Very recently, Hoang in [**40**] proposed a second-order NSFD method that simultaneously preserves the positivity and LAS of one-dimensional autonomous dynamical systems. Besides, many results focusing on NSFD schemes preserving the positivity and LAS of specific models in biology, ecology, epidemiology and other fields are also remarkable (see, for example, [**1, 2, 8, 80, 81, 89**] and references therein).

Motivated and inspired by the above reasons, we will apply the NSFD methodology proposed by Mickens to construct a dynamically consistent NSFD scheme for the model (1.1). This scheme is constructed based on a novel non-local approximation for the right-side functions. It is proved by rigorous mathematical analyses that the NSFD scheme preserves the positivity, asymptotic stability and conservation laws for all the values of the step size. Moreover, the constructed NSFD scheme is easy to be implemented and can be used to solve a broad range of mathematical problems arising in real-life. This claim is also confirmed by a set of illustrative numerical examples.

The plan of this work is as follows:

Section 2 provides some basic concepts and preliminaries. The NSFD scheme is introduced and analyzed in Section 3. A set of numerical experiments is investigated and reported in Section 4. Finally, conclusions, remarks and some open problems are given in Section 5.

2. Definitions and preliminaries

In this section, we recall from literature some essential concepts and preliminaries related to qualitative theory of continuous and discrete dynamical systems, conservation laws and NSFD schemes.

2.1. Stability of equilibria of continuous-time dynamical systems. Consider the continuous-time dynamical system (1.1). We recall that a point $y^* \in \mathbb{R}^n$ is called an equilibrium point of System (1.1) if $f(y^*) = 0$ (see [**49, 85**]). Without loss of generality, we can always assume that $y^* = 0$.

DEFINITION 2.1 ([**49, 85**]). The equilibrium point $y^* = 0$ of (1.1) is said to be:

- stable, if, for each $\epsilon > 0$, there is $\delta = \delta(\epsilon)$ such that

$$\|y(0)\| < \delta \quad \text{implies that} \quad \|y(t)\| < \epsilon, \quad \forall t \geq 0;$$

- unstable if it is not stable;
- (locally) asymptotically stable if it is stable and δ can be chosen such that

$$\|y(0)\| < \delta \quad \text{implies that} \quad \lim_{t\to\infty} y(t) = 0.$$

The following theorem is known as the Lyapunov's indirect method [**49**, **52**, **53**, **85**].

THEOREM 2.2. *Let $y^* = 0$ be an equilibrium point for the system* (1.1), *where $f : D \to \mathbb{R}^n$ is continuously differentiable and D is a neighborhood of the origin. Let*

$$A = \frac{\partial f}{\partial y}(y)\bigg|_{y=0}.$$

Then,

- y^* is LAS if $Re\lambda_i < 0$ for all eigenvalues of A;
- y^* is unstable if $Re\lambda_i > 0$ for one or more the eigenvalues of A.

REMARK 2.3. An equilibrium point y^* of (1.1) is said to be hyperbolic if none of the eigenvalues of the matrix A lies on the imaginary axis and non-hyperbolic otherwise (see [**85**]). Hence, Theorem 2.2 is only applicable for hyperbolic equilibrium points.

THEOREM 2.4 ([**45**, **84**]). *Suppose that f in* (1.2) *has the property that solutions of initial value problems $y(0) = y_0 \geq 0$ are unique and, for all i, $f_i(y) \geq 0$ whenever $y \geq 0$ satisfies $y_i = 0$. Then $y(t) \geq 0$ for all $t \geq 0$ for which it is defined, provided $y(0) \geq 0$.*

REMARK 2.5. We infer from Theorem 2.4 that the set $\mathbb{R}^n_+$ is a positively invariant set of the model (1.1) if the condition (1.2) holds.

2.2. Conservation laws. We consider three classes of conservation laws for the model (1.1), that are direct conservation, generalized conservation and sub-conservation laws [**65**].

For the system (1.1), let us denote by $P_T(t) = y_1(t) + y_2(t) + \ldots + y_n(t)$. We say that the system (1.1) satisfies

- a direct conservation law (DCL) if
$$\frac{dP_T}{dt} = 0,$$
which implies that P_T is constant;
- a generalized conservation law (GCL) if
$$\frac{dP_T}{dt} = a_1 - b_1 P_T, \quad a_1, b_1 > 0,$$
which implies that $P_T(t)$ monotonically converges to a_1/b_1 as $t \to \infty$;
- a sub-conservation law (SCL) if a sub-set of the populations, $P_S(t)$, where
$$P_S(t) = P_{1S}(t) + P_{2S}(t) + \ldots + P_{mS}(t), \quad 2 \leq m \leq n,$$
has the property
$$\frac{dP_T}{dt} = 0,$$
or
$$\frac{dP_T}{dt} = a_1 - b_1 P_T.$$

In [**65**], Mickens and Washington introduced NSFD discretizations of some mathematical models satisfying the above conservation laws.

2.3. Stability of equilibria of discrete-time dynamical systems. Consider a general discrete-time dynamical system governed by first-order difference equations of the form

$$y_{k+1} = g(y_k), \quad y_0 \in \mathbb{R}^n, \tag{2.1}$$

where $g : D \to \mathbb{R}$ and $D \subset \mathbf{R}^n$ is the domain of the function g. The vector y_0 is called the initial value or initial data. Note that the uniqueness of the sequence $\{y_k\}_{k=0}^{\infty}$ is automatic if it exists.

A point $y^* \in \mathbf{R}^n$ is called a fixed point or an equilibrium point of the system (2.1) if $g(y^*) = y^*$ (see [**34, 85**]).

DEFINITION 2.6 ([**34, 85**]). Let y^* be an equilibrium point of the system (2.1). Then, y^* is said to be:

- stable if given $\epsilon > 0$ there exists $\delta > 0$ such that
$$\|y_0 - y^*\| < \delta \quad \text{implies} \quad \|y_k - y^*\| < \epsilon,$$
for all $k > 0$.
- unstable if it is not stable;
- attracting if there exists $\eta > 0$ such that
$$\|y_0 - y^*\| < \eta \quad \text{implies} \quad \lim_{k\to\infty} y_k = y^*;$$
- (locally) asymptotically stable if it is stable and attracting.

THEOREM 2.7 ([**34, 85**]). *Let $f \in \mathcal{C}^2(\mathbb{R}^n, \mathbb{R}^n)$. Then a fixed point y^* of the system* (2.1) *is locally asymptotically stable if the eigenvalues of $df(y^*)$ lie strictly inside the unit circle. If any of the eigenvalues lie outside the unit circle the fixed point is unstable.*

REMARK 2.8. A fixed point y^* is said to be hyperbolic if none of the eigenvalues of the matrix $df(y^*)$ lies on the unit circle and non-hyperbolic if otherwise (see [**85**]). Hence, Theorem 2.7 only applicable for hyperbolic equilibrium points.

2.4. Nonstandard finite difference schemes. Consider a finite difference scheme that approximates solutions of (1.1) in the form

$$D_{\Delta t}(y_k) = F_{\Delta t}(f; y_k), \tag{2.2}$$

where $D_{\Delta t}(y_k) \approx dy/dt$, $F_{\Delta t}(f; y_k) \approx f(y)$ and $t_k = k\Delta t$, Δt is the step size.

According to the Mickens' methodology [**59–64**], an NSFD scheme for the equation (1.1) is a discrete model constructed based on a set of six rules. In particular, NSFD schemes for first-order differential equations can be defined as follows [**6, 7, 29**].

DEFINITION 2.9. The finite difference scheme (2.2) is called an NSFD scheme if at least one of the following conditions is satisfied:

- $D_{\Delta t}(y_k) = \dfrac{y_{k+1} - y_k}{\phi(\Delta t)}$, where $\phi(\Delta t) = \Delta t + \mathcal{O}(\Delta t^2)$ is a non-negative function and is called a nonstandard denominator function;
- $F_{\Delta t}(f; y_k) = g(y_k, y_{k+1}, \Delta t)$, where $g(y_k, y_{k+1}, \Delta t)$ is a non-local approximation of the right-hand side of the system (1.1).

Definition 2.10 ([**7**,**8**]). Assume that the solutions of the equation (1.1) satisfy some property $\mathcal{P}$. The numerical scheme (2.2) is called (qualitatively) stable with respect to property $\mathcal{P}$ (or $\mathcal{P}$-stable), if for every value of $\Delta t > 0$ the set of solutions of (2.2) satisfies property $\mathcal{P}$.

Definition 2.11 ([**6**,**7**,**64**]). Consider the differential equation $dy/dt = f(y)$. Let a finite difference scheme for the equation be $y_{k+1} = F(y_k; \Delta t)$. Let the differential equation and/or its solutions have property $\mathcal{P}$. The discrete model equation is dynamically consistent with the differential equation if it and/or its solutions also have property $\mathcal{P}$.

Definition 2.12. (see [**88**]) The NSFD method (2.2) is called *positive* or *positivity-preserving*, if, for any value of the step size Δt, and $y_0 \in \mathbb{R}^n_+$ its solution remains positive, i.e., $y_k \in \mathbb{R}^n_+$ for all $k \in \mathbb{N}$.

3. Construction of the NSFD scheme

In this section, we construct a generalized NSFD scheme for the model (1.1) and investigate its qualitative dynamical properties.

3.1. Derivation of the NSFD scheme and its positivity. First, suppose that the function f satisfies the condition **(C1)** with a real number α. We use the following non-local approximation for the function f

$$f(y(t_k)) = f(y(t_k)) + \big(my(t_k) - my(t_k)\big) \approx f(y_k) + my_k - my_{k+1}, \tag{3.1}$$

where m is a real number. By combining (3.1) with the nonstandard approximation for the first-order derivative dy/dt, we obtain the following NSFD model for the system (1.1):

$$\frac{y_{k+1} - y_k}{\phi(\Delta t)} = f(y_k) + my_k - my_{k+1}, \tag{3.2}$$

where $\phi(\Delta t) = \Delta t + \mathcal{O}(\Delta t^2)$ as $\Delta t \to 0$. The explicit form of the scheme (3.2) is given by

$$y_{k+1} = \frac{y_k + \phi f(y_k) + \phi m y_k}{1 + m\phi}, \tag{3.3}$$

or equivalently,

$$y_{k+1} = y_k + \frac{\phi}{1 + m\phi} f(y_k). \tag{3.4}$$

Remark 3.1. Although the rule of nonlocal approximation of the NSFD methodology states that nonlinear terms in the right-hand side $f(y)$ should be approximated in a nonlocal way, in many cases linear terms with negative coefficients, namely, $-cy$ ($c > 0$) can be also nonlocally approximated by $-cy_{k+1}$ to guarantee the positivity of the corresponding NSFD schemes (see, for instance, [**1**,**8**,**12**,**22**,**43**,**44**,**55**,**56**,**65**]). Hence, the approximation (3.1) can be considered as a nonlocal approximation, which follows the NSFD methodology.

Lemma 3.2. *The NSFD sheme* (3.2) *preserves the positivity of the model* (1.1) *for all the values of the step size* Δt *whenever*

$$m \geq m_P := \max\{\alpha, 0\}. \tag{3.5}$$

PROOF. We use mathematical induction to prove the positivity of the NSFD scheme (3.2). Indeed, assume that $y_k \geq 0$. Then,

$$f(y_k) + my_k \geq f(y_k) + \alpha y_k \geq 0.$$

Combining this estimate with (3.3), we obtain

$$y_{k+1} = \frac{y_k + \phi(f(y_k) + my_k)}{1 + m\phi} \geq 0.$$

This is the desired conclusion. The proof is complete. □

REMARK 3.3. The discretization (3.1) guarantees that the NSFD scheme (3.2) is consistent. Similarly to convergence analysis presented in [**19**], we can conclude that the NSFD scheme (3.2) is convergent of order 1.

REMARK 3.4. It is easy to see from (3.4) that the NSFD scheme (3.2) can be rewritten in the form

$$\frac{y_{k+1} - y_k}{\Phi(\Delta t)} = f(y_k),$$

where

$$\Phi(\Delta t) = \frac{\phi(\Delta t)}{1 + m\phi(\Delta t)} = \phi(\Delta t) + \mathcal{O}(\phi^2(\Delta t)) = \Delta t - \mathcal{O}(\Delta t^2).$$

Hence, the NSFD scheme (3.2) can be equivalently derived by using renormalization of the denominator. Here, it is important to note that conditions for the NSFD scheme (3.2) to be dynamically consistent with the model (1.1) will be set to m.

3.2. Equilibria and their asymptotic stability. Let us denote by $\mathcal{E}_C$ and $\mathcal{E}_D$ the sets of equilibrium points of the continuous-time dynamical system (1.1) and the discrete-time model (3.2), respectively. It should be emphasized that $\mathcal{E}_C \subset \mathcal{E}_D$ for most standard finite difference schemes such as the Taylor's and Runge-Kutta schemes since they often generate spurious equilibrium points, which depend on the step size [**59–64**]. However, we infer from the system (3.4) that:

LEMMA 3.5. *$\mathcal{E}_C = \mathcal{E}_D$ for all the finite values of the step size Δt. In other words, the NSFD scheme* (3.2) *preserves the set of equilibrium points of the model* (1.1).

Let us denote by $\mathcal{E}_C^S$ the set of all asymptotically stable equilibrium points of the system (1.1). To analyze the stability of the NSFD scheme (3.2), we introduce the following notation

$$\begin{aligned} \mathcal{F} &= \bigcup_{y^* \in \mathcal{E}_C^S} \sigma(J_C(y^*)), \\ m_S &= \max_{\lambda \in \mathcal{F}} \left\{ -\frac{|\lambda|^2}{2Re\lambda} \right\}, \end{aligned} \tag{3.6}$$

where $J_C(y^*)$ is the Jacobian matrix of the system (1.1) evaluating at y^* and $\sigma(J_C(y^*))$ is the set of eigenvalues of $J_C(y^*)$. Note that $Re\lambda < 0$ for all $\lambda \in \mathcal{F}$ (see Theorem 2.2).

THEOREM 3.6. *Let m be a real number with the property that*

$$m \geq m_S, \tag{3.7}$$

where m_S is given in (3.6). *Then, the NSFD scheme* (3.2) *preserves the LAS of the system* (1.1) *for all the values of the step size Δt.*

PROOF. Let y^* be any equilibrium point of the continuous system (1.1). Let us denote by $J_C(y^*)$ and $J_D(y^*)$ the Jacobian matrices of the systems (1.1) and (3.2) evaluating at y^*, respectively. Then, it follows from the equation (3.4) that

$$J_D = I + \frac{\phi}{1+m\phi} J_C, \tag{3.8}$$

where I is the identity matrix. The relation (3.8) implies that

$$\begin{aligned}\det(xI - J_C) &= \det\left[\left(x + \frac{1+m\phi}{\phi}\right)I - \frac{1+m\phi}{\phi}J_D\right]\\ &= \left(\frac{1+m\phi}{\phi}\right)^n \det\left[\left(1 + \frac{\phi x}{1+m\phi}\right)I - J_D\right].\end{aligned}$$

Hence, λ is an eigenvalue of J_C if and only if $\mu = 1 + \dfrac{\phi}{1+m\phi}\lambda$ is an eigenvalue of J_D. This implies that

$$\begin{aligned}|\mu_i|^2 = \left|1 + \frac{\phi}{1+m\phi}\lambda_i\right|^2 &= \left(1 + \frac{\phi}{1+m\phi}Re\lambda_i\right)^2 + \left(\frac{\phi}{1+m\phi}Im\lambda_i\right)^2\\ &= 1 + \frac{2\phi}{1+m\phi}Re\lambda_i + \left(\frac{\phi}{1+m\phi}\right)^2 |\lambda_i|^2.\end{aligned} \tag{3.9}$$

We now consider the following two cases of the stability of y^*.

Case 1. y^* is an unstable equilibrium point of the system (1.1). In this case, it follows from Theorem 2.2 that there exists $\lambda_i \in \sigma(J_C(y^*))$ such that $Re\lambda_i > 0$. Then, we infer from (3.9) that the corresponding eigenvalue μ_i of $J_D(y^*)$ satisfies

$$|\mu_i|^2 = 1 + \frac{2\phi}{1+m\phi}Re\lambda_i + \left(\frac{\phi}{1+m\phi}\right)^2 |\lambda_i|^2 > 1,$$

which implies that y^* is an unstable equilibrium point of the system (3.2) (see Theorem 2.7).

Case 2. y^* is an asymptotically stable equilibrium point of the system (1.1). In this case, $Re\lambda < 0$ for all $\lambda \in \sigma(J_C(y^*))$. By using (3.9), we obtain $|\mu|^2 < 1$ if and only if

$$(2mRe\lambda + |\lambda|^2)\phi < -2Re\lambda.$$

This condition is always satisfied if $m \geq -|\lambda|^2/(2Re\lambda)$. Hence, we obtain from the condition $m \geq m_s$ and Theorem 2.7 that y^* is an asymptotically stable equilibrium point of the NSFD scheme (3.2).

Finally, the proof is completed by combining Cases 1 and 2. $\square$

REMARK 3.7. In many cases, it is not easy to obtain conditions for NSFD schemes to be dynamically consistent with respect to the asymptotic stability of dynamical systems (see, for instance, [**19, 22, 24, 26, 30, 43, 44, 87**]). However, we can obtain the stability threshold m_S of the NSFD scheme (3.2) by simple calculations or a simple numerical algorithm.

3.3. Conservation laws. Assume that the system (1.1) satisfies conservation laws. We show that the NSFD scheme (3.2) has the ability to preserve this property.

LEMMA 3.8. *The NSFD scheme* (3.2) *preserves the DCL for all finite step sizes.*

PROOF. Assume that the system (1.1) satisfies a DCL, that is the total population satisfies

$$\frac{dP_T}{dt} = \frac{dy_1}{dt} + \frac{dy_2}{dt} + \ldots + \frac{dy_n}{dt} = 0,$$

which implies that $\sum_{i=1}^n y_i(t)$ is constant, and hence, $\sum_{i=1}^n f_i(y(t)) = 0$ for all $t \geq 0$. By using (3.4) we obtain

$$P_{T,k+1} := \sum_{i=1}^n y_{i,k+1} = \sum_{i=1}^n y_{i,k} + \frac{\phi}{1+m\phi}\sum_{i=1}^n f_i(y_k) = \sum_{i=1}^n y_{i,k+1} := P_{T,k}.$$

This implies that the DCL is preserved automatically by the NSFD scheme (3.2). The proof is complete. □

Next, assume that a GCL exists for the system (1.1), that is

$$\frac{dP_T}{dt} = a_1 - b_1 P_T, \quad a_1, b_1 > 0. \tag{3.10}$$

Note that (3.10) has an exact finite difference scheme of the form (see [**65**])

$$\frac{P_{T,k+1} - P_{T,k}}{\phi^*(\Delta t)} = a_1 - b_1 P_{T,k}, \quad \phi^*(\Delta t) = \frac{1 - e^{-b_1 \Delta t}}{b_1}. \tag{3.11}$$

Applying the NSFD scheme (3.2) we obtain

$$P_{T,k+1} = P_{T,k} + \frac{\phi}{1+m\phi}(a_1 - b_1 P_{T,k}). \tag{3.12}$$

LEMMA 3.9. *The following assertions hold for the difference equation* (3.12)*:*

(1) *the sequence* $\{P_{T,k}\}_k$ *defined in* (3.12) *monotonically converges to* a_1/b_1 *as* $k \to \infty$*, provided that*

$$m \geq m_{GCL} := b_1; \tag{3.13}$$

(2) *The scheme* (3.12) *is an exact finite difference scheme for* (3.10) *if*

$$\phi(\Delta t) = \phi_E(\Delta t) := \frac{\phi^*(\Delta t)}{1 - m\phi^*(\Delta t)}, \tag{3.14}$$

where ϕ^* *is given in* (3.11).

PROOF. **Proof of Part (1).** It follows from (3.12) that

$$\begin{aligned}
P_{T,k+1} &= P_{T,k} + \frac{\phi}{1+m\phi}(a_1 - b_1 P_{T,k}) = \frac{a_1\phi}{1+m\phi} + \left(1 - \frac{b_1\phi}{1+m\phi}\right)P_{T,k}\\
&= \frac{a_1\phi}{1+m\phi} + \left(1 - \frac{b_1\phi}{1+m\phi}\right)\left[\frac{a_1\phi}{1+m\phi} + \left(1 - \frac{b_1\phi}{1-m\phi}\right)P_{T,k-1}\right]\\
&= \frac{a_1\phi}{1+m\phi} + \frac{a_1\phi}{1+m\phi}\left(1 - \frac{b_1\phi}{1+m\phi}\right) + \left(1 - \frac{b_1\phi}{1+m\phi}\right)^2 P_{T,k-1}.
\end{aligned}$$

By continuing the above process, we have

$$P_{T,k+1} = \frac{a_1\phi}{1+m\phi}\sum_{i=0}^k \left(1 - \frac{b_1\phi}{1+m\phi}\right)^i + \left(1 - \frac{b_1\phi}{1+m\phi}\right)^k P_{T,0}.$$

Since $m \geq b_1$, $1 - \dfrac{b_1\phi}{1+m\phi} \in (-1,1)$. This implies that

$$\lim_{t\to\infty} P_{T,k} = \frac{a_1\phi}{1+m\phi}\left(\frac{b_1\phi}{1+m\phi}\right)^{-1} = \frac{a_1}{b_1}. \tag{3.15}$$

On the other hand, we infer from $m \geq b_1$ that

$$\frac{\partial P_{T,k+1}}{\partial P_{T,k}} = 1 + \frac{\phi}{1+m\phi}(-b_1) \geq 0,$$

which implies the monotonocity of $\{P_{T,k}\}_k$. The proof is completed.

Proof of Part (2). The exact scheme (3.11) can be written in the form

$$P_{T,k+1} = P_{T,k} + \phi^*(a_1 - b_1 P_{T,k}). \tag{3.16}$$

We infer from (3.16) and (3.12) that (3.12) is an exact scheme for (3.10) if

$$\phi^* = \frac{\phi}{1+m\phi}.$$

This condition is equivalent to (3.14). The proof is complete. □

From Lemma 3.9 we have:

THEOREM 3.10. *The NSFD scheme* (3.2) *satisfies the GCL if one of the following conditions holds:*

(1) $m \geq m_{GCL} := b_1$.
(2) $\phi(\Delta t) = \phi_E(\Delta t)$, *where* ϕ_E *is given in* (3.14).

REMARK 3.11. The function ϕ_E given in (3.14) satisfies $\phi(\Delta t) = \Delta t + \mathcal{O}(\Delta t^2)$ and is defined for all $\Delta t > 0$ if $b_1 \geq m$. When $b_1 < m$, it is defined only if $\Delta t \neq -\dfrac{\ln(m-b_1)}{b_1}$.

REMARK 3.12. The sub-conservation law will be satisfied if the DCL and GCL are satisfied.

3.4. Dynamics consistency. For the sake of convenience, we summarize the results constructed in Subsections 3.1–3.4 in the following theorem.

THEOREM 3.13 (Dynamic consistency). *The NSFD scheme* (3.2) *is*

- *dynamically consistent with the positivity of the model* (1.1) *if the condition* (3.5) *is satisfied;*
- *dynamically consistent with the LAS of the model* (1.1) *if the condition* (3.7) *is satisfied;*
- *dynamically consistent with the DCL of the model* (1.1) *for all the values of* m*;*
- *dynamically consistent the GCL the model* (1.1) *if*

$$m \geq m_{GCL} := b_1 \tag{3.17}$$

or

$$\phi(\Delta t) = \phi_E(\Delta t), \tag{3.18}$$

where ϕ_E *is given in* (3.14).

REMARK 3.14.

- The positivity, LAS and conservation laws are independent of each other. So, they may not occur simultaneously. Therefore, the threshold parameter m should be selected depending on properties that the model (1.1) can have.
- The determination of m_P in (3.5) and m_{GCL} in (3.17) is trivial while it is also easy to write a numerical algorithm to compute m_S.
- Since the conditions for the NSFD scheme (3.2) to be dynamically consistent with respect to the positiivity, LAS and conservation laws of the model (1.1) are set to the parameter m, we can choose the denominator function $\phi(\Delta t)$ to be Δt. This shows the role of the parameter m for the dynamic consistency of the NSFD scheme (3.2).

3.5. A note on the nonstandard Euler scheme. As emphasized in Remark 3.4, the NSFD scheme (3.2) can be considered as a nonstandard Euler method. For the sake of convenience, we now consider a special case of the parameter m, namely, $m = 0$. In this case, the NSFD scheme (3.2) becomes

$$y_{k+1} = y_k + \phi(\Delta t)f(y_k). \tag{3.19}$$

This can be considered as a "weakly" NSFD scheme [**28**].

Because of the absence of parameter m, conditions for the dynamic consistency of the scheme (3.19) will be set to the denominator function ϕ. Below, we will give conditions for the scheme (3.19) to be dynamically consistent with the model (1.1).

We infer from (3.19) that $y_{k+1} \geq 0$ for all $\phi > 0$ if $\alpha \leq 0$. In this case, the nonstandard Euler scheme is unconditionally positive. Otherwise, if $\alpha > 0$, then it follows from (1.2) that

$$y_{k+1} = y_k + \phi f(y_k) \geq y_k + (-\alpha y_k) = (1 - \phi\alpha)y_k.$$

Consequently, $y_{k+1} \geq 0$ if $\phi < (\alpha)^{-1}$. So, we have the following result on the positivity of the nonstandard Euler scheme.

LEMMA 3.15. *The nonstandard Euler scheme* (3.19) *preserves the positivity of the model* (1.1) *for all the values of the step size* Δt *whenever the denominator function satisfies the following condition for all* $\Delta t > 0$

$$\phi(\Delta t) < \phi_P := \begin{cases} \infty, & \alpha \leq 0, \\ (\alpha)^{-1}, & \alpha > 0. \end{cases} \tag{3.20}$$

The following result is a direct consequence of the results presented in [**28, 29**].

COROLLARY 3.16. The nonstandard Euler scheme (3.19) preserves the LAS of the model (1.1) for all the values of the step size Δt if the denominator function satisfies the following condition for all $\Delta t > 0$

$$\phi(\Delta t) < \phi_S := \min_{\lambda \in \mathcal{F}} \left\{ -\frac{2Re\lambda}{|\lambda|^2} \right\}, \tag{3.21}$$

where the set $\mathcal{F}$ is defined in (3.6).

It is easily seen that the nonstandard Euler scheme always preserves the DCL. On the other hand, when applying the nonstandard Euler scheme for systems satisfying the GLC, we obtain

$$\frac{P_{T,k+1} - P_{T,k}}{\phi(\Delta t)} = a_1 - b_1 P_{T,k}. \tag{3.22}$$

Hence, the GLC is preserved if

$$\phi < \phi_{GCL} = (b_1)^{-1}. \tag{3.23}$$

Furthermore, the scheme (3.22) becomes an exact scheme for (3.10) if

$$\phi(\Delta t) = \frac{1 - e^{-b_1 \Delta t}}{b_1}. \tag{3.24}$$

Hence, we obtain the following assertion.

COROLLARY 3.17. The nonstandard Euler scheme (3.19) is always dynamically consistent with respect to the DCL and is dynamically consistent with respect to the GCL if the condition (3.23) or the condition (3.24) holds.

REMARK 3.18.

- By choosing suitable denominator functions, the nonstandard Euler scheme (3.19) can preserve the positivity, LAS and conservation laws as the NSFD scheme (3.2). Although the schemes (3.2) and (3.19) are all convergent of order 1, in the next section we will provide a numerical example in which errors of Scheme (3.2) are better than errors of Scheme (3.19).
- The condition (3.24) can conflict with the conditions (3.20) and (3.21).

4. Numerical experiments

In this section, we perform some numerical examples to illustrate the theoretical findings and to show advantages of the proposed NSFD scheme.

EXAMPLE 4.1 (Numerical simulation of single-species population models). Consider single-species population models, which describe the population size changes in the absence of disease and of maturation delay [**18**]

$$\frac{dN}{dt} = B(N)N - dN, \tag{4.1}$$

where $N(t)$ stands for the population size at the time t, $d > 0$ is the death rate constant, and $B(N)N$ is a birth rate function with $B(N)$ has the following properties for all $N \in (0, \infty)$

(A1) $B(N) > 0$;
(A2) $B(N)$ is continuously differentiable with $B'(N) < 0$;
(A3) $B(0^+) > d > B(\infty)$.

The following functions are examples of birth functions $B(N)$ that are found in biological systems and satisfy (A1)-(A3) (see [**18**])

(B1) $B_1(N) = be^{-aN}$, with $a > 0$ and $b > d$;
(B2) $B_2(N) = \dfrac{p}{q + N^n}$, with $p, q, n > 0$ and $p > qd$;
(B3) $B_3(N) = \dfrac{A}{N} + c$, with $A > 0$, $d > c > 0$.

It is clear that the model (4.1) satisfies the condition **(C1)** with $\alpha = d$. On the other hand, it possesses two equilibrium points, that are $N^0 = 0$ and $N^* = B^{-1}(d)$. Moreover, N^0 is always unstable while N^* is always asymptotically stable. Therefore, the condition **(C2)** also satisfies. If $B(N) = B_3(N)$, then the model (4.1) becomes linear and satisfies the GCL.

We consider the model (4.1) with

$$B(N) = B_1(N) = be^{-aN}N, \quad a = d = 1, \quad b = 12$$

as a test problem for numerical schemes. In this case, the model has a unique equilibrium point

$$N^* = -\ln(d/b)/a \approx 2.4849,$$

which is locally asymptotically stable. Note that the conservation laws are absent in this example.

For the NSFD scheme (3.2), we compute the parameter m as

$$m \geq \max\{m_P,\ m_S\} = \max\{1,\ 1.2425\} = 1.2425.$$

For the nonstandard Euler scheme, the condition for the denominator function can be computed as

$$\phi(\Delta t) < \min\{\phi_P,\ \phi_S\} = \min\{1,\ 0.8049\} = 0.8049$$

Numerical solutions generated by the NSFD scheme (3.2) with $m = 2.5$ and $\Delta t \in \{2.0,\ 0.1\}$ and by the nonstandard Euler scheme with $\phi(\Delta t) = (1 - e^{-2.5\Delta t})/2.5$ and $\Delta t \in \{1,\ 0.1\}$ are depicted in Figure 1. In this figure, each blue curve represents a solution corresponding to a specific initial value and the green arrows represents the evolution of the continuous model. We observe from Figure 1 that the numerical solutions generated by the NSFD model (3.2) are positive and stable. Furthermore, the NSFD model replicates the dynamics of the continuous model and its behaviour does not depend on the chosen step sizes. Similarly, the nonstandard Euler scheme (3.19) also preserves the positivity and stability of the continuous model regardless of the chosen step sizes.

We now use the standard Euler scheme and a well-known two-stage Runge-Kutta (RK2) scheme, namely, the explicit trapezoidal scheme (see [**13**]) for the purpose of comparing them with the NSFD schemes. Approximations obtained from these standard schemes are sketched in Figure 2. It is clear that the standard Euler scheme generates an unstable numerical solution, which oscillates around the equilibrium position. Hence, the dynamics of the continuous model cannot be preserved. The similar unusual phenomena occur when using the standard RK2 scheme with step sizes $\Delta t = 1$ and $\Delta t = 1.2$. More clearly, the RK2 scheme using $\Delta t = 1.2$ generates the approximation oscillating around the equilibrium point and when $\Delta t = 1.0$, the corresponding numerical solution converges to a spurious equilibrium point. This result is completely consistent with analysis performed by Mickens in [**59**–**63**].

Before ending this example, we compute errors obtained from the NSFD scheme (3.2), the nonstandard Euler scheme (3.19) and the standard Euler scheme. Note that theses schemes are all convergent of order 1 since $\phi(\Delta t) = \Delta t + \mathcal{O}(\Delta t^2)$. For this purpose, we consider $N(0) = 2$ and use the approximation generated by the classical four-stage Runge-Kutta scheme (see [**13**]) with $\Delta t = 10^{-5}$ as a benchmark (reference) solution since it is impossible to find the exact solution. Errors provided by the three numerical schemes are shown in Tables 1 and 2. In these tables, *error* and $error_T$ are computed by

$$error = \max_{1 \leq k \leq N} |N(t_k) - N_k|, \quad error_T = \sum_{i=1}^{N} |N(t_k) - N_k|.$$

TABLE 1. The values of the *error* of the NSFD scheme (3.2) with $m_s = 1.5$, the nonstandard Euler scheme with $\phi(h) = (1 - e^{-1.5\Delta t})/1.5$ and the standard Euler scheme

Δt	NSFD *error*	N. Euler *error*	S. Euler *error*
1	0.053229803837052	0.200391028658851	1.895818024330166
10^{-1}	0.002547712779198	0.010718671314132	0.025728835277831
10^{-2}	$3.474294297345359e-004$	$9.794823732640623e-004$	0.002320869273385
10^{-3}	$3.568404980169859e-005$	$9.709160784732163e-005$	$2.298145180823497e-004$
10^{-4}	$3.577852512925972e-006$	$9.700655808853043e-006$	$2.295902635651714e-005$
10^{-5}	$3.578797715952931e-007$	$9.699803174001431e-007$	$2.295678681374369e-006$

TABLE 2. The values of the $error_T$ of the NSFD scheme (3.2) with $m_s = 1.5$, the nonstandard Euler scheme with $\phi(h) = (1 - e^{-1.5\Delta t})/1.5$ and the standard Euler scheme

Δt	NSFD $error_T$	N. Euler $error_T$	S. Euler $error_T$
1	0.056862533841534	0.257861518971072	15.530365282265436
10^{-1}	0.030063654051754	0.109862865600443	0.256967636704650
10^{-2}	0.039734990275286	0.103390564043856	0.247233541153000
10^{-3}	0.040703612013705	0.102744510888389	0.246264374931792
10^{-4}	0.040800485618428	0.102679929285270	0.246167506109657
10^{-5}	0.040810166884441	0.102673437145564	0.246157825305468

From Tables 1 and 2, we see that the errors provided by the NSFD scheme (3.2) are better. The errors of the nonstandard Euler scheme are better than the ones provided by the standard Euler scheme. Hence, advantages of the nonstandard schemes in term of errors are shown.

EXAMPLE 4.2 (Numerical simulation of a predator-prey system). Consider predator-prey model with a Beddington-DeAngelis functional response and linear intrinsic growth rate of the prey population, which is derived from a general predator-prey model proposed by DeAngelis et al in [**27**]. The model under consideration is given by

$$\begin{aligned} \frac{dx}{dt} &= x - \frac{Axy}{1+x+y}, \\ \frac{dy}{dt} &= \frac{Exy}{1+x+y} - Dy, \end{aligned} \tag{4.2}$$

where x and y are the prey and predator population sizes, respectively, and all the parameters are assumed to be positive. We refer the readers to [**31**] for details of this model.

Clearly, the model (4.2) satisfies the condition (**C1**) with $\alpha = \max\{A-1, D\}$. On the other hand, stability analysis in [**31**] implies that **(C2)** holds if $A \neq E$. Numerical simulation of the model (4.2) by NSFD schemes was also examined in [**29**].

We now consider the predator-prey model (4.2) with the following parameters

$$A = 6, \quad D = 5, \quad E = 7.$$

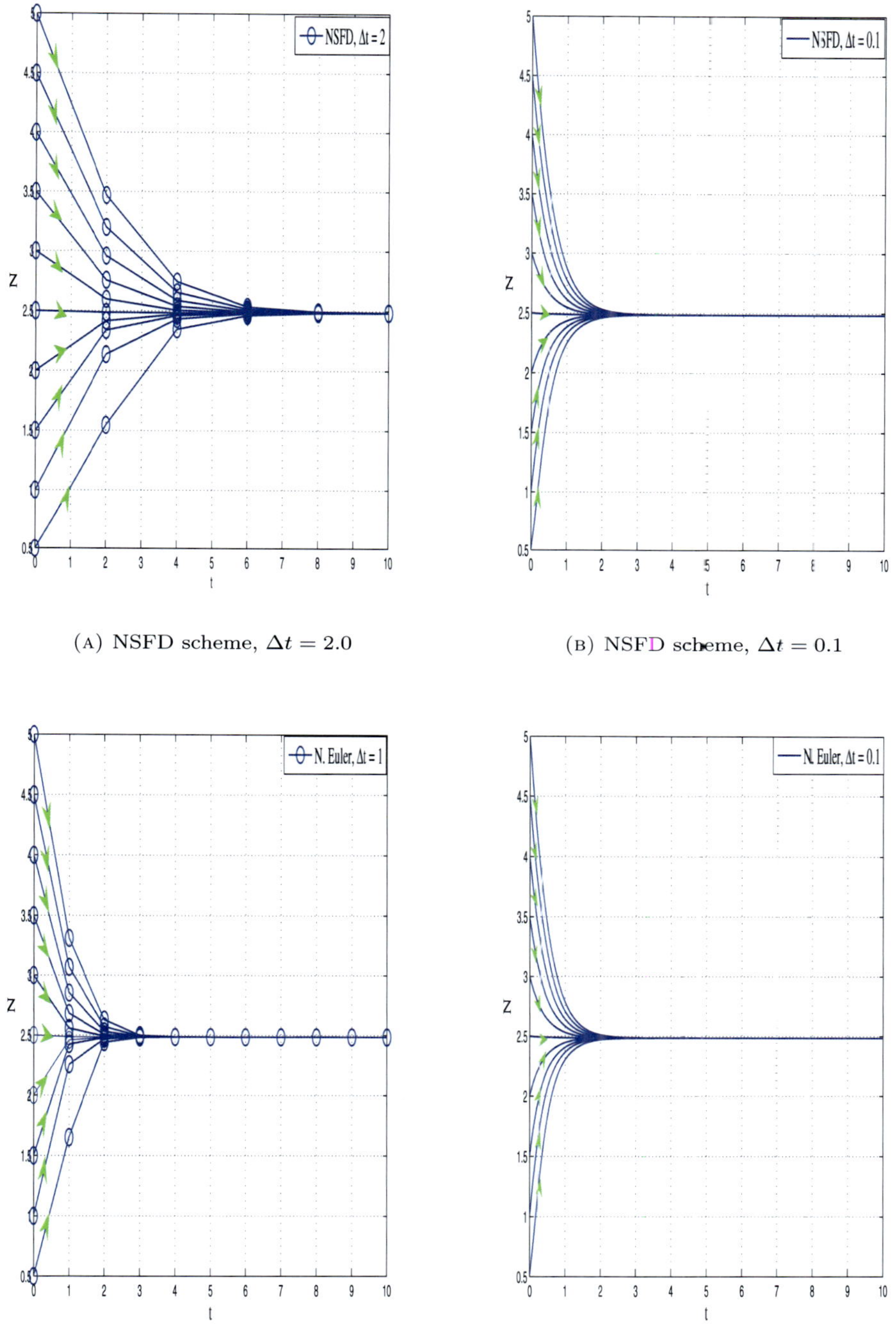

(A) NSFD scheme, $\Delta t = 2.0$

(B) NSFD scheme, $\Delta t = 0.1$

(C) Nonstandard Euler scheme, $\Delta t = 1.0$

(D) Nonstandard Euler scheme, $\Delta t = 0.1$

FIGURE 1. Numerical approximations of the model (4.1) generated by the NSFD scheme (3.2) with $m = 2.5$ and $\Delta t \in \{2.0, 0.1\}$ and by the nonstandard Euler scheme (3.19) with $\phi(\Delta t) = (1 - e^{-2.5\Delta t})/2.5$ and $\Delta t \in \{1, 0.1\}$.

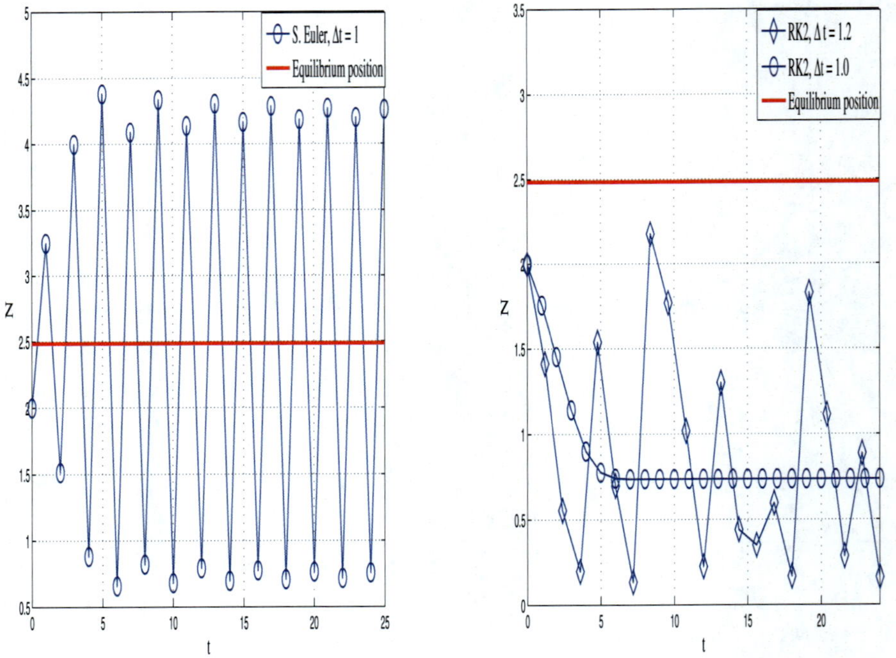

(A) Euler scheme using $\Delta t = 1.0$.

(B) RK2 scheme using $\Delta t = 1.0$ and $\Delta t = 1.2$

FIGURE 2. Numerical approximation of the model (4.1) generated by the standard Euler and RK2 schemes.

In this case, the model has a unique positive equilibrium point $(x^*, y^*) = (6,\ 1.4)$ and the Jacobian matrix evaluating at (x^*, y^*) is

$$J(x^*, y^*) = \begin{pmatrix} 0.7143 & -3.5714 \\ 0.3333 & -0.8333 \end{pmatrix}.$$

Consequently, its eigenvalues are

$$\lambda_{1,2} = -0.0595 \pm 0.7692i,$$

which implies that (x^*, y^*) is locally asymptotically stable.

By simple calculations, we determine the condition for the parameter m of the NSFD scheme (3.2) as

$$m \geq \max\{m_P, m_S\} = 5.$$

Therefore, we can take $m = 5.1$.

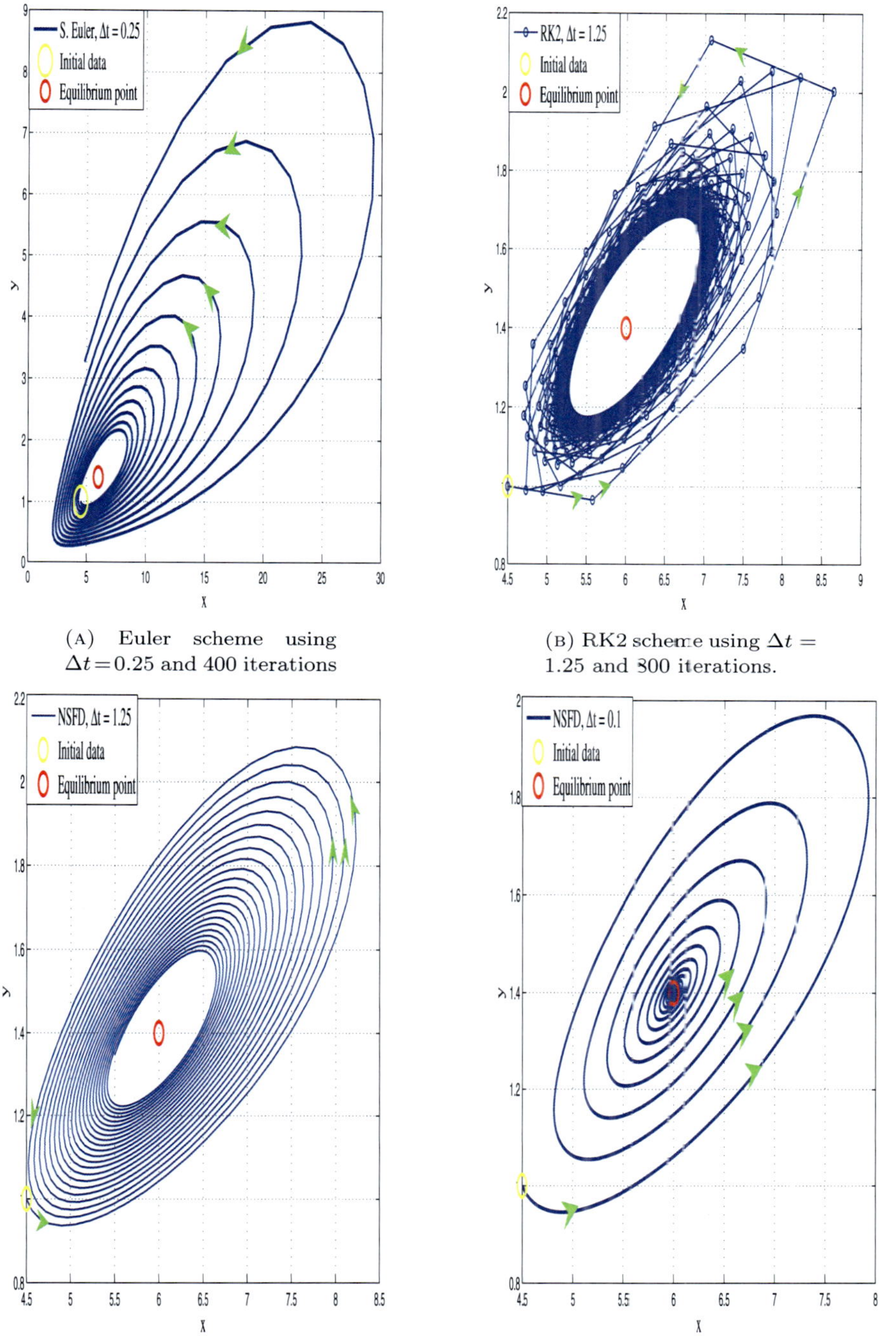

(A) Euler scheme using $\Delta t = 0.25$ and 400 iterations

(B) RK2 scheme using $\Delta t = 1.25$ and 800 iterations.

(C) NSFD scheme using $\Delta t = 1.25$ and 800 iterations

(D) NSFD scheme using $\Delta t = 0.1$ and 10^5 iterations

FIGURE 3. Numerical approximations of the model (4.2) generated by the NSFD scheme (3.2) and the standard Euler and RK2 schemes.

Numerical approximations provided by the Euler and RK2 schemes and the NSFD scheme (3.2) using $(x(0), y(0)) = (4.5, 1)$ are depicted in Figure 3. We see that the phase plane generated by the Euler scheme is unstable and spirals out from the equilibrium position; the RK2 scheme provides an approximation which oscillates around the equilibrium point; meanwhile, the approximations obtained from the NSFD scheme (3.2) correctly preserves the dynamics of the predator-prey model.

Note that the condition for the denominator function ϕ of the nonstandard Euler scheme (3.19) is

$$\phi < \min\{\phi_P, \phi_S\} = \frac{1}{5}.$$

Hence, we can take $\phi(\Delta t) = (1 - e^{-\tau\Delta t})/\tau$, where $\tau > 5$. Then, the dynamic consistency of the nonstandard Euler scheme is guaranteed.

EXAMPLE 4.3 (Numerical simulation of a vaccination model). Consider an infectious disease with a preventive (prophylactic) vaccine and an effective therapeutic treatment proposed by Gumel and Moghadas in [**38**]. This model monitors the temporal dynamics of susceptible individuals (S), vaccinated individuals (V) and infected individuals (I) and is described by

$$\begin{aligned} \frac{dS}{dt} &= \Pi - \frac{c\beta_1 I}{1+I}S - \xi S + \alpha^* I - \mu S, \\ \frac{dV}{dt} &= \xi S - \frac{c\beta_2 I}{1+I}V - \mu V, \\ \frac{dI}{dt} &= \frac{c\beta_1 I}{1+I}S + \frac{c\beta_2 I}{1+I}V - \alpha^* I - \mu I. \end{aligned} \tag{4.3}$$

Here, all the parameters are positive. Mathematical formulation and qualitative study for the model (4.3) were performed in [**38**].

It is easy to check that the model (4.3) satisfies the condition **(C1)** with $\alpha = \max\{c\beta_1 + \xi + \mu, c\beta_2 + \mu, \alpha^* + \mu\}$. Meanwhile, the condition **(C2)** holds if the basic reproduction number (BRN) of the model satisfies

$$\mathcal{R}_0 = \frac{c\beta_1\Pi}{(\mu + \alpha^*)(\mu + \xi)} + \frac{c\beta_2\Pi\xi}{\mu(\mu + \alpha^*)(\mu + \xi)} \neq 1.$$

Also, the GCL is satisfied because

$$\frac{d(S + V + I)}{dt} = \Pi - \mu(S + V + I).$$

In this example, we consider the vaccination model (4.3) with the following parameters

$$\begin{aligned} &\Pi = 700, \quad \beta_1 = 10^{-4}, \quad \beta_2 = 10^{-6}, \quad \mu = 0.03 \qquad \text{(see [38])}, \\ &\alpha^* = 0.95, \qquad c = 8, \qquad \xi = 0.95, \qquad\qquad \text{(assumed)}. \end{aligned}$$

In this case, the BRN is $\mathcal{R}_0 = 0.7677 < 1$ and the model has a unique disease-free (uninfected) equilibrium point $E^0 = (S^0, V^0, I^0) = (714.2857,\ 2.2619 \times 10^4,\ 0)$, which is locally asymptotically stable (see [**38**]).

We now compute the parameter m for the NSFD scheme (3.2). First, it is easy to see that

$$m_P = \max\{c\beta_1 + \xi + \mu, c\beta_2 + \mu, \alpha + \mu\} = 0.98.$$

Since the Jacobian matrix of the model evaluating at E^* has three eigenvalues, that are

$$\lambda_1 = -0.98, \quad \lambda_2 = -0.03, \quad \lambda_3 = -0.23,$$

we obtain

$$m_S = 0.49.$$

Note that the model satisfying a GCL, namely,

$$\frac{d(S+V+I)}{dt} = \Pi - \mu(S+V+I).$$

Consequently, the condition for the preservation of the GCL is

$$m \geq m_{GCL} = \mu = 0.03.$$

Therefore, the condition for the parameter m is

$$m \geq \max\{m_P, m_S, m_{GCL}\} = 0.98.$$

In a similar way, we can determine the condition for the nonstandard Euler scheme as

$$\phi(\Delta t) = \frac{1 - e^{-\tau \Delta t}}{\tau}, \quad \tau > (0.98)^{-1} \approx 1.02.$$

Numerical solutions obtained from the standard Euler, RK2 and NSFD scheme are depicted in Figure 4, respectively. In these figures, each blue curve represents a phase space corresponding to a specific initial value, the green arrows show the evolution of the continuous model and the red circle indicates the position of the stable equilibrium point.

We observe from Figures 4a and 4b that the standard Euler scheme generates a negative and unstable numerical approximation, which oscillates around the equilibrium point with an increasing amplitude; the solution provided by the standard RK2 scheme is not positive and converges to a spurious equilibrium point; conversely, the NSFD scheme correctly preserves the dynamics of the continuous model regardless of the chosen step sizes (see Figures 4c and 4d).

Note that the vaccination model satisfies a GCL. For this reason, we now consider approximations of the total population generated by the Euler, RK2 and NSFD scheme (3.2). From numerical approximations in Figure 5, we see that the NSFD schemes preserves the GCL but this property is not preserved by the standard schemes.

5. Conclusions, discussions and some open problems

In this work, we have applied the NSFD methodology proposed by Mickens to construct a generalized NSFD method for a class of autonomous dynamical systems that satisfies positivity, stability and conservation laws. This NSFD method is constructed based on a novel non-local approximation for the right-side functions of the dynamical systems. By rigorous mathematical analyses, we have shown that the NSFD method is dynamically consistent with respect to the positivity, asymptotic stability and three classes of conservation laws, including direct conservation, generalized conservation and sub-conservation laws. Additionally, a set of numerical experiments has been also performed to illustrate the theoretical findings and to show advantages of the proposed NSFD method over some well-known standard

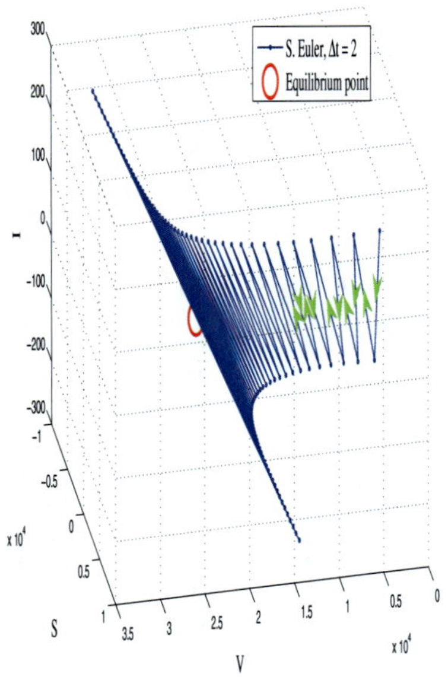

(A) Euler scheme using $(S(0), V(0), I(0)) = (500, 2000, 100)$ and $\Delta t = 2$ after 100 iterations

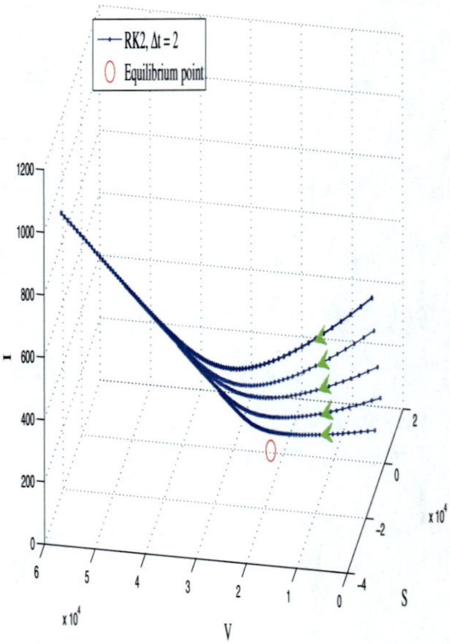

(B) RK2 scheme using $\Delta t = 2$ after 100 iterations

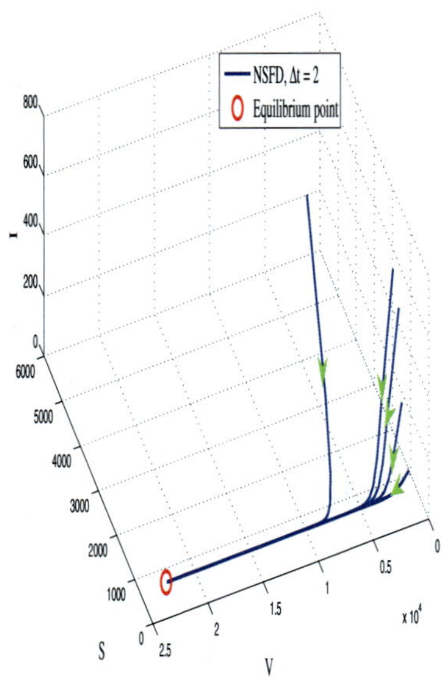

(C) NSFD scheme with $\Delta t = 2.0$ after 250 iterations

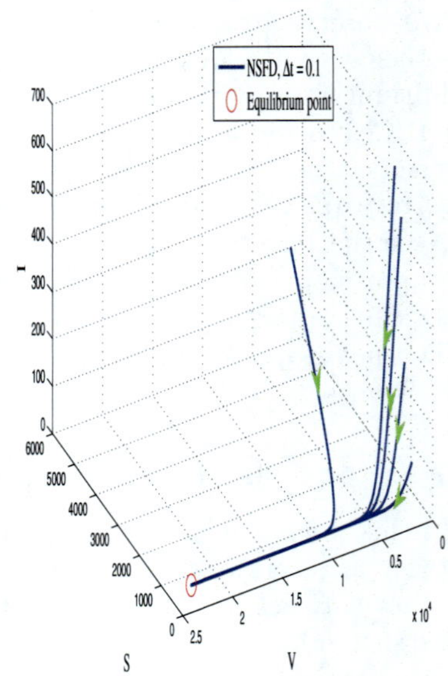

(D) NSFD scheme with $\Delta t = 0.1$ after 5000 iterations

FIGURE 4. Numerical approximations of the model (4.3) generated by the standard Euler and RK2 schemes and by the NSFD scheme (3.2).

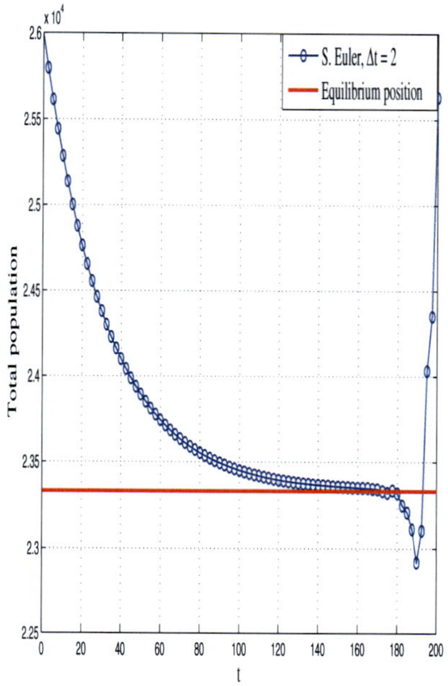

(A) Euler scheme using $\Delta t = 2.5$

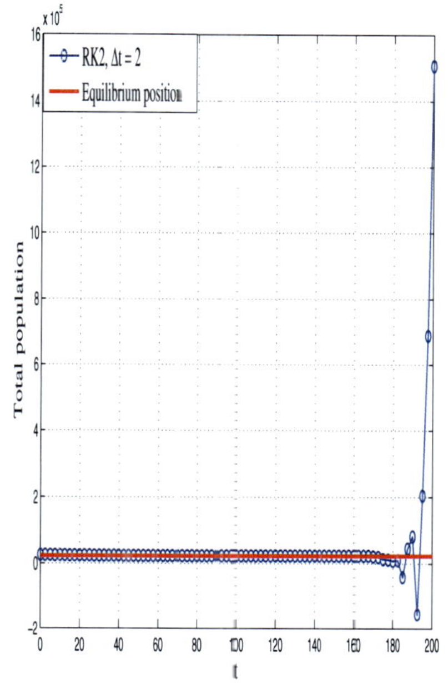

(B) RK2 scheme using $\Delta t = 2.5$

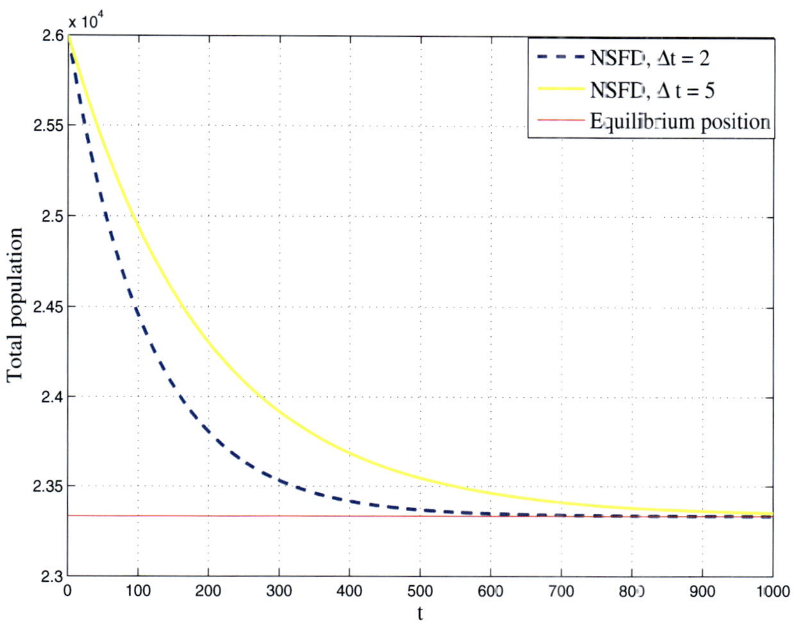

(C) NSFD scheme using $\Delta t = 2.5$ and $\Delta t = 5.0$

FIGURE 5. Approximations of the total population generated by the standard Euler and RK2 schemes and the NSFD scheme (3.2) using $(S(0), V(0), I(0)) = (5000, 20000, 1000)$.

numerical schemes. The numerical results are not only consistent with the theoretical assertions but also indicate that the NSFD method is easy to be implemented and can be used to solve a broad range of mathematical models arising in real-life.

It worth noting that the threshold parameters (m_S, m_P, m_{GCL}) for the NSFD method (3.2) and $(\phi_P, \phi_S, \phi_{GCL})$ for the nonstandard Euler method (3.19) can be computed easily. More clearly, these threshold parameters can be obtained by simple calculations or simple numerical algorithms.

Similarly to the NSFD methods formulated in [**7**, **28**, **29**], the NSFD method (3.2) also provides a stability-preserving numerical scheme for autonomous dynamical systems in general contexts. However, the positivity of the NSFD schemes in [**7**, **28**, **29**] is not guaranteed automatically while the NSFD method (3.2) has the ability to preserve not only the positivity but also the conservation laws of the system (1.1).

In [**88**], Wood and Kojouharov introduced a nonstandard numerical method preserving the positivity and LAS of autonomous dynamical systems. This method is computationally efficient, easy to be implemented and can be used to solve a broad range of problems in science and engineering. However, it requires the determination of the sign of the right-hand side functions at each iteration, and hence, the computational time is increased. For the NSFD method (3.2), we only need to compute the parameter m exactly once and then use it for all iterations. Furthermore, the computation of m is easy and can be done by a simple algorithm.

Before ending this section, we mention some open problems and future directions that research in NSFD methods can take.

(1) The NSFD method (3.2) requires the determination of the number α in the condition (1.2). Although α can be determined easily for most of the models in literature, it is necessary to construct an algorithm for computing α in general cases. Even though many important mathematical models in real-word applications satisfy the condition (1.2), this condition leads to a big restriction of the application of the proposed NSFD method. Hence, it is important to study extensions of the NSFD method (3.2) for mathematical models having the positivity and asymptotic stability but not satisfying Condition (1.2).

(2) NSFD methods preserving the asymptotic stability of non-hyperbolic equilibrium points: Most stability-preserving NSFD methods including the NSFD method (3.2) require equilibrium points of dynamical systems to be hyperbolic. The construction of generalized NSFD methods preserving the LAS of dynamical systems even when they have non-hyperbolic equilibrium points has been studied in some previous work (see, for instance, [**11**]) but is still an important open problem.

(3) Global asymptotic stability (GAS) of NSFD methods: It is well-known that analyzing the GAS of NSFD schemes for mathematical models having the GAS property is very important but not a simple task in general. This problem has been mentioned and considered in many previous works (see, for instance [**3**, **8**, **22**–**26**, **42**, **81**, **82**]). We observe from the numerical examples in Section 4 that the NSFD method can preserve not only the LAS but also the GAS of the continuous models under consideration. Therefore, it is suitable to predict that there exists a positive number m_{GAS} that plays as a GAS threshold of the NSFD method (3.2), i.e.,

the NSFD method (3.2) preserves the GAS of the system (1.1) whenever $m \geq m_{GAS}$. This prediction will be useful in studying the construction of NSFD methods preserving the GAS of dynamical systems.

(4) High-order NSFD methods: High-order NSFD methods for differential equations have attracted the attention of many researchers with various approaches (see, for example [**17, 21, 37, 40, 50, 55, 56**]); hence, the improvement accuracy of the NSFD method (3.2) is essential with many practice applications. A simple approach for improving accuracy of the NSFD method (3.2) is to combine it with Richardson's extrapolation method or variable step size strategies. This is easy to do because (3.2) is simple and fully explicit.

(5) New NSFD methods for differential equation models arising in real-world applications: Although NSFD methods have been strongly developed and achieved many important results, fields of theory and practice always pose new and complex problems; hence, constructing effective NSFD methods for these problems is very essential. Recently, NSFD methods for fractional differential equations have been widely studied (see, for instance [**12, 20**]). It was proved that NSFD methods for fractional-order systems are efficient and powerful; however, except for the positivity, other properties of fractional-order NSFD methods have not been intensively discussed. It is easy to extend the NSFD method (3.2) for systems in the context of fractional derivative operators. Then, the asymptotic stability in particular and qualitative study in general for the fractional-order NSFD schemes will be posed. It is appropriate to predict that the parameter m under suitable conditions will guarantee the dynamic consistency of the proposed NSFD schemes.

Acknowledgments

We would like to thank the editor and anonymous referees for useful and valuable comments that led to a great improvement of the paper.

References

[1] Elias M. Adamu, Kailash C. Patidar, and Andriamihaja Ramanantoanina, *An unconditionally stable nonstandard finite difference method to solve a mathematical model describing visceral leishmaniasis*, Math. Comput. Simulation **187** (2021), 171–190, DOI 10.1016/j.matcom.2021.02.007. MR4227114

[2] Oluwaseye Adekanye and Talitha Washington, *Nonstandard finite difference scheme for a Tacoma Narrows Bridge model*, Appl. Math. Model. **62** (2018), 223–236, DOI 10.1016/j.apm.2018.05.027. MR3834137

[3] H. Al-Kahby, F. Dannan, and S. Elaydi, *Non-standard discretization methods for some biological models*, Applications of nonstandard finite difference schemes (Atlanta, GA, 1999), World Sci. Publ., River Edge, NJ, 2000, pp. 155–180, DOI 10.1142/9789812813251_0004. MR1831405

[4] L. J. S. Allen, *Introduction to Mathematical Biology*, Pearson Education, Inc, Prentice Hall, 2007.

[5] P. Amarasekare and H. Possingham, *Patch dynamics and metapopulation theory: the case of successional species*, J. Theor. Biol. **209** (2001)333-344.

[6] Roumen Anguelov and Jean M.-S. Lubuma, *Nonstandard finite difference method by nonlocal approximation*, Math. Comput. Simulation **61** (2003), no. 3-6, 465–475, DOI 10.1016/S0378-4754(02)00106-4. MODELLING 2001 (Pilsen). MR1984145

[7] Roumen Anguelov and Jean M.-S. Lubuma, *Contributions to the mathematics of the nonstandard finite difference method and applications*, Numer. Methods Partial Differential Equations **17** (2001), no. 5, 518–543, DOI 10.1002/num.1025. MR1849163

[8] R. Anguelov, Y. Dumont, J. M.-S. Lubuma, and M. Shillor, *Dynamically consistent nonstandard finite difference schemes for epidemiological models*, J. Comput. Appl. Math. **255** (2014), 161–182, DOI 10.1016/j.cam.2013.04.042. MR3093413

[9] Roumen Anguelov, Jean M.-S. Lubuma, and Simon K. Mahudu, *Qualitatively stable finite difference schemes for advection-reaction equations*, J. Comput. Appl. Math. **158** (2003), no. 1, 19–30, DOI 10.1016/S0377-0427(03)00468-0. Selected papers from the Conference on Computational and Mathematical Methods for Science and Engineering (Alicante, 2002). MR2013601

[10] R. Anguelov, P. Kama, and J. M.-S. Lubuma, *On non-standard finite difference models of reaction-diffusion equations*, J. Comput. Appl. Math. **175** (2005), no. 1, 11–29, DOI 10.1016/j.cam.2004.06.002. MR2105666

[11] R. Anguelov, J. M.-S. Lubuma, and M. Shillor, *Topological dynamic consistency of nonstandard finite difference schemes for dynamical systems*, J. Difference Equ. Appl. **17** (2011), no. 12, 1769–1791, DOI 10.1080/10236198.2010.488226. MR2854823

[12] Abraham J. Arenas, Gilberto González-Parra, and Benito M. Chen-Charpentier, *Construction of nonstandard finite difference schemes for the SI and SIR epidemic models of fractional order*, Math. Comput. Simulation **121** (2016), 48–63, DOI 10.1016/j.matcom.2015.09.001. MR3425284

[13] Uri M. Ascher and Linda R. Petzold, *Computer methods for ordinary differential equations and differential-algebraic equations*, Society for Industrial and Applied Mathematics (SIAM), Philadelphia, PA, 1998, DOI 10.1137/1.9781611971392. MR1638643

[14] Fred Brauer and Carlos Castillo-Chavez, *Mathematical models in population biology and epidemiology*, 2nd ed., Texts in Applied Mathematics, vol. 40, Springer, New York, 2012, DOI 10.1007/978-1-4614-1686-9. MR3024808

[15] Vincenzo Capasso and Gabriella Serio, *A generalization of the Kermack-McKendrick deterministic epidemic model*, Math. Biosci. **42** (1978), no. 1-2, 43–61, DOI 10.1016/0025-5564(78)90006-8. MR529097

[16] Michael Chapwanya, Jean M.-S. Lubuma, and Ronald E. Mickens, *Positivity-preserving nonstandard finite difference schemes for cross-diffusion equations in biosciences*, Comput. Math. Appl. **68** (2014), no. 9, 1071–1082, DOI 10.1016/j.camwa.2014.04.021. MR3269960

[17] Benito M. Chen-Charpentier, Dobromir T. Dimitrov, and Hristo V. Kojouharov, *Combined nonstandard numerical methods for ODEs with polynomial right-hand sides*, Math. Comput. Simulation **73** (2006), no. 1-4, 105–113, DOI 10.1016/j.matcom.2006.06.008. MR2307094

[18] K. Cooke, P. van den Driessche, and X. Zou, *Interaction of maturation delay and nonlinear birth in population and epidemic models*, J. Math. Biol. **39** (1999), no. 4, 332–352, DOI 10.1007/s002850050194. MR1727839

[19] Jacky Cresson and Frédéric Pierret, *Non standard finite difference scheme preserving dynamical properties*, J. Comput. Appl. Math. **303** (2016), 15–30, DOI 10.1016/j.cam.2016.02.007. MR3479269

[20] Jacky Cresson and Anna Szafrańska, *Discrete and continuous fractional persistence problems—the positivity property and applications*, Commun. Nonlinear Sci. Numer. Simul. **44** (2017), 424–448, DOI 10.1016/j.cnsns.2016.07.016. MR3554828

[21] Quang A. Dang and Manh Tuan Hoang, *Positive and elementary stable explicit nonstandard Runge-Kutta methods for a class of autonomous dynamical systems*, Int. J. Comput. Math. **97** (2020), no. 10, 2036–2054, DOI 10.1080/00207160.2019.1677895. MR4145346

[22] Quang A. Dang and Manh Tuan Hoang, *Nonstandard finite difference schemes for a general predator-prey system*, J. Comput. Sci. **36** (2019), 101015, 14, DOI 10.1016/j.jocs.2019.07.002. MR3987494

[23] Quang A Dang and Manh Tuan Hoang, *Complete global stability of a metapopulation model and its dynamically consistent discrete models*, Qual. Theory Dyn. Syst. **18** (2019), no. 2, 461–475, DOI 10.1007/s12346-018-0295-y. MR3982749

[24] Quang A. Dang and Manh Tuan Hoang, *Lyapunov direct method for investigating stability of nonstandard finite difference schemes for metapopulation models*, J. Difference Equ. Appl. **24** (2018), no. 1, 15–47, DOI 10.1080/10236198.2017.1391235. MR3750528

[25] Quang A Dang and Manh Tuan Hoang, *Positivity and global stability preserving NSFD schemes for a mixing propagation model of computer viruses*, J. Comput. Appl. Math. **374** (2020), 112753, 12, DOI 10.1016/j.cam.2020.112753. MR4061809

[26] Quang A. Dang and Manh Tuan Hoang, *Dynamically consistent discrete metapopulation model*, J. Difference Equ. Appl. **22** (2016), no 9, 1325–1349, DOI 10.1080/10236198.2016.1197213. MR3589595

[27] D. L. DeAngelis, R. A. Goldstein, and R. V. O'Neill, *A Model for Tropic Interaction*, Ecology **56** (1975) 881-892.

[28] Dobromir T. Dimitrov and Hristo V. Kojouharov, *Nonstandard finite-difference schemes for general two-dimensional autonomous dynamical systems*, Appl. Math. Lett. **18** (2005), no. 7, 769–774, DOI 10.1016/j.aml.2004.08.011. MR2144725

[29] Dobromir T. Dimitrov and Hristo V. Kojouharov, *Stability-preserving finite-difference methods for general multi-dimensional autonomous dynamical systems*, Int. J. Numer. Anal. Model. **4** (2007), no. 2, 280–290. MR2287608

[30] Dobromir T. Dimitrov and Hristo V. Kojouharov, *Dynamically consistent numerical methods for general productive-destructive systems*, J. Difference Equ. Appl. **17** (2011), no. 12, 1721–1736, DOI 10.1080/10236191003781947. MR2854820

[31] Dobromir T. Dimitrov and Hristo V. Kojouharov, *Complete mathematical analysis of predator-prey models with linear prey growth and Beddington-DeAngelis functional response*, Appl. Math. Comput. **162** (2005), no. 2, 523–538, DOI 10.1016/j.amc.2003.12.106. MR2111853

[32] Leah Edelstein-Keshet, *Mathematical models in biology*, Classics in Applied Mathematics, vol. 46, Society for Industrial and Applied Mathematics (SIAM), Philadelphia, PA, 2005. Reprint of the 1988 original, DOI 10.1137/1.9780898719147. MR2131632

[33] Matthias Ehrhardt and Ronald E. Mickens, *A nonstandard finite difference scheme for convection-diffusion equations having constant coefficients*, Appl. Math. Comput. **219** (2013), no. 12, 6591–6604, DOI 10.1016/j.amc.2012.12.068. MR3027827

[34] Saber Elaydi, *An introduction to difference equations*, 3rd ed., Undergraduate Texts in Mathematics, Springer, New York, 2005. MR2128146

[35] H. Fatoorehchi and M. Ehrhardt, *Numerical and semi-numerical solutions of a modified Thévenin model for calculating terminal voltage of battery cells*, J Energy Storage **145**(2022) 103746.

[36] S. M. Garba, A. B. Gumel, A. S. Hassan, and J. M.-S. Lubuma, *Switching from exact scheme to nonstandard finite difference scheme for linear delay differential equation*, Appl. Math. Comput. **258** (2015), 388–403, DOI 10.1016/j.amc.2015.01.088. MR3323077

[37] Gilberto González-Parra, Abraham J. Arenas, and Benito M. Chen-Charpentier, *Combination of nonstandard schemes and Richardson's extrapolation to improve the numerical solution of population models*, Math. Comput. Modelling **52** (2010), no. 7-8, 1030–1036, DOI 10.1016/j.mcm.2010.03.015. MR2718414

[38] A. B. Gumel and S. M. Moghadas, *A qualitative study of a vaccination model with nonlinear incidence*, Appl. Math. Comput. **143** (2003), no. 2-3, 409–419, DOI 10.1016/S0096-3003(02)00372-7. MR1981706

[39] Herbert Hethcote, Ma Zhien, and Liao Shengbing, *Effects of quarantine in six endemic models for infectious diseases*, Math. Biosci. **180** (2002), 141–160, DOI 10.1016/S0025-5564(02)00111-6. John A. Jacquez memorial volume. MR1935292

[40] Manh Tuan Hoang, *A novel second-order nonstandard finite difference method for solving one-dimensional autonomous dynamical systems*, Commun. Nonlinear Sci. Numer. Simul. **114** (2022), Paper No. 106654, 13, DOI 10.1016/j.cnsns.2022.106654. MR4447711

[41] Manh Tuan Hoang, *Positivity and boundedness preserving nonstandard finite difference schemes for solving Volterra's population growth model*, Math. Comput. Simulation **199** (2022), 359–373, DOI 10.1016/j.matcom.2022.04.003. MR4412235

[42] Manh Tuan Hoang, *Dynamically consistent nonstandard finite difference schemes for a virus-patch dynamic model*, J. Appl. Math. Comput. **68** (2022), no. 5, 3397–3423, DOI 10.1007/s12190-021-01673-z. MR4487815

[43] Manh Tuan Hoang, *Reliable approximations for a hepatitis B virus model by nonstandard numerical schemes*, Math. Comput. Simulation **193** (2022), 32–56, DOI 10.1016/j.matcom.2021.09.020. MR4330944

[44] Manh Tuan Hoang, *Dynamical analysis of a generalized hepatitis B epidemic model and its dynamically consistent discrete model*, Math. Comput. Simulation **205** (2023), 291–314, DOI 10.1016/j.matcom.2022.10.006. MR4499405

[45] Zoltán Horváth, *On the positivity step size threshold of Runge-Kutta methods*, Appl. Numer. Math. **53** (2005), no. 2-4, 341–356, DOI 10.1016/j.apnum.2004.08.026. MR2128530

[46] W. O. Kermack and A. G. McKendrick, *Contributions to the mathematical theory of epidemics - I*, Proc. R. Soc. Lond. **115** (1927) 700-721.

[47] W. O. Kermack and A. G. McKendrick, *Contributions to the mathematical theory of epidemics - II*, Proc. R. Soc. Lond. **138** (1932) 55-83.

[48] J. E. Keymer, P.A. Marquet, J. X. Velasco-Hernandez, and S. A. Levin, *Extinction thresholds and metapopulation persistence in dynamic landscapes*, Am. Nat. **156** (2000) 478-494.

[49] H. K. Khalil, *Nonlinear systems*, Third Edition, Prentice Hall, 2002.

[50] Hristo V. Kojouharov, Souvik Roy, Madhu Gupta, Fawaz Alalhareth, and John M. Slezak, *A second-order modified nonstandard theta method for one-dimensional autonomous differential equations*, Appl. Math. Lett. **112** (2021), Paper No. 106775, 6, DOI 10.1016/j.aml.2020.106775. MR4154331

[51] C. M.Kribs-Zaleta and J. X. Velasco-Hernández, *A simple vaccination model with multiple endemic states*, Math. Biosci. **164** (2000) 183-201.

[52] J. P. LaSalle, *The stability of dynamical systems*, Regional Conference Series in Applied Mathematics, Society for Industrial and Applied Mathematics, Philadelphia, Pa., 1976. With an appendix: "Limiting equations and stability of nonautonomous ordinary differential equations" by Z. Artstein. MR0481301

[53] A. M. Lyapunov, *The general problem of the stability of motion*, Internat. J. Control **55** (1992), no. 3, 521–790, DOI 10.1080/00207179208934253. Translated by A. T. Fuller from Édouard Davaux's French translation (1907) of the 1892 Russian original; With an editorial (historical introduction) by Fuller, a biography of Lyapunov by V. I. Smirnov, and the bibliography of Lyapunov's works collected by J. F. Barrett; Lyapunov centenary issue. MR1154209

[54] Maia Martcheva, *An introduction to mathematical epidemiology*, Texts in Applied Mathematics, vol. 61, Springer, New York, 2015, DOI 10.1007/978-1-4899-7612-3. MR3409181

[55] J. Martín-Vaquero, A. Martín del Rey, A. H. Encinas, J. D. Hernández Guillén, A. Queiruga-Dios, and G. Rodríguez Sánchez, *Higher-order nonstandard finite difference schemes for a MSEIR model for a malware propagation*, J. Comput. Appl. Math. **317** (2017), 146–156, DOI 10.1016/j.cam.2016.11.044. MR3606071

[56] J. Martín-Vaquero, A. Queiruga-Dios, A. Martín del Rey, A. H. Encinas, J. D. Hernández Guillén, and G. Rodríguez Sánchez, *Variable step length algorithms with high-order extrapolated non-standard finite difference schemes for a SEIR model*, J. Comput. Appl. Math. **330** (2018), 848–854, DOI 10.1016/j.cam.2017.03.031. MR3717635

[57] Robert Mattheij and Jaap Molenaar, *Ordinary differential equations in theory and practice*, Classics in Applied Mathematics, vol. 43, Society for Industrial and Applied Mathematics (SIAM), Philadelphia, PA, 2002. Reprint of the 1996 original, DOI 10.1137/1.9780898719178. MR1946758

[58] Jaime Mena-Lorca and Herbert W. Hethcote, *Dynamic models of infectious diseases as regulators of population sizes*, J. Math. Biol. **30** (1992), no. 7, 693–716, DOI 10.1007/BF00173264. MR1175099

[59] Ronald E. Mickens, *Nonstandard finite difference models of differential equations*, World Scientific Publishing Co., Inc., River Edge, NJ, 1994. MR1275372

[60] R. E. Mickens, *Applications of Nonstandard Finite Difference Schemes*, World Scientific, 2000.

[61] Ronald E. Mickens (ed.), *Advances in the applications of nonstandard finite difference schemes*, World Scientific Publishing Co. Pte. Ltd., Hackensack, NJ, 2005, DOI 10.1142/9789812703316. MR2213290

[62] Ronald E. Mickens, *Nonstandard finite difference schemes for differential equations*, J. Difference Equ. Appl. **8** (2002), no. 9, 823–847, DOI 10.1080/1023619021000000807. In honor of Professor Allan Peterson on the occasion of his 60th birthday. MR1919887

[63] Ronald E. Mickens, *Nonstandard finite difference schemes—methodology and applications*, World Scientific Publishing Co. Pte. Ltd., Hackensack, NJ, [2021] ©2021. Expanded second edition of [1275372]. MR4274594

[64] Ronald E. Mickens, *Dynamic consistency: a fundamental principle for constructing nonstandard finite difference schemes for differential equations*, J. Difference Equ. Appl. **11** (2005), no. 7, 645–653, DOI 10.1080/10236190412331334527. American Mathematical Society Special Session on Difference Equations and Discrete Dynamics. MR2173250

[65] Ronald E. Mickens and Talitha M. Washington, *NSFD discretizations of interacting population models satisfying conservation laws*, Comput. Math. Appl. **66** (2013), no. 11, 2307–2316, DOI 10.1016/j.camwa.2013.06.011. MR3125375

[66] Ronald E. Mickens, *A nonstandard finite difference scheme for a PDE modeling combustion with nonlinear advection and diffusion*, Math. Comput. Simulation **69** (2005), no. 5-6, 439–446, DOI 10.1016/j.matcom.2005.03.008. MR2169718

[67] Ronald E. Mickens, *A nonstandard finite difference scheme for the diffusionless Burgers equation with logistic reaction*, Math. Comput. Simulation **62** (2003), no. 1-2, 117–124, DOI 10.1016/S0378-4754(02)00180-5. Nonlinear waves: computation and theory, II (Athens, GA, 2001). MR1983580

[68] R. E. Mickens, *A nonstandard finite difference scheme for a Fisher PDE having nonlinear diffusion*, Comput. Math. Appl. **45** (2003), no. 1-3, 429–436, DOI 10.1016/S0898-1221(03)80028-7. Numerical methods in physics, chemistry, and engineering. MR1991379

[69] Ronald E. Mickens, *A nonstandard finite difference scheme for a nonlinear PDE having diffusive shock wave solutions*, Math. Comput. Simulation **55** (2001), no. 4-6, 549–555, DOI 10.1016/S0378-4754(00)00309-8. Nonlinear waves: computation and theory (Athens, GA, 1999). MR1821679

[70] Ronald E. Mickens, *A nonstandard finite-difference scheme for the Lotka-Volterra system*, Appl. Numer. Math. **45** (2003), no. 2-3, 309–314, DOI 10.1016/S0168-9274(02)00223-4. MR1967579

[71] Ronald E. Mickens and Isom H. Herron, *Approximate rational solutions to the Thomas-Fermi equation based on dynamic consistency*, Appl. Math. Lett. **116** (2021), Paper No. 106994, 7, DOI 10.1016/j.aml.2020.106994. MR4197233

[72] Ronald E. Mickens, Kale Oyedeji, and Sandra Rucker, *Exact finite difference scheme for second-order, linear ODEs having constant coefficients*, J. Sound Vibration **287** (2005), no. 4-5, 1052–1056, DOI 10.1016/j.jsv.2005.03.012. MR2197734

[73] R. E. Mickens and I. Ramadhani, *Finite-difference schemes having the correct linear stability properties for all finite step-sizes. III*, Comput. Math. Appl. **27** (1994), no. 4, 77–84, DOI 10.1016/0898-1221(94)90056-6. MR1256770

[74] R. E. Mickens and A. B. Gumel, *Construction and analysis of a non-standard finite difference scheme for the Burgers-Fisher equation*, J. Sound Vibration **257** (2002), no. 4, 791–797, DOI 10.1006/jsvi.2001.4240. MR1941103

[75] R. E. Mickens and A. B. Gumel, *Numerical study of a non-standard finite-difference scheme for the van der Pol equation*, J. Sound Vibration **250** (2002), no. 5, 955–963, DOI 10.1006/jsvi.2001.3783. MR1903060

[76] Ronald E. Mickens, *NSFD scheme for acoustic propagation with the linearized Euler equations*, Math. Comput. Simulation **127** (2016), 189–193, DOI 10.1016/j.matcom.2013.11.008. MR3501297

[77] S. M. Moghadas and A. B. Gumel, *Global stability of a two-stage epidemic model with generalized non-linear incidence*, Math. Comput. Simulation **60** (2002), no. 1-2, 107–118, DOI 10.1016/S0378-4754(02)00002-2. MR1916901

[78] Kailash C. Patidar, *On the use of nonstandard finite difference methods*, J. Difference Equ. Appl. **11** (2005), no. 8, 735–758, DOI 10.1080/10236190500127471. MR2156651

[79] K. C. Patidar, *Nonstandard finite difference methods: recent trends and further developments*, J. Difference Equ. Appl. **22** (2016), no. 6, 817–849, DOI 10.1080/10236198.2016.1144748. MR3520261

[80] Lih-Ing W. Roeger, *Dynamically consistent discrete Lotka-Volterra competition models derived from nonstandard finite-difference schemes*, Discrete Contin. Dyn. Syst. Ser. B **9** (2008), no. 2, 415–429, DOI 10.3934/dcdsb.2008.9.415. MR2373389

[81] Lih-Ing Wu Roeger and Glenn Lahodny Jr., *Dynamically consistent discrete Lotka-Volterra competition systems*, J. Difference Equ. Appl. **19** (2013), no. 2, 191–200, DOI 10.1080/10236198.2011.621894. MR3021781

[82] Lih-Ing W. Roeger, *Dynamically consistent discrete-time SI and SIS epidemic models*, Discrete Contin. Dyn. Syst. **Dynamical systems, differential equations and applications. 9th AIMS Conference. Suppl.** (2013), 653–662, DOI 10.3934/proc.2013.2013.653. MR3462410

[83] Lih-Ing W. Roeger, *Exact finite-difference schemes for two-dimensional linear systems with constant coefficients*, J. Comput. Appl. Math. **219** (2008), no. 1, 102–109, DOI 10.1016/j.cam.2007.07.015. MR2437699

[84] Hal L. Smith and Paul Waltman, *The theory of the chemostat: Dynamics of microbial competition*, Cambridge Studies in Mathematical Biology, vol. 13, Cambridge University Press, Cambridge, 1995, DOI 10.1017/CBO9780511530043. MR1315301

[85] A. M. Stuart and A. R. Humphries, *Dynamical systems and numerical analysis*, Cambridge Monographs on Applied and Computational Mathematics, vol. 2, Cambridge University Press, Cambridge, 1996. MR1402909

[86] Amit Kumar Verma, Mukesh Kumar Rawani, and Carlo Cattani, *A numerical scheme for a class of generalized Burgers' equation based on Haar wavelet nonstandard finite difference method*, Appl. Numer. Math. **168** (2021), 41–54, DOI 10.1016/j.apnum.2021.05.019. MR4268650

[87] Daniel T. Wood, Dobromir T. Dimitrov, and Hristo V. Kojouharov, *A nonstandard finite difference method for n-dimensional productive-destructive systems*, J. Difference Equ. Appl. **21** (2015), no. 3, 240–254, DOI 10.1080/10236198.2014.997228. MR3316778

[88] Daniel T. Wood and Hristo V. Kojouharov, *A class of nonstandard numerical methods for autonomous dynamical systems*, Appl. Math. Lett. **50** (2015), 78–82, DOI 10.1016/j.aml.2015.06.008. MR3378952

[89] Daniel T. Wood, Hristo V. Kojouharov, and Dobromir T. Dimitrov, *Universal approaches to approximate biological systems with nonstandard finite difference methods*, Math. Comput. Simulation **133** (2017), 337–350, DOI 10.1016/j.matcom.2016.04.007. MR3575285

Department of Mathematics, FPT University, Hoa Lac Hi-Tech Park, Km29 Thang Long Blvd, Hanoi, Viet Nam

Email address: tuanhm14@fe.edu.vn

Email address: hmtuan01121990@gmail.com

Contemporary Mathematics
Volume **793**, 2024
https://doi.org/10.1090/conm/793/15883

Higher-order modified nonstandard finite difference methods for autonomous dynamical systems

Fawaz K. Alalhareth, Madhu Gupta, Hristo V. Kojouharov, and Souvik Roy

In celebration of the 80th birthday of Dr. Ronald E. Mickens.

ABSTRACT. Nonstandard finite difference (NSFD) methods are known to offer numerous advantages over traditional techniques for numerically solving various problems in science and engineering. However, in general, they are only of first-order accuracy. In this chapter, we present four new classes of modified NSFD methods, which are not only elementary stable and preserve the positivity of solutions, but are also of second-order accuracy. We also present a series of numerical simulations that support the theoretical results and demonstrate the advantages of the new modified NSFD methods when compared to other classical standard and nonstandard numerical methods.

1. Introduction

Nonstandard finite difference (NSFD) methods provide an accurate and efficient way to numerically solve many problems in science and engineering, and are also known to offer various advantages over traditional techniques. One of their most important properties is the complete absence of the elementary numerical instabilities that plague the classical finite difference methods.

Ronald E. Mickens introduced the nonstandard finite difference methods about three decades ago to numerically solve ordinary differential equations while preserving the essential properties of their exact solutions [**23–25, 29**]. Since then, NSFD methods have been extensively developed to approximate the solutions of various types of differential equations. Since the onset, several NSFD methods have been constructed for specific ODE and PDE systems used to model physical processes and chemical reactions along with biological interactions such as food-chain and epidemic models (e.g. [**5–7, 14, 15, 20, 22, 23, 26–28, 30, 31, 35, 38**]). In particular, Dimitrov and Kojouharov designed a series of *elementary stable nonstandard* (ESN) numerical methods for a variety of special classes of autonomous dynamical systems [**8, 10–13**] culminating with the *positive and elementary stable nonstandard* (PESN) numerical methods for general multi-dimensional productive-destructive systems [**13, 36**] as well as for solving general autonomous dynamical

2020 *Mathematics Subject Classification.* Primary 65L05, 65L20; Secondary 65L06, 65L12.
Key words and phrases. Nonstandard, Finite difference, Positivity, Elementary stable, Second-order, NSFD, ESN, PESN.

The third author was the corresponding author.

systems with positive solutions [37]. The majority of these studies have focused on preserving key properties of the corresponding dynamical system, however the proposed nonstandard numerical methods are usually only of first-order accuracy.

Motivated by the earlier works on ESN and PESN numerical methods, we develop a methodology for designing second-order ESN and PESN numerical methods for solving general autonomous dynamical systems of the form:

$$\vec{x}'(t) = \vec{f}(\vec{x}); \quad \vec{x}(t_0) = \vec{x}_0, \tag{1.1}$$

where $\vec{x}$ represents the vector-function $[x_1(t), \ldots, x_n(t)]^T$, $x_i : [t_0, T) \to \mathbb{R}$, $\vec{f} = [f_1, \ldots, f_n]^T \in C^2(\mathbb{R}^n; \mathbb{R}^n)$ is differentiable, $\vec{x}_0 \in \mathbb{R}^n$. We assume that System (1.1) has a finite number of only hyperbolic equilibria. A general finite difference method which approximates the solution of System (1.1) on the interval $[t_0, T]$ can be written as:

$$D_{i,h}(\vec{x}^k) = F_{i,h}(f_i; \vec{x}^k), \; k = 0, \cdots, N_t, \tag{1.2}$$

where $D_{i,h}(\vec{x}^k) \approx x_i'\big|_{t=t_k}$, $F_{i,h}(f_i; \vec{x}^k) \approx f_i(\vec{x})$, $\vec{x}^k \approx \vec{x}(t_k)$, $t_k = t_0 + kh$, $k = 0, \cdots, N_t$, $i = 1, \cdots, n$, with mesh size $h > 0$. This chapter presents four new classes of second-order modified nonstandard finite difference methods, which have been previously introduced and analyzed in [**1**–**3**, **16**]. The new nonstandard numerical methods satisfy the two main properties of NSFD methods, as formalized in [**5**], which are that the denominator function $\varphi(h)$ from the discretization of the derivative, i.e., $D_h(x^k) = \dfrac{x^{k+1} - x^k}{\varphi(h)}$ is a non-negative function of the form $\varphi(h) = h + \mathcal{O}(h^2)$, and the right-hand side function is discretized non-locally, i.e., $F_h(f; x^k) = g(x^k, x^{k+1}, h)$. All of the new numerical methods are elementary stable [**5**, **9**, **38**], meaning that the fixed points of the methods are the same as the equilibria of Equation (1.1) and vice-versa, and also that their local stability properties are the same for any value of the step-size h. In addition, two of the new modified NSFD methods are also positivity-preserving [**13**, **36**], meaning that their numerical solutions remain positive for any value of the step-size h and any positive initial conditions. Furthermore, the proposed new NSFD methods use a novel modified nonstandard denominator function $\varphi_i(h, \vec{x}^k)$ in the discretization of the derivative, which is dependent not only on the step-size h but also on the numerical solution $\vec{x}^k$, ensuring the second-order accuracy of the numerical methods.

The chapter is organized as follows: First, we present the modified nonstandard one-stage and two-stage theta methods, followed by a discussion of the modified nonstandard two-stage explicit Runge-Kutta method, all of which are elementary stable and of second-order accuracy. Next, we introduce the two second-order modified nonstandard finite difference methods that are both positive and elementary stable. We also present numerical simulations, in Section 4, of two classical mathematical models in biology to illustrate the performance of the proposed new NSFD numerical methods. In the last section, concluding remarks and future research directions are outlined.

2. Second-order modified ESN numerical methods

In this section, we present two new classes of second-order modified NSFD methods for Equation (1.1) that are elementary stable. The first class consists of two modified nonstandard theta methods, and the second is a second-order improvement on the nonstandard two-stage explicit Runge-Kutta (ERK2) method.

2.1. Second-order modified ESN (SOESN) theta method for general autonomous dynamical systems. The following result holds for the new second-order modified nonstandard theta method [**2, 16**]:

THEOREM 2.1. *Let $\vec{f} \in C^2(\mathbb{R}^n;\mathbb{R}^n)$ and let $\varphi_i : \mathbb{R}^+ \times \mathbb{R}^n \to \mathbb{R}^+$, for $i = 1,\dots,n$, satisfies the following conditions:*

(I) $\varphi_i(h,\vec{x}) = h + (1-2\theta)\dfrac{\langle \nabla_x f_i(\vec{x}), \vec{f}(\vec{x})\rangle}{f_i(\vec{x})}\dfrac{h^2}{2} + \mathcal{O}(h^3)$, *for all* $1 \le i \le n$,

(II) $0 < \varphi_i(h,\vec{x}) < \dfrac{2|\mathrm{Re}(\lambda_i)|}{|2\theta-1||\lambda_i|^2}$, $0 \le \theta \le 1$, $\theta \ne \frac{1}{2}$, *for all hyperbolic equilibria* $\vec{x}^*$ *of Equation* (1.1) *with* $h > 0$ *and for all* $\vec{x} \in \mathbb{R}^n$.

Then, the following modified nonstandard one-stage theta method:

$$\frac{x_i^{k+1} - x_i^k}{\varphi_i(h,\vec{x}^k)} = f_i\left(\theta\vec{x}^{k+1} + (1-\theta)\vec{x}^k\right), \tag{2.1}$$

and the modified nonstandard two-stage theta method [**21**]:

$$\frac{x_i^{k+1} - x_i^k}{\varphi_i(h,\vec{x}^k)} = \theta f_i(\vec{x}^{k+1}) + (1-\theta) f_i(\vec{x}^k), \tag{2.2}$$

$i = 1,\dots,n$, *for approximating the solution of Equation* (1.1), *are both elementary stable and second-order accurate.*

PROOF. A sketch of the proof is presented below, with more details available in [**2, 16**].

First, the second-order accuracy of the modified nonstandard one-stage theta method (2.1) is easily proven by using a Taylor series expansion about t_k and enforcing Condition (I), which yields:

$$x_i(t_{k+1}) - \left[x_i(t_k) + \varphi_i(h,\vec{x}(t_k)) f_i\left(\theta\vec{x}(t_{k+1}) + (1-\theta)\vec{x}(t_k)\right)\right] = \mathcal{O}(h^3)$$

for each $i = 1, 2, \dots n$.

To prove the elementary stability of the method, let $\vec{x}^*$ be a hyperbolic equilibrium of System (1.1) and $J = J(x^*)$ be the Jacobian evaluated at $\vec{x}^*$ with eigenvalues $\lambda_1, \lambda_2, \dots, \lambda_n$. If x^* is a stable equilibrium, then $Re(\lambda_i) < 0$ for all $i = 1, 2, \dots, n$, and if x^* is an unstable equilibrium, then there exists an i_0 such that $Re(\lambda_{i_0}) > 0$. Let Λ denote the Jordan form of J, then $J = S^{-1}\Lambda S$, where S is a non-singular complex $n \times n$ matrix, and using the substitution $\vec{y} = S\vec{\epsilon}$, with $\vec{\epsilon} = \vec{x} - \vec{x}^*$, yields the following linear system:

$$\vec{y}' = \Lambda\vec{y}. \tag{2.3}$$

Applying Method (2.1) to System (2.3) yields the following stability function:

$$L(\varphi_i) = \frac{1 + \lambda_i(1-\theta)\varphi_i(h,\vec{y}^k)}{1 - \lambda_i\theta\varphi_i(h,\vec{y}^k)}, \quad i = 1, 2, \dots n. \tag{2.4}$$

Next, imposing Condition (II) and using the fact that x^* is a stable equilibrium, results in $|L(\varphi_i)| < 1$ for $i = 1, 2, \dots, n$, which implies that x^* is a stable fixed point. Similarly, one can show that if x^* is an unstable equilibrium then there exist an i_0 such that $|L(\varphi_{i_0})| > 1$, and therefore, x^* is an unstable fixed point, which proves that x^* is stable equilibrium of System (1.1) if and only if it is a stable fixed point of Method (2.1).

The proof of second-order accuracy and elementary stability of the modified nonstandard two-stage theta method (2.2) is done using similar arguments. □

The following lemma provides a recipe for constructing a general class of nonstandard denominator functions which satisfy the conditions of Theorem (2.1) [**2**, **16**].

LEMMA 2.2. *Let $\phi_{i_1} : \mathbb{R}_+ \to \mathbb{R}_+$ and $\phi_{i_2} : \mathbb{R} \to \mathbb{R}_+$ satisfy the following conditions:*

(a) $0 < \phi_{i_1}(h) < 1$, *for all* $h > 0$, *and* $\phi_{i_1}(h) = h - \dfrac{h^2}{2} + \mathcal{O}(h^3)$.

(b) $0 < \phi_{i_2}(h) < M$, *for all* $h \in \mathbb{R}$ *and some* $M > 0$, *and* $\phi_{i_2}(h) = 1 + h + \mathcal{O}(h^3)$.

Then, the functions

$$\varphi_i(h, \vec{x}) = \frac{\phi_{i_1}(\alpha h)}{\alpha}\phi_{i_2}\left(\frac{\alpha - q_i(\vec{x})}{2}h\right), \quad i = 1, \dots, n, \tag{2.5}$$

with $\alpha > \dfrac{M}{2}\max_\Omega \dfrac{|2\theta - 1||\lambda|^2}{|Re(\lambda)|}$, *where* $\Omega = \bigcup_{\vec{x}^* \in \Gamma} \sigma(J(\vec{x}^*))$ *and* Γ *denotes the set of all equilibria of System* (1.1), *and* $q_i(\vec{x}) = -(1 - 2\theta)\dfrac{\langle \nabla_x f_i(\vec{x}), \vec{f}(\vec{x})\rangle}{f_i(\vec{x})}$, *satisfy the conditions* (*I*) *and* (*II*) *of Theorem* (2.1).

REMARK 2.3. There exists a variety of functions ϕ_{i_1} and ϕ_{i_2} that satisfy the conditions (a) and (b) of Lemma 2.2, with one such set of functions being

$$\phi_{i_1}(h) = 1 - e^{-h} \text{ and } \phi_{i_2}(h) = 1 + \tanh(h).$$

It is also important to note that the nonstandard denominator functions (2.5) in Lemma 2.2 involve the product of two functions, where one, $\dfrac{\phi_{i_1}(\alpha h)}{\alpha}$, accounting for the stability properties of the numerical method, due to the choice of the parameter α, while the other, $\phi_{i_2}\left(\dfrac{\alpha - q_i(\vec{x})}{2}h\right)$, accounting for the second-order accuracy of the numerical method, due to the the form of the parameter $q_i(\vec{x})$.

REMARK 2.4. In the one-dimensional case [**18**, **21**], a general class of nonstandard denominator functions $\varphi(h, x^k)$ that satisfies the conditions of Theorem 2.1 can be constructed of the form $\varphi(h, x^k) = \dfrac{\phi(hq)}{q}$, with $q = -f_x(x^k) \neq 0$, using, for example, the functions $\phi(h) = 1 - e^{-h}$. Here, the denominator function $\varphi(h, x^k)$ depends only on the derivative of the right-hand side function f of the autonomous differential equation.

2.2. Second-order modified ESN (SOESN) ERK2 method for general autonomous dynamical systems.

The following result holds for the new second-order modified nonstandard two-stage ERK2 method [**2**, **16**]:

THEOREM 2.5. *Let $\vec{f} = [f_1, \dots, f_n] \in C^2(\mathbb{R}^n; \mathbb{R}^n)$ and let $\varphi : \mathbb{R}^+ \to \mathbb{R}^+$ satisfy the following conditions*

(I) $\varphi(h) = h + \mathcal{O}(h^3)$,

(II) $0 < \varphi(h) < \frac{1}{q}$, *for all* $\lambda \in \Omega$, *where* $q > \max_{\Omega} \frac{|\lambda|^2}{2|Re(\lambda)|}$, $\Omega = \bigcup_{\vec{x}^* \in \Gamma} \sigma(J(\vec{x}^*))$ *and* Γ *denotes the set of all hyperbolic equilibria* $\vec{x}^*$ *of System* (1.1).

Then the following modified nonstandard two-stage ERK2 method for approximating the solution of Equation (1.1)*:*

$$
(2.6)\quad x_i^{k+1} = x_i^k + \varphi(h)\left\{(1-\omega)f_i(\vec{x}^k) + \omega f_i\left(\vec{x}^k + \frac{1}{2\omega}\vec{f}(\vec{x}^k)\varphi(h)\right)\right\},\ 0 < \omega \leq 1,
$$

$i = 1, \ldots, n$, *is conservative, elementary stable, and second-order accurate, provided the method does not introduce additional fixed points other than the equilibria of Equation* (1.1).

PROOF. The second-order accuracy of the modified nonstandard two-stage ERK2 method (2.6) can be shown using the same approach as in the proof of Theorem (2.1). Similarly, to prove the elementary stability of the method, yields the following stability function:

$$
(2.7)\qquad L(\varphi(h)) = 1 + \varphi(h)\lambda_i + \frac{1}{2}\lambda_i^2\varphi^2(h).
$$

Next, imposing Condition (II) and using the fact that x^* is a stable equilibrium, results in $|L(\varphi)| < 1$ for $i = 1, 2, \ldots, n$, which implies that x^* is a stable fixed point. Likewise, one can show that if x^* is an unstable equilibrium then there exist a λ_{i_0}, with $Re(\lambda_{i_0}) > 0$, such that $|L(\varphi_{i_0})| > 1$, and therefore, x^* is an unstable fixed point of Method (2.6).

Finally, because the denominator function $\varphi(h)$ is independent of x and thus the same for each component $i = 1, \ldots, n$, it is simple to show that $\sum_{i=1}^{n} x_i^{k+1} = \sum_{i=1}^{n} x_i^k$, which proves that the modified nonstandard two-stage ERK2 method (2.6) is also conservative. □

REMARK 2.6. There exists a variety of functions $\varphi(h)$ which satisfy the conditions of Theorem 2.5, and hence ensure the second-order accuracy and elementary stability of the method (2.6). One such denominator function is $\varphi(h) = \frac{\tanh(qh)}{q}$, where $q > \max_{\Omega} \frac{|\lambda|^2}{2|Re(\lambda)|}$.

3. Second-order modified PESN numerical methods

In this section, we present the SOPESN1 and SOPESN2 methods, two new classes of modified NSFD methods that are not only elementary stable and second-order accurate, but also positivity-preserving. The SOPESN1 methods are intended to approximate the solutions of productive-destructive systems, whereas the SOPESN2 methods can be used to solve general autonomous dynamical systems.

3.1. Second-order modified PESN (SOPESN1) method for productive-destructive systems. In this section, we consider general productive-destructive systems of the form:

$$
(3.1)\qquad \vec{x}' = \vec{P}(\vec{x}) - \vec{D}(\vec{x});\quad \vec{x}(t_0) = \vec{x}_0 \in \mathbb{R}_+^n,
$$

where $\vec{P} = [P_1, \ldots, P_n]$, $\vec{D} = [D_1, \ldots, D_n]$, with $P_i, D_i \in C^2(\mathbb{R}^n_+; \mathbb{R}_+)$. Equation (3.1) is commonly used for models in science and engineering, with the functions $\vec{P}$ and $\vec{D}$ representing the production and destruction processes, respectively. The following result holds for the new SOPESN1 numerical method [**1**, **3**, **4**]:

THEOREM 3.1. *Let $\vec{P}, \vec{D} \in C^2(\mathbb{R}^n_+; \mathbb{R}^n_+)$ and $\varphi_i : \mathbb{R}_+ \times \mathbb{R}^n_+ \to \mathbb{R}_+$ satisfies the following condition:*

$$\varphi_i(h, \vec{x}) = h - q_i(\vec{x})\frac{h^2}{2} + \mathcal{O}(h^3), \tag{3.2}$$

where

$$q_i(\vec{x}) = -\sum_{j=1}^{n} \left(\frac{\partial P_i(\vec{x})}{\partial x_j} - \frac{\partial D_i(\vec{x})}{\partial x_j}\right)\left(\frac{P_j(\vec{x}) - D_j(\vec{x})}{P_i(\vec{x}) - D_i(\vec{x})}\right) - \frac{2D_i(\vec{x})}{x_i}, \tag{3.3}$$

whenever $D_i(\vec{x}) \neq P_i(\vec{x})$. Then, the SOPESN1 numerical method with denominator function (3.2)–(3.3)*:*

$$\frac{x_i^{k+1} - x_i^k}{\varphi_i(h, \vec{x}^k)} = P_i(\vec{x}^k) - D_i(\vec{x}^k)\frac{x_i^{k+1}}{x_i^k}, \tag{3.4}$$

preserves the positivity of solutions of Equation (3.1)*, is elementary stable, and second-order accurate.*

PROOF. A sketch of the proof is presented below, with more details available in [**1**, **3**, **4**]. The positivity of the SOPESN1 method follows from the fact that when $\vec{x}^k \in \mathbb{R}^n_+$, then $P_i(\vec{x}^k) > 0$ and $D_i(\vec{x}^k) > 0$ for all $i = 1, \ldots, n$, and therefore,

$$x_i^{k+1} = \frac{x_i^k\left(x_i^k + \varphi_i(h, \vec{x}^k)P_i(\vec{x}^k)\right)}{x_i^k + \varphi_i(h, \vec{x}^k)D_i(\vec{x}^k)} > 0,$$

since $\varphi_i(h, \vec{x}^k) > 0$.

To prove the SOPESN1 method is of second-order accuracy, we consider the two cases of $D_i(\vec{x}^k) \neq P_i(\vec{x}^k)$ and $D_i(\vec{x}^k) = P_i(\vec{x}^k)$ separately. First, note that when $D_i(\vec{x}^k) = P_i(\vec{x}^k)$, then the SOPESN1 method (3.4) reduces to $x_i^{k+1} = x_i^k$, which represents an exact method for solving Equation (3.1) at $t = t_k$. When $D_i(\vec{x}^k) \neq P_i(\vec{x}^k)$, note that since $\varphi_i(h, \vec{x}) = h + \mathcal{O}(h^2)$ and $x_i \neq 0$, then there exists $\delta_i > 0$ such that

$$\frac{1}{1 + \dfrac{D_i(\vec{x})\varphi_i(h, \vec{x})}{x_i}} = 1 - \frac{D_i(\vec{x})\varphi_i(h, \vec{x})}{x_i} + \frac{D_i^2(\vec{x})\varphi_i^2(h, \vec{x})}{x_i^2} + \mathcal{O}(h^3), \tag{3.5}$$

whenever $0 < h < \delta_i$. Using Equation (3.5), the second-order accuracy of the method is easily proven by a Taylor series expansion about t_k which yields:

$$x_i(t_{k+1}) - \frac{x_i(t_k) + \varphi_i(h, \vec{x}(t_k))P_i(\vec{x}(t_k))}{1 + \dfrac{\varphi_i(h, \vec{x}(t_k))D_i(\vec{x}(t_k))}{x_i(t_k)}} = \mathcal{O}(h^3).$$

Finally, to show that the numerical method (3.4) is elementary stable, we first note that all fixed points x^* of the SOPESN1 method are equilibria of Equation (3.1) and vice versa. Next, similarly to the proof in Theorem (2.1), we apply the SOPESN1 method to the system:

$$\dot{\vec{y}} = \Lambda\vec{y}, \tag{3.6}$$

where Λ is the Jordan form of the Jacobian J evaluated at the equilibrium $\vec{x}^*$, with eigenvalues $\lambda_1, \lambda_2, \ldots, \lambda_n$, $J = S^{-1}\Lambda S$, and $\vec{y} = S\vec{\epsilon}$, with $\vec{\epsilon} = \vec{x} - \vec{x}^*$. For illustrative purposes, we outline the proof only for the case when all eigenvalues are real. The proof of elementary stability for the other cases is presented in details in [**1, 3, 4**].

If x^* is stable equilibrium of System (3.1), then applying Method (3.4) to System (3.6) yields:

$$y_i^{k+1} = \left(\frac{1}{1 - \varphi_i(h, \vec{y}^k)\lambda_i}\right) y_i^k.$$

To show that x^* is a stable fixed point of the numerical method (3.4) is equivalent to show that $y_i^k \to 0$ for all $i = 1, \ldots, n$. This, in turn, is equivalent to

$$\left|\frac{1}{1 - \varphi_i(h, \vec{y}^k)\lambda_i}\right| < 1,$$

which is clearly satisfied given that x^* is a stable equilibrium of System (3.1). If x^* is an unstable fixed point of System (3.1), then there exists an eigenvalue $\lambda_{i_0} > 0$. Applying the SOPESN1 method to System (3.6) yields

$$y_{i_0}^{k+1} = \left(1 + \varphi_{i_0}(h, \vec{y}^k)\lambda_{i_0}\right) y_{i_0}^k.$$

Since $\varphi_{i_0}(h, \vec{y}^k) > 0$, then clearly $\left|1 + \varphi_{i_0}(h, \vec{y}^k)\lambda_{i_0}\right| > 1$, and therefore, $\vec{x}^*$ is unstable fixed point of Method (3.4), which completes the proof. □

3.2. Second-order modified PESN (SOPESN2) method for general autonomous dynamical systems. The following result holds for the new SOPESN2 numerical method for approximating the solutions of Equation [**1, 3, 4**]:

$$\vec{x}' = \vec{f}(\vec{x}); \quad \vec{x}(t_0) = \vec{x}_0 \in \mathbb{R}_+^n, \tag{3.7}$$

where $\mathbb{R}_+^n = \{(x_1, x_2, \ldots, x_n) \in \mathbb{R}^n : x_i > 0,\ \forall i = 1, \ldots, n\}$, the vector-function $\vec{x} = [x_1, \ldots, x_n]$ is unknown, and the vector-function $\vec{f} = [f_1, \ldots, f_n] \in C^2(\mathbb{R}_+^n; \mathbb{R}^n)$ such that $\mathbb{R}_+^n$ is positively invariant.

THEOREM 3.2. *Let $\vec{f} \in C^2(\mathbb{R}_+^n; \mathbb{R}^n)$ and let $\varphi_i : \mathbb{R}_+ \times \mathbb{R}_+^n \to \mathbb{R}_+$ satisfy the following:*

$$\varphi_i(h, \vec{x}) = h - q_i(\vec{x})\frac{h^2}{2} + \mathcal{O}(h^3), \tag{3.8}$$

where

$$q_i(\vec{x}) = \begin{cases} -\displaystyle\sum_{j=1}^{n} \frac{\partial f_i(\vec{x})}{\partial x_j}\frac{f_j(\vec{x})}{f_i(\vec{x})}, & f_i(\vec{x}) > 0 \\ \dfrac{2f_i(\vec{x})}{x_i} - \displaystyle\sum_{j=1}^{n} \frac{\partial f_i(\vec{x})}{\partial x_j}\frac{f_j(\vec{x})}{f_i(\vec{x})}, & f_i(\vec{x}) < 0 \end{cases}. \tag{3.9}$$

Then, the following SOPESN2 numerical method with denominator function (3.8)–(3.9):

$$\frac{x_i^{k+1} - x_i^k}{\varphi_i(h, \vec{x}^k)} = w_i^k f_i(\vec{x}^k), \tag{3.10}$$

where

$$w_i^k = \begin{cases} 1, & \text{if } f_i(\vec{x}^k) \geq 0 \\ \dfrac{x_i^{k+1}}{x_i^k}, & \text{if } f_i(\vec{x}^k) < 0 \end{cases},$$

preserves the positivity of solutions of Equation (1.1), *is elementary stable, and second-order accurate.*

PROOF. The second-order accuracy, positivity-preserving, and elementary stability properties of the SOPESN2 method (3.10) can be shown using the same approach as in the proof of Theorem 3.1 for the SOPESN1 method, with details presented in [**1**, **3**, **4**]. The main difference between the proofs is that the SOPESN2 method takes on two different forms due to the parameter w_i^k, which depend on the sign of the value of the right-hand side function components $f_i(\vec{x}^k)$ at each time-step k. □

REMARK 3.3. There exists many functions that satisfy the hypothesis of Theorems 3.1 and 3.2, e.g., one can choose

$$\varphi_i(h, \vec{x}^k) = \begin{cases} \dfrac{1 - e^{-q_i(\vec{x}^k)h}}{q_i(\vec{x}^k)}, & \text{if } q_i(\vec{x}^k) \neq 0 \\ h, & \text{if } q_i(\vec{x}^k) = 0 \end{cases}. \tag{3.11}$$

with $q_i(\vec{x}^k)$ as in Equation (3.3) and as in Equation (3.9), respectively.

4. Numerical simulations

To demonstrate the performance of the proposed new methods, we present a series of numerical simulations using the second-order modified nonstandard theta methods (2.1) and (2.2) with $\theta = 0$ (SOESN EE method [**17**]), the second-order modified nonstandard two-stage ERK2 method (2.6) with $\omega = 1/2$ (SOESN ERK2 method), and the two SOPESN methods. We consider two autonomous ODEs from mathematical biology with well-known asymptotic behavior of their positive solutions, namely a 1-D modified logistic growth model and a 2-D predator-prey model. The two specific models are chosen in order to easily highlight the numerical differences between the four new modified NSFD methods.

We first consider the following modified logistic growth equation [**34**, p. 124]:

$$\frac{dx}{dt} = ax\left(\frac{x}{L} - 1\right)\left(1 - \frac{x}{K}\right), \tag{4.1}$$

with $a > 0$ and $0 < L < K$. Equation (4.1) has two stable equilibria $x^* = 0$ and $x^* = L$, and one unstable equilibrium $x^* = K$. Here, for the right-hand side function of Equation (4.1)

$$f(x) = ax\left(\frac{x}{L} - 1\right)\left(1 - \frac{x}{K}\right),$$

the production and destruction functions have the forms

$$P(x) = ax^2\left(\frac{K + L}{KL}\right) \text{ and } D(x) = ax\left(\frac{x^2 + KL}{KL}\right),$$

respectively, with $f(x) = P(x) - D(x)$.

Applying the four second-order modified NSFD methods for solving Equation (4.1) yields the following expressions:

- SOPESN1 method (3.4):

$$\frac{x^{k+1}-x^k}{\varphi^{[I]}(h,x^k)} = a\left(x^k\right)^2\left(\frac{K+L}{KL}\right) - ax^k\left(\frac{\left(x^k\right)^2-KL}{KL}\right)\frac{x^{k+1}}{x^k}$$

- SOPESN2 method (3.10):

$$\frac{x^{k+1}-x^k}{\varphi^{[II]}(h,x^k)} = \begin{cases} ax^k\left(\dfrac{x^k}{L}-1\right)\left(1-\dfrac{x^k}{K}\right), & \text{if } f(x^k)\geq 0\\ ax^k\left(\dfrac{x^k}{L}-1\right)\left(1-\dfrac{x^k}{K}\right)\dfrac{x^{k+1}}{x^k}, & \text{if } f(x^k)<0\end{cases}$$

- SOESN EE method (2.2), with $\theta = 0$ [**17**]:

$$\frac{x^{k+1}-x^k}{\varphi^{[III]}(h,x^k)} = ax^k\left(\frac{x^k}{L}-1\right)\left(1-\frac{x^k}{K}\right)$$

- SOESN ERK2 method (2.6), with $\omega = 1/2$:

$$\frac{x^{k+1}-x^k}{\varphi^{[IV]}(h)} = ax^k\left(\frac{x^k}{L}-1\right)\left(1-\frac{x^k}{K}\right)$$

$$+\frac{1}{2}a\left(x^k+f\left(x^k\right)\varphi^{[IV]}(h)\right)\left(\frac{x^k+f\left(x^k\right)\varphi^{[IV]}(h)}{L}-1\right)\left(1-\frac{x^k-f\left(x^k\right)\varphi^{[IV]}(h)}{K}\right).$$

The corresponding nonstandard denominator functions in the above methods are chosen as

$$\varphi^{[J]}(h,x^k) = \frac{1-e^{-q^{[J]}(x^k)h}}{q^{[J]}(x^k)},\quad J = I, II, III,$$

where

$$q^{[I]}(x) = \frac{a\left(x^2-KL-2(K+L)x\right)}{KL},$$

$$q^{[II]}(x) = \begin{cases}\dfrac{a(3x^2+KL-2(K+L)x)}{KL}, & f(x)>0\\ a\left(\dfrac{x^2}{LK}-1\right), & f(x)<0\end{cases},$$

$$q^{[III]}(x) = \frac{a(3x^2+KL-2(K+L)x)}{KL},$$

as in Equation (3.3), Equation (3.9), and [**17**], respectively and

$$\varphi^{[IV]}(h) = \frac{\tanh(qh)}{q},$$

where $q = 0.5 > 0.25 = \max\left\{\dfrac{|f_x(0)|}{2}, \dfrac{|f_x(K)|}{2}, \dfrac{|f_x(L)|}{2}\right\}$, as in Theorem 2.5.

First, we consider the following set of parameters values: $a = 0.5$, $L = 0.9$, and $K = 1$ with initial condition $x(0) = 0.5$. As shown in Figure 1(A), all modified second-order NSFD methods behave very well and converge to the exact solution

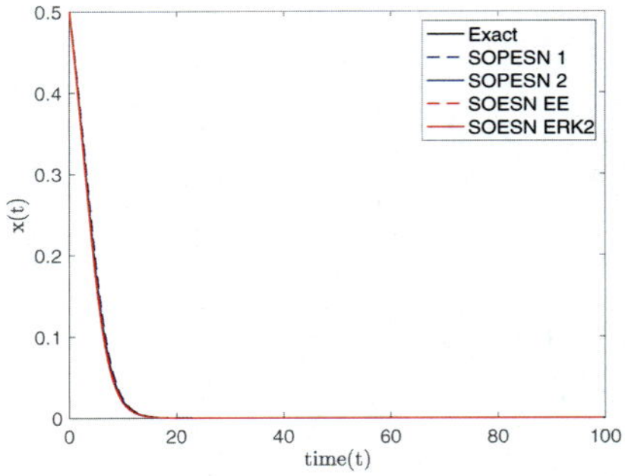

(A) $h = 0.6$

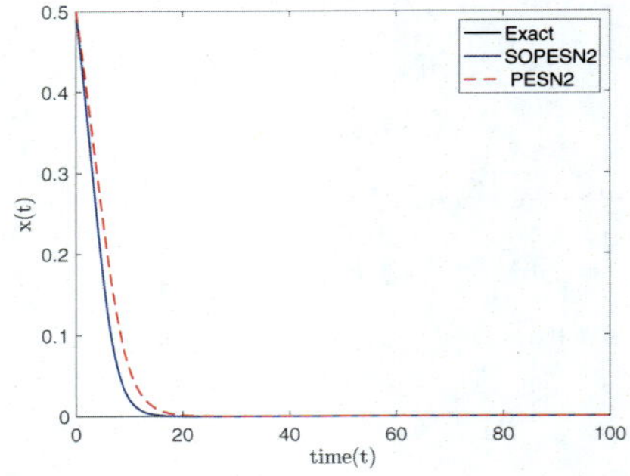

(B) $h = 0.6$

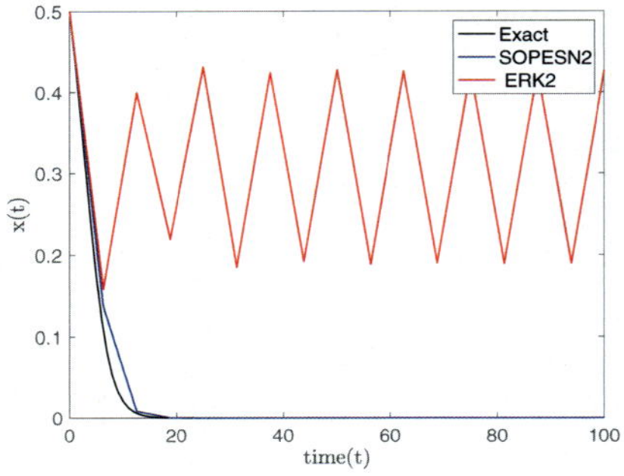

(C) $h = 6.25$

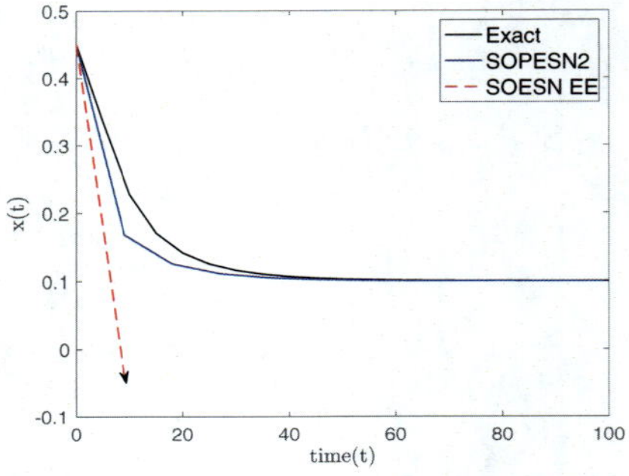

(D) $h = 9$

FIGURE 1. Comparison of the new modified second-order nonstandard numerical methods with other methods for solving Equation (4.1) with $a > 0$, (A)-(C), and $a < 0$, (D), respectively, using different values of the step-size h.

for $h = 0.6$. Figure1(B) shows the second-order SOPESN2 method clearly outperforming the first-order PESN2 method [**37**], with $\varphi(h) = (1 - e^{-0.5h})/0.5$, for $h = 0.6$. When the accuracy of the presented second-order modified NSFD methods is compared to other first-order ESN and PESN methods [**1**–**3**, **16**], similar numerical behavior is observed. Figure1(C) compares the SOPESN2 method to the classical second-order two-stage explicit Runge-Kutta method (ERK2) [**33**] for $h = 6.25$. The SOPESN2 method preserves the positivity and converges to the exact solution, whereas the ERK2 method not only produces oscillatory solutions but also introduces new artificial equilibria for large values of the time-step [**17**].

Next, we consider the case of $a < 0$ with the following set of parameters values: $a = -0.1$, $L = 0.5$, and $K = 0.1$. In this case, Equation (4.1) has two unstable equilibria $x^* = 0$ and $x^* = L$, and one stable equilibrium $x^* = K$. Figure1(D) depicts how sometimes, for large values of the time-step h, the numerical solution produced by the modified NSFD methods that are not positivity-preserving, such as the SOESN EE and NSFD ERK2 methods, can initially jump the unstable equilibrium $x^* = 0$ and thus go negative, leading to an eventual solution blow-up, whereas the numerical solution produced by the positivity-preserving SOPESN2

TABLE 1. Comparison of discrete L^2 error (top) and the rates of convergence (bottom) of the new modified nonstandard numerical methods and the PESN2 method for solving Equation (4.1).

Discrete L^2 error					
h	SOPESN1	SOPESN2	PESN2	SOESN EE	SOESN ERK2
0.5000	0.0838637	0.0008975	0.3439226	0.0015533	0.0398884
0.2500	0.0210937	0.0002305	0.1769097	0.0003931	0.0100924
0.1250	0.0052714	0.0000584	0.0896342	0.0000989	0.0025330
0.0625	0.0013166	0.0000147	0.0451006	0.0000248	0.0006342
0.0312	0.0003289	0.0000037	0.0226195	0.0000062	0.0001586
0.0156	0.0000822	0.0000009	0.0113268	0.0000016	0.0000397
0.0078	0.0000205	0.0000002	0.0056676	0.0000004	0.0000099
Rates of convergence					
0.5000	-	-	-	-	-
0.2500	1.9912	1.9611	0.9591	1.9824	1.9827
0.1250	2.0006	1.9807	0.9809	1.9909	1.9944
0.0625	2.0014	1.9902	0.9909	1.9956	1.9978
0.0312	2.0011	1.9902	0.9956	2.0000	1.9995
0.0156	2.0004	2.0395	0.9978	1.9542	1.9982
0.0078	2.0035	2.1699	0.9989	2.0000	2.0036

method always remains positive and converges to the exact solution. This is not surprising given that the elementary stable SOESN EE method is not designed to preserve the positivity of the solution for arbitrary large step-sizes and only locally preserves the stability of the equilibria.

To further illustrate the accuracy of the new modified NSFD methods, we perform a set of numerical simulations of Equation (4.1) with different values of the step-size $h = 1/2^n$ for $n = 1, \ldots, 7$. Table 1 displays the discrete L^2 error for different values of the step-size h and the corresponding rates of convergence of the modified NSFD methods and the PESN2 method. The rate of convergence for step-size h is computed using the relative L^2 error data E_h and E_{2h} at successive step-sizes h and $2h$, respectively, and is defined as

$$r(h) = \frac{\ln\left(E_{2h}/E_h\right)}{\ln(2)}.$$

As we can see in the table, all the presented new modified NSFD methods are second-order accurate, while the PESN2 method is only of first-order accuracy.

Next, we discuss the performance of the four proposed new second-order modified NSFD methods when applied to the following Rosenzweig–MacArthur predator–prey system with a Holling-type II predator functional response [**10**, **19**]:

$$\begin{aligned} \frac{dx}{dt} &= bx(1-x) - \frac{axy}{c+x} \\ \frac{dy}{dt} &= \frac{xy}{c+x} - dy. \end{aligned} \tag{4.2}$$

In the numerical simulations, we use the parameter values $a = 2.0, b = 1.0, c = 0.5$ and $d = 6$. In this case, stability analysis of System (4.2) shows that there exist two equilibria $E_0 = (0,0)$ and $E_1 = (1,0)$, with E_0 unstable and E_1 globally stable [**10**]. The eigenvalues of $J(E_0)$ are $\lambda_1 = 1$ and $\lambda_2 = -6.0$, while of $J(E_1)$ are $\lambda_3 = -1$ and $\lambda_4 = -\frac{16}{3}$. For solving System (4.2), the production and destruction functions used in the SOPESN2 method have the following form:

$$P_1(x) = bx,\ D_1(x) = bx^2 + \frac{axy}{c+x},\ \text{and}\ P_2(x) = \frac{xy}{c+x},\ D_2(x) = dy.$$

The denominator functions used in the SOPESN1 and SOPESN2 methods are

$$\varphi_i^{[J]}(h, x^k, y^k) = \frac{1 - e^{-q_i^{[J]}(x^k,y^k)h}}{q_i^{[J]}(x^k, y^k)},\quad J = I, II,$$

with $q_i^{[I]}(x^k, y^k)$ as in Theorem 3.1 for the SOPESN1 method, and $q_i^{[II]}(x^k, y^k)$ as in Theorem 3.2 for the SOPESN2 method. The denominator function in the SOESN EE method is a product of two functions, Equation (2.5):

$$\varphi_i^{[III]}(h, x, y) = \frac{\phi_{i_1}(\alpha h)}{\alpha}\phi_{i_2}\left(\frac{\alpha - q_i^{[III]}(x,y)}{2}h\right),$$

where we use $\phi_{i_1}(h) = 1 - e^{-h}$ and $\phi_{i_2}(h) = 1 - \sin h$. We have chosen

$$\alpha = 5.67 > \frac{M}{2}\max\left\{\frac{|\lambda_1|^2}{|Re(\lambda_1)|} = 1, \frac{|\lambda_2|^2}{|Re(\lambda_2)|} = \frac{16}{3}\right\},$$

with $M > 2$ and $q_i^{[III]}(x, y)$ calculated according to Lemma 2.2. The denominator function used in the SOESN ERK2 method is $\varphi^{[IV]}(h) = \tanh(qh)/q$, where

$$q = 3.1 > \frac{1}{2}\max\left\{\frac{|\lambda_1|^2}{|Re(\lambda_1)|} = 1, \frac{|\lambda_2|^2}{|Re(\lambda_2)|} = \frac{16}{3}\right\}.$$

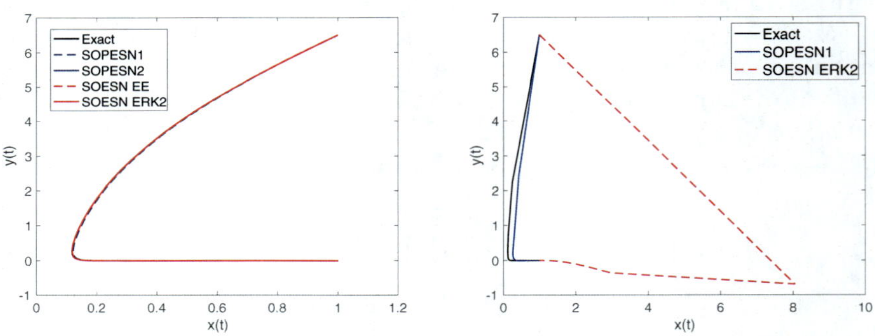

(A) $h = 0.001$ (B) $h = 0.1924$

FIGURE 2. Comparison of the new modified second-order nonstandard numerical methods with each other, applied to System (4.2), for different values of the step-size h.

In the previous example of Equation (4.1), we demonstrated the behavior of some of the classical standard and first-order nonstandard numerical methods. We obtained similar numerical results when applying those same methods to solve System (4.2). Therefore, to avoid repetition, this example will only compare and contrast the benefits of the proposed new modified NSFD methods. Figure 2(A) shows the excellent performance of all four second-order modified NSFD methods when solving System (4.2), with an initial condition $(x(0), y(0)) = (1, 6.5)$ and a time-step size $h = 0.001$, as all numerical solutions very closely resemble the exact solution. In Figure 2(B), a numerical comparison is presented of the SOESN ERK2 method and the positivity-preserving SOPESN2 method for solving System (4.2) using a much larger step-size $h = 0.1924$. In this case, the SOESN ERK2 solution, represented by the red dashed line in Figure 2(B), initially falls below the x-axis, indicating that positivity is violated. This is because the SOESN ERK2 method is not unconditionally positivity-preserving and, in some case, produces negative solution values when very large step-sizes h are used. Despite the initial negative values, the SOESN ERK2 solution quickly recovers and eventually converges to the stable equilibrium point $E_1 = (1, 0)$, because the numerical method is designed to preserve the stability of the local stability properties of the equilibria for any time-step size h when the solution is close to an equilibrium. It is also important to highlight how much more computationally efficient the SOESN ERK2 method is than the SOPESN2 method. This is because the denominator function $\varphi^{[IV]}(h)$ in the SOESN ERK2 method is the same for all solution components and does not change over time (Theorem 2.5), whereas the denominator function $\varphi_i^{[II]}(h, x^k)$ in the SOPESN2 method (Theorem 3.2) must be re-calculated for each solution component i and at every time step. In particular, the SOESN ERK2 method approximates the exact solution 312% faster than the SOPESN2 method, in the numerical simulations shown in Figure 2(A), making it the preferred modified NSFD numerical method in that case. The SOESN ERK2 method is therefore much more appealing than the SOPESN2 method in situations where the second-order modified ESN numerical methods naturally produce positive solutions, when short-term solution dynamics are not of practical importance, or when the positivity of the numerical solution is not required, for example, when solving ODEs arising as adjoint equations in an ODE-constrained optimal control problem that have non-positive solutions but require numerical methods which are elementary stable.

5. Discussion and conclusions

This chapter presented four new classes of second-order modified nonstandard methods: the second-order modified nonstandard theta methods (2.1) and (2.2), the second-order modified nonstandard ERK2 method (2.6), the second-order modified PESN method for productive-destructive systems (3.4), and the second-order modified PESN method for general autonomous dynamical systems (3.10). Table 2 provides a visual summary of the research findings.

The fundamental idea underlying the development of the proposed NSFD numerical methods is the use of a novel modified nonstandard denominator function in the discretization of the derivative. In the case of the modified NSFD theta methods, the denominator function is a product of two functions, where one of them satisfies the second-order accuracy property of the methods, while the other satisfies the stability criteria of Theorem 2.1. For the modified nonstandard ERK2 method, a single denominator function suffices to satisfy the conditions for second-order accuracy and elementary stability. Furthermore, because the nonstandard

TABLE 2. Summary of all four new classes of modified nonstandard finite difference methods.

Method	Autonomous ODEs	Order of accuracy	Positivity preserving	Elementary stable
SOESN θ	General	Second	No	Yes
SOESN ERK2	General	Second	No	Yes
SOPESN1	Productive-Destructive	Second	Yes	Yes
SOPESN2	General	Second	Yes	Yes

denominator function is constant for all solution components, the SOESN ERK2 method is conservative. The nonstandard denominator functions in the SOPESN1 and SOPESN2 methods are different for each solution component, however there exist many such functions that can be used to guarantee the second-order accuracy as well as the elementary stability and positivity-preserving properties of the methods. In addition, each of the new numerical methods can be written in an explicit form, which makes them easy to implement and computationally efficient.

The new modified NSFD methods presented here can be used to solve a wide range of applied problems, from models in physics, chemistry, and biology to economics and engineering [**31**, **32**]. To demonstrate the advantages and differences between the new methods, specific applications from population dynamics for solving the modified logistic growth equation and the Rosenzweig-MacArthur predator-prey system with a Holling-type II predator functional response were presented. A comparison with classical standard and nonstandard finite difference methods revealed the superior performance of the proposed modified NSFD methods. The simulations also numerically demonstrated the significance of the elementary stability and positivity-preserving properties of the proposed new numerical methods, as well as their second-order accuracy, making them a preferred choice for solving autonomous dynamical systems in science, engineering, and economics.

A future goal of this research is to develop a general framework for improving the order of accuracy of other classes of dynamically consistent nonstandard finite difference methods.

References

[1] Fawaz Karhan Alalhareth, *Higher-Order Nonstandard Finite Difference Methods for Autonomous Differential Equations with Applications in Mathematical Ecology*, ProQuest LLC, Ann Arbor, MI, 2022. Thesis (Ph.D.)–The University of Texas at Arlington. MR4512117

[2] Fawaz K Alalhareth, Madhu Gupta, Hristo V Kojouharov, and Souvik Roy, *Second-order modified nonstandard theta and Runge-Kutta methods for n-dimensional autonomous differential equations*, Preprint, 2022.

[3] Fawaz K Alalhareth, Madhu Gupta, Hristo V Kojouharov, and Souvik Roy, *Second-order modified positive and elementary stable nonstandard numerical methods for n-dimensional autonomous differential equations*, Preprint, 2022.

[4] Fawaz K Alalhareth, Madhu Gupta, Souvik Roy, and Hristo V Kojouharov, *Second-order modified positive and elementary stable nonstandard numerical methods for n n-dimensional autonomous differential equations*, Mathematical Methods in the Applied Sciences, `http://doi.org/10.1002/mma.9560`.

[5] Roumen Anguelov and Jean M.-S. Lubuma, *Contributions to the mathematics of the nonstandard finite difference method and applications*, Numer. Methods Partial Differential Equations **17** (2001), no. 5, 518–543, DOI 10.1002/num.1025. MR1849163

[6] Abraham J. Arenas, Gilberto González-Parra, and Blas Melendez Caraballo, *A nonstandard finite difference scheme for a nonlinear Black-Scholes equation*, Math. Comput. Modelling **57** (2013), no. 7-8, 1663–1670, DOI 10.1016/j.mcm.2011.11.009. MR3033392

[7] Michael Chapwanya, Jean M.-S. Lubuma, and Ronald E. Mickens, *Nonstandard finite difference schemes for Michaelis-Menten type reaction-diffusion equations*, Numer. Methods Partial Differential Equations **29** (2013), no. 1, 337–360, DOI 10.1002/num.21733. MR3003114

[8] Dobromir T. Dimitrov and Hristo V. Kojouharov, *Analysis and numerical simulation of phytoplankton-nutrient systems with nutrient loss*, Math. Comput. Simulation **70** (2005), no. 1, 33–43, DOI 10.1016/j.matcom.2005.03.001. MR2165762

[9] Dobromir T. Dimitrov and Hristo V. Kojouharov, *Nonstandard finite-difference schemes for general two-dimensional autonomous dynamical systems*, Appl. Math. Lett. **18** (2005), no. 7, 769–774, DOI 10.1016/j.aml.2004.08.011. MR2144725

[10] Dobromir T. Dimitrov and Hristo V. Kojouharov, *Positive and elementary stable nonstandard numerical methods with applications to predator-prey models*, J. Comput. Appl. Math. **189** (2006), no. 1-2, 98–108, DOI 10.1016/j.cam.2005.04.003. MR2202966

[11] Dobromir T. Dimitrov and Hristo V. Kojouharov, *Stability-preserving finite-difference methods for general multi-dimensional autonomous dynamical systems*, Int. J. Numer. Anal. Model. **4** (2007), no. 2, 280–290. MR2287608

[12] Dobromir T. Dimitrov and Hristo V. Kojouharov, *Nonstandard finite-difference methods for predator-prey models with general functional response*, Math. Comput. Simulation **78** (2008), no. 1, 1–11, DOI 10.1016/j.matcom.2007.05.001. MR2414152

[13] Dobromir T. Dimitrov and Hristo V. Kojouharov, *Dynamically consistent numerical methods for general productive-destructive systems*, J. Difference Equ. Appl. **17** (2011), no. 12, 1721–1736, DOI 10.1080/10236191003781947. MR2854820

[14] Deqiong Ding, Qiang Ma, and Xiaohua Ding, *An unconditionally positive and global stability preserving NSFD scheme for an epidemic model with vaccination*, Int. J. Appl. Math. Comput. Sci. **24** (2014), no. 3, 635–646, DOI 10.2478/amcs-2014-0046. MR3243520

[15] Gilberto González-Parra, Abraham J. Arenas, and Benito M. Chen-Charpentier, *Combination of nonstandard schemes and Richardson's extrapolation to improve the numerical solution of population models*, Math. Comput. Modelling **52** (2010), no. 7-8, 1030–1036, DOI 10.1016/j.mcm.2010.03.015. MR2718414

[16] Madhu Gupta, *On Some Problems in Sparse Hybrid Imaging, Non-Standard Finite Difference Methods, and Fokker-Planck Frameworks in Esophageal Cancer*, ProQuest LLC, Ann Arbor, MI, 2022. Thesis (Ph.D.)–The University of Texas at Arlington. MR4512116

[17] Madhu Gupta, John M Slezak, Fawaz K Alalhareth, Souvik Roy, and Hristo V Kojouharov, *Second-order nonstandard explicit Euler method*, AIP Conference Proceedings **2302** (2020), no. 1, 110003.

[18] Madhu Gupta, John M. Slezak, Fawaz K. Alalhareth, Souvik Roy, and Hristo V. Kojouharov, *Second-order modified nonstandard explicit Runge-Kutta and theta methods for one-dimensional autonomous differential equations*, Appl. Appl. Math. **16** (2021), no. 2, Art. 1, 788–803. MR4389214

[19] Crawford Stanley Holling, *The functional response of predators to prey density and its role in mimicry and population regulation*, The Memoirs of the Entomological Society of Canada **97** (1965), no. S45, 5–60.

[20] Aditya Kaushik, *Nonstandard perturbation approximation and traveling wave solutions of nonlinear reaction diffusion equations*, Numer. Methods Partial Differential Equations **24** (2008), no. 1, 217–238, DOI 10.1002/num.20244. MR2371356

[21] Hristo V Kojouharov, Souvik Roy, Madhu Gupta, Fawaz Alalhareth, and John M Slezak, *A second-order modified nonstandard theta method for one-dimensional autonomous differential equations*, Applied Mathematics Letters **112** (2021), 106775.

[22] Liping Liu, Dominic P. Clemence, and Ronald E. Mickens, *A nonstandard finite difference scheme for contaminant transport with kinetic Langmuir sorption*, Numer. Methods Partial Differential Equations **27** (2011), no. 4, 767–785, DOI 10.1002/num.20551. MR2830660

[23] Ronald E. Mickens, *Nonstandard finite difference models of differential equations*, World Scientific Publishing Co., Inc., River Edge, NJ, 1994. MR1275372

[24] Ronald E. Mickens, *Applications of nonstandard finite difference schemes*, World Scientific, 2000.

[25] Ronald E. Mickens, *Nonstandard finite difference methods, in advances in the applications of nonstandard finite difference schemes*, World Scientific, River Edge, NJ, 2005.

[26] Ronald E. Mickens, *A nonstandard finite difference scheme for a PDE modeling combustion with nonlinear advection and diffusion*, Math. Comput. Simulation **69** (2005), no. 5-6, 439–446, DOI 10.1016/j.matcom.2005.03.008. MR2169718

[27] Ronald E. Mickens, *A numerical integration technique for conservative oscillators combining nonstandard finite-difference methods with a Hamilton's principle*, J. Sound Vibration **285** (2005), no. 1-2, 477–482, DOI 10.1016/j.jsv.2004.09.027. MR2213110

[28] Ronald E. Mickens, *Determination of denominator functions for a NSFD scheme for the Fisher PDE with linear advection*, Math. Comput. Simulation **74** (2007), no. 2-3, 190–195, DOI 10.1016/j.matcom.2006.10.006. MR2307856

[29] Ronald E. Mickens, *Nonstandard finite difference schemes: Methodology and applications*, World Scientific, 2020.

[30] Ronald E. Mickens and Pedro M. Jordan, *A new positivity-preserving nonstandard finite difference scheme for the DWE*, Numer. Methods Partial Differential Equations **21** (2005), no. 5, 976–985, DOI 10.1002/num.20073. MR2154229

[31] Kailash C. Patidar, *On the use of nonstandard finite difference methods*, J. Difference Equ. Appl. **11** (2005), no. 8, 735–758, DOI 10.1080/10236190500127471. MR2156651

[32] Kailash C. Patidar, *Nonstandard finite difference methods: recent trends and further developments*, J. Difference Equ. Appl. **22** (2016), no. 6, 817–849, DOI 10.1080/10236198.2016.1144748. MR3520261

[33] Alfio Quarteroni, Riccardo Sacco, and Fausto Saleri, *Numerical mathematics*, 2nd ed., Texts in Applied Mathematics, vol. 37, Springer-Verlag, Berlin, 2007, DOI 10.1007/b98885. MR2265914

[34] Ronald W. Shonkwiler and James Herod, *Mathematical biology: An introduction with Maple and Matlab*, 2nd ed., Undergraduate Texts in Mathematics, Springer, New York, 2009, DOI 10.1007/978-0-387-70984-0. MR2549073

[35] Yuanyuan Wang, *Numerical dynamics of a nonstandard finite-difference-θ method for a red blood cell survival model*, Adv. Difference Equ., posted on 2015, 2015:86, 12, DOI 10.1186/s13662-015-0432-8. MR3321024

[36] Daniel T. Wood, Dobromir T. Dimitrov, and Hristo V. Kojouharov, *A nonstandard finite difference method for n-dimensional productive-destructive systems*, J. Difference Equ. Appl. **21** (2015), no. 3, 240–254, DOI 10.1080/10236198.2014.997228. MR3316778

[37] Daniel T. Wood and Hristo V. Kojouharov, *A class of nonstandard numerical methods for autonomous dynamical systems*, Appl. Math. Lett. **50** (2015), 78–82, DOI 10.1016/j.aml.2015.06.008. MR3378952

[38] Daniel T. Wood, Hristo V. Kojouharov, and Dobromir T. Dimitrov, *Universal approaches to approximate biological systems with nonstandard finite difference methods*, Math. Comput. Simulation **133** (2017), 337–350, DOI 10.1016/j.matcom.2016.04.007. MR3575285

Department of Mathematics, College of Arts & Sciences, Najran University, Najran, Kingdom of Saudi Arabia

Email address: `fkalalhareth@nu.edu.sa`

Mathematical Sciences Department, George Mason University, Fairfax, Virginia 22030

Email address: `mgupta25@gmu.edu`

Department of Mathematics, The University of Texas at Arlington, Arlington, Texas 76019-0408

Email address: `hristo@uta.edu`

Department of Mathematics, The University of Texas at Arlington, Arlington, Texas 76019-0408

Email address: `souvik.roy@uta.edu`

Contemporary Mathematics
Volume **793**, 2024
https://doi.org/10.1090/conm/793/15907

A simple NSFD inspired method for Monod kinetics with small half saturation constants in the chemostat setting

Hermann J. Eberl

ABSTRACT. Monod kinetics is an important nonlinearity that appears in mathematical modelling of microbial systems, but (under different names) also in many other applications in Mathematical Biology and Process Engineering. Although seemingly innocuous, for some extreme parameter values (notably very small half saturation concentrations and large decay rates), sophisticated high order solvers for ordinary differential equations have been known to fail. We explore this breakdown situation and suggest a simple, low order, easy to implement method that is inspired by so-called *Nonstandard Finite Difference* or *Mickens* schemes. We find that these can be a viable alternative to modern initial value problem solvers, in the problematic cases of extreme parameter values.

1. Introduction

Implementations of many sophisticated and often highly accurate modern methods for the numerical integration of ordinary differential equations are available in a variety of multi-purpose software for numerical computation, such as `R/deSolve`, `python/scipy`, and in `julia` via its package `DifferentialEquations.jl`, to name just a few non proprietary ones; see also [**16**]. For most users, the appeal of such packages is that for a huge class of problems these solvers work "out of the box" and with the methods' default parameters. That is, often it is not required to have an in depth knowledge of the algorithms and how to control them. Moreover, modern ODE solvers are complex, not easy to understand for the non-expert, and it is difficult to efficiently program them *de novo* in any programming language and to thoroughly test them. When the "out of the box" approach does not work, one of the challenges in using such packages can be to decide which of the available methods is a good candidate for the problem at hand. This requires the user to understand the differences between the algorithms and which method parameters have which effect on stability and accuracy of the method. Considering the technical sophistication behind these methods, this is often difficult and requires expert

2020 *Mathematics Subject Classification.* Primary 37N25, 65L12.

Key words and phrases. Chemostat, Mickens scheme, Monod kinetics, nonstandard finite difference method, numerical solution.

Supported by the Natural Sciences and Engineering Research Council of Canada (NSERC) under Discovery Grant RGPIN-2019-05003.

knowledge that many users of such multipurpose numerical software do not have. The advantage of such user friendly packages is then lost.

In mathematical models of microbial population and resource dynamics, the term Monod kinetics refers to a substrate dependent growth rate that marks a smooth transition between the constant 0th order kinetics where the growth limiting substrate is available in abundance, and linear 1st order kinetics when the resource becomes limited [**7**]. If the concentration of the growth limiting substrate is denoted by C then the substrate dependent Monod growth rate is given by the function

$$g(C) = \mu_{max} \frac{C}{K + C} \tag{1.1}$$

where μ_{max} is the maximum specific growth rate and K is the half saturation concentration, i.e the concentration value at which the growth rate is half the maximum, i.e. $g(K) = \mu_{max}/2$. This parameter controls when substrate is limited or abundant. If $C \gg K$ then $g(C) \approx \mu_{max}$, i.e. the system is saturated and we have an approximately constant growth rate [0th order kinetics]. If $C \ll K$ then substrate is limited and we have $g(C) \approx \frac{\mu_{max}}{K} C$, i.e. the growth rate is proportional to the available substrate [1st order kinetics].

This nonlinearity also appears in other applications in Mathematical Biology and Environmental Engineering, under different names: Holling Type 2 functional/numerical response in ecology [**14**], Michaelis-Menten in biochemistry and enzyme kinetics [**14**], hyperbolic force of infection (or incidence rate) in mathematical epidemiology [**2, 5**]. The function is known as Langmuir isotherm in the modeling of sorption processes [**10**]. In river quality modeling it appears with the dependent variable *biological oxygen demand* as argument in the Braun-Bertheoux model [**18**] (see also [**6**]), and with the dependent variable *dissolved oxygen concentration* as argument in a nonlinear version of the Streeter-Phelps equation [**6**]. In operations management this nonlinearity sometimes is used as a clearing function, cf [**19**].

For the most part, models involving Monod kinetics are numerically well behaved and do not pose a challenge for such modern ODE solver packages. However, over the years we have in several applications encountered numerical difficulties in cases where the concentrations of the growth limiting substrates can be large relative to K initially but is quickly degraded to become limited, for example in cases where initially $C \gg K$ and where μ_{max} is sufficiently large. In such cases the transition from 0th order to 1st order kinetics can be very rapid, or almost instantaneous.

In this brief essay we want to explore the numerical difficulties for Monod kinetics with small half saturation concentration that can practically be observed. We will also propose a very simple and easy to implement low order method that is inspired by so-called *Non-standard Finite Difference* or *Mickens* schemes that avoids the problems that much more sophisticated modern high order methods might encounter in such situations.

2. Problem formulation

2.1. Monod decay. To illustrate the numerical problems that can arise in the case of small K it suffices to study the simplest setting, namely degradation of the growth limiting substrate as described by the differential equation

$$\dot{C} = -\mu_{max}\frac{C}{K+C} \tag{2.1}$$

with positive parameters μ_{max} and K. It is easy to verify that this equation preserves positivity and possesses a unique solution that exists for all $t > 0$ and converges to 0 as $t \to \infty$. That is, from a mathematical perspective this equation does not pose any challenges other than the non-availability of a closed form solution in terms of elementary functions. For illustration purposes we consider here the simplest ODE solver, namely the first order forward Euler method, but we point out that the same issues arise in higher order methods as well (albeit often delayed, i.e. at smaller time steps or smaller half saturation concentrations).

Let $C_p > 0$ be an approximation of $C(t_p)$, where $t_p = p \cdot \tau$, at the pth time step, $p \geq 0$. Then the Forward Euler Method

$$C_{p+1} = C_p - \tau g(C_p) = C_p \cdot \left(1 - \frac{\tau\mu_{max}}{K + C_p}\right). \tag{2.2}$$

does not necessarily preserve positivity. Indeed, if $C_p < \mu_{max}\tau - K$ then $C_{p+1} < 0$, and if $C_p \ll \mu_{max}\tau - K$, then we even can have $C_{p+1} \ll -K$. In that later case $g(C_{p+1}) \approx \mu_{max}$, thus the numerical approximation behaves like $C_{p+1} \approx C_p - \tau \cdot \mu_{max}$, i.e C_p becomes a strictly decreasing sequence that diverges at constant rate such that $\lim_{p\to\infty} C_p = -\infty$. In this case, the method approximates the correct solution of the wrong differential equation $\dot{C} = -\mu_{max}$, and therefore no breakdown situations due to instability, etc. might occur.

Similar observations can be made for more advanced classes of solvers. To discuss this, we will look at the Linearly Implicit Euler Method. This is a very simple first order stiff solver which forms the basis of more sophisticated higher order Rosenbrock and Rosenbrock-Wanner methods and their embedded time step controlled variants. Starting point here is the Implicit Euler Method with subsequent linearisation:

$$C_{p+1} = C_p - \tau g(C_{p+1}) \approx C_p - \tau\left[g(C_p) + (C_{p+1} - C_p)g'(C_p)\right] \tag{2.3}$$

Replacing $g(C)$ by (1.1) and isolating for C_{p+1} we obtain the recurrence

$$C_{p+1} = C_p \cdot \left(1 - \frac{\frac{\tau\mu_{max}}{K+C_p}}{1 + \frac{\tau\mu_{max}K}{(K+C_p)^2}}\right) \tag{2.4}$$

In this case, if C_p is positive, then C_{p+1} becomes negative for

$$\tau > \frac{K + C_p}{\mu_{max}} / \left(1 - \frac{K}{K + C_p}\right)$$

and if

$$\tau \gg \frac{K + C_p}{\mu_{max}} / \left(1 - \frac{K}{K + C_p}\right)$$

then we can find even $C_{p+1} \ll -K$. We point out that for given K, μ_{max} and C_p the critical time step for the Forward Euler method to become negative is smaller than that of the Linearly Implicit Euler method, but that for small K, i.e. for $K \ll C_p$ (which is the situation that we investigate here) they both become practically indistinguishable.

At the core of the matter in both cases is the observation that for very small K and sufficiently large μ_{max} the numerical method within one time step can go from $C^p \gg K$ to $C^{p+1} \ll -K$ due to the steep decay of $C(t)$. The observation discussed here carries over to more complex models that involve Monod kinetics. An important such model which we will study below is the chemostat model, in which degradation of the substrate C is coupled to the growth of bacterial biomass.

2.2. The chemostat setting.

2.2.1. *The single species, single substrate chemostat.* The single species, single substrate chemostat model for the dependent variables biomass density X [mass/volume] and concentration C [mass/volume] of the growth controlling substrate reads

$$\begin{cases} \dot{C} &= Q(C_0 - C) - N\frac{XC}{K+C}, \\ \dot{X} &= -QX + M\frac{XC}{K+C}. \end{cases} \tag{2.5}$$

The derivatives on the left hand side are taken with respect to the independent variable T [time] that does not explicitly occur in the equations beyond that. Here, the positive constant Q [1/time] is the reactor's dilution rate, C_0 [mass/volume] is the substrate concentration on inflow, M [1/time] is the maximum specific growth rate, K [mass/volume] is the half saturation concentration, and Nu [1/time] is the maximum substrate utilization rate that is related to M via the yield coefficient that is determined by the stoichiometry of the system.

To simplify notation and some calculations going forward we re-cast this model in dimensionless form using the new variables and parameters

$$c := \frac{C}{C_0}, \quad x := \frac{NX}{QC_0}, \quad t := QT \tag{2.6}$$

and parameters

$$\mu := \frac{M}{Q}, \quad k := \frac{K}{C_0}. \tag{2.7}$$

This leads to

$$\begin{cases} c' &= 1 - c - \frac{xc}{k+c}, \\ x' &= -x + \frac{\mu xc}{k+c}. \end{cases} \tag{2.8}$$

where the derivatives on the left hand side are taken with respect to t. The following summarizes the well-known features of this model. The proofs are straightforward and utilize only standard arguments. They can be found in many books, e.g. in [**7**].

PROPOSITION 2.1. *Model* (2.8) *with non-negative initial data and positive parameters* μ, k *possesses a unique non-negative solution that exists for all* $t > 0$.

PROPOSITION 2.2. *For all positive parameters* μ, k, *the model* (2.8) *has the washout equilibrium* $(c^*, x^*) = (1, 0)$. *It is asymptotically stable if* $\frac{\mu}{k+c^*} < 1$ *and unstable if the inequality is reversed. In that case, a second, asymptotically stable positive equilibrium exists,* $(c^{**}, x^{**}) = \left(\frac{k}{\mu-1}, \mu(1 - c^{**})\right)$, *to which all solutions in the positive cone converge.*

There is a simple but important interpretation of this last result: if under ideal growth condition (i.e., $c = 1$) production of new biomass is slower than washout from the reactor, then the system will converge to complete washout, i.e. the absence of biomass. If, on the other hand, the washout from the system is slower than maximum biomass production (i.e. production at inflow conditions), then the system will attain a positive steady state with biomass present. The attained concentration value $c^{**} = \frac{k}{\mu-1}$ is called the break-even concentration. At this concentration biomass growth is balanced by hydrodynamic washout, $\frac{\mu c^{**}}{k-c^{**}} = 1$.

2.2.2. *A Mickens inspired semi-implicit time integrator.* NSFD methods for (2.1) and some related models have been introduced and discussed in some detail in [**2**]. Some of those ideas we will use here for the chemostat model (2.8). To construct our numerical method, we use two of the rules for the derivation of Non-standard Finite Difference (NSFD) methods after Mickens in [**12**]: (i) the order of the approximation of the derivative is the same as the order of the derivative itself (Rule No. 1), and (ii) we express nonlinearities nonlocal in time, by writing products of dependent variables as products of terms evaluated at different time instances (Rule No 3; see also part 2 of Definition 5 in [**1**]).

More specifically, we use the forward 1st order difference approximation for the derivative, as in the explicit and linearly implicit Euler method, we approximate the linear washout term at the new time level, and we approximate xc nonlocally in time at time $t = t_p$ by $x(t_p)c(t_{p+1})$ or by $x(t_{p+1})c(t_p)$, such that the resulting discrete equation is linear in the dependent variable at the new time step. Then we have

$$\begin{cases} c_{p+1} &= c_p + \tau(1 - c_{p+1} - \frac{x_p c_{p+1}}{k+c_p}), \\ x_{p+1} &= x_p + \tau(-x_{p+1} + \frac{\mu x_{p+1} c_p}{k+c_p}). \end{cases} \tag{2.9}$$

where $\tau > 0$ is the discrete timestep size such that $t_p = p\tau$ for $p = 0, 1, 2, \ldots$, and $x_p \approx x(t_p)$, $c_p \approx c(t_p)$ are the numerical approximation of the exact solution at time t_p.

By rearranging terms, (2.9) can be rewritten as the recursion

$$\begin{cases} c_{p+1} &= \frac{c_p+\tau}{1+\tau(1+\frac{x_p}{k+c_p})} \\ x_{p+1} &= \frac{x_p}{1+\tau(1-\frac{\mu c_p}{k+c_p})} \end{cases} \tag{2.10}$$

We will first show that this discrete approximation reflects the properties of Propositons 2.1 and 2.2 of the continuous model, i.e. that the method is dynamically consistent with the model that it approximates, cf. also Rule 4 in [**12**].

PROPOSITION 2.3. (a) *If* $\mu\frac{c_p}{k+c_p} < 1$ *then* $c_p > 0, x_p > 0$ *implies* $c_{p+1} > 0, x_{p+1} > 0$ *for all* $\tau > 0$. (b) *If* $\mu\frac{c_p}{k+c_p} > 1$ *then* $c_p > 0, x_p > 0$ *implies* $c_{p+1} > 0, x_{p+1} > 0$ *for all* $0 < \tau < \frac{1}{\frac{\mu c_p}{k+c_p}-1}$.

PROOF. The assertion is easily verified by straightforward calculations. □

In the above Proposition, (a) is the case in which biomass washout dominates growth. Case (b) is the reverse scenario. The time step constraint here is to guarantee positivity of x_{p+1}; for c_{p+1} the assertion holds for all $\tau > 0$. The time step constraint $\tau < \frac{1}{\frac{\mu c_p}{k+c_p}-1}$ for stability is practically not a severe restriction, as it simply states that the simulation time step should be chosen smaller than the characteristic time scale of net growth. This is an obvious accuracy requirement if one is interested in simulation of biomass growth. Both cases (a) and (b) give

COROLLARY 2.4. *If* $0 < \tau < \frac{1}{\frac{\mu}{k+1}-1}$ *then* $c_0 > 0, x_0 > 0$ *implies that* $c_p > 0, x_p > 0$ *for all* $p > 0$*, and that the solution* (c_p, x_p) *for all* $p > 0$ *is unique.*

PROOF. Positivity for all $p > 0$ follows by induction from Proposition 2.3. Uniqueness of the solution (c_p, x_p), $p > 0$ follows then directly from the construction of the recursion in (2.10). □

Before we study the stability of the discrete system at steady states we confirm

LEMMA 2.5. *The steady states* (c^*, x^*) *and* (c^{**}, x^{**}) *of the ODE model* (2.8) *are also steady states of its discrete approximation* (2.10).

PROOF. The proof is direct. Substituting $(c^*, x^*) = (1, 0)$ for (c_p, x_p) into the RHS of (2.10) returns on the LHS $(c_{p+1}, x_{p+1}) = (1, 0) = (c_p, x_p)$. Similarly, substituting (c^{**}, x^{**}) for (c_p, x_p) into the RHS of (2.10) and using that $\mu\frac{c^{**}}{k+c^{**}} - 1 = 0$ immediately returns $x_{p+1} = x_p = x^{**}$. Using also that $x^{**} = \mu(1 - c^{**})$, we find

$$c_{p+1} = \frac{c_p + \tau}{1 + \frac{\tau}{c_p}}, \quad \text{and thus} \quad \frac{c_{p+1}}{c_p} = 1$$

from which we have indeed $c_{p+1} = c_p = c^{**}$, as required. □

PROPOSITION 2.6. (a) *If the washout steady state* $(1, 0)$ *of the continuous model* (2.8) *is linearly asymptotically stable, then it is also linearly asymptotically stable for the NSFD recursion* (2.10) *for all time steps choices* $\tau > 0$.

(b) *If the washout steady state* $(1, 0)$ *of the continuous model* (2.8) *is linearly unstable, then it is also linearly unstable for the NSFD recursion* (2.10) *for all time steps choices* $0 < \tau < \frac{1}{\frac{\mu c_p}{k+c_p}-1}$.

PROOF. The assertion follows in the usual manner from linearisation arguments. The Jacobian of the discrete system (2.10) evaluated in the washout equilibrium $(1, 0)$ takes the simple form

$$J(1,0) = \begin{bmatrix} \frac{1}{1+\tau} & * \\ 0 & \frac{1}{1-\tau(\frac{\mu}{k+1}-1)} \end{bmatrix}$$

where entry $*$ in position [1,2] denotes a non-zero value that does not affect the eigenvalue calculations. This matrix has the diagonal elements as its two eigenvalues. We find

$$0 < \lambda_1 = \frac{1}{1+\tau} < 1 \quad \text{for all} \quad \tau > 0.$$

For $\lambda_2 = \frac{1}{1-\tau(\frac{\mu}{k+1}-1)}$) we find $0 < \lambda_2 < 1$ for all $\tau > 0$ if $\frac{\mu}{k+1} - 1 < 0$, i.e. if the washout equilibrium of (2.8) is asymptotically stable, i.e. in case (a). If $\frac{\mu}{k+1} - 1 > 0$, i.e. in case (b), we have $\lambda_2 > 1$, i.e. the washout equilibrium is unstable. □

We remark that from the view point of stability the time step range for part (*b*) can be increased to include also $-1 < \lambda_2 < 0$. In this case, however, the solutions will show spurious oscillations around the equilibrium.

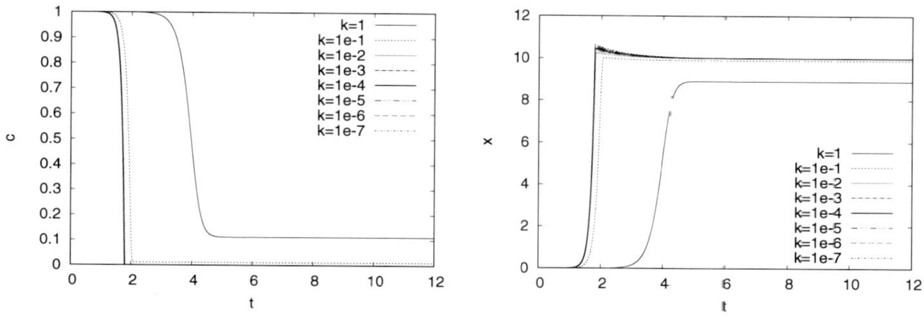

FIGURE 1. Solution of the chemostat model (2.8) computed with the NSFD recursion (2.10) for various values of k.

3. Numerical results

In the basic Monod decay model of Section 2.1 the true solution of the initial value problem converges to 0 and we have seen that in a single time step, depending on parameters, traditional ODE solvers can overshoot into the negatives and then get lost. In the chemostat model from Section 2.2.1 the substrate concentration converges to a positive value c^{**} if the washout equilibrium is unstable. Therefore, it is not *a priori* clear whether the numerical solution might get close enough to 0 (and hence to the singularity at $-K$) for this overshooting problem to occur. In that sense, the more complex chemostat model might be expected to potentially be better behaved. However, note that as k becomes smaller, also c^{**} moves closer to 0. We investigate this more in a series of numerical experiments.

3.1. Implementation. All simulations that we report here were carried out in `Julia-1.7.3` on a Lenovo ThinkPad X1 Carbon with Intel Core i7-8650U CPU, under Ubuntu 20.04.4. The NSFD recursion (2.10) was *de novo* implemented as a single loop with a time step τ that is fixed for each model simulation.

For comparison purposes we also simulated the chemostat model using somewhat arbitrarily selected algorithms of Julia's vast `DifferentialEquations.jl` package [**3**, **15**]. These methods were used as is, without fine tuning their parameters, in accordance with the overall purpose of our study.

3.2. Model solution in dependence of k. The simulations that we carried out were run for fixed parameters $\mu = 10$, and initial data $c(0) = 1$, $x(0) = 10^{-6}$. The simulation interval was $0 \le t \le 20$, large enough for the system to reach steady state. The half saturation concentration k was varied as $k = 10^{-j}$ with $j = 0, .., 7$. In all cases the nontrivial equilibrium (c^{**}, x^{**}) as per Proposition 2.2 is asymptotically stable and should be attained. The washout equilibrium $(1, 0)$ is unstable.

In Figure 1 we display the solutions computed with the recursion (2.10). Within plotting accuracy, the model attains steady state at about $t = 10 \sim 12$ in all cases. Therefore, the plots are restricted to $0 < t < 12$. The solution component $c(t)$ remains approximately at initial/inflow concentration for some time and then drops rapidly to the steady state value c^{**}. The smaller k, the sooner this drop happens and the steeper it is. Note that per Proposition 2.2 the steady state values c^{**} decreases as k decreases. For x the image is similar but reversed: after an initial lag phase we observe a steep increase in x and then a leveling of at x^{**}, which becomes the larger, the smaller k is. For the smaller k values x^{**} becomes indistinguishable within plotting accuracy, in accordance with Proposition 2.2. For small k we observe that in the transients the numerical simulation for x overshoots to a maximum and then declines from there to x^{**}. This is indicative of the chosen timestep being too large for the method to capture the transients with sufficient accuracy. The convergence of the NSFD recursion in dependence of time step choice is investigated in the next section.

The methods tested from the `julia` solver package, either did not converge for the smaller k that we tested, or gave obviously erroneous solutions. In most cases for small enough k a method would abort with an error message "instability detected" after c attaining large negative numbers. In some instances we observed that for k in an intermediate range a method took a long time but eventually found a proper solution. These methods would fail then for smaller k.

Among the many methods we surveyed (see the julia documentation [**3**] for a description of methods), the simple *ImplicitEuler* worked best. The smallest k for which the method returned the correct steady state value was $k = 10^{-5}$ and it would fail for smaller k (simulation data not shown). Method `Hairer42` failed for $k < 1$. Methods `Hairer4` and `KenCarp4` failed for $k < 10^{-1}$. Methods `Trapezoid`, `Tsit5`, `AutoTsit5(Rosenbrock23())`, as well as the default without solver specified failed for $k < 10^{-2}$. Methods `Rodas5`, `BS3`,`Velds4` failed for $k < 10^{-3}$. The methods `Veldd4`, `RadauIIA5`, `Rosenbrock23` failed for $k < 10^{-4}$. The method `ROS3P`, for all values of k, seems to have converged to the unstable equilibrium $(c^*, x^*) = (1, 0)$. From this brief *ad hoc* survey of methods applied to a single test case, it appears that there is no indication that stiff solvers would perform better than non-stiff solvers.

The NSFD recursion (2.10) returned the correct steady state quickly for all choices of $k = 10^{-j}$, with $j = 1, ..8$.

3.3. Convergence of the NSFD recursion. To assess the convergence properties of the NSFD recursion (2.10), we simulated for $k = 10^{-1}$, $k = 10^{-3}$, $k = 10^{-5}$ the model over the time interval $[0, t_{end} = 20]$ with stepsizes $\tau_j := t_{end} \cdot 2^{-j}$, for $j = 8, \ldots, 20$. We compared the solution components c and x for τ_j and τ_{j-1} in the 1-norm .

More specifically, let c_p^j, $p = 0, \ldots, 2^j$, denote the approximation of $c(t)$ at time $t = p\tau_j$. That is c_p^j approcimates $c(p \cdot \tau_j)$ computed with time step size $\tau_j = t_{end} \cdot 2^{-j}$. Then we have that for two subsequent grids with step sizes τ_j and τ_{j-1} the approximations c_{2p}^j and c_p^{j-1} are taken at the same time $t = 2p\tau_j = p\tau_{j-1}$.

Then we calculate the 1-norm-difference (scaled by the number of data points) between the solutions for two subsequent time steps τ_j and τ_{j-1} as

$$\frac{1}{2^j} \sum_{i=0}^{2^{j-1}} \left| c_i^{j-1} - c_{2i}^j \right| . \tag{3.1}$$

For x the calculations are performed in the same manner.

These data are plotted in Figure 2. We found for all three tested values of k and both solution components c and x that halving the time steps from τ_{j-1} to τ_j leads to a decrease of the difference between the two corresponding solutions by a factor 2. This establishes and verifies the expected first order of the convergence,

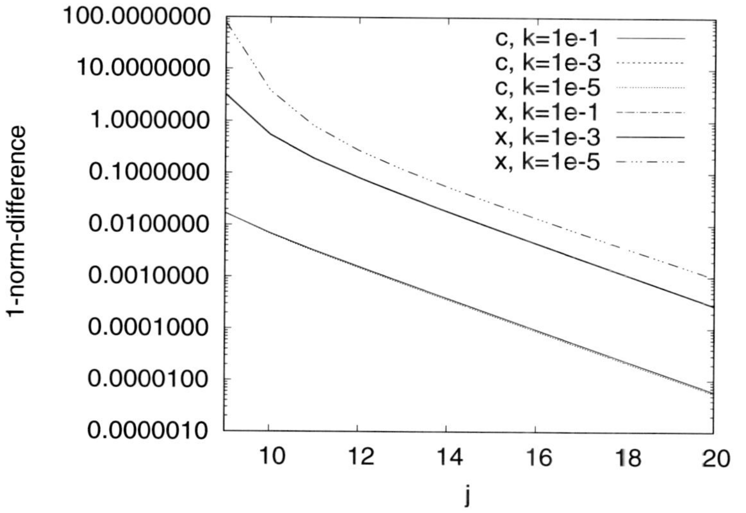

FIGURE 2. Convergence test for the NSFD recursion (2.10). The model was simulated for three different values of k with timesteps $\tau_j := 20 \cdot 2^{-j}$, for $j = 8, \ldots, 20$. Plotted is the difference between the solution components c and x computed with timesteps τ_j and τ_{j-1}, measured in the 1-norm that has been calculated per (3.1).

i.e we find the same convergence order the Implicit Euler method, which as per the results reported in the previous section was among the traditional methods the one that was successful for the largest range of k values. The corresponding results for the maximum-norm are very similar, data not shown.

4. Discussion, limitations and outlook

4.1. Summary. The NSFD inspired semi-implicit recursion for the numerical approximation of the solution of the chemostat model that we studied here is, by construction, a low order method. While many implementations of modern, sophisticated high order ODE solvers are available, we have seen that these can go astray when applied to Monod kinetics with small half saturation concentrations and relatively large substrate utilization rates. In these cases too large time steps might lead to overstepping the singularity of the Mono equation. As we have seen, also error controlled methods with adaptive time stepping are not absolutely protected from this dilemma. The NSFD recursion does not suffer from this problem from the perspective of stability for moderate time steps. The method is easy to implement, it is derivative free, and does not require the solution of nonlinear equations.

4.2. Alternative formulation of Monod kinetics. The solutions of the underlying ordinary differential equation remain positive. The breakdown of numerical methods happens, when the discrete approximation drops into the negative range, close to, or below the singularity at $-K$, and when the method cannot recover from this. This suggests that improvement might be achieved, by replacing $g(C)$ with another function, with which it coincides for $C \geq 0$ and which remains non-negative throughout. A candidate that is sometimes used is the piecewise defined $\tilde{g}(C) = g(\max\{0, C\})$, which cuts off g at 0 for negative arguments but it is continuous at $C = 0$. The resulting ODE still satisfies the Lipschitz condition, but the right hand side of (2.1), (2.5) and other equations involving this term are not differentiable anymore at $C = 0$. Therefore, the differential equation does not have enough regularity for the convergence theory of higher order methods to apply and one must expect at least an order reduction, even if only in an intermediate step the numerical approximation dips into the negatives, i.e. in the cases in which we have seen that ODE solvers with (1.1) might fail. In the case of the chemostat model, we know that c^{**} is the closer to 0 the smaller k is, i.e for very small k we might expect such order reduction to occur.

To investigate this further we repeated the simulation of Section 3.2 with the same methods from `julia`'s package `DifferentialEquations.jl`, but now using $\tilde{g}(C)$ instead of $g(C)$. The vast majority of solvers (but not all) converged to steady state for a wider range of k, some even for all ranges tested. In many instances, however, we observed a loss in accuracy of the computed steady state values of several leading digits, compared to the exact, theoretical one. Hence, also the work around to replace $g(C)$ by $\tilde{g}(C)$ does not provide a reliable safeguard against breakdown of, or inaccuracies introduced by the used numerical method.

4.3. Possible modifications of the NSFD recursion. In (2.10) both c_{p+1} and x_{p+1} are computed from c_p and x_p. An alternative idea that can be investigated is to compute first c_{p+1} from c_p and x_p and then x_{p+1} from c_{p+1} and x_p, or *vice versa*. These methods are equally easy to implement, but lead to more involved algebraic calculations for the analysis of the resulting methods, for example in regards to dynamic consistency of the method (Rule No. 4 in [**12**]). For a comparison of the performance of several such approaches for (2.1) and some related equations we refer to [**2**].

4.4. Related problems. The simple NSFD inspired recursion that we studied here is based on Rules No. 1 and 3 of [**12**], which can be also applied to many other types of equations which might be much more involved than the simple Monod decay and chemostat models that we discussed here. The same issues around Monod kinetics that we investigated here can arise in much more larger models of microbial processes, such as [**8, 9**], which consist of dozens of ODEs. In many applications of such models a sensitivity analysis is routinely carried out, in which thousands of simulations are run for randomly chosen parameters sets that are drawn from some probability distributions. This could result in very small half saturation concentrations, even if the default parameter values are not. The general idea behind our NSFD inspired recursion could be applicable to these more involved systems as well.

Similarly, the ideas to generate low order, derivative free numerical methods can also be used for the time integration of parabolic PDE systems of type

$$u_t = \nabla\big(D(u)\nabla u\big) - F(u)u,$$

where $u = u(t, x)$ is the solution vector, and $D(u), F(u)$ are matrices. D encodes the spatial diffusion processes. It is diagonal in the absence of cross-diffusion. F encodes the reactions and may include Monod kinetics in many. One has then

$$u_{p+1} = u_p + \tau\nabla\big(D(u_p)\nabla u_{p+1}\big) - F(u_p)u_{p+1}$$

After discretisation of the spatial terms, say with Finite Differences, Finite Volumes or Finite Elements, in every time step the solution of a (usually sparse and diagonally dominant) linear system is required to determine the solution vector u_{p+1} at the new time step. For this task efficient solvers exist that are straightforward to parallelize. As it is the case for ODE models, this time integration scheme is Jacobian free (unlike, say, Rosenbrock-Wanner methods) and it does not require the solution of nonlinear systems (unlike, say fully implicit methods). Methods like or very similar to this type have been used successfully, but not yet thoroughly investigated from a theoretical/performance perspective, for highly degenerate biofilm models with a singularity in the diffusion coefficient for one biomass species and Monod kinetics in the reaction terms [**4**, **11**, **20**, **21**], and later extended to models with several species [**13**, **17**].

4.5. Accuracy. Although, as we have seen from the view point of stability, that the NSFD inspired scheme (2.10) only requires moderate time steps, practical observations have shown that to obtain also accurate results some very small time steps might be required. In fact we have seen that the for given time step the accuracy of the method might depend on k, cf also Figure 1, and that by extension the smaller k (i.e. the steeper the transition between 0th and 1st order kinetics), the finer a numerical resolution is required. This is not surprising as the method, by construction and intent is a low order method. Therefore, in cases where efficient traditional ODE solvers work, the NSFD recursion cannot be recommended as an alternative. One question that might be worth exploring is whether some form of time step control can be implement, say one that is based on comparing solutions based on calculations with time steps τ and $\tau/2$, i.e. making the step from t_p to t_{p+1} with a single step of (2.10), and in two steps of length $\tau/2$ and then comparing the computed values. If they are sufficiently small together one can try increasing the time step for $t_p \to t_{p+1}$, if they are too far apart, one resorts to a smaller time step instead. Another possible avenue to improve accuracy and to permit larger time step is to use nonstandard denominator functions $\phi(\tau)$, as in Rule No. 2 for the construction of Mickens schemes in [**12**]. This is relatively straightforward to do for linear differential equations or other equations for which much is known about the exact solution, but it seems that there are no immediate good candidates for the problem at hand, so we have not yet explored this.

5. Conclusion

Nowadays, a broad selection of more or less easy to use, high order and efficient software tools exist for the simulation of mathematical models that are cast as initial value problems of ordinary differential equations. These are designed to work "out-of-the-box" in many or even most cases with little input from the user beyond the actual model description.

However, there are also situations where such packaged methods might fail. In this contribution we explored one of these breakdown scenarios that we have encountered over the years several times in models of bacterial systems, namely models that involve Monod kinetics and an extreme choice of parameters, specifically low half saturation concentrations coupled with a relatively high maximum substrate utilization rate. The same nonlinearity, which is a smooth transition from 0th order to 1st order kinetics, arises under different names also in many other application areas of Mathematical Biology.

In applications, in which modern higher order methods work well these should be preferred as they are designed to be more accurate. However, in more critical situations in which these sophisticated modern methods fail, the low order method that we have studied here provides an easy to implement and easy to use alternative.

References

[1] Roumen Anguelov and Jean M.-S. Lubuma, *Contributions to the mathematics of the nonstandard finite difference method and applications*, Numer. Methods Partial Differential Equations **17** (2001), no. 5, 518–543, DOI 10.1002/num.1025. MR1849163

[2] Michael Chapwanya, Jean M.-S. Lubuma, and Ronald E. Mickens, *From enzyme kinetics to epidemiological models with Michaelis-Menten contact rate: Design of nonstandard finite difference schemes*, Comput. Math. Appl. **64** (2012), no. 3, 201–213, DOI 10.1016/j.camwa.2011.12.058. MR2944803

[3] *DifferentialEquations.jl: Scientific Machine Learning (SciML) Enabled Simulation and Estimation*, `https://docs.sciml.ai/DiffEqDocs/stable/`, Accessed on 2022-12-19

[4] Hermann J. Eberl and Laurent Demaret, *A finite difference scheme for a degenerated diffusion equation arising in microbial ecology*, Proceedings of the Sixth Mississippi State–UBA Conference on Differential Equations and Computational Simulations, Electron. J. Differ. Equ. Conf., vol. 15, Southwest Texas State Univ., San Marcos, TX, 2007, pp. 77–96. MR2316131

[5] Hermann J. Eberl, Harry J. Gaebler, and Yrjö T. Gröhn, *A brief note on a multistrain SIR model with complete cross-protection and nonlinear force of infection*, Commun. Nonlinear Sci. Numer. Simul. **103** (2021), Paper No. 106001, 14, DOI 10.1016/j.cnsns.2021.106001. MR4302634

[6] A. V. Gotosev, *Modification of the Streeter–Phelps System with the Aim to Account for the Feedback between Dissolved Oxygen Concentration and Organic Matter Oxidation Rate*, Wat. Resources, **37** (2010) no. 2, 245–251.

[7] J. Harmand, C. Lobry, A. Rapaport, T. Sari, *The Chemostat: Mathematical Theory of Microorganism Cultures*, Wiley, 2017

[8] IWA Task Group for Mathematical Modelling of Anaerobic Wastewater Processes, *Anaerobic Digestion Model No.1 (ADM1)*, IWA Publishing, London, 2002

[9] IWA Task Group on Mathematical Modelling for the Design and Operation of Biological Wastewater Treatment, *Activated Sludge Models ASM1, ASM2, ASM2d and ASM3*, IWA Publishing, London, 2000

[10] J. D. Logan, *Transport modeling in hydrogeochemical systems*, Springer, New York, 2001

[11] J. E. Macías-Díaz, R. E. Landry, and A. Puri, *A finite-difference scheme in the computational modelling of a coupled substrate-biomass system*, Int. J. Comput. Math. **91** (2014), no. 10, 2199–2214, DOI 10.1080/00207160.2013.870661. MR3264886

[12] Ronald E. Mickens, *Nonstandard finite difference schemes for differential equations*, J. Difference Equ. Appl. **8** (2002), no. 9, 823–847, DOI 10.1080/1023619021000000807. In honor of Professor Allan Peterson on the occasion of his 60th birthday. MR1919887

[13] N. Muhammad, H. J. Eberl. *OpenMP Parallelization of a Mickens Time-Integration Scheme for a Mixed-Culture Biofilm Model and its Performance on Multi-core and Multi-processor Computers*, LNCS, **5976** (2010) 180-195.

[14] J. Müller, C. Kuttler, *Methods and Models in Mathematical Biology Deterministic and Stochastic Approaches*, Springer, Berlin, Heilderberg, 2015

[15] C. Rackauckas, Q. Nie, `Differentialequations.jl`–a performant and feature-rich ecosystem for solving differential equations in `julia`, *J. Open Research Software*, 5(1), 2017

[16] C. Rackauckas, *A Comparison Between Differential Equation Solver Suites In MATLAB, R, Julia, Python, C, Mathematica, Maple, and Fortran*, The Winnower, 9:e153459.98975, 2018

[17] Kazi Rahman and Hermann J. Eberl, *Numerical treatment of a cross-diffusion model of biofilm exposure to antimicrobials*, Parallel processing and applied mathematics. Part I, Lecture Notes in Comput. Sci., vol. 8384, Springer, Heidelberg, 2014, pp. 134–144, DOI 10.1007/978-3-642-55224-3_14. MR3218368

[18] S. Rinaldi, R. Soncini-Sessa, H. Stehfest, H. Tamura, *Modeling and Control of River Quality*, McGrah-Hill, New York, 1978

[19] B. Selcuk, J. C. Fransoo, A. G. de Kok, *Work-in-process clearing in supply chain operations planning*, IIE Transactions **40** (2008) 206-220

[20] S. Sirca, M. Horvat, *Computational Methods for Physicists*, Springer, Berlin and Heidelberg, 2012

[21] Y. O. Tijani, A. R. Appadu, A. A. Aderogba, *Some Finite Difference Methods to Model Biofilm Growth and Decay: Classical and Non-Standard*, Computation, **9** (2021) 123

DEPT. MATHEMATICS AND STATISTICS, UNIVERSITY OF GUELPH, 50 STONE RD E, GUELPH ONTARIO, CANADA N1G 2W1

Email address: heberl@uoguelph.ca

Contemporary Mathematics
Volume **793**, 2024
https://doi.org/10.1090/conm/793/15879

Dynamics preserving nonstandard finite difference scheme for a microbial population model incorporating environmental stress

Michael Chapwanya and Phindile Dumani

ABSTRACT. Microbial populations depend on their environment but can also modify it. In addition to breaking down complex nutrients for their growth, microbes can exhibit negative behavior by engineering the environment in ways that are detrimental to their proliferation. In this work, a deterministic mathematical model is formulated to include the environmental noise from the toxin-producing microbial population accounting for the switch of cells to dormancy at high concentrations. A dynamically consistent nonstandard finite difference scheme is designed. Theoretical and numerical investigation of the proposed model is presented to provide insight into the conditions that may lead to the extinction of the microbial population.

1. Introduction

Bacterial biofilms are multicellular communities embedded in a self-produced polymeric matrix comprised of polysaccharides, proteins, and extracellular matrix [**25**]. Self-organization and positioning of bacteria biofilms, both in space and time, have been observed in natural environments and experimental studies leading to, for example, regularly repeating patterns, see [**19, 23**]. Cell density and signaling are two crucial components in the formation of a spatially organised biofilm [**21**]. Biofilm formation occurs as a result of environmental cues such as starvation, chemical concentrations or change in temperature. Although cell signaling and distinct cell genes involved in the formation of biofilms differ between species, there exist other commonalities. Simple cells like the bacterium *E. coli* [**23**], *S. typhimurium* [**26**], to name a few, present regulated cellular anatomy resulting from the dynamic interactions between proteins and the cell membrane. It is noted in [**26**] that different bacteria present different patterns, in particular, during the formation stage. For example, *S. typhimurium* present structured law, while swarm ring appears for *E. coli.*

A review on cooperation in microbial populations can be found in [**5**]. The authors provided an ecological overview where ensembles of local subpopulations

2020 *Mathematics Subject Classification.* Primary 37N25, 65L12.

Key words and phrases. microbial population oscillation; dynamic consistency; environmental stress; nonstandard finite difference.

The authors were supported by the South African DSI-NRF SARChI Chair in Mathematical Models and Methods in Bioengineering and Biosciences.

continuously emerge, grow and disappear again. The subpopulations vary in size from a few initial cells to billions in established communities. It has been reported that environmental noise such as the presence of toxins, produced by the microbes themselves, can be catastrophic. For example, metabolic waste products such as lactic acid, acetate and hydrogen, change the pH levels of an environment [**7**]. In their natural habitat, bacteria constantly compete with rivaling species found in the same ecological niche for survival and growth-limiting resources. Biological warfares occur where one species produces antibiotics or 'chemical weapons' to suppress the growth or eradicate competing microorganisms. Some antimicrobial compounds do not only affect the competitors but also act against their own as well as identical bacteria [**10, 17**]. In some cases, such metabolic waste can be a resource for another species, allowing both polluters (waste producers) and detoxifiers (cross feeders) to potentially benefit [**7, 9**]. On the other hand, the spreading of phages can change the fitness landscape [**22, 24**].

It has been observed that bacteria modify the environmental pH to such a degree that they make their surroundings so acidic leading to the swift and complete extinction of the whole population. Termed 'ecological suicide' in their experiments, extinction of soil bacteria species *Paenibacillus sp.*, was observed in well-mixed batch culture in media containing glucose as the main carbon source. See also [**18, 20**] and references therein. The observation is that the denser the bacteria, the more waste is produced and thus the risk of population collapse is higher at high population densities. This was observed in about 25% of the soil microbes tested. In this case, low-density populations thrive while high density leads to population collapse. A study supporting the ability of bacteria to change media pH in their environment, indirectly rendering it more or less hospitable for other species, can be found in [**6**]. The authors in [**18**] observed that hostile environment can also cause oscillations in the population size over time. Although these experiments are useful for providing insights into specific mechanisms of interactions, it is unclear to what extent findings from these studies scale up to predict the generic properties of large microbial communities or the interactions therein.

Despite the ubiquity of these environment-mediated population collapses, mathematical models designed to understand these processes are lacking in the literature. One of the models is proposed in [**16**] to explain how microbes interacting through their chemical environment can achieve coexistence in a continuous growth setup. The authors observed that, in addition to consumption/degradation, facilitation and self-restraint interactions played a critical role in allowing species coexistence. Using pH as a model parameter, in [**20**] the authors proposed a deterministic model supported by experiments to understand how the bacteria manipulate their environment and also react to it.

Motivated by the existing literature, we propose a system of ordinary differential equations representing a general microbial population consisting of active and inactive cells capable of producing waste products modifying their environment to pH levels that could lead to either population oscillation ('boom and bust'), [**8**] or 'ecological suicide' [**18**]. The model is generic and can help to test hypotheses about these unknown processes. It is always important to bear in mind that the model is only an approximation of reality, and its predictions are subject to some uncertainty. In Section 2.1 we investigate a reduced model of microbial populations that either die or become dormant in an ecosystem with stressful environmental conditions. Furthermore, in Section 2.2, a full model representing microbes that switch states depending on the toxins' concentration levels in the environment is investigated. We design a nonstandard finite difference scheme (in Section 3) and present numerical results and discussions related to the presented models in Sections 2.1 and 2.2. We discuss our findings and conclude in Section 5.

2. Model formulation

Mathematical models taking into account the consumption of a limiting nutrient in the presence of a chemoattractant have been considered extensively in the literature. See for example, [**3**, **14**, **15**, **23**], and references therein. In [**15**], the authors summarised model mechanisms for *E. coli* experiments in a semi-solid medium. In addition, it is assumed that non-motile cells are dead and do not contribute to the dynamics of the model. Chemotaxis was taken into account to represent the directed movement of organisms in response to the concentration gradient. However, as indicated above, mathematical models including the negative feedback from the microbe's activities are lacking in the literature.

In this work, we consider an ecosystem comprising a general microbial population capable of modifying its environment to pH levels that are not hospitable to themselves. We begin by denoting $b(t)$, the density of active microorganisms, and $c(t)$ the concentration of the waste product in the surrounding environment. In the absence of $c(t)$, the bacteria growth follows a simple Logistic law

$$\frac{db}{dt} = k_3 b(k_4 - b), \tag{2.1}$$

where k_4 is the carrying capacity and k_3k_4 is the density independent growth rate. The equation governing the change in concentration $c(t)$ is given by

$$\frac{dc}{dt} = g(b) - k_5 c - bf(c), \tag{2.2}$$

where k_5 denotes the rate of natural degradation of c, see for example [**6**]. Thus, we allow for the consumption of metabolic waste, see for example [**7**] for further details. Moreover, we generalise the consumption function, $f(c)$, and assume it satisfies the following hypotheses: $f(c) = 0$ if and only if $c = 0$ and $f'(c) \geq 0$, for all c. The production function $g(b)$ saturates and includes input from both external sources and active microbes, hence we assume it satisfies the following conditions: $g(b) > 0$, $g'(b) > 0$ for all $b \geq 0$. Readers can consult, for example, the work [**3**, **4**],[**18**, Supplementary Information] and the references therein.

It has been reported that organisms that fail to adapt to this stress would either die or become dormant [**11**]. Considering these ecological processes, our model takes the form

$$\begin{aligned}
\frac{da}{dt} &= k_1 q(c) b - k_2 p(c) a - k_8 a, \\
\frac{db}{dt} &= k_3 b(k_4 - b) - k_1 q(c) b + k_2 p(c) a, \\
\frac{dc}{dt} &= g(b) - k_5 c - bf(c),
\end{aligned} \tag{2.3}$$

where $a(t)$ denotes the density of dormant bacteria. The functions $q(c)$ and $p(c)$, each of them $\in (0,1)$, denote the hibernation and reawakening of bacteria in response to the concentration c, respectively, and are thought of as cell-specific probabilities of switching states. See for example [**8**] and the literature therein for further discussion on these switch functions. In our work, we invoke assumptions on the transition functions $p(c)$ and $q(c)$.

$\mathcal{H}_1 \ : p : \mathbb{R}_+ \to$ (0,1] is C^1, $p'(c) \leq 0$ for all c and $\lim_{c\to\infty} p(c) = 0$,

$\mathcal{H}_2 \ : q : \mathbb{R}_+ \to$ (0,1] is C^1, $q'(c) \geq 0$ for all c and $\lim_{c\to\infty} q(c) = 1$.

Here we have assumed the concentration of energy source is large compared to the concentration of cells (negligible nutrient consumption). In addition, c is not included in the proliferation - which is consistent with the work of [**15**, Section 5.5]. The model is rendered dimensionless by choosing the following scales

$$w = \frac{a}{k_4}, \quad u = \frac{b}{k_4}, \quad v = \frac{c}{k_4}, \quad t \sim \frac{1}{k_3 k_4},$$

to obtain

$$\begin{aligned} \frac{dw}{dt} &= rq(v)u - sp(v)w - \mu w, \\ \frac{du}{dt} &= u(1-u) - rq(v)u + sp(v)w, \\ \frac{dv}{dt} &= \alpha(\beta g(u) - v) - \gamma u f(v), \end{aligned} \tag{2.4}$$

where $r = k_1/(k_3k_4)$, $s = k_2/(k_3k_4)$, $\beta = [g]/(k_4k_5)$, $\gamma = [f]/(k_3k_4)$, $\alpha = k_5/(k_3k_4)$ and $\mu = k_8/(k_4k_3)$. The notation $[*]$ refers to the scale for the given term. The system (2.4) is appended with the following initial conditions

$$w(0) = w_0 \geq 0, \quad u(0) = u_0 \geq 0, \quad v(0) = v_0 \geq 0. \tag{2.5}$$

The following result summarises some of the features of this model.

PROPOSITION 1. *System* (2.4) *defines a dynamical system on the biologically feasible region*

$$\Omega = \left\{ (w,u,v) \in \mathbb{R}^3_+ : w + u + v \leq \frac{2 + \alpha\beta \max_{u\geq 0}[g(u)]}{\alpha_m} \right\},$$

where $\alpha_m = \min\{1, \alpha, \mu\}$.

The proof of the result utilizes straightforward arguments and will not be included here. In the next sections, we will study system (2.4) together with its submodels to gain insight into the model dynamics.

2.1. Model without bacteria resuscitation. Observing that once active bacteria fail to adapt to the stress level $v(t)$, they would either die or become dormant (see for example [**11**]), we propose the following reduced model

$$\begin{aligned} \frac{du}{dt} &= u(1-u) - rq(v)u, \\ \frac{dv}{dt} &= \alpha(\beta g(u) - v) - \gamma u f(v). \end{aligned} \tag{2.6}$$

Under this setup, the term $rq(v)u$ represents the bacteria switching into a dormant state, or bacterial mortality due to environmental stress. We remark that $1/(rq(v))$

describes the resistance or residence time of microbes under inhospitable conditions and it is dependent on the concentration $v(t)$. The assumption on $q(0) > 0$ represents the existence of random switching or a result of paracrine signaling.

Setting the right-hand side of $(2.6)_1$ to zero we see that $u = 0$ or $u = 1 - rq(v)$. That is, $u = 0$ together with the right-hand side of $(2.6)_2$ set to zero, we get the following boundary equilibrium $P^0 = (0, \beta g(0))$, representing the extinction state. Using the linearisation technique, P^0 is locally asymptotically stable provided $1 - rq(\beta g(0)) < 0$, otherwise, it is unstable. Essentially we require that $r > 1/q(\beta g(0))$ for the population to derive itself to extinction.

On the other hand, assuming $u^* \neq 0$ or equivalently $u^* = 1 - rq(v^*)$, then from setting the right hand of $(2.6)_2$ to zero, v^* must satisfy the equation

$$H(v^*) = \alpha(\beta g(u^*) - v^*) - \gamma u^* f(v^*) = 0,$$

with $\lim_{v^* \to 0^+} H(v^*) = \alpha\beta g(rq(0) - 1) > 0$ and

$$H'(v^*) = -rq(v^*)\left[\alpha\beta g'(u^*) - \gamma f(v^*)\right] - \gamma u^* f'(v^*) - \alpha,$$

where $\lim_{v^* \to 0^+} H'(v^*) = -\alpha - \gamma f'(0)(1 - rq(0)) < 0$. First we see that if $\alpha\beta g'(u^*) - \gamma f(v^*) \geq 0$, then $H(v^*)$ is monotonically decreasing and a unique coexistence equilibrium point $P^* = (u^*, v^*)$ exists provided $1 - rq(v^*) > 0$.

On the other hand, if $\alpha\beta g'(u^*) - \gamma f(v^*) < 0$, together with $\lim_{v^* \to 0^+} H(v^*) > 0$ and $\lim_{v^* \to 0^+} H'(v^*) < 0$, in addition to $\lim_{v^* \to \infty} H'(v^*) = -\alpha - \gamma f'(1 - rq(v^*)) < 0$, suggests the existence of at least one coexistence equilibrium point $P^* = (u^*, v^*)$ provided $1 - rq(v^*) > 0$. This will be illustrated numerically in Figs. 4 and 5.

Next, we consider the local stability of the co-existence equilibrium, P^*, whose Jacobian at P^* is given by

$$J(u^*, v^*) = \begin{pmatrix} -u^* & -rq'(v^*)u^* \\ \alpha\beta g'(u^*) - \gamma f(v^*) & -\alpha - \gamma u^* f'(v^*) \end{pmatrix}.$$

The trace and determinant of $J(u^*, v^*)$ are given by

$$-u^* - \alpha - \gamma u^* f'(v^*) < 0, \text{ and } rq(v^*)u^*[\alpha\beta g'(u^*) - \gamma f(v^*)] + u^*[\alpha + \gamma u^* f'(v^*)],$$

respectively. Clearly, we see that if $\alpha\beta g'(u^*) - \gamma f(v^*) \geq 0$, a unique co-existence equilibrium state exists and it is locally asymptotically stable. It is not clear what happens if $\alpha\beta g'(u^*) - \gamma f(v^*) < 0$. However, the following result rules out the possibility of periodic orbits.

PROPOSITION 2. *The uniform steady state system* (2.3) *has no periodic orbits.*

PROOF. To prove this result, we apply the Dulac-Bendixon criterion. Let $Z \subset X$ be open and simply connected in the first quadrant. Let $h_u(u, v) = u(1 - u) - rq(v)u$ and $h_v(u, v) = \alpha(\beta g(u) - v) - \gamma u f(v)$. We need to show that there exists a function $\phi : Z \to \mathbb{R}$, continuously differentiable on Z, such that

$$\frac{\partial(\phi h_u)}{\partial u} + \frac{\partial(\phi h_v)}{\partial v}$$

is either strictly negative or strictly positive everywhere on Z. Let $\phi(u, v) = \dfrac{1}{u}$. Taking Z to be open in the first quadrant, clearly, ϕ is continuously differentiable in Z. Furthermore,

$$\frac{\partial(\phi h_u)}{\partial u} + \frac{\partial(\phi h_v)}{\partial v} = -\frac{1}{u}\left(u + \alpha + \gamma u f'(v)\right) < 0.$$

Hence, the system has no periodic orbits in the open first quadrant. □

REMARK 2.1. The initial investigation centered on how microbial populations can derive themselves to extinction. It is clear from the proposed reduced model that this depends on the resistance of the population to the toxicity of the environment they reside in. The better the microbes can resist environmental toxic changes, the better they can avoid extinction. This can also be linked to the use of antibiotics as a treatment strategy for infections as alluded to in [**12**]. It is crucial to understand antibiotic resistance before administering it.

REMARK 2.2. The proposed reduced model (2.6) recovers the dynamics where bacteria acidify the environment inducing negative feedback on their survival. Under this setup, dormant cells are assumed to be dead [**11**]. The qualitative results indicated the nonexistence of a limit cycle, a conclusion which agrees with the discussion in [**18**, Fig. 2], and a suggestion that the oscillations may be a result of experimental conditions.

2.2. The full model. We now focus on the case where bacteria switch between states (active and inactive) depending on the change in concentration $v(t)$ in the system. In addition, we assume dormant bacteria do not die. We begin by recalling the model system (2.4) and present some mathematical results.

PROPOSITION 3. *System* (2.4) *admits the following equilibria: (a) the microbial-free state* $P^0 = (0, 0, \beta g(0))$, *and (b) a unique co-existence state* $P^* = (w^*, u^*, v^*)$, *where* $u^* = 1$, $w^* = \dfrac{rq(v^*)}{sp(v^*)}$ *and* v^* *satisfies* $\alpha(\beta g(1) - v^*) - \gamma f(v^*) = 0$.

PROOF. First, setting the right hand side of equations (2.4) to zero, we get P^0 and P^* as the only equilibria where v^* satisfies $\alpha(\beta g(1) - v^*) - \gamma f(v^*) = 0$. Defining $H(v^*) = \alpha(\beta g(1) - v^*) - \gamma f(v^*)$, then $H(v^*)$ is a monotonically decreasing function since $H'(v^*) = -\alpha - \gamma f'(v^*) < 0$, where we have used the properties of functions f and g outlined in Section 2. Furthermore, $\lim_{v^* \to 0^+} H(v^*) = \alpha\beta g(1) > 0$, hence, a unique solution v^* satisfying $H(v^*) = \alpha(\beta g(1) - v^*) - \gamma f(v^*)$ exists. □

The Jacobian of system (2.4) at the microbial-free state, P^0, is given by

$$J(P^0) = \begin{pmatrix} -sp(\beta g(0)) & rq(\beta g(0)) & 0 \\ sp(\beta g(0)) & 1 - rq(\beta g(0)) & 0 \\ 0 & \alpha\beta g'(0) & -\alpha \end{pmatrix}.$$

Clearly, $-\alpha$ is one of the eigenvalues of $J(P^0)$. The remaining eigenvalues are obtained from the reduced matrix

$$\begin{pmatrix} -sp(\beta g(0)) & rq(\beta g(0)) \\ sp(\beta g(0)) & 1 - rq(\beta g(0)) \end{pmatrix},$$

whose trace is $(1 - rq(\beta g(0))) - sp(\beta g(0))$ and the determinant given by $-sp(\beta g(0))$. Under the hypotheses $\mathcal{H}_1$ and $\mathcal{H}_2$, the microbial-free equilibrium state is unstable. This result, in comparison with the reduced model results, clearly support the idea that microbial populations are capable of surviving hostile environment if they can switch state depending on the concentration $v(t)$.

Next, we consider the local stability of the co-existence equilibrium, P^*. The Jacobian matrix of system (2.4) at the co-existence equilibrium is

$$J(P^*) = \begin{pmatrix} -sp(v^*) & rq(v^*) & rq'(v^*) - sp'(v^*)w^* \\ sp(v^*) & -1 - rq(v^*) & -rq'(v^*) + sp'(v^*)w^* \\ 0 & \alpha\beta g'(1) - \gamma f(v^*) & -\alpha - \gamma f'(v^*) \end{pmatrix}.$$

The characteristic equation of the Jacobian matrix $J(P^*)$ is given by

$$Q(\lambda) := \lambda^3 + s_1\lambda^2 + s_2\lambda + s_3 = 0,$$

the coefficients are given as

$$\begin{aligned} s_1 &= -\text{trace } J(P^*) = sp(v^*) + (1 + rq(v^*)) + (\alpha + \gamma f'(v^*)), \\ s_2 &= (\alpha + \gamma f'(v^*))(1 + rq(v^*) + sp(v^*)) \\ &\quad + (\gamma f(v^*) - \alpha\beta g'(1))(-rq'(v^*) + sp'(v^*)w^*) + sp(v^*), \\ s_3 &= -\det J(P^*) = sp(v^*)\,(\alpha + \gamma f'(v^*)) \,. \end{aligned}$$

Clearly, s_1 and s_3 are unconditionally positive and s_2 is positive provided $\gamma f(v^*) - \alpha\beta g'(1) \leq 0$. Thus, the first assumption of *Rough–Hurwitz* is verified. It remains to check if Δ is positive. We have

$$\begin{aligned} \Delta &= s_1 s_2 - s_3 \\ &= (sp(v^*) + rq(v^*) + 1) \times s_2 + sp(v^*)(\alpha + \gamma f'(v^*))^2 + (\alpha + \gamma f'(v^*))^2\,(rq(v^*) + 1) \\ &\quad + (\alpha + \gamma f'(v^*))(\gamma f(v^*) - \alpha\beta g'(1))(-rq'(v^*) + sp'(v^*)w^*). \end{aligned}$$

However, it is unclear if the complete statement is positive. Assuming $\gamma f(v^*) - \alpha\beta g'(1) \leq 0$, then the coefficient s_2 and the second-*Hurwitz* determinant, $\Delta = s_1 s_2 - s_3$, are positive. However, under $\gamma f(v^*) - \alpha\beta g'(1) > 0$, Δ can be zero, meaning that the polynomial $Q(\lambda)$ may have pure imaginary conjugate roots and one real root.

Choosing β as the bifurcation parameter, then $Q(\lambda)$ has a pair of complex conjugate roots, $a \pm bi$ where $a, b \in \mathbb{R}$, which cross the real axis as β passes through β^*, then $\Delta(\beta)$ changes sign as β passes through β^*. Rewriting Δ, we have

$$\begin{aligned} \Delta(\beta) &= -\alpha\beta g'(1)(-rq'(v^*) + sp'(v^*)w^*)\,[(\alpha + \gamma f'(v^*)) + (1 + sp(v^*) + rq(v^*))] \\ &\quad + (1 + sp(v^*) + rq(v^*))\,\left[sp(v^*) + (\alpha + \gamma f'(v^*))(1 + sp(v^*) + rq(v^*)) + (\alpha - \gamma f'(v^*))^2\right] \\ &\quad + \gamma f(v^*)(-rq'(v^*) + sp'(v^*)w^*)\,[(\alpha + \gamma f'(v^*)) + (1 + sp(v^*) + rq(v^*))] \,. \end{aligned}$$

For Hopf bifurcation, we need an additional property to verify the so-called transversality condition that indicates that the eigenvalues cross the x−axis with a non-zero velocity. In other words

$$\frac{da}{d\beta}(\beta^*) = a'(\beta^*) \neq 0. \tag{2.7}$$

Verifying (2.7) is equivalent to verifying $\Delta'(\beta^*) \neq 0$. We proceed to show numerically that $\Delta'(\beta^*) < 0$, see Fig. 1(b). We deduce that a Hopf bifurcation related to the bifurcation parameter, β, may occur at (P^*, β^*) provided the following conditions are met. To summarise, Hopf bifurcation occurs if

REMARK 2.3.

- There exists a β^* such that $\Delta(\beta^*) = 0$, such Jacobian matrix at the coexistence state P^* has a pair of imaginary eigenvalues as well as one real negative eigenvalue,
- and $\Delta'(\beta^*) \neq 0$.

The remark is supported numerically in Fig. 1 to illustrate the existence of Hopf bifurcation and oscillatory solutions.

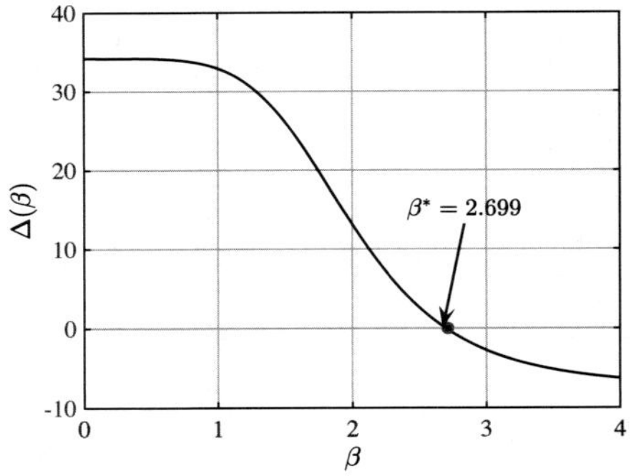

(a) Plot for $\Delta(\beta)$ with β.

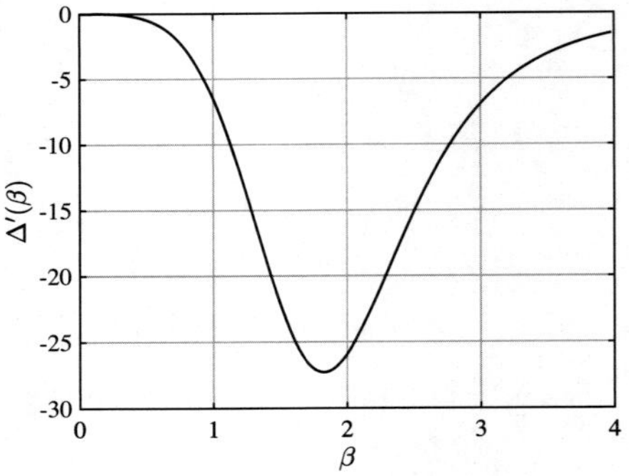

(b) Plot for $\Delta'(\beta)$ with β.

FIGURE 1. Illustrating the existence of Hopf bifurcation with $r = s = \alpha = 1$, and $\gamma = 2$. There exists a $\beta^* = 2.699$ such that $\Delta(\beta^*) = 0$.

3. Nonstandard finite difference method

In this section, we propose a nonstandard-finite difference scheme to solve the system of equations. We first recall the differential model (2.4) and let $(w^n, u^n, v^n)^T$ denote an approximation of $(w(t_n), u(t_n), v(t_n))^T$ where $t_n = n\Delta t$, with $n \in \mathbb{N}$ and $\Delta t > 0$. We propose the following scheme constructed via the nonlocal approximation (see [**13**]) of the nonlinear terms

$$\begin{aligned} \frac{w^{n+1} - w^n}{\Delta t} &= rq(v^n)u^n - sp(v^n)w^{n+1} - \mu w^{n+1}, \\ \frac{u^{n+1} - u^n}{\Delta t} &= \left(1 - u^{n+1}\right) u^n - rq(v^n)u^{n+1} + sp(v^n)w^{n+1}, \\ \frac{v^{n+1} - v^n}{\Delta t} &= \alpha\left(\beta g(u^n) - v^{n+1}\right) - \gamma v^{n+1} h(v^n)u^{n+1}, \end{aligned} \tag{3.1}$$

where, for convenience, we have used $f(v) = vh(v)$. The numerical scheme (3.1) is called nonstandard finite difference method because nonlinear terms are approximated in a nonlocal way.

Consider system (2.4), written in the compact form

$$\frac{dy}{dt} = G(y).$$

Equivalently, system (3.1) is given in the form

$$y^{n+1} = F(\Delta t)\left(y^n\right), \quad y(0) = y_0,$$

where $y^n = (w^n, u^n, v^n)$ and

$$F(\Delta t)(y^n) = \begin{pmatrix} \dfrac{w^n + \Delta t r q(v^n) u^n}{1 + \Delta t \mu + \Delta t s p(v^n)} \\ \dfrac{u^n + \Delta t u^n + \Delta t s p(v^n) w^{n+1}}{1 + \Delta t u^n + \Delta t r q(v^n)} \\ \dfrac{v^n + \Delta t \alpha \beta g(u^n)}{1 + \alpha \Delta t + \Delta t \gamma h(v^n) u^{n+1}} \end{pmatrix}. \tag{3.2}$$

Using standard techniques [**2**], we now verify the dynamic consistency of the scheme. For consistency, F must satisfy the following requirements:

$$F(0)(y) = y, \quad \frac{dF(0)}{d\Delta t}(y) = G(y).$$

Clearly, the first condition is satisfied. We have that $\dfrac{dF(0)}{d\Delta t}(y)\Big|_{\Delta t=0}$ is given by

$$\begin{pmatrix} \dfrac{rq(v)u - sp(v)w - \mu w}{(1 + \Delta t \mu + \Delta t s p(v))^2} \\ \dfrac{u(1-u) - rq(v)u + sp(v)w}{(1 + \Delta t u + \Delta t r q(v))^2} \\ \dfrac{\alpha(\beta g(u) - v) - \gamma u v h(v)}{(1 + \alpha \Delta t + \Delta t \gamma h(v) u)^2} \end{pmatrix}_{\Delta t = 0} = \begin{pmatrix} rq(v)u - sp(v)w - \mu w \\ u(1-u) - rq(v)u + sp(v)w \\ \alpha(\beta g(u) - v) - \gamma u f(v) \end{pmatrix}.$$

The above system (3.2) can be rewritten in the explicit form

$$A(\Delta t, y^n) y^{n+1} = B(\Delta t, y^n) y^n + \psi(0, 0, 1 + \beta g(0))^t, \tag{3.3}$$

where matrices $A(\Delta t, y^n)$ and $B(\Delta t, y^n)$ are given by

$$\begin{pmatrix} 1 + \Delta t \mu + \Delta t s p(v^n) & 0 & 0 \\ 0 & 1 + \Delta t u^n + \Delta t r q(v^n) & 0 \\ 0 & 0 & 1 + \alpha \Delta t + \Delta t \gamma h(v^n) u^{n+1} \end{pmatrix}$$

and

$$\begin{pmatrix} 1 & \Delta t r q(v^n) & 0 \\ \Delta t s p(v^n) & 1 + \Delta t & 0 \\ 0 & 0 & 1 + \alpha \Delta t \beta g(u^n) \end{pmatrix},$$

respectively. We see that $A(\Delta t, y^n)$ is an *M-matrix* since it is a diagonal matrix. Hence, its inverse, $A(\Delta t, y^n)^{-1}$, is non-negative. We have that $B(\Delta t, y^n) \geq 0$, which is sufficient for the scheme to preserve the positivity of solutions. We conclude that scheme (3.1) defines a discrete dynamical system on the same domain as the continuous model (2.4). See [**1**] and references therein for more details. Setting $y^{n+1} = y^n$, then the scheme preserves the same fixed points, $P^0 = (0, 0, \beta g(0))$ and $P^* = (w^*, u^*, v^*)$ as the continuous model (2.4).

Following (3.2) and (3.3), then $F(\Delta t)(y) = A^{-1}(\Delta t, y)B(\Delta t, y)$ so that $\dfrac{dF(\Delta t)}{dy}(y)$ is given by

$$\begin{pmatrix} \dfrac{1}{1+\Delta t\mu+\Delta tsp(v)} & \dfrac{\Delta trq(v)}{1+\Delta t\mu+\Delta tsp(v)} & J_{1,3} \\ \dfrac{\Delta tsp(v)}{1+\Delta tu+\Delta trq(v)} & \dfrac{1+\Delta t}{1+\Delta tu+\Delta trq(v)} - \dfrac{\Delta t\,(u+\Delta tu+\Delta tsp(v)w)}{(1+\Delta tu+\Delta trq(v))^2} & J_{2,3} \\ 0 & \dfrac{\Delta t\alpha\beta g'(u)}{1+\alpha\Delta t+\Delta t\gamma h(v)u} - \dfrac{(\alpha v+\Delta t\alpha\beta g(u))\,\Delta t\gamma h(v)}{(1+\alpha\Delta t+\Delta t\gamma h(v)u)^2} & J_{3,3} \end{pmatrix},$$

where

$$J_{1,3} = \frac{\Delta trq'(v)u}{1+\Delta t\mu+\Delta tsp(v)} - \frac{(w+\Delta trq(v)u)\,\Delta tsp'(v)}{(1+\Delta t\mu+\Delta tsp(v))^2},$$

$$J_{2,3} = \frac{\Delta tsp'(v)w}{1+\Delta tu+\Delta trq(v)} - \frac{\Delta trq'(v)\,(u+\Delta tu+\Delta tsp(v)w)}{(1+\Delta tu+\Delta trq(v))^2},$$

$$J_{3,3} = \frac{1}{1+\alpha\Delta t+\Delta t\gamma h(v)u} - \frac{\Delta t\gamma h'(v)u\,(\alpha v+\Delta t\alpha\beta g(u))}{(1+\alpha\Delta t+\Delta t\gamma h(v)u)^2}.$$

We proceed to show that all eigenvalues of the Jacobian matrix evaluated at the fixed points lie in a unit circle.

Evaluating $\dfrac{dF(\Delta t)}{dy}(y)$ at the bacteria-free equilibrium, we obtain

$$\begin{pmatrix} \dfrac{1}{1+\Delta t\mu+\Delta tsp(\beta g(0))} & \dfrac{\Delta trq(\beta g(0))}{1+\Delta t\mu+\Delta tsp(\beta g(0))} & 0 \\ \dfrac{\Delta tsp(\beta g(0))}{1+\Delta trq(\beta g(0))} & \dfrac{1+\Delta t}{1+\Delta trq(\beta g(0))} & 0 \\ 0 & J_{3,2}(P^0) & \dfrac{1}{1+\alpha\Delta t} \end{pmatrix},$$

where

$$J_{3,2}(P^0) = \frac{\Delta t\alpha\beta g'(0)}{1+\alpha\Delta t} - \frac{\alpha(\beta g(0)+\Delta t\beta g(0))\Delta t\gamma h(\beta g(0))}{(1+\alpha\Delta t)^2}.$$

Now, we need to verify that all the eigenvalues of the Jacobian matrix at P^0 lie inside a unit circle. First, we see that $\dfrac{1}{1+\alpha\Delta t}$ is the first eigenvalue of the matrix $\dfrac{dF(\Delta t)}{dy}(0,0,\beta g(0))$ which is strictly less than one. The remaining eigenvalues are obtained from the reduced matrix

$$J_r = \begin{pmatrix} \dfrac{1}{1+\Delta t\mu+\Delta tsp(\beta g(0))} & \dfrac{\Delta trq(\beta g(0))}{1+\Delta t\mu+\Delta tsp(\beta g(0))} \\ \dfrac{\Delta tsp(\beta g(0))}{1+\Delta trq(\beta g(0))} & \dfrac{1+\Delta t}{1+\Delta trq(\beta g(0))} \end{pmatrix}.$$

To show whether the remaining eigenvalues lie in a unit circle, we use the well-known Jury's conditions for $J_r \in \mathbb{R}^2 \times \mathbb{R}^2$. In this case, we need to show that

$$(i)\quad 1-\det(J_r) > 0,$$

$$(ii)\quad 1-\operatorname{trace}(J_r)+\det(J_r) > 0,$$

$$(iii)\quad 0 < J_{11} < 1,\ 0 < J_{22} < 1.$$

The first condition, $1 - \det(J_r) > 0$, is equivalent to

$$\Delta t[2\Delta tsp(\beta g(0))rq(\beta g(0)) + \mu(1 + \Delta trq(\beta g(0))) + (sp(\beta g(0)) + rq(\beta g(0)) - 1)],$$

which is strictly positive provide $sp(\beta g(0))+rq(\beta g(0))-1 > 0$. The second condition is equivalent to

$$(\Delta t)^2[\mu(rq(\beta g(0)) - 1) - sp(\beta g(0))],$$

which is strictly positive provided $\mu(rq(\beta g(0)) - 1) - sp(\beta g(0)) > 0$. The last condition is clearly not satisfied provided $rq(\beta g(0)) - 1 < 0$, that is, there is a possibility that the fixed point P^0 lies inside a unit circle provided $rq(\beta g(0))-1 > 0$.

Similarly, for the stability of P^*, we proceed to evaluate $\frac{dF(\Delta t)}{dy}(y)$ at the co-existence fixed point and we obtain the characteristic equation

$$\lambda^3 + s_2\lambda^2 + s_1\lambda + s_0 = 0, \tag{3.4}$$

where the roots of equation (3.4) lie inside the unit circle provided

$$|s_0 + s_2| < 1 + s_1, \quad \text{and} \quad |s_1 - s_0 s_2| < 1 - s_0^2. \tag{3.5}$$

Theoretical verification of these conditions result in an overlong expression, thus we proceed numerically. In the results shown in Fig. 2, we choose $r = 1$, $s = 1$, $\gamma = 0.1$, $\alpha = 1$, $\beta = 1.1$ so that $\gamma f(v^*) - \alpha\beta g'(1) \leq 0$. Clearly, for $\Delta t > 0$, conditions (3.5) are satisfied provided $\gamma f(v^*) - \alpha\beta g'(1) \leq 0$ and we conclude that all eigenvalues lie in a unit circle. Hence, the coexistence fixed point for the discrete model (3.1) is locally asymptotically stable provided $\gamma f(v^*) - \alpha\beta g'(1) \leq 0$.

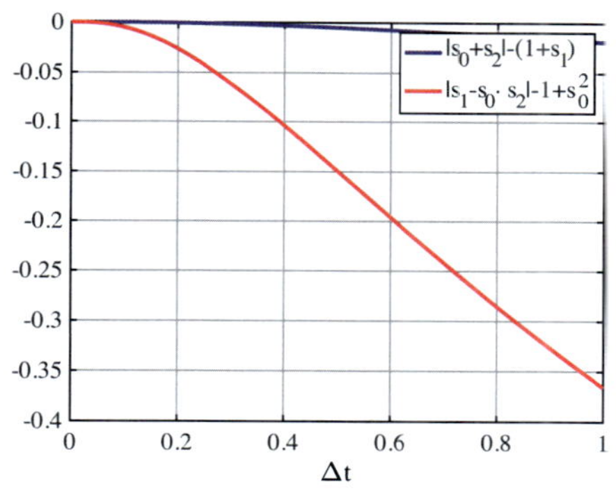

FIGURE 2. Numerical illustration for the local stability of the co-existence fixed point P^* provided $\gamma f(v^*) - \alpha\beta g'(1) = -0.1250 < 0$

In addition, in Fig. 3 we observe that provided $\gamma f(v^*) - \alpha\beta g'(1) > 0$, there is a possibility that P^* can be unstable through a Hopf bifurcation leading to oscillatory solutions and this is also illustrated in Fig. 7. In Fig. 3 we choose $r = 1$, $s = 1$, $\gamma = 2$, $\alpha = 1$ with $\gamma f(v^*) - \alpha\beta g'(1) > 0$. It is clear that when $\beta = 2.8$ the first condition of (3.5) is not satisfied but when $\beta = 2$, P^* remains locally asymptotically stable.

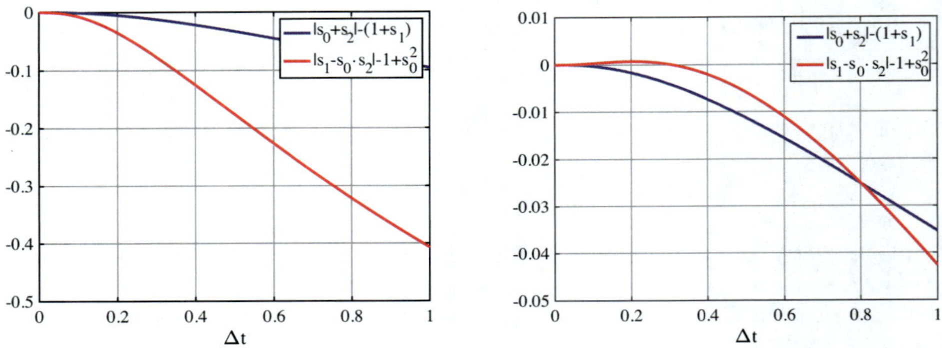

(a) $\beta = 2.0$ (b) $\beta = 2.8$

FIGURE 3. Numerical illustration for: (a) the local stability of the co-existence fixed point P^* when $\gamma f(v^*) - \alpha\beta g'(1) = 1.5000$ and (b) possibility of Hopf bifurcation when $\gamma f(v^*) - \alpha\beta g'(1) = 2.1000$.

4. Numerical simulations

In this section, we present numerical simulations in support of the theoretical results derived above. Throughout this chapter, we choose $f(v) = v$ and

$$q(v) = \frac{\omega + (v)^m}{k^m + (v)^m}, \tag{4.1}$$

with $m \geq 1$ and $\omega = 0.2 > 0$, satisfying the assumptions on $f(v)$ and hypothesis $\mathcal{H}_2$, respectively. A comprehensive discussion on the choice of $q(v)$ can be found in [**8**]. As discussed further in their work, the choice of $f(v)$ has 'cosmetic' effect and will not be discussed further here. In addition to the choice of $q(v)$ in (4.1), we also have

$$p(v) = \frac{k^m}{k^m + (v)^m}, \tag{4.2}$$

with $m \geq 1$ which satisfies hypothesis $\mathcal{H}_1$.

4.1. Model without bacteria resuscitation. We have shown in Section 2.1 that the system without a quiescent stage does not admit periodic orbits. Numerical simulations to support the results and the effect of parameter changes will be investigated in this section. We integrate the system of equations using the following reduced nonstandard finite difference scheme corresponding to scheme (3.1).

$$\begin{aligned} \frac{u^{n+1} - u^n}{\Delta t} &= \left(1 - u^{n+1}\right) u^n - rq(v^n)u^{n+1}, \\ \frac{v^{n+1} - v^n}{\Delta t} &= \alpha\left(\beta g(u^n) - v^{n+1}\right) - \gamma v^{n+1} h(v^n) u^{n+1}. \end{aligned} \tag{4.3}$$

In Fig. 4(a) we illustrate the stability of the extinction equilibrium point when the parameters are chosen such that $1 - rq(\beta g(0)) = -0.100 < 0$. In this case, only a unique boundary equilibrium P^0 exists. If $1 - rq(v^*) = 0.292 > 0$ together with $\gamma f(v^*) - \alpha\beta g'(u^*) = -0.368 \leq 0$, we have a unique coexistence equilibrium point $(u^*, v^*) = (0.2918, 1.1583)$ which is stable locally. This is illustrated in Fig. 4(b).

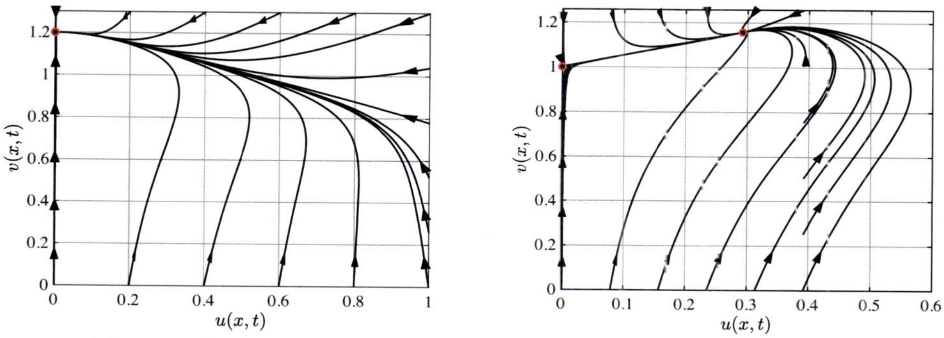

(a) $1 - rq(\beta g(0)) < 0$.

(b) $\gamma f(v^*) - \alpha\beta g'(u^*) \leq 0$.

FIGURE 4. Phase planes corresponding to system (2.6). The parameters are chosen such that $r = 1.0$ with: $\beta = 1.2$, $\gamma = 1.0$ in figure (a), and $\beta = 1.0$, $\gamma = 0.2$ in figure (b).

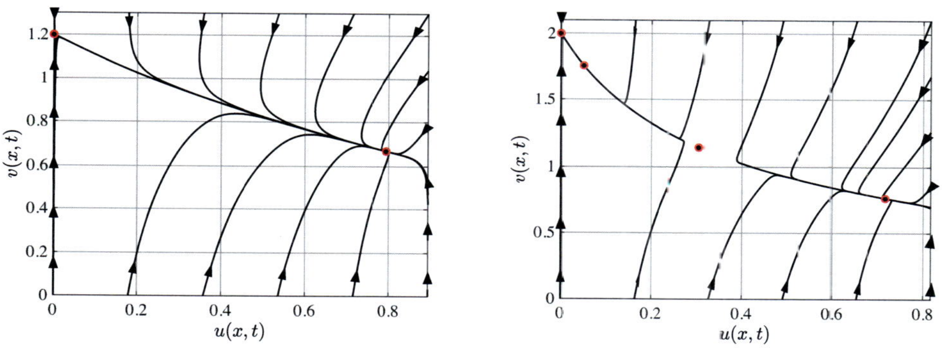

(a) $\gamma f(v^*) - \alpha\beta g'(u^*) > 0$.

(b) $\gamma f(v^*) - \alpha\beta g'(u^*) > 0$.

FIGURE 5. Phase planes corresponding to system (2.6). The parameters are chosen such that $r = 1$, $\alpha = 1$ with: $\beta = 1.2$, $\gamma = 2.0$ in figure (a), and $\beta = 2.0$, $\gamma = 3.8$ in figure (b).

We further investigate the case $1 - rq(v^*) > 0$ together with $\gamma f(v^*) - \alpha\beta g'(u^*) < 0$. As illustrated in Fig. 5, we either have a unique coexistence equilibrium point which is locally asymptotically stable Fig. 5(a), or a bistable case Fig. 5(b). This is supported by experimental observations reported in [**16**, **20**].

4.2. The full model. Similar to the previous reduced system (2.6), we solve the full system using scheme (3.1). In Figs. 6, and 7 we illustrate the properties of model (2.4). Unless stated differently under the figure caption, the parameters are chosen as follows: $r = 1.0$, $s = 1.0$, $\gamma = 2.0$, $\alpha = 1$ and $\beta = 3.0$.

Microbial population oscillations have been observed in both the field and experiments, supported by mathematical models, see for example [**8**, **18**] to name a few. Based on their experiments, the authors in [**18**, Supplementary information] proposed a mathematical model consisting of two differential equations representing the change in bacterial density in response to the change in proton concentration (pH). An ad hoc function to account for a pH dependence of the bacterial growth

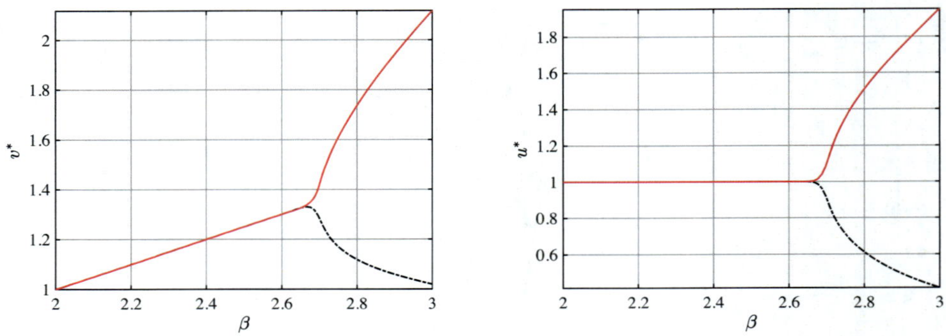

(a) Bifurcation on v^*.

(b) Bifurcation on u^*.

FIGURE 6. Bifurcation with respect to β illustrating the existence of population oscillations. The figure shows the minimal (broken line) and maximal (solid line) values of the environment (figure (a)) and active microbial population (figure (b)).

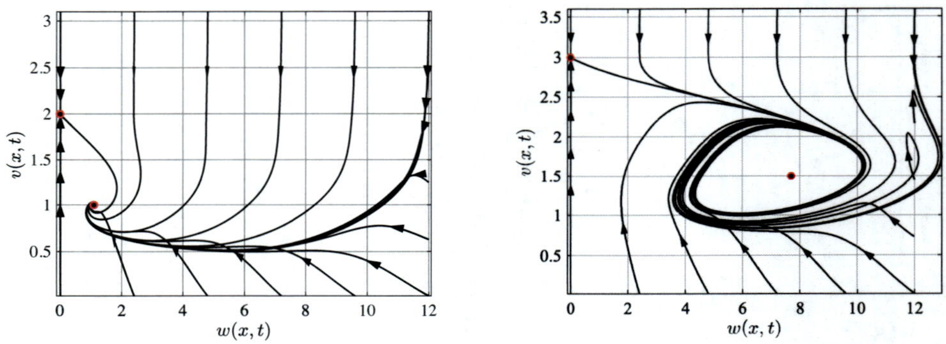

(a) $\beta = 2.0$.

(b) $\beta = 3.0$.

FIGURE 7. Phase plane for model (2.4) illustrating the existence of oscillatory solutions and local stability.

was coupled to the standard Logistic growth model. The rate of change of proton concentration was assumed to depend on the rate of change of bacteria density to account for the observed non-monotonic time profiles. In this work, we proposed a generic model to account for the change in bacteria density in response to the change in its environment. We allow for the bacteria to switch state (active or dormant) in response to their surroundings without going into the molecular details, see Section 2.2.

Similar to the experimental work of [**18**], coexistence and population oscillations were observed in the model proposed here. In particular, from the experiment in [**18**] population extinction was observed at low buffer levels, population oscillation at medium buffer levels and stable coexistence at high buffer levels . This phenomenon is replicated here via the production term $\beta g(u)$, where a decrease in β is used to emulate an increase in buffer concentration as in experiments [**18**]. When $\beta = 2.0$, representing high buffer levels, a stable coexistence state is observed, see

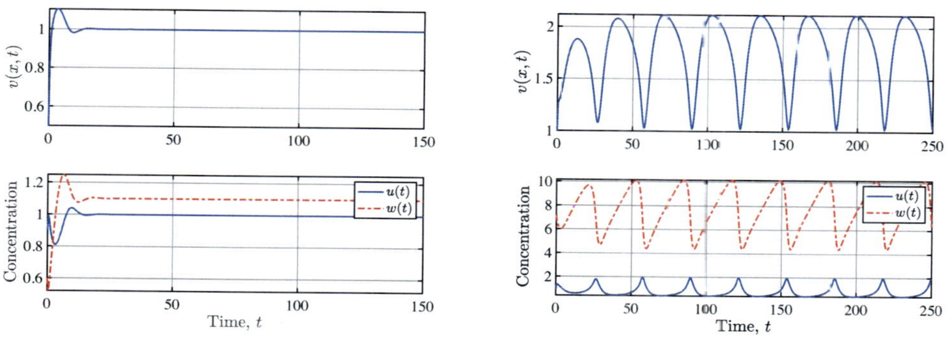

(a) $\beta = 2.0$. (b) $\beta = 3.0$.

FIGURE 8. Time series solution of (2.4) corresponding to parameter selection: $r = 1.0$, $s = 1.0$, $\gamma = 2.0$ and $\alpha = 1.0$.

Fig. 8(a). For medium buffer, $\beta = 3.0$, both population oscillations and coexistence are observed, see Figs. 7 and 8(b).

5. Conclusions

Understanding bacterial responses to physiological conditions is an important priority for combating opportunistic infections. The proposed model has underpinnings of antibiotic resistance. For example, the work [**12**] studied the dynamics of *E. faecalis* communities exposed to antibiotics revealing scenarios where increasing population size or delaying drug exposure can promote population collapse. In [**20**], the authors showed that the way microbes modify their environment and react to it sets the interactions within single-species populations and also between different species.

Microbial growth and survival depend on the environment and interactions within. These populations are capable of changing their environment by producing chemicals, thereby changing the pH of their environment. The proposed model takes into account the dynamics of bacteria transitioning from one state to another depending on the concentration levels of the toxin in their surroundings. Our results support the observed rapid eradication, also named 'ecological suicide', of the microbial population as a result of changes in their environment [**18**]. In addition, we observed the occurrence of self-sustained oscillations in the microbial population mediated by the change in concentration levels of the toxin. In Fig. 8, we see that as the toxin concentration levels increase, the number of active cells declines which supports the negative interaction between active microbe density and toxin concentrations. We can conclude that as the toxin concentration levels increase causing the deterioration of the environment, this leads to population collapse. Our results support the experimental observations in [**18**] showing population collapse as a result of environmental deterioration leading to oscillatory dynamics.

The role of interactions between genotypes and their consequences for spatial structure and functional outcomes is increasingly being studied to better understand the ecology and evolution of microbial communities. As reported in [**19**], organisms are not randomly dispersed but can be found in spatial aggregates at many scales, leading to spatial heterogeneity. They further argued that generic factors intrinsic to the populations such as interactions between the organisms play a major role in the emergence of these patterns. Regular spatial patterns and heterogeneity can be observed in different microbial ecosystems. Experimental observations ranged from bistability, successive growth, extended suicide, and stabilization. In future work, we will explore the variety of spatial patterns that could possibly emerge from the system and the ecological consequences that follow.

References

[1] R. Anguelov, Y. Dumont, J. M.-S. Lubuma, and M. Shillor, *Dynamically consistent non-standard finite difference schemes for epidemiological models*, J. Comput. Appl. Math. **255** (2014), 161–182, DOI 10.1016/j.cam.2013.04.042. MR3093413

[2] R. Anguelov, J. M.-S. Lubuma, and M. Shillor, *Topological dynamic consistency of non-standard finite difference schemes for dynamical systems*, J. Difference Equ. Appl. **17** (2011), no. 12, 1769–1791, DOI 10.1080/10236198.2010.488226. MR2854823

[3] M. Chen, M. Fan, R. Liu, X. Wang, X. Yuan, and H. Zhu, *The dynamics of temperature and light on the growth of phytoplankton*, Journal of Theoretical Biology, **385** (2015) 8–19.

[4] M. Chen, M. Fan, X. Yuan, and H. Zhu, *Effect of seasonal changing temperature on the growth of phytoplankton*, Math. Biosci. Eng. **14** (2017), no. 5-6, 1091–1117, DOI 10.3934/mbe.2017057. MR3657119

[5] J. Cremer, A. Melbinger, K. Wienand, T. Henriquez, H. Jung, and E. Frey, *Cooperation in microbial populations: theory and experimental model systems*, Journal of Molecular Biology, **431** (2019) no. 23, 4599–4644.

[6] M. Dohi and A. Mougi, *A coexistence theory in microbial communities*, Royal Society Open Science, **5** (2018) no. 9, 180476.

[7] S. Estrela, E. Libby, J. Van Cleve, F. Débarre, M. Deforet, W. R. Harcombe, J. Peña, S. P. Brown, and M. E. Hochberg, *Environmentally mediated social dilemmas*, Trends in Ecology & Evolution, **34** (2019) no. 1, 6–18.

[8] A. Fowler and H. Winstanley, *Microbial dormancy and boom-and-bust population dynamics under starvation stress*, Theoretical Population Biology, **120** (2018), 114–120.

[9] J. E. Goldford, N. Lu, D. Bajić, S. Estrela, M. Tikhonov, A. Sanchez-Gorostiaga, D. Segrè, P. Mehta, and A. Sanchez, *Emergent simplicity in microbial community assembly*, Science, **361** (2018) no. 6401, 469–474.

[10] J. E. González-Pastor, *Cannibalism: a social behavior in sporulating bacillus subtilis*, FEMS Microbiology Reviews, **35** (2011) no. 3, 415–424.

[11] Y. Guan, N. Jiang, Y. Wu, Z. Yang, A. Bello, and W. Yang, *Disentangling the role of salinity-sodicity in shaping soil microbiome along a natural saline-sodic gradient*, Science of the Total Environment, **765** (2021) no. 142738.

[12] K. M. Hallinen, J. Karslake, and K. B. Wood, *Delayed antibiotic exposure induces population collapse in enterococcal communities with drug-resistant subpopulations*, Elife, **9** (2020), e52813.

[13] R. E. Mickens, *Nonstandard finite difference models of differential equations*, World Scientific Publishing Co., Inc., River Edge, NJ, 1994. MR1275372

[14] M. Mimura, H. Sakaguchi, and M. Matsushita, *Reaction-diffusion modelling of bacterial colony patterns*, Physica A: Statistical Mechanics and its Applications, **282** (2000) no. 1-2, 283–303.

[15] J. D. Murray, *Mathematical biology. II: Spatial models and biomedical applications*, 3rd ed., Interdisciplinary Applied Mathematics, vol. 18, Springer-Verlag, New York, 2003. MR1952568

[16] L. Niehaus, I. Boland, M. Liu, K. Chen, D. Fu, C. Henckel, K. Chaung, S. E. Miranda, S. Dyckman, M. Crum, et al, *Microbial coexistence through chemical-mediated interactions*, Nature Communications, **10** (2019) no. 1, 1–12.

[17] P. F. Popp and T. Mascher, *Coordinated cell death in isogenic bacterial populations: sacrificing some for the benefit of many?*, Journal of Molecular Biology, **431** (2019) no. 23, 4656–4669.

[18] C. Ratzke, J. Denk, and J. Gore, *Ecological suicide in microbes*, Nature Ecology & Evolution, **2** (2018) no. 5, 867–872.

[19] C. Ratzke and J. Gore, *Self-organized patchiness facilitates survival in a cooperatively growing Bacillus subtilis population*, Nature Microbiology, **1** (2016) no. 5, 1–5.

[20] C. Ratzke and J. Gore, *Modifying and reacting to the environmental pH can drive bacterial interactions*, PLoS Biology, **16** (2018) no. 3, e2004248.

[21] D. Romero, M. F. Traxler, D. López, and R. Kolter, *Antibiotics as signal molecules*, Chemical Reviews, **111** (2011) no. 9, 5492–5505.

[22] M. Simmons, K. Drescher, C. D. Nadell, and V. Bucci, *Phage mobility is a core determinant of phage-bacteria coexistence in biofilms*, The ISME Journal, **12** (2018) no. 2, 531–543.

[23] L. Tsimring, H. Levine, I. Aranson, E. Ben-Jacob, I. Cohen, O. Shochet, and W. N. Reynolds, *Aggregation patterns in stressed bacteria*, Physical Review Letters, **75** (1995) no. 9, 1859.

[24] M. Vasse, C. Torres-Barceló, and M. E. Hochberg, *Phage selection for bacterial cheats leads to population decline*, Proceedings of the Royal Society B: Biological Sciences, **282** (2015) no. 1818, 20152207.

[25] H. F. Winstanley, M. Chapwanya, M. J. McGuinness, and A. C. Fowler, *A polymer-solvent model of biofilm growth*, Proc. R. Soc. Lond. Ser. A Math. Phys. Eng. Sci. **467** (2011), no. 2129, 1449–1467, DOI 10.1098/rspa.2010.0327. MR2782165

[26] D. E. Woodward, R. Tyson, M. Myerscough, J. Murray, E. Budrene, and H. Berg, *Spatio-temporal patterns generated by Salmonella typhimurium*, Biophysical Journal, *68* (1995) no. 5, 2181–2189.

Department of Mathematics & Applied Mathematics, University of Pretoria, Pretoria 0002, South Africa
Email address: `m.chapwanya@up.ac.za`

Department of Mathematics & Applied Mathematics, University of Pretoria, Pretoria 0002, South Africa
Email address: `phindile.dumani@up.ac.za`

Contemporary Mathematics
Volume **793**, 2024
https://doi.org/10.1090/conm/793/15906

Using unity approximations to construct nonstandard finite difference schemes for Bernoulli differential equations

Treena S. Basu, Ron Buckmire, Zaheer Coovadia, Mayra Diaz, David A. Iniguez, and Alexandra Scott

ABSTRACT. We explore the use of nonstandard finite difference (NSFD) methods in numerically approximating solutions to Bernoulli ordinary differential equations (BDEs) with constant coefficients. Specifically, we investigate the numerical performance of NSFD schemes that deploy unity approximations. A unity approximation can be created by introducing and replacing the number 1 with an approximation using non-local terms of the form $\dfrac{2x_{k+1}}{x_{k+1}+x_k}$ where $x_k \approx x_{k+1}$ on a discrete grid. Our NSFD schemes derived from unity approximations produce solutions that compare favorably with numerical solutions obtained using standard finite difference (SFD) schemes. The NSFD schemes presented here demonstrate their utility by possessing one or more of the following properties: enhanced accuracy, preserving positivity, and maintaining dynamic consistency (i.e., other essential qualities of the ODE pertaining to its domain and the asymptotic properties of the equilibria such as stability).

1. Introduction

Ordinary differential equations (ODEs) and partial differential equations (PDEs) are frequently used to model complex laws of nature and other phenomena. For instance, the linear ODE $\dfrac{dx}{dt} = -\lambda x, x(0) = x_0 > 0$ is used to model exponential decay of radioactive isotopes for carbon dating and other uses. A pair of first-order nonlinear differential equations known as the Lotka–Volterra equations [**1**] can be used to model the dynamics of biological systems in which two species interact, one as a predator and the other as prey. The dynamics of an epidemic can be modeled with a system of ODEs using the famous SIR model [**2**], where S represents the susceptible population and I and R represent the infected population and the removed population respectively. However, even simple and well-known ODE models like the Lotka-Volterra model and the SIR model do not have analytic closed-form solutions [**3**, **4**]. Thus, in order to produce solutions to these ODE models and many others one must often use numerical approximation techniques.

2020 *Mathematics Subject Classification.* Primary 65L05, 65L12.

The first and second authors are the corresponding authors.

For undergraduate students, affiliations shown are correct when they did the work in this chapter.

Although standard finite difference (SFD) methods are among the most popular methods used to numerically approximate solutions to differential equations [**5–9**], there can be difficulties. In some cases, SFD schemes are unstable [**10**] or only capable of preserving the important properties of the differential equation if the step size is small enough [**11–16**].

Ronald E. Mickens, Clark Atlanta University Professor Emeritus, has been proposing non-standard finite difference (NSFD) schemes, also known as Mickens-type finite differences, since the 1970s. Many, many schemes created by him or citing his work show improved accuracy compared to SFD methods. Furthermore, numerical solutions generated through NSFD methods can preserve essential properties of the ODE and their solutions, oftentimes when SFD methods can not. Although the field of numerical analysis that includes NSFD schemes has long been and continues to be an area of active research, there are still many opportunities for exploration, discovery and contribution. We are excited and honored to be invited to submit this summary of our recent work in this area for inclusion in this volume in celebration of Prof. Micken's 80th birthday (February 7, 2023). We note with pride that many of the results presented here were accomplished in collaboration with undergraduate student researchers (Zaheer Coovadia, Mayra Diaz, David A. Iniguez, and and Alexandra Scott) in summer 2021.

1.1. NSFD Schemes for ODEs. Consider an ODE of the form $\dfrac{dx}{dt} = f(x,t)$ defined on $a \le t \le b$ with initial condition (IC) $x(a) = x_0$ where $a, b, x_0 \in \mathbb{R}$ are known parameters. Our goal is to obtain a function $x(t)$ that satisfies the ODE and the IC, which together are known as an intial value problem (IVP). However, for most IVPs, only a numerical approximation to the solution $x(t)$ can be found. In these cases, the solution technique is to discretize the interval $[a, b]$ into N subintervals, define a grid separation parameter $h = \frac{b-a}{N}$, let $t_k = a + kh$ for all $k = 0, \cdots, N$ and approximate the derivative to convert the differential equation to a difference equation. Generating a numerical solution to the ODE means that we can find values $\{x_k\}_0^N$ where x_k is arbitrarily close to $x(t_k)$ as N increases to infinity (and h decreases to 0). In this scenario, typically one solves a difference equation to obtain values of $\{x_k\}_0^N$.

A finite difference scheme is considered to be a NSFD scheme if at least one of the following criteria are met

(1) The discrete derivative has a more general form than that used in conventional discretization [**17, 18**]. For instance, in the case of discretizing a first-order derivative one would consider the form

$$\frac{dx}{dt} = \frac{x_{k+1} - \psi(h)x_k}{\phi(h)}$$

where $\psi(h)$ and $\phi(h)$ have the following properties

$$\psi(h) = 1 + O(h), \text{ and } \quad \phi(h) = h + O(h)$$

and h is the grid separation parameter (or step size).

(2) Linear and non-linear terms involving the dependent variable are modelled non-locally [**17**]. For example, $x_k \approx 2x_k - x_{k+1}$, and $x_k^2 \approx x_k x_{k+1}$ are potential ways in which a linear term such as x and a non-linear term x^2 could be discretized non-locally.

More details regarding the criteria for NSFD schemes can be found in the literature [**17**, **19**–**21**]. NSFD schemes are often more desirable than SFDs schemes because of:

- the ability of NSFD schemes to sometimes numerically outperform SFD schemes in various ways
- the capability of NSFD schemes to preserve important properties of the solutions of the differential equation, such as positivity, monotonicity, periodicity, stability, etc.

1.2. Motivation. Following in the footsteps of Prof. Mickens, many researchers have tried (and succeeded) in finding NSFD schemes that can outperform SFDs in the solution of ODEs. In this subsection we will discuss the prior research which motivates the results we present in this chapter and summarize the problems we chose to investigate. Specifically, we will be looking at a subset of NSFD schemes that are obtained using unity approximations. We restrict our investigations to ODEs that are Bernoulli differential equations (BDEs) with constant coefficients of the form:

$$\frac{dx}{dt} = \lambda_1 x^p - \lambda_2 x, \quad x(0) = x_0 > 0, \quad \lambda_1\lambda_2 \neq 0, \quad p \in \mathbb{Q},\, 0 < p < 1. \tag{1.1}$$

Lubuma and Patidar [**24**] presented an exact finite difference scheme for the BDE of the form

$$\frac{dy}{dt} = \lambda y(1 - y^{m-1}), \quad y(0) = y_0, \quad \lambda \neq 0, \quad m \in \mathbb{R} \tag{1.2}$$

by implementing a change of variable $v = y^{1-m}$, thus transforming Equation (1.2) into the linear ODE

$$\frac{dv}{dt} = (1-m)\lambda v - \lambda(1-m), \quad v(0) = y_0^{1-m} \tag{1.3}$$

whose exact scheme is known from [**17**, **23**]. Equation (1.2) is related to the BDEs in (1.1) that we will be focusing on in this chapter when p is replaced by m and $\lambda_1 = \lambda_2 = \lambda$.

Roeger and Mickens [**23**] studied IVPs of the form

$$\frac{dx}{dt} = -\lambda x^\alpha, \quad x(t_0) = x_0 > 0, \quad \lambda > 0, \quad \alpha > 0 \tag{1.4}$$

and such equations often appear in the modelling of a broad range of physical and engineering phenomena [**17**, **20**]. Roeger and Mickens provided [**23**] an exact finite difference scheme for Equation 1.4 in two cases: (i) $0 < \alpha < 1$ (ii) $\alpha \geq 1$ when α is rational. Equation (1.4) is also a special case of the BDEs in (1.1), this time with $\lambda_2 = 0$ and p replaced by α.

The work presented in this chapter was motivated by the research described above, but our work can be distinguished from these prior results in the following ways:

(1) While Lubuma and Patidar [**24**] present an exact finite difference scheme for the BDE in Equation (1.2), their work uses a transformation of this nonlinear ODE into a linear ODE and relies upon knowledge of the exact FD scheme for this linear ODE to produce the exact finite difference scheme for the original (untransformed) BDE. The work presented below provides examples of using a unity scheme to arrive at the exact NSFD

scheme for the BDE without having to transform it into linear form first. Using this example as a proof-of-concept, our unity approximation method may be applied to other ODEs that can't benefit from being transformed to a linear ODE in order to produce the exact FD scheme.

(2) Recall, Roeger and Mickens gave the form of an exact NSFD scheme for Equation (1.4) for $0 < \alpha < 1$ when α is rational. The work presented in this chapter deploys unity approximations to replicate this exact Roeger-Mickens scheme when $\alpha = 1/2$ case and in the general $0 < \alpha < 1$ case.

(3) Similarly, we also show that we can use unity NSFD approximations to replicate the exact FD scheme that Lubuma-Patidar provided for the BDE. We do this for the special case of $p = 1/2$ and in the general case where $p = n/m$ with $m > n$.

(4) We also look at the simple decay equation $\dfrac{dx}{dt} = -\lambda x$ since in some sense it is a special case of the Roeger-Mickens equation in Equation (1.4) with $\alpha = 1$ as well as the generalized BDE where $\lambda_1 = 0$. We show that one can find NSFD schemes which use unity approximations that numerically outperform SFD schemes for this elementary IVP. This corresponds to both the $p = 1$ and $\lambda_2 = 0$ (with $\lambda_1 < 0$) **and** the $\lambda_1 = 0$ case for our BDE in Equation (1.1).

Table 1 below outlines the specific BDEs that are investigated in this paper along with the name we refer to it by, the parameter values that distinguish it from the others and then the form of the equation.

TABLE 1. BDEs: $\dfrac{dx}{dt} = \lambda_1 x^p - \lambda_2 x$ and $p \in \mathbb{Q}$.

Equation Name	BDE Parameters	BDE
Simple Decay	$\lambda_1 = 0,\ \lambda_2 > 0,\ p$ any	$\dfrac{dx}{dt} = -\lambda x$
Root Decay	$\lambda_1 < 0,\ \lambda_2 = 0,\ p = \dfrac{1}{2}$	$\dfrac{dx}{dt} = -\lambda\sqrt{x}$
Generalized Root Decay	$\lambda_1 < 0,\ \lambda_2 = 0,\ 0 < p < 1$	$\dfrac{dx}{dt} = -\lambda x^{n/m},\ n < m$
Root BDE	$\lambda_1 \neq 0,\ \lambda_2 \neq 0, p = \dfrac{1}{2}$	$\dfrac{dx}{dt} = \lambda_1\sqrt{x} - \lambda_2 x$
Generalized BDE	$\lambda_1 \neq 0,\ \lambda_2 \neq 0,\ 0 < p < 1$	$\dfrac{dx}{dt} = \lambda_1 x^p - \lambda_2 x$

1.3. Organization. The rest of this paper is organized as follows. In Section 2, we examine two NSFD schemes developed through unity approximations applied to the simple decay equation. We present numerical results that demonstrate the NSFD unity schemes are second-order and outperform SFD schemes. We start with this problem because it is a special case of both of the problems that motivated this work (Roeger-Mickens and Lubuma-Patidar).

Next, in Section 3 we show that we can use unity NSFD approximations to replicate the Roeger-Mickens exact FD scheme for a special case (the Root Decay Equation). Then we conclude this section by applying a unity NSFD approximation to replicate the exact FD scheme for the Generalized Root Decay equation, i.e. Equation (1.1) when p is rational and $0 < p < 1$.

In Section 4 we again show that a unity approximation can generate an exact NSFD scheme for first a special case and then the generalized problem. This time the general form is the generalized BDE in Equation (1.1) where $\lambda_1\lambda_2 \neq 0$ and $0 < p < 1$ and the special case is $p = 1/2$ (the Root BDE).

We conclude the chapter by discussing some ways in which our work may be extended in Section 6.

2. NSFD schemes using unity approximations for the simple decay equation

We begin by analyzing the following IVP:

$$\frac{dx}{dt} = -\lambda x, \quad x(0) = x_0 > 0, \quad \lambda > 0 \tag{2.1}$$

and refer to the ODE in Equation (2.1) as the *Simple Decay Equation* [**17**]. Note that Equation (2.1) is a special case of

- the Bernoulli ODE given in Equation (1.1) where $\lambda_1 = 0$, $\lambda_2 = \lambda > 0$;
- the Roeger-Mickens Equation found in (1.4) with $\alpha = 1$.

The exact analytic solution $x(t)$ to the decay equation IVP is given by

$$x(t) = x_0 \exp^{-\lambda(t-t_0)} \tag{2.2}$$

and the solution decreases asymptotically to 0 for $t \in [0, \infty)$ and remains positive. We summarize the properties of Equation (2.1) below [**17**]

- $x(t) = 0$ is a constant solution;
- the fixed point $x = 0$ is globally stable;
- the solution $x(t)$ maintains the sign of the initial condition, i.e., $x(t) > 0$ for $t > 0$;
- $x(t)$ monotonically decreases in magnitude to zero for any $x_0 > 0$.

An exact finite difference scheme (EFDS) to Equation (2.1) can be written as [**17**]

$$x_{k+1} = \exp^{-\lambda h} x_k. \tag{2.3}$$

From Equation (2.3) it can be inductively established that

$$x_k = \exp^{-\lambda hk} x_0. \tag{2.4}$$

The idea behind the NSFD schemes explored in this section is to use the "unity approximation" as a ratio of $1 = \dfrac{2x}{2x}$ and then include non-local approximations of x in different parts of the finite-difference expression. More specifically, we view the approximations of unity as $1 \approx \dfrac{x_A + x_B}{x_C + x_D}$, where A, B, C, and D are either k or $k + 1$. Since there are 2 possible values for each of the 4 variables (A, B, C and D) in the expression, there are $2^4 = 16$ possibilities for this particular unity approximation. Of course, due to the commutativity of addition and trivial cases there will not be 16 distinct schemes to consider. By ignoring schemes that are (1) numerically unstable finite difference schemes or (2) dynamically consistent with the Simple Decay Equation we end up with two schemes that we are interested in investigating further.

In the next subsection, we analyze these two remaining NSFD schemes created through unity approximations.

2.1. NSFD Scheme 2.1. We can view the Simple Decay Equation in (2.1) as

$$\frac{dx}{dt} = -\lambda x = -\lambda x \cdot \mathbf{1} = -\lambda x \cdot \left(\frac{2x}{2x}\right). \tag{2.5}$$

Using $1 = \frac{2x}{2x} \approx \frac{x_{k+1} + x_{k+1}}{x_{k+1} + x_k}$ as an approximation of unity we produce the following scheme

$$\frac{x_{k+1} - x_k}{h} = -\lambda x_k \left(\frac{x_{k+1} + x_{k+1}}{x_{k+1} + x_k}\right). \tag{2.6}$$

Simplifying yields a quadratic equation in x_{k+1}

$$x_{k+1}^2 + 2\lambda h x_k x_{k+1} - x_k^2 = 0.$$

Solving for x_{k+1} results in

$$x_{k+1} = \frac{-2\lambda h x_k \pm \sqrt{4x_k^2(\lambda^2 h^2 + 1)}}{2}.$$

x_{k+1} has a negative and a positive solution. For $x_0 > 0$ only the positive solution is physically meaningful. Thus, the **NSFD Scheme 2.1** is given by

$$x_{k+1} = \left(-\lambda h + \sqrt{\lambda^2 h^2 + 1}\right) x_k. \tag{2.7}$$

We can show that this scheme can be re-written as

$$x_k = \left(-\lambda h + \sqrt{\lambda^2 h^2 + 1}\right)^k x_0. \tag{2.8}$$

One can observe that $x = 0$ is a stable fixed point of Equation (2.7). By inspection, one can see that this scheme maintains positivity.

Next, we implement Scheme 2.1 given by Equation (2.7) using parameter values of $\lambda = 1$, $x_0 = 1$ and $h = \{0.1, 0.01, 0.001, 0.0001\}$. For all values of h the numerical solution is real and maintains positivity.

TABLE 2. Error analysis NSFD Scheme 2.1 for the decay equation

Scheme 2.1	h Ratio	L^∞ Error	Ratio of Errors	Order
$h = 0.1000$	$\sim$	$6.10896281e-04$	$\sim$	$\sim$
$h = 0.0100$	10	$6.13109921e-06$	99.6389	1.9984
$h = 0.0010$	10	$6.13132385e-08$	99.9963	2.0000
$h = 0.0001$	10	$6.13105622e-10$	100.00436	2.0000

From the Table 2 we observe that when the ratio of subsequent step sizes h_k/h_{k+1} is $\frac{1}{10}$ the L^∞ error between the true solution to the Simple Decay Equation and the numerical solution obtained through NSFD Scheme 2.1 decreases by a factor of approximately $\frac{1}{100}$. This indicates that Scheme 2.1 is a second-order scheme. Further, the fifth column in Table 2 computes

$$\frac{\ln(\text{Ratio of Errors})}{\ln(h_k/h_{k+1})} = \frac{\ln(\text{Ratio of Errors})}{\ln(10)}$$

which approaches 2 and thus establishes that the order of Scheme 2.1 is second-order.

2.2. NSFD Scheme 2.2. This time we view the Simple Decay Equation (2.1) using a different version of an approximation of unity, i.e., $\frac{x_{k+1}+x_k}{2x_k}$ to produce

$$\frac{x_{k+1}-x_k}{h} = -\lambda x_k \cdot \mathbf{1} = -\lambda x_k \left(\frac{x_{k+1}+x_k}{2x_k}\right). \tag{2.9}$$

Solving for x_{k+1} results in the following **NSFD Scheme 2.2**

$$x_{k+1} = \left(\frac{1-\lambda h/2}{1+\lambda h/2}\right) x_k. \tag{2.10}$$

which can be re-written as

$$x_k = \left(\frac{1-\frac{\lambda h}{2}}{1+\frac{\lambda h}{2}}\right)^k x_0. \tag{2.11}$$

Furthermore, we note that $x = 0$ is a stable fixed point of Equation (2.10). By inspection, one can see that this scheme maintains positivity when $h < \frac{2}{\lambda}$.

The results obtained from implementing NSFD Scheme 2.2 using parameter values of $\lambda = 1$, $x_0 = 1$ and $h = \{0.1, 0.01, 0.001, 0.0001\}$ are displayed in Table 3. These numerical results show that Scheme 2.2 is second-order accurate. For all values of h the numerical solution is real and maintains positivity.

TABLE 3. Error analysis NSFD Scheme 2.2 for the decay equation

Scheme 2.2	h Ratio	L^∞ Error	Ratio of Errors	Order
$h = 0.1000$	$\sim$	$3.06898789e-04$	$\sim$	$\sim$
$h = 0.0100$	10	$3.06569522e-06$	100.1074	2.0005
$h = 0.0010$	10	$3.06565833e-08$	100.0012	2.0000
$h = 0.0001$	10	$3.06931480e-10$	99.8808	1.9995

Thus, to summarize, both NSFD Scheme 2.1 and NSFD Scheme 2.2 are second-order schemes and therefore, more desirable than standard first-order schemes such as implicit and explicit Euler. Note that $x = 0$ is a fixed point for both NSFD schemes presented in this Section 2. Also, numerical solutions from NSFD Schemes 2.1 and NSFD Scheme 2.2 are real and maintain positivity. Thus, both NSFD schemes are dynamically consistent with the Simple Decay Equation (2.1).

3. NSFD schemes for the Root Decay Equation and the Generalized Root Decay Equation

In this section we consider NSFD schemes that exploit two different ways the number "1" can be approximated algebraically and utilized to discretize Equation

(3.1) which we refer to as the *Root Decay Equation.* Note that the Root Decay Equation is equivalent to

- the Roeger-Mickens model given in Equation (1.4) for the case when $\alpha = \frac{1}{2}$
- our BDE given in Equation (1.1) with $\lambda_2 = 0, \lambda_1 = -\lambda < 0$ and $p = \frac{1}{2}$.

The Root Decay IVP being considered is

$$\frac{dx}{dt} = -\lambda\sqrt{x}, \quad x(0) = x_0 > 0, \quad \lambda > 0. \tag{3.1}$$

The exact solution $x(t)$ is given by

$$x(t) = \begin{cases} \left(\sqrt{x_0} - \frac{\lambda}{2}t\right)^2, & 0 < t \leq t_c = \frac{2\sqrt{x_0}}{\lambda} \\ 0 & , \quad t > t_c. \end{cases} \tag{3.2}$$

Note the solution is nonnegative, bounded and monotonically decreasing for all $t \geq 0$. This is important because we will want to produce NSFD schemes that possess these qualities for dynamic consistency.

Since Equation (3.1) is the special case when $\alpha = 1/2$ from Equation (1.4), the corresponding exact finite difference scheme adapted from [**23**] is

$$x_{k+1} = \left(\sqrt{x_k} - \frac{\lambda h}{2}\right)^2. \tag{3.3}$$

The exact scheme in Equation (3.3) is clearly positivity-preserving since it is a square. From Equation (3.3) it can be inductively established that

$$x_k = \left(\sqrt{x_0} - \frac{k\lambda h}{2}\right)^2. \tag{3.4}$$

We summarize the properties of the Root Decay Equation (3.1) below

- $x(t) = 0$ is a constant solution to Equation (3.1) and $x = 0$ is a fixed-point of Equation (3.1).
- Since $x(0) > 0$ and $\frac{dx}{dt} < 0$, the solution $x(t)$ can only decrease from an initial bounded value x_0.
- $x(t)$ remains non-negative to maintain real solutions of Equation (3.1). Clearly, there exists a t value equal to $t_c = \frac{2\sqrt{x_0}}{\lambda}$ after which $x(t)$ is identically equal to zero for all $t > t_c$.

In this section, we demonstrate that we can use NSFD schemes from unity approximations that produce exact FD schemes for the generalized root decay equation and a special case when $p = 1/2$. Note that in light of the property of the exact solution $x(t)$ discussed in the last bullet point above, when computing numerical solutions for the Root Decay and the Generalized Root Decay Equation, we restrict our calculations to the domain $0 \leq t \leq t_c$ after choosing values of the parameters x_0, λ and h.

3.1. NSFD Scheme 3.1: an exact scheme. We propose the following unity approximation in a NSFD for the Root Decay Equation (3.1)

$$\frac{x_{k+1} - x_k}{h} = -\lambda\sqrt{x_k} \cdot \mathbf{1} = -\lambda\sqrt{x_k}\left(\frac{\sqrt{x_{k+1}} + \sqrt{x_k}}{2\sqrt{x_k}}\right). \tag{3.5}$$

Making the substitution $u_{k+1} = \sqrt{x_{k+1}} \Leftrightarrow u_k^2 = x_k$ results in the following quadratic equation in u_{k+1}

$$2u_{k+1}^2 + \lambda h u_{k+1} + \lambda h u_k - 2u_k^2 = 0. \tag{3.6}$$

Solving for u_{k+1} we obtain

$$u_{k+1} = u_k - \frac{\lambda h}{2}. \tag{3.7}$$

Thus, we obtain **NSFD Scheme 3.1** as

$$x_{k+1} = \left(\sqrt{x_k} - \frac{\lambda h}{2}\right)^2. \tag{3.8}$$

Note this is the same as the exact finite difference scheme presented in Equation (3.3).

3.2. NSFD Scheme 3.2: an exact scheme for the Generalized Root Decay Equation. In this subsection we consider a more general form of the Root Decay Equation given in (3.1). Specifically, we consider the IVP

$$\frac{dx}{dt} = -\lambda x^{\frac{n}{m}}, \quad x(0) = x_0 > 0, \quad \lambda > 0, \quad m, n \in \mathbb{Z}^+, m > n. \tag{3.9}$$

We refer to Equation (3.9) as the *Generalized Root Decay Equation.* Note that Equation (3.9) is equivalent to

- Our BDE in Equation (1.1) when $\lambda_1 = -\lambda < 0$, $\lambda_2 = 0$ and p is rational with $0 < p < 1$;
- Roeger-Mickens IVP in (1.4) when $0 < \alpha < 1$.

The exact solution $x(t)$ is given by

$$x(t) = \begin{cases} \left(x_0^{1-\frac{n}{m}} - \lambda t(1 - \frac{n}{m})\right)^{\frac{1}{1-\frac{n}{m}}} & 0 < t \le t_c = \dfrac{x_0^{1-\frac{n}{m}}}{\lambda(1-\frac{n}{m})} \\ 0, & t > t_c. \end{cases} \tag{3.10}$$

Equation (3.9) has a known exact finite difference scheme provided in [**17**]

$$x_{k+1} = \left(x_k^{1-\frac{n}{m}} - \lambda h(1 - \frac{n}{m})\right)^{\frac{1}{1-\frac{n}{m}}} \quad n \neq m \tag{3.11}$$

i.e., the solution will decrease and stay positive until it reaches 0 at t_c and stays at 0.

In order to generate an NSFD scheme we propose the following approximation to unity

$$1 \approx \frac{m-n}{m} \frac{\sum_{i=1}^{m} x_{k+1}^{\frac{(m-n)(m-i)}{m}} x_k^{\frac{(m-n)(i-1)}{m}}}{\sum_{i=1}^{m-n} x_{k+1}^{m-n-i} x_k^{\frac{n}{m}+i-1}}. \tag{3.12}$$

In the computations below, as with all unity approximations, we assume that if the discretization is fine enough, i.e., $x_{k+1} \approx x_k$, then the RHS of Equation (3.12) is approximately 1. The details of the calculation are given here:

$$\begin{aligned}
&\left[\frac{m-n}{m}\cdot\frac{\sum_{i=1}^{m} x_{k+1}^{\left(\frac{(m-n)(m-i)}{m}\right)}\cdot x_k^{\left(\frac{(m-n)(i-1)}{m}\right)}}{\sum_{i=1}^{m-n} x_{k+1}^{(m-n-i)}\cdot x_k^{\left(\frac{n}{m}+i-1\right)}}\right]\\
&\approx\left[\frac{m-n}{m}\cdot\frac{\sum_{i=1}^{m} x_k^{\left(\frac{(m-n)(m-i)}{m}+\frac{(m-n)(i-1)}{m}\right)}}{\sum_{i=1}^{m-n} x_k^{\left(m-n-i+\frac{n}{m}+i-1\right)}}\right]\\
&=\left[\frac{m-n}{m}\cdot\frac{\sum_{i=1}^{m} x_k^{\left(\frac{m^2-mi-nm+ni+mi-m-ni+n}{m}\right)}}{\sum_{i=1}^{m-n} x_k^{\left(m-n+\frac{n}{m}-1\right)}}\right]\\
&=\left[\frac{m-n}{m}\cdot\frac{\sum_{i=1}^{m} x_k^{\left(\frac{m^2-nm-m+n}{m}\right)}}{\sum_{i=1}^{m-n} x_k^{\left(m-n-1+\frac{n}{m}\right)}}\right]\\
&=\left[\frac{m-n}{m}\cdot\frac{\sum_{i=1}^{m} x_k^{\left(m-n-1+\frac{n}{m}\right)}}{\sum_{i=1}^{m-n} x_k^{\left(m-n-1+\frac{n}{m}\right)}}\right]\\
&=\frac{m-n}{m}\cdot\frac{m}{m-n}\cdot\frac{x_k^{\left(m-n-1+\frac{n}{m}\right)}}{x_k^{\left(m-n-1+\frac{n}{m}\right)}}\\
&=1
\end{aligned}$$

We now put forward the following unity approximation in a NSFD for Equation (3.9).

$$\frac{x_{k+1}-x_k}{h}=-\lambda x_k^{\frac{n}{m}}\cdot\mathbf{1}=-\lambda x_k^{\frac{n}{m}}\,\frac{m-n}{m}\,\frac{\sum_{i=1}^{m} x_{k+1}^{\frac{(m-n)(m-i)}{m}}\,x_k^{\frac{(m-n)(i-1)}{m}}}{\sum_{i=1}^{m-n} x_{k+1}^{m-n-i}\,x_k^{\frac{n}{m}+i-1}}. \tag{3.13}$$

By solving for x_{k+1} we will show that Equation (3.13) is equivalent to the exact discretization of Equation (3.9) given in Equation (3.11). We make use of the following standard factorizations

$$u-v=(u^{\frac{1}{m}}-v^{\frac{1}{m}})\sum_{i=1}^{m} u^{\frac{m-i}{m}}v^{\frac{i-1}{m}},\quad m>1 \tag{3.14}$$

and

$$x_{k+1}^{m-n}-x_k^{m-n}=(x_{k+1}-x_k)\sum_{i=1}^{m-n} x_{k+1}^{m-n-i}x_k^{i-1},\quad m>n. \tag{3.15}$$

Substituting $u = x_{k+1}^{m-n}$ and $v = x_k^{m-n}$ in Equation (3.14) results in

$$x_{k+1}^{m-n} - x_k^{m-n} = \left(x_{k+1}^{\frac{m-n}{m}} - x_k^{\frac{m-n}{m}}\right) \sum_{i=1}^{m} x_{k+1}^{\frac{(m-n)(m-i)}{m}} x_k^{\frac{(m-n)(i-1)}{m}}. \tag{3.16}$$

Equating (3.15) and (3.16) gives

$$(x_{k+1}-x_k) \sum_{i=1}^{m-n} x_{k+1}^{m-n-i} x_k^{i-1} = \left(x_{k+1}^{\frac{m-n}{m}} - x_k^{\frac{m-n}{m}}\right) \sum_{i=1}^{m} x_{k+1}^{\frac{(m-n)(m-i)}{m}} x_k^{\frac{(m-n)(i-1)}{m}} \tag{3.17}$$

which implies

$$\frac{x_{k+1}^{\frac{m-n}{m}} - x_k^{\frac{m-n}{m}}}{x_{k+1} - x_k} = \frac{\sum_{i=1}^{m-n} x_{k+1}^{m-n-i} x_k^{i-1}}{\sum_{i=1}^{m} x_{k+1}^{\frac{(m-n)(m-i)}{m}} x_k^{\frac{(m-n)(i-1)}{m}}}. \tag{3.18}$$

We multiply the LHS of Equation (3.13) by the LHS of Equation (3.18), and similarly we multiply the RHS of Equation (3.13) by the RHS of Equation (3.18), we obtain

$$\begin{aligned} &\frac{x_{k+1} - x_k}{h} \cdot \frac{x_{k+1}^{\frac{m-n}{m}} - x_k^{\frac{m-n}{m}}}{x_{k+1} - x_k} \\ &= -\lambda x_k^{\frac{n}{m}} \frac{m-n}{m} \frac{\sum_{i=1}^{m} x_{k+1}^{\frac{(m-n)(m-i)}{m}} x_k^{\frac{(m-n)(i-1)}{m}}}{\sum_{i=1}^{m-n} x_{k+1}^{m-n-i} x_k^{\frac{n}{m}+i-1}} \frac{\sum_{i=1}^{m-n} x_{k+1}^{m-n-i} x_k^{i-1}}{\sum_{i=1}^{m} x_{k+1}^{\frac{(m-n)(m-i)}{m}} x_k^{\frac{(m-n)(i-1)}{m}}}. \end{aligned} \tag{3.19}$$

Simplification and factoring out an $x_k^{\frac{n}{m}}$ in the denominator of the second fraction of the RHS of Equation (3.19) results in

$$\frac{x_{k+1}^{\frac{m-n}{m}} - x_k^{\frac{m-n}{m}}}{h} = -\lambda x_k^{\frac{n}{m}} \frac{m-n}{m} \frac{\sum_{i=1}^{m-n} x_{k+1}^{m-n-i} x_k^{i-1}}{x^{\frac{n}{m}} \sum_{i=1}^{m-n} x_{k+1}^{m-n-i} x_k^{i-1}} \tag{3.20}$$

which becomes

$$\frac{x_{k+1}^{\frac{m-n}{m}} - x_k^{\frac{m-n}{m}}}{h} = -\lambda \frac{m-n}{m}. \tag{3.21}$$

Solving for x_{k+1} results in **NSFD Scheme 3.2**

$$x_{k+1} = \left(x_k^{\frac{m-n}{m}} - \lambda h \frac{m-n}{m}\right)^{\frac{m}{m-n}}. \tag{3.22}$$

It is clear the NSFD Scheme 3.2 presented in Equation (3.22) is equivalent to the exact FD scheme in Equation (3.11) that was presented in [**17**]. Thus, NSFD Scheme 3.2 is exact.

In this section we have shown that one can use unity approximations to regenerate exact NSFD schemes for the generalized root decay equation and the special case $p = \frac{1}{2}$ (Root Decay equation).

4. NSFD Schemes for the Bernoulli Differential Equation (BDE)

In this section we are considering the BDE given in Equation (1.1) and restate it below for convenience.

$$\frac{dx}{dt} = \lambda_1 x^p - \lambda_2 x, \quad x(0) = x_0 > 0, \quad \lambda_1, \lambda_2 \neq 0, \quad p \quad \text{is rational}, \quad 0 < p < 1.$$

Here again we will use two different unity approximations to generate exact FD schemes for two differet versions of the BDE. The key result presented in this section is that when p is rational and $0 < p < 1$ there exists an approximation of unity that leads to an exact finite difference scheme for the BDE. The exact analytical solution to Equation (1.1) is given by

$$x(t) = \left(\left(x_0^{1-p} - \frac{\lambda_1}{\lambda_2} \right) e^{\lambda_2 t(p-1)} + \frac{\lambda_1}{\lambda_2} \right)^{\frac{1}{1-p}}. \tag{4.1}$$

We summarize the properties of Equation (1.1) below

- $x(t) = 0$ is a constant solution and $x = 0$ is a fixed point;
- $x(t) = 1$ is a constant solution and $x = 1$ is a fixed point provided $\lambda_1 = \lambda_2$;
- $x(t) = -1$ is also a constant solution and $x = -1$ is also a fixed point only when $\lambda_1 = \lambda_2$ and n, m are both odd, where $p = \dfrac{n}{m}$;
- the fixed-point $x = 0$ is unstable, the fixed-point $x = 1$ is globally stable and $x = -1$ is globally stable when n and m are both odd.

Lubuma and Patidar [**24**], showed the existence of an exact FD scheme for the BDE by transforming it to a linear ODE first, and then expressing the EFDS in terms of the linear ODE. Using their idea we use the substitution $v = x^{1-p}$ which results in the linear ODE

$$\frac{dv}{dt} = -\lambda_2(1-p)v + \lambda_1(1-p).$$

Thus, an exact finite difference scheme to Equation (1.1) can be written as [**17,24**]

$$x_{k+1} = \left(e^{\lambda_2 h(p-1)} (x_k)^{1-p} + \frac{\lambda_1}{\lambda_2} - \frac{\lambda_1}{\lambda_2} e^{\lambda_2 h(p-1)} \right)^{\frac{1}{1-p}} \tag{4.2}$$

From Equation (4.2) it can be inductively established that

$$x_{k+1} = \left(e^{\lambda_2 kh(p-1)} (x_0)^{1-p} + \frac{\lambda_1}{\lambda_2} - \frac{\lambda_1}{\lambda_2} e^{\lambda_2 kh(p-1)} \right)^{\frac{1}{1-p}}. \tag{4.3}$$

In this section we will show how to use a unity approximation to generate an exact NSFD scheme for the Root BDE (i.e. special case of the BDE when $p = \dfrac{1}{2}$) first and then also for the generalized BDE.

4.1. NSFD Scheme 4.1: an exact scheme for the Bernoulli ODE with $p = \frac{1}{2}$.

In this subsection, we specifically focus on using unity approximations to develop NSFD schemes for the Bernoulli ODE Equation (1.1) with $\lambda_1 = \lambda_2 = \lambda \neq 0$, $p = \dfrac{1}{2}$ which is the IVP

$$\frac{dx}{dt} = \lambda(\sqrt{x} - x), \quad x(0) = x_0 > 0. \tag{4.4}$$

We will show that it is possible to derive an exact finite difference scheme for Equation (4.4) using the technique of unity approximations. The exact analytic solution to equation (4.4) is given by

$$x(t) = \left((\sqrt{x_0} - 1)e^{-\frac{\lambda t}{2}} + 1\right)^2. \tag{4.5}$$

Lubuma et al. [**24**] showed that an exact finite difference scheme for the Bernoulli ODE in (4.4) can be developed through performing a substitution $v = x^{\frac{1}{2}}$ thereby converting Equation (4.4) to the linear ODE

$$\frac{dv}{dt} = -\frac{\lambda}{2}v + \frac{\lambda}{2}. \tag{4.6}$$

The analytic solution to Equation (4.6) can be found through separation of variables and is given by

$$v(t) = e^{-\frac{\lambda}{2}t}v_0 + 1 - e^{-\frac{\lambda}{2}t}. \tag{4.7}$$

The exact finite difference scheme for Equation (4.6) can be found in a way similar to Equation (2.3) in [**17**] and is given by

$$v_{k+1} = e^{-\frac{\lambda}{2}h}v_k + 1 - e^{-\frac{\lambda}{2}h}. \tag{4.8}$$

From Equation (4.8) it can be inductively established that

$$v_k = e^{-\frac{\lambda}{2}kh}v_0 + 1 - e^{-\frac{\lambda}{2}kh}. \tag{4.9}$$

Thus, an exact finite difference scheme for the Bernouli ODE in Equation (4.4) can be obtained from Equation (4.8) by recalling the substitution $v_{k+1} = \sqrt{x_{k+1}}$ and $v_k = \sqrt{x_k}$

$$\sqrt{x_{k+1}} = e^{-\frac{\lambda}{2}h}\sqrt{x_k} + 1 - e^{-\frac{\lambda}{2}h}. \tag{4.10}$$

From Equation (4.10) it can be established through induction that

$$\sqrt{x_k} = e^{-\frac{\lambda}{2}kh}\sqrt{x_0} + 1 - e^{-\frac{\lambda}{2}kh}. \tag{4.11}$$

We have shown that the exact NSFD scheme for the generalized BDE can be rewritten as Equation (4.11). We will show below that when we use a unity approximation to find the exact FD scheme for the generalized BDE it will be identical to the expression in Equation (4.11).

We now propose the following unity approximation in an exact NSFD scheme for Equation (4.4)

$$\frac{x_{k+1} - x_k}{\frac{1-e^{-\frac{\lambda}{2}h}}{\frac{\lambda}{2}}} = \lambda\sqrt{x_k}\cdot\mathbf{1} - \lambda x_k\cdot\mathbf{1} = \lambda\sqrt{x_k}\cdot\left(\frac{\sqrt{x_{k+1}} + \sqrt{x_k}}{2\sqrt{x_k}}\right) - \lambda x_k\cdot\left(\frac{\sqrt{x_{k+1}} + \sqrt{x_k}}{2\sqrt{x_k}}\right). \tag{4.12}$$

Solving Equation (4.12) for x_{k+1}, and using the fact that $x_{k+1} - x_k$ can be factored as $(\sqrt{x_{k+1}} - \sqrt{x_k})(\sqrt{x_{k+1}} + \sqrt{x_k})$ we obtain **NSFD Scheme 4.1** as

$$\sqrt{x_{k+1}} = e^{-\frac{\lambda}{2}}\sqrt{x_k} + 1 - e^{-\frac{\lambda}{2}}. \tag{4.13}$$

To show that this is the exact scheme for Equation (4.4), we will use induction on Equation (4.13) and writing out $\sqrt{x_0}, \sqrt{x_1}, \ldots, \sqrt{x_{k-1}}$ in terms of $\sqrt{x_0}$ to obtain

$$\sqrt{x_k} = (\sqrt{x_0} - 1)e^{-kh/2} + 1. \tag{4.14}$$

Note squaring the Equation (4.14) yields

$$x_k = \left(\sqrt{x_0} - 1)e^{-kh/2} + 1\right)^2 \tag{4.15}$$

which is analogous to Equation (4.5) and is the square of Equation (4.11).

Therefore, we have shown that the NSFD scheme 4.1 created through a unity approximation is an exact discretization of the solution of the Bernoulli ODE in Equation (4.4).

4.2. NSFD Scheme 4.2: an exact unity NSFD scheme for the generalized BDE. In this section we present an exact NSFD scheme derived using unity approximations, named NSFD Scheme 4.2, for the generalized BDE in Equation (1.1) for the case p is rational and $0 < p < 1$. That is, we consider p to be of the form, $p = \dfrac{n}{m}$, where m and n are co-prime integers with $n < m$. The resulting IVP is

$$\frac{dx}{dt} = \lambda_1 x^{\frac{n}{m}} - \lambda_2 x, \quad x(0) = x_0 > 0, \lambda_1\lambda_2 \neq 0, n < m. \tag{4.16}$$

The core result presented in this section is that when p is rational and $0 < p < 1$, there exists an approximation of unity that leads to an exact finite difference scheme for the Bernoulli ODE in Equation (4.16). The exact analytic solution to Equation (4.16) is given in Equation (4.1).

We propose the following unity approximation resulting in an exact NSFD scheme for Equation (4.16) using the denominator function $\phi(h) = \dfrac{1 - e^{-(\frac{m-n}{m})\lambda_2 h}}{(\frac{m-n}{m})\lambda_2}$:

$$\frac{x_{k+1} - x_k}{\phi(h)} = \lambda_1 x_k^{\frac{n}{m}} \cdot 1 - \lambda_2 x_k \cdot 1 \tag{4.17}$$

$$= \lambda_1 x_k^{\frac{n}{m}} \left[\left(\frac{m-n}{m}\right) \cdot \frac{\sum_{i=1}^{m} x_{k+1}^{\frac{(m-n)(m-i)}{m}} \cdot x_k^{\frac{(m-n)(i-1)}{m}}}{\sum_{i=1}^{m-n} x_{k+1}^{m-n-i} \cdot x_k^{\frac{n}{m}+i-1}} \right]$$

$$- \lambda_2 x_k \left[\left(\frac{m-n}{m}\right) \cdot \frac{\sum_{i=1}^{m} x_{k+1}^{\frac{(m-n)(m-i)}{m}} \cdot x_k^{\frac{(m-n)(i-1)}{m}}}{\sum_{i=1}^{m-n} x_{k+1}^{m-n-i} \cdot x_k^{\frac{n}{m}+i-1}} \right]. \tag{4.18}$$

This is equivalent to

$$\frac{x_{k+1} - x_k}{\phi(h)} = \left(\lambda_1 x_k^{\frac{n}{m}} - \lambda_2 x_k\right) \left[\left(\frac{m-n}{m}\right) \cdot \frac{\sum_{i=1}^{m} x_{k+1}^{\frac{(m-n)(m-i)}{m}} \cdot x_k^{\frac{(m-n)(i-1)}{m}}}{\sum_{i=1}^{m-n} x_{k+1}^{m-n-i} \cdot x_k^{\frac{n}{m}+i-1}} \right]. \tag{4.19}$$

Note that the Taylor series expansion of $\phi(h)$ is given by

$$\phi(h) \approx h - \frac{h^2(\lambda_2(m-n))}{2m} = h + O(h^2),$$

and therefore this is a valid denominator function satisfying Mickens rules for NSFD schemes [**17**]. One can see that if the discretization is fine enough, that is, $x_{k+1} \approx x_k$

then the expression $\left[\frac{m-n}{m}\cdot\frac{\sum_{i=1}^{m} x_{k+1}^{\left(\frac{(m-n)(m-i)}{m}\right)}\cdot x_k^{\left(\frac{(m-n)(i-1)}{m}\right)}}{\sum_{i=1}^{m-n} x_{k+1}^{(m-n-i)}\cdot x_k^{\left(\frac{n}{m}+i-1\right)}}\right]$ is an approximation of unity.

Next, we show that Equation (4.17) is equivalent to the exact discretization of the analytical solution in Equation (4.3). To do this we make use of the factorization in Equation (3.15) and Equation (3.16).

If we multiply the LHS of Equation (4.19) by the LHS of the Equation (3.18), and similarly we multiply the RHS of Equation (4.19) by the RHS of (3.18), we obtain

$$\frac{x_{k+1}-x_k}{\phi(h)}\cdot\frac{x_{k+1}^{(m-n)/m}-x_k^{(m-n)/m}}{x_{k+1}-x_k}=\left(\lambda_1 x_k^{\frac{n}{m}}-\lambda_2 x_k\right)\cdot \tag{4.20}$$

$$\left[\left(\frac{m-n}{m}\right)\cdot\frac{\sum_{i=1}^{m} x_{k+1}^{\frac{(m-n)(m-i)}{m}}\cdot x_k^{\frac{(m-n)(i-1)}{m}}}{\sum_{i=1}^{m-n} x_{k+1}^{m-n-i}\cdot x_k^{\frac{n}{m}+i-1}}\right]\cdot\left[\frac{\sum_{i=1}^{m-n} x_{k+1}^{m-n-i}x_k^{i-1}}{\sum_{i=1}^{m}\left(x_{k+1}^{(m-n)(m-i)/m}x_k^{(m-n)(i-1)/m}\right)}\right].$$

Note that we can factor out $x_k^{\frac{n}{m}}$ from the denominator of the first term in brackets as follows:

$$\frac{x_{k+1}^{(m-n)/m}-x_k^{(m-n)/m}}{\phi(h)}=\left(\lambda_1 x_k^{\frac{n}{m}}-\lambda_2 x_k\right)\left(\frac{m-n}{m}\right)\cdot \tag{4.21}$$

$$\left[\frac{\sum_{i=1}^{m} x_{k+1}^{\frac{(m-n)(m-i)}{m}}\cdot x_k^{\frac{(m-n)(i-1)}{m}}}{x_k^{\frac{n}{m}}\sum_{i=1}^{m-n} x_{k+1}^{m-n-i}\cdot x_k^{i-1}}\right]\cdot\left[\frac{\sum_{i=1}^{m-n} x_{k+1}^{m-n-i}x_k^{i-1}}{\sum_{i=1}^{m}\left(x_{k+1}^{(m-n)(m-i)/m}x_k^{(m-n)(i-1)/m}\right)}\right]. \tag{4.22}$$

The remaining the summations in Equation (4.21) cancel and we are left with the following

$$\begin{aligned}\frac{x_{k+1}^{(m-n)/m}-x_k^{(m-n)/m}}{\phi(h)}&=\left(\lambda_1 x_k^{\frac{n}{m}}-\lambda_2 x_k\right)\cdot\left(\frac{m-n}{m}\right)\cdot\frac{1}{x_k^{\frac{n}{m}}}\\&=\left(\frac{m-n}{m}\right)\left(\lambda_1-\lambda_2 x_k^{1-\frac{n}{m}}\right)\\&=\left(\frac{m-n}{m}\right)\cdot\lambda_1-\left(\frac{m-n}{m}\right)\lambda_2 x_k^{1-\frac{n}{m}}.\end{aligned} \tag{4.23}$$

Recalling that $\phi(h)=\dfrac{1-e^{-(\frac{m-n}{m})\lambda_2 h}}{(\frac{m-n}{m})\lambda_2}$, we solve Equation (4.23) for x_{k+1} which results in

$$x_{k+1}=\left(e^{-(\frac{m-n}{m})\lambda_2 h}x_k^{\frac{m-n}{m}}+\frac{\lambda_1}{\lambda_2}-\frac{\lambda_1}{\lambda_2}e^{-(\frac{m-n}{m})\lambda_2 h}\right)^{\frac{m}{m-n}}. \tag{4.24}$$

Note Equation (4.24) is equivalent to Equation (4.3) with $p=\dfrac{n}{m}$, and therefore we have shown that our proposed NSFD Scheme 4.2 that utilizes a unity approximation given in Equation (3.12) is an exact scheme for Equation (4.16).

5. Summary of results

In this paper we analyzed NSFD schemes generated through approximations of unity for various parameter values of the general BDE given below:

$$\frac{dx}{dt} = \lambda_1 x^p - \lambda_2 x, \quad x(0) = x_0 > 0, \quad \lambda_1\lambda_2 \neq 0, \quad p \in \mathbb{Q}, 0 < p < 1.$$

For the Simple Decay Equation, we used approximations of unity to produce second-order NSFD schemes (Scheme 2.1 and Scheme 2.2) which outperform basic SFD schemes while maintaining essential properties of the analytic solution to the ODE. Numerical verification of the second-order convergence of NSFD Schemes 2.1 and 2.2 are provided. This example shows the utility of the unity approximation idea to produce useful numerical methods that surpass SFD methods.

For the generalized Root Decay equation and the special case $p = \frac{1}{2}$ we generate an exact finite difference scheme (Scheme 3.2 and 3.1, respectively) for each case using an approximation of unity. Similarly, for the generalized BDE and the special case when $p = \frac{1}{2}$ we again generated an exact finite difference scheme (Scheme 4.2 and 4.1, respectively) for both cases using an approximation of unity.

We also separately conducted numerical experiments using many other unity approximations for the root decay and the root BDE which often resulted in second-order FD schemes but didn't maintain dynamic consistency so we decided not to include them in this paper.

Through our work presented here we have demonstrated that unity approximations are extremely useful. They can be used to (re)generate exact NSFD schemes previously found using other techniques and NSFD unity approximations can often be used to generate NSFD schemes that typically outperform SFD schemes.

Below, in Table 4 below we summarize the unity approximations that have appeared in this paper along with scheme they resulted in and the equation they were associated with.

TABLE 4. Summary of Unity Approximations in this Chapter

Name	Unity Approximation	Scheme	Exact Status
Simple Decay	$\frac{x_{k+1}+x_{k+1}}{x_{k+1}+x_k}$	2.1	NO
Simple Decay	$\frac{x_{k+1}+x_k}{2x_k}$	2.2	NO
Root Decay	$\frac{\sqrt{x_{k+1}}+\sqrt{x_k}}{2\sqrt{x_k}}$	3.1	YES
Generalized Root Decay	$\frac{m-n}{m}\frac{\sum_{i=1}^{m} x_{k+1}^{\frac{(m-n)(m-i)}{m}} x_k^{\frac{(m-n)(i-1)}{m}}}{\sum_{i=1}^{m-n} x_{k+1}^{m-n-i} x_k^{\frac{n}{m}+i-1}}$	3.2	YES
Root BDE	$\frac{\sqrt{x_{k+1}}+\sqrt{x_k}}{2\sqrt{x_k}}$	4.1	YES
Generalized BDE	$\frac{m-n}{m}\frac{\sum_{i=1}^{m} x_{k+1}^{\frac{(m-n)(m-i)}{m}} x_k^{\frac{(m-n)(i-1)}{m}}}{\sum_{i=1}^{m-n} x_{k+1}^{m-n-i} x_k^{\frac{n}{m}+i-1}}$	4.2	YES

6. Future work

There are many other interesting autonomous ODEs unrelated to BDEs, such as systems of nonlinear ODEs, that could be the subject of future work where we generate NSFD schemes using unity approximations and investigate how (or whether) they outperform basic SFD schemes.

Although this chapter mostly focused on the use of unity NSFD approximations to regenerate exact FD schemes, the authors have also been investigating many other ODEs, typically ones where there is an exact solution known but NOT a known exact FD scheme. In these cases one can very often show the unity-based NSFD schemes generate solutions that outperform SFD schemes. Our summer 2022 cohort of undergraduate research students looked at NSFD schemes for problems of the form $\frac{dx}{dt} = -x^3 + F(x)$ where $F(x)$ is a low-degree polynomial. In fact, future work could include explorations of the use of NSFD schemes generated from unity approximations applied to other related ODEs, such as $\frac{dx}{dt} = \lambda x^p + F(x)$. Note that when $F(x) = x$ this is a BDE but $p > 1$ and $p \neq 3$ would include cases that we have not yet examined.

References

[1] Quang A. Dang and Manh Tuan Hoang, *Nonstandard finite difference schemes for a general predator-prey system*, J. Comput. Sci. **36** (2019), 101015, 14, DOI 10.1016/j.jocs.2019.07.002. MR3987494

[2] Ronald E. Mickens, *An exactly solvable model for the spread of disease*, College Math. J. **43** (2012), no. 2, 114–121, DOI 10.4169/college.math.j.43.2.114. MR2897472

[3] Lih-Ing W. Roeger, *Nonstandard finite-difference schemes for the Lotka-Volterra systems: generalization of Mickens's method*, J. Difference Equ. Appl. **12** (2006), no. 9, 937–948, DOI 10.1080/10236190600909380. MR2262331

[4] A. R. Yaghoubi and H. Saberi Najafi, *Non-standard finite difference schemes for investigating stability of a mathematical model of virus therapy for cancer*, Appl. Appl. Math. **14** (2019), no. 2, 805–819. MR4044146

[5] S. M. Moghadas, M. E. Alexander, and B. D. Corbett, *A non-standard numerical scheme for a generalized Gause-type predator-prey model*, Phys. D **188** (2004), no. 1-2, 134–151, DOI 10.1016/S0167-2789(03)00285-9. MR2045311

[6] Deqiong Ding, Qiang Ma, and Xiaohua Ding, *A non-standard finite difference scheme for an epidemic model with vaccination*, J. Difference Equ. Appl. **19** (2013), no. 2, 179–190, DOI 10.1080/10236198.2011.614606. MR3021780

[7] A. B. Gumel, S. M. Moghadas, and R. E. Mickens, *Effect of a preventive vaccine on the dynamics of HIV transmission*, Commun. Nonlinear Sci. Numer. Simul. **9** (2004), no. 6, 649–659, DOI 10.1016/S1007-5704(03)00024-8. MR2070525

[8] Ron Buckmire, *Applications of Mickens finite differences to several related boundary value problems*, Advances in the applications of nonstandard finite difference schemes, World Sci. Publ., Hackensack, NJ, 2005, pp. 47–87, DOI 10.1142/9789812703316_0003. MR2212198

[9] Ron Buckmire, Karl McMurtry, and Ronald E. Mickens, *Numerical studies of a nonlinear heat equation with square root reaction term*, Numer. Methods Partial Differential Equations **25** (2009), no. 3, 598–609, DOI 10.1002/num.20361. MR2510749

[10] K. C. Patidar, *Nonstandard finite difference methods: recent trends and further developments*, J. Difference Equ. Appl. **22** (2016), no. 6, 817–849, DOI 10.1080/10236198.2016.1144748. MR3520261

[11] Ronald E. Mickens, *Nonstandard finite difference schemes for differential equations*, J. Difference Equ. Appl. **8** (2002), no. 9, 823–847, DOI 10.1080/1023619021000000807. In honor of Professor Allan Peterson on the occasion of his 60th birthday. MR1919887

[12] Michael Chapwanya, Jean M.-S. Lubuma, and Ronald E. Mickens, *Positivity-preserving nonstandard finite difference schemes for cross-diffusion equations in biosciences*, Comput. Math. Appl. **68** (2014), no. 9, 1071–1082, DOI 10.1016/j.camwa.2014.04.021. MR3269960

[13] Jacky Cresson and Frédéric Pierret, *Non standard finite difference scheme preserving dynamical properties*, J. Comput. Appl. Math. **303** (2016), 15–30, DOI 10.1016/j.cam.2016.02.007. MR3479269

[14] Dobromir T. Dimitrov and Hristo V. Kojouharov, *Dynamically consistent numerical methods for general productive-destructive systems*, J. Difference Equ. Appl. **17** (2011), no. 12, 1721–1736, DOI 10.1080/10236191003781947. MR2854820

[15] Ronald E. Mickens and P. M. Jordan, *A new positivity-preserving nonstandard finite difference scheme for the DWE*, Numer. Methods Partial Differential Equations **21** (2005), no. 5, 976–985, DOI 10.1002/num.20073. MR2154229

[16] Ronald E. Mickens, *Numerical integration of population models satisfying conservation laws: NSFD methods*, J. Biol. Dyn. **1** (2007), no. 4, 427–436, DOI 10.1080/17513750701605598. MR2363352

[17] Ronald E. Mickens, *Nonstandard finite difference models of differential equations*, World Scientific Publishing Co., Inc., River Edge, NJ, 1994. MR1275372

[18] Ronald E. Mickens and Arthur Smith, *Finite-difference models of ordinary differential equations: influence of denominator functions*, J. Franklin Inst. **327** (1990), no. 1, 143–149, DOI 10.1016/0016-0032(90)90062-N. MR1032894

[19] Ronald E. Mickens (ed.), *Applications of nonstandard finite difference schemes*, World Scientific Publishing Co., Inc., River Edge, NJ, 2000. Papers from the Minisymposium on Nonstandard Finite Difference Schemes: Theory and Applications held at the SIAM Annual Meeting, Atlanta, GA, 1999, DOI 10.1142/9789812813251. MR1831401

[20] Ronald E. Mickens (ed.), *Advances in the applications of nonstandard finite difference schemes*, World Scientific Publishing Co. Pte. Ltd., Hackensack, NJ, 2005, DOI 10.1142/9789812703316. MR2213290

[21] Ronald E. Mickens, *Difference equations: Theory, applications and advanced topics*, 3rd ed., Monographs and Research Notes in Mathematics, CRC Press, Boca Raton, FL, 2015. MR3362807

[22] Ronald E. Mickens, *Difference equation models of differential equations*, Math. Comput. Modelling **11** (1988), 528–530, DOI 10.1016/0895-7177(88)90549-3. Mathematical modelling in science and technology (St. Louis, MO, 1987). MR959933

[23] Lih-Ing Wu Roeger and Ronald E. Mickens, *Exact finite difference and non-standard finite difference schemes for* $\mathrm{d}x/\mathrm{d}t = -\lambda x^\alpha$, J. Difference Equ. Appl. **18** (2012), no. 9, 1511–1517, DOI 10.1080/10236198.2011.574622. MR2974134

[24] *Advances in the Applications of Nonstandard Finite Difference Schemes*. Chapter 12, World Scientific: Singapore (2005)

[25] K. F. Gurski, *A simple construction of nonstandard finite-difference schemes for small nonlinear systems applied to SIR models*, Comput. Math. Appl. **66** (2013), no. 11, 2165–2177, DOI 10.1016/j.camwa.2013.06.034. MR3125364

[26] Lisha Wang and L.-I. W. Roeger, *Nonstandard finite difference schemes for a class of generalized convection-diffusion-reaction equations*, Numer. Methods Partial Differential Equations **31** (2015), no. 4, 1288–1309, DOI 10.1002/num.21951. MR3343609

Mathematics Department, Occidental College, 1600 Campus Road, Los Angeles, California 90041

Email address: basu@oxy.edu

Mathematics Department, Occidental College, 1600 Campus Road, Los Angeles, California 90041

Email address: ron@oxy.edu

Mathematics Department, Boston College, 140 Commonwealth Ave, Chestnut Hill, Massachusetts 02467

Email address: coovadia.zaheer@gmail.com

Mathematics Department, Heritage University, 3240 Fort Road, Toppenish, Washington 98948

Email address: diazmayra137@gmail.com

Mathematics Department, California State Polytechnic University, Pomona, California 91768

Email address: dainiguez@cpp.edu

Department of Mathematics, Tufts University, 419 Boston Ave, Medford, Massachusetts 02155

Email address: alexandra.scott29@gmail.com

Contemporary Mathematics
Volume **793**, 2024
https://doi.org/10.1090/conm/793/15903

On the structure and evolution of poroacoustic solitary waves: Finite-time gradient catastrophe under the Darcy–Jordan model

C. L. Shimp, J. V. Lambers, and P. M. Jordan

This chapter is dedicated to Prof. Ronald E. Mickens on the occasion of his 80th birthday

ABSTRACT. We revisit and extend the poroacoustic-based investigation carried out by Jordan in [Wave Motion **94** (2020), 102498], wherein the topic of wave overturning was the central focus. In particular, we develop a numerical method that accurately solves the Cauchy problem for the 1D, uni-directional, Darcy–Jordan model, even after the solution profile becomes multi-valued. Through the analysis of the behavior of the solution over time, the spatial discretization is adapted so that all solution values at each point (x, t) in the space-time grid can be computed through targeted application of root-finding methods, and without any knowledge of an analytical solution. The effectiveness of the method is then demonstrated using a variety of physically-relevant initial conditions.

1. Introduction

In his 2020 study of poroacoustic solitary waves propagating in homogeneous gases, Jordan [**1**] considered the (dimensionless) initial-value problem (IVP):

$$u_t + (1 + \epsilon\beta u)u_x + \tfrac{1}{2}\delta u = 0, \quad x \in \mathbb{R}, \quad t > 0, \tag{1a}$$

$$u(x, 0) = u_0(x), \quad x \in \mathbb{R}, \tag{1b}$$

i.e., the Cauchy problem on $\mathbb{R}$ for a special case of the damped Riemann equation, for the following pulse-type initial conditions (IC)s:

$$u_0(x) = \begin{cases} \dfrac{\varkappa^2}{\varkappa^2 + x^2}, & \text{Lorentzian,} \\ \\ \exp(-\varkappa|x|), & \text{symmetric-exponential,} \end{cases} \tag{2}$$

where $\varkappa(> 0)$ is a free (dimensionless) parameter.

As shown in Ref. [**1**, § 2], Eq. (1a) is the *right-running* approximation to the weakly-nonlinear equation of motion [**2**, Eq. (2.16)]

$$\phi_{xx} - \phi_{tt} - \delta\phi_t = \epsilon\partial_t[(\phi_x)^2 + (\beta - 1)(\phi_t)^2], \tag{3}$$

Key words and phrases. Cauchy problem, finite-amplitude poroacoustics, gradient catastrophe, Kepler's equation, solitary waves.

J.V.L. and P.M.J. were supported by U.S. Office of Naval Research (ONR) funding.

The third author is the corresponding author.

which follows on applying the finite-amplitude approximation scheme to the 1D version of the poroacoustic formulation that has come to be known as the Darcy–Jordan model[1] (DJM). In Eq. (3), which is expressed in terms of the dimensionless variables defined in Ref. [**2**], $\phi = \phi(x,t)$ is the (1D) scalar velocity potential, which is related to $u = u(x,t)$ via $u = \phi_x$, where the velocity vector is assumed to be of the form $\mathbf{u} = (u(x,t), 0, 0)$; the coefficient of nonlinearity [**5**] is given by

$$\beta = \begin{cases} (\gamma + 1)/2, & \text{gases}, \\ 1 + B/(2A), & \text{liquids}, \end{cases} \tag{4}$$

where γ denotes the ratio of specific heats of the fluid phase when the fluid under consideration is a gas (see Section 2.1 below) and B/A denotes the parameter of nonlinearity [**5**]; $\epsilon \propto 1/c_e$ is the Mach number, where $c_e(> 0)$ denotes the (constant) speed of sound in the undisturbed fluid [**6**]; and $\delta \propto \chi\nu/K$ is the dimensionless Darcy coefficient, where the constants $\chi \in (0,1)$ and $K(> 0)$ denote, respectively, the porosity and permeability of the porous solid and $\nu(> 0)$ is the (constant) kinematic viscosity of the fluid.

It should be noted that the derivation of Eq. (3) is based, in part, on the assumptions $0 < \epsilon \ll 1$ and $\delta \sim \mathcal{O}(\epsilon)$, which are central to the application of the finite-amplitude approximation scheme in Ref. [**2**], and the neglect of nonlinear terms of $\mathcal{O}(\epsilon^2)$. Going further in simplifying, the linearization of Eq. (3) is achieved by neglecting *only* nonlinear terms of $\mathcal{O}(\epsilon)$, i.e., setting $\epsilon := 0$. This final order-of-magnitude based approximation reduces Eq. (3) to the well known *damped wave equation*, a PDE which describes a wide range of natural phenomena; see, e.g., Ref. [**7**] and those cited therein.

By employing the method of characteristics (see Section 2.2 below), Jordan was able to obtain exact, explicit, solutions, valid for all $t > 0$, for both cases of u_0 considered. As such, he was able to analytically examine the (mathematical) process of "wave overturning", by which a single-valued solution of a nonlinear IVP becomes *multi-valued*; see, e.g., Ref. [**8**, pp. 37–46]. The process of wave overturning occurs after the solution profile in question suffers finite-time *gradient catastrophe*; see Section 2.3 below. In the case of IVP (1), this means the development of a vertical tangent at a point on the leading (i.e., right-hand) side of the profile at some time $t = t^\star(> 0)$. Said behavior will always be *possible* in the cases studied below since only positive, pulse-type ICs, for which $|u_0'(x)|$ is bounded on the entire real line, are considered. [In this chapter, a prime denotes the derivative of a function of a single (real) variable.] For compact ICs, this means that $u_0(x) > 0$, $|u_0'(x)| \leq M_b$ for all $x \in \operatorname{supp}(u_0)$, where $M_b(> 0)$ is some constant and $\operatorname{supp}(\mathfrak{F})$ denotes the set-theoretic support of the function $\mathfrak{F}$.

From the physical standpoint, the primary value of determining the multi-valued solution is so that the *equal-area rule* (EAR) can be applied to calculate the resulting shock wave's location along the x-axis. Knowing the shock's location as a function of t not only allows us to perform *shock fitting* [**8**, § 2.8], the purpose of which is to replace the multi-valued waveform with one that is discontinuous but *single-valued*, but it also allows us to determine the shock's velocity, knowledge of which is necessary (but *not* sufficient) if an expression for the shock's amplitude is to be derived; see Section 3.1 below, wherein the mathematical simplicity of the

[1] Also known to as the "Jordan–Darcy" model; see Ref. [**3**], as well as Ref. [**4**, § 8.1].

resulting solution profile allows the aforementioned quantities to be determined analytically. Unfortunately, when it comes to yielding solutions that can be expressed *both* exactly and explicitly, for all $t > 0$, very few ICs are as amenable as those in Eq. (2); i.e., for only a small subset of physically relevant ICs are we able to analytically examine the process of wave overturning. Consequently, we are forced to turn to numerical methods if progress is to be made in the case of, e.g., the cos-pulse and Gaussian ICs.

Generally, a numerical method for solving an IVP such as ours produces a solution that, at a fixed time, is a function defined on a spatial grid. Unfortunately, once gradient catastrophe occurs, the exact solution, at a fixed time, is no longer a function of the spatial variable, as its graph does not pass the vertical line test. By contrast, the numerical algorithm presented in this paper computes solutions that account for this overturning effect. This is accomplished by seeking all solutions of an equation that characterizes the wave variable, which arises from applying the method of characteristics. The process of finding all solution values for a given (x, t) combines the efficiency of the secant method and the reliability of bisection. With each point in the spatial grid, we associate an interval within which this root-finding process is performed, which is updated at each time step in such a way as to ensure that no roots are missed in the case of gradient catastrophe, while also preserving efficiency. The spatial grid itself is also updated, so that grid points are concentrated in areas of interest.

In the investigation carried out below, our primary aims are three-fold: introduce, test, and then utilize the aforementioned numerical algorithm. In achieving such, we computationally extend the analysis performed in Refs. [**1**] and [**9**] to the case of other, physically relevant, ICs that also yield multi-valued solutions—solutions whose mathematical complexity, however, precludes any form of in-depth treatment by currently available analytical methods/tools.

To these ends, the present chapter is organized as follows. In Section 2, we review the physics of perfect gases, the method of characteristics as applied to IVP (1), and the mathematics of jump of phenomena. In Section 3, we present/examine explicit, analytically derived, solutions to IVP (1) that evolve from certain pulse-type ICs. Then, in Section 4, details on our numerical scheme and its construction are presented. In Section 5, we employ our scheme to numerically generate post-gradient catastrophe solution profiles based on the ICs given in Refs. [**1**] and [**9**] and those considered in Section 3; we also compare/contrast our scheme with/to one from the WENO family. Lastly, in Section 6, our main result is summarized and possible extensions of the present analysis are proposed.

REMARK 1. We note for later reference that Eq. (1a) can also be expressed as the (dimensionless) scalar balance law

$$u_t + \mathfrak{q}_x = -\tfrac{1}{2}\delta u, \tag{5}$$

where the associated flux is given by

$$\mathfrak{q} = u + \tfrac{1}{2}\epsilon\beta u^2; \tag{6}$$

see also Ref. [**1**]. Here, we observe that $\mathfrak{q}$ is strongly convex, which means that it is also *strictly convex* [**10**], with strong convexity constant $\epsilon\beta$. The fact that $\mathfrak{q}$ is strictly convex is a critical element in ensuring that the inequality stated in Eq. (20) below is satisfied under all of the ICs we shall consider.

2. Physical and mathematical preliminaries

2.1. Perfect gases: a review. With the exception of γ, which is rendered dimensionless by way of its defining relation, all variables and parameters discussed in this subsection are expressed in *dimensional* form.

In the case of a *perfect gas*, to which we hereafter limit our focus, p, ρ, and θ satisfy the following special case of the *ideal gas law* [**11**, § 2.5]:

$$p = (c_p - c_v)\rho\theta \qquad (c_p, c_v := \text{const.}). \tag{7}$$

Here, $p(>0)$ is the thermodynamic pressure; $\rho(>0)$ is the mass density of the gas; $\theta(>0)$ is the absolute temperature; $c_p > c_v > 0$ are the specific heats at constant pressure and volume, respectively; and $\gamma = c_p/c_v$. In addition, the thermodynamic variable $c(>0)$, which represents the instantaneous speed of sound in the gas, is given by $c = \sqrt{\gamma p/\rho}$ in the case of perfect gases, and $\gamma \in (1, 5/3]$, where for the *diatomic* sub-class of perfect gases (e.g., air), $\gamma = 1.40(\implies \beta = 1.2)$ [**5, 11**].

It should be noted that IVP (1) was formulated under the assumption that the gas under consideration is in its equilibrium state (i.e., is *homogeneous-quiescent* [**6**, p. 14]) at $x = \mp\infty$. That is, while p, ρ, and θ, assume their respective equilibrium state values p_e, ρ_e, and θ_e, all of which are *positive* constants related via Eq. (7), at $x = \mp\infty$, u, on the other hand, is *zero* at $x = \mp\infty$. It should also be noted that c_e is given by $c_e = \sqrt{\gamma p_e/\rho_e}$ in the case of perfect gases.

2.2. General solution via the method of characteristics. By applying the method of characteristics (see, e.g., Ref. [**12**, p. 196]), one finds that the exact solution of IVP (1) can always be expressed as

$$u(x,t) = u_0(\xi)\exp(-\delta t/2), \tag{8a}$$

$$\eta - \xi = (\alpha^\star)^{-1}u_0(\xi)g(t). \tag{8b}$$

In this pair of coupled (transcendental) equations, $\xi = \xi(x,t)$ is the wave variable, we let $\eta = x - t$,

$$g(t) := 1 - \exp(-\delta t/2), \tag{9}$$

and

$$\alpha^\star = \frac{\delta}{2\epsilon\beta}, \tag{10}$$

where $\alpha^\star \in (0,1)$ is the critical amplitude value for poroacoustic acceleration waves under the DJM [**2**].

To obtain an *explicit* (exact) solution to IVP (1), it is necessary and sufficient that Eq. (8a) be solvable for ξ in terms of x and t. As shown in Ref. [**9**], however, not all explicit solutions remain valid/applicable beyond the time of gradient catastrophe; see also Section 3.2 below.

2.3. Gradient catastrophe, jump definition, and shock conditions. In the present (1D) setting, gradient catastrophe [**13**] refers to the situation in which u_x blows up at some particular time, which we denote by $t^\star$; here, i.e., in the case of IVP (1),

$$u_x(x,t) = \frac{u_0'(\xi)\exp(-\delta t/2)}{1 + u_0'(\xi)g(t)/\alpha^\star}. \tag{11}$$

The physical interpretation of this blow-up is that a (right-propagating) *shock wave*, i.e., a jump in u, has begun to form at time $t = t^\star$; see, e.g., Refs. [**8**, **14**].

In this chapter, we define the amplitude of the jump in a function $\mathfrak{F} = \mathfrak{F}(x,t)$ across a propagating *singular surface*, e.g., the plane $x = \mathfrak{S}(t)$, as

$$[\![\mathfrak{F}]\!](t) := \mathfrak{F}^- - \mathfrak{F}^+; \tag{12}$$

here, employing the usual notation convention of singular surface theory [**15**, § 173], we have set

$$\mathfrak{F}^\mp := \lim_{x \to \mathfrak{S}(t)\uparrow\downarrow} \mathfrak{F}(x,t), \tag{13}$$

respectively, where both limits are assumed to exist. Physically, $x = \mathfrak{S}(t)$ represents a *wavefront* propagating to the right along, and perpendicular to, the x-axis; the velocity of this wavefront along the x-axis is, of course, given by

$$V(t) = \frac{\mathrm{d}\mathfrak{S}(t)}{\mathrm{d}t}, \tag{14}$$

where $V(t) > 0$ follows from the right-running-only nature of $\mathfrak{S}(t)$'s motion.

If $t^\star \in \mathbb{R}^+$, then, post gradient catastrophe, u and $\mathfrak{q}$ satisfy

$$[\![\mathfrak{q}]\!]_{\text{sh}}(t) = V_{\text{sh}}(t)[\![u]\!]_{\text{sh}}(t) \qquad (t > t^\star), \tag{15}$$

which of course is the (1D) *Rankine–Hugoniot condition*; see, e.g., Refs. [**14**] and [**16**, § 6.2]. Here and below, we employ a "sh" subscript to denote a shock-related quantity; i.e., a quantity that is of relevance only for $t > t^\star$.

In the case of Eq. (1a), it is easily established, on taking jumps in Eq. (6), that

$$[\![\mathfrak{q}]\!]_{\text{sh}}(t) = [\![u]\!]_{\text{sh}}(t) + \tfrac{1}{2}\epsilon\beta\left\{2u^+[\![u]\!]_{\text{sh}}(t) + [\![u]\!]^2_{\text{sh}}(t)\right\}, \tag{16}$$

where the quadratic term has been simplified using the formula for the jump of a product, viz.,

$$[\![\mathfrak{F}\mathfrak{G}]\!] = \mathfrak{F}^+[\![\mathfrak{G}]\!] + \mathfrak{G}^+[\![\mathfrak{F}]\!] + [\![\mathfrak{F}]\!][\![\mathfrak{G}]\!]. \tag{17}$$

If $x = \mathfrak{S}_{\text{sh}}(t)$ can be determined, say, via the EAR (see, e.g., Refs. [**12**, § 5] and [**8**, **14**]), then $V_{\text{sh}}(t)$ follows from Eq. (14); thus, from Eqs. (15) and (16), the shock amplitude we seek is the *nonzero* solution of the quadratic

$$(V_{\text{sh}}(t) - 1)[\![u]\!]_{\text{sh}}(t) - \tfrac{1}{2}\epsilon\beta\left\{2u^+[\![u]\!]_{\text{sh}}(t) + [\![u]\!]^2_{\text{sh}}(t)\right\} = 0, \tag{18}$$

assuming of course that u^+ is also known. If $u^+ = 0$, then our shock amplitude is given by the particularly simple expression

$$[\![u]\!]_{\text{sh}}(t) = \frac{2\,(V_{\text{sh}}(t) - 1)}{\epsilon\beta} \qquad (u^+ = 0). \tag{19}$$

And, along with the Rankine–Hugoniot condition, all shocks studied below satisfy that *Lax entropy condition* [**10**], which herein takes the form

$$[\![u]\!]_{\text{sh}}(t) > 0; \tag{20}$$

consequently, all are admissible from the physical standpoint.

2.4. Acceleration waves and higher order discontinuities. If the IC under consideration is at best C^0-smooth (e.g., the symmetric-exponential case), then we will have, for $t \in (0, t^\star)$ at *minimum*, $[\![u_x]\!](t) \neq 0$ across at least one surface. When this occurs, the (1D) *Maxwell compatibility condition* tells us that u_t also suffers a jump across the same surface(s) and, moreover, that, at each such surface, the two jumps are related via

$$[\![u_t]\!](t) = -V(t)[\![u_x]\!](t). \tag{21}$$

This result, it should be noted, follows from *Hadamard's lemma* [**4**] when $[\![u]\!](t) =$ const., in particular, zero, across said surface(s); see also Refs. [**15**, § 180] and [**16**, § 6.9]. If $[\![u_x]\!](t), [\![u_t]\!](t) \neq 0$ across a surface $x = \mathfrak{S}(t)$, then this surface is termed an *acceleration wave* [**15**, § 190]; see also Straughan [**4**, § 8.1.3], who discusses such jumps in the present (i.e., poroacoustic) context.

Lastly, we shall refer to waves in which $[\![u]\!](t) = [\![u_x]\!](t) = 0$, but $[\![u_{xx}]\!](t)$, $[\![u_{xt}]\!](t)$, etc., are *nonzero*, as second order *weaker waves*, where in this regard we have adopted the terminology of Lindsay and Straughan [**17**, § 10]. It should be noted that these weaker waves also satisfy the compatibility condition; e.g., $[\![u_x]\!](t) = 0$ implies that $[\![u_{tx}]\!](t) = -V(t)[\![u_{xx}]\!](t)$.

3. Selected ICs that yield explicit solutions

All numerical results presented in this section were computed/plotted using the software package MATHEMATICA (ver. 11.2).

3.1. Compact, piecewise linear IC: the witch's hat. Consider first the simplest possible (C^0-smooth) pulse-type IC:

$$u_0(\xi) = \begin{cases} 1 - \varkappa|\xi|, & |\xi| < 1/\varkappa, \\ 0, & |\xi| \geq 1/\varkappa, \end{cases} \tag{22}$$

where $\varkappa(> 0)$ is a free parameter. Under this IC, the functional form of which is known by a number of different names [**18**, pp. 250, 254],

$$t^\star = -2\delta^{-1} \ln(1 - \alpha^\star/\varkappa). \tag{23}$$

Since $t^\star \in \mathbb{R}^+$ only for $\alpha^\star < \varkappa$, we shall assume that $\alpha^\star < \varkappa$ holds throughout the present subsection.

It is readily shown, on substituting Eq. (22) into Sys. (8), that the exact, explicit, *tri-valued* solution of IVP (1) is given by

$$u(x,t) = \exp(-\delta t/2)\hat{u}(x,t) \qquad (t > 0), \tag{24}$$

where in the present subsection we have

$$
(25)\quad \hat{u}(x,t) = \\
= \begin{cases} \begin{cases} 0, & x \le t - 1/\varkappa, \\ 1 + \dfrac{\varkappa(x-t) - (\varkappa/\alpha^\star)g(t)}{1 + (\varkappa/\alpha^\star)g(t)}, & t - 1/\varkappa < x < t + g(t)/\alpha^\star, \\ 1, & x = t + g(t)/\alpha^\star, \\ 1 - \dfrac{\varkappa(x-t) - (\varkappa/\alpha^\star)g(t)}{1 - (\varkappa/\alpha^\star)g(t)}, & t + 1/\varkappa > x > t + g(t)/\alpha^\star, \\ 0, & x \ge t + 1/\varkappa, \end{cases} & t \in (0, t^\star), \\ \begin{cases} 0, & x \le t - 1/\varkappa, \\ 1 + \dfrac{\varkappa(x-t) - (\varkappa/\alpha^\star)g(t)}{1 + (\varkappa/\alpha^\star)g(t)}, & t - 1/\varkappa < x < t + g(t)/\alpha^\star, \\ 1, & x = t + g(t)/\alpha^\star, \\ 1 + \dfrac{\varkappa(x-t) - (\varkappa/\alpha^\star)g(t)}{|1 - (\varkappa/\alpha^\star)g(t)|}, & t + g(t)/\alpha^\star > x > t + 1/\varkappa, \\ 0, & x \ge t + 1/\varkappa, \end{cases} & t > t^\star. \end{cases}
$$

As noted above, and shown (numerically) in Ref. [**1**], use of the EAR allows us to determine the shock's location, i.e., $x = \mathfrak{S}_{\rm sh}(t)$, along the x-axis. Here, this is accomplished by solving the following integral equation for $\mathfrak{S}_{\rm sh}$:

$$
(26)\quad \int_{t+1/\varkappa}^{\mathfrak{S}_{\rm sh}(t)} \hat{u}(x,t)\Big|_{\rm Green} \mathrm{d}x = \int_{\mathfrak{S}_{\rm sh}(t)}^{t+g(t)/\alpha^\star} \hat{u}(x,t)\Big|_{\rm Blue} \mathrm{d}x \\
- \int_{\mathfrak{S}_{\rm sh}(t)}^{t+g(t)/\alpha^\star} \hat{u}(x,t)\Big|_{\rm Green} \mathrm{d}x \quad (t > t^\star),
$$

which can be immediately reduced to

$$
(27)\quad \int_{t+1/\varkappa}^{t+g(t)/\alpha^\star} \hat{u}(x,t)\Big|_{\rm Green} \mathrm{d}x = \int_{\mathfrak{S}_{\rm sh}(t)}^{t+g(t)/\alpha^\star} \hat{u}(x,t)\Big|_{\rm Blue} \mathrm{d}x \qquad (t > t^\star);
$$

see Fig. 1 for the references to color. As in Ref. [**1**][2], the quantity on which our EAR calculation is based is the (dimensionless) momentum density, as approximated under the finite-amplitude scheme [**12**, § 5].

Omitting the details, it is a straightforward matter to solve[3] Eq. (27) and show that, under Eq. (22),

$$
(28)\quad \mathfrak{S}_{\rm sh}(t) = (t - 1/\varkappa) + \varkappa^{-1}\sqrt{2(1 + \varkappa g(t)/\alpha^\star)} \qquad (t > t^\star);
$$

see the red line (i.e., the shock-front) in Fig. 1, and note that the two small triangles formed by its "fitting" are of *equal* area. Here, we observe that

$$
(29)\quad t + 1/\varkappa < \mathfrak{S}_{\rm sh}(t) < t + g(t)/\alpha^\star \qquad (t > t^\star),
$$

where we note that $\mathfrak{S}_{\rm sh}(t)$ does not exist for $t \in (0, t^\star)$. We also have

$$
(30)\quad V_{\rm sh}(t) = 1 + \frac{\epsilon\beta\exp(-\delta t/2)}{\sqrt{2(1 + \varkappa g(t)/\alpha^\star)}} \qquad (t > t^\star),
$$

[2]With regard to the use of the EAR in Ref. [**1**], see also Ref. [**9**, § 4, Item (ii)].

[3]While we are able to determine $\mathfrak{S}_{\rm sh}(t)$ analytically under Eq. (22), as alluded to in Section 1, it must be stressed that ICs under which this is possible are the rare exception, not the norm.

which easily follows on substituting Eq. (28) into Eq. (14).

On substituting Eq. (30) into Eq. (19), taking note of the fact that $u^+ = 0$ under the present IC, we find, after simplifying, that

$$\llbracket u \rrbracket_{\text{sh}}(t) = \frac{2\exp(-\delta t/2)}{\sqrt{2(1+\varkappa g(t)/\alpha^\star)}} \qquad (t > t^\star), \tag{31}$$

recalling that $\llbracket u \rrbracket_{\text{sh}}(t)$ occurs across $x = \mathfrak{S}_{\text{sh}}(t)$.

From the physical standpoint, then, the $t > t^\star$ case of Eq. (25) is to be replaced by the shock profile

$$\begin{aligned} u_{\text{sh}}(x,t) &= \exp(-\delta t/2) \\ &\times \begin{cases} 0, & x \le t - 1/\varkappa, \\ 1 + \dfrac{\varkappa(x-t) - (\varkappa/\alpha^\star)g(t)}{1+(\varkappa/\alpha^\star)g(t)}, & t - 1/\varkappa < x < \mathfrak{S}_{\text{sh}}(t), \quad t > t^\star, \\ 0, & x > \mathfrak{S}_{\text{sh}}(t), \end{cases} \end{aligned} \tag{32}$$

a discontinuous but singled-valued waveform for which

$$\sup_{x\in\mathbb{R}}[u_{\text{sh}}(x,t)] = \llbracket u \rrbracket_{\text{sh}}(t) \qquad (t > t^\star). \tag{33}$$

where $\sup(\cdot)$ denotes the supremum of its argument.

REMARK 2. For $t \in (0, t^\star)$, Eq. (25) admits three acceleration waves (recall Section 2.4); they occur across the surfaces $x = t \mp 1/\varkappa$ and $x = t + g(t)/\alpha^\star$. For $t > t^\star$, i.e., under Eq. (32), only the one across $x = t - 1/\varkappa$ remains; its amplitude is given by

$$\llbracket u_x \rrbracket(t)\Big|_{x=t-1/\varkappa} = -\frac{\varkappa\exp(-\delta t/2)}{1+(\varkappa/\alpha^\star)g(t)} \qquad (t > 0), \tag{34}$$

while its velocity is $V(t) = 1$. Qualitatively, the "fate" of the acceleration wave given by Eq. (34) is the same as that of its counterpart $x = \mathfrak{S}_1(t)$ in Ref. [**9**, § 2.1], wherein the cos-pulse IC is considered.

REMARK 3. For $t \in (0, t^\star)$, as exemplified by Eq. (25), the C- (i.e., differentiability) class of a given solution profile is the same as that of the IC from which it arose. For $t > t^\star$, however, all solution profiles exhibit a (propagating) jump discontinuity, i.e., a shock wave, as exemplified by Eq. (32).

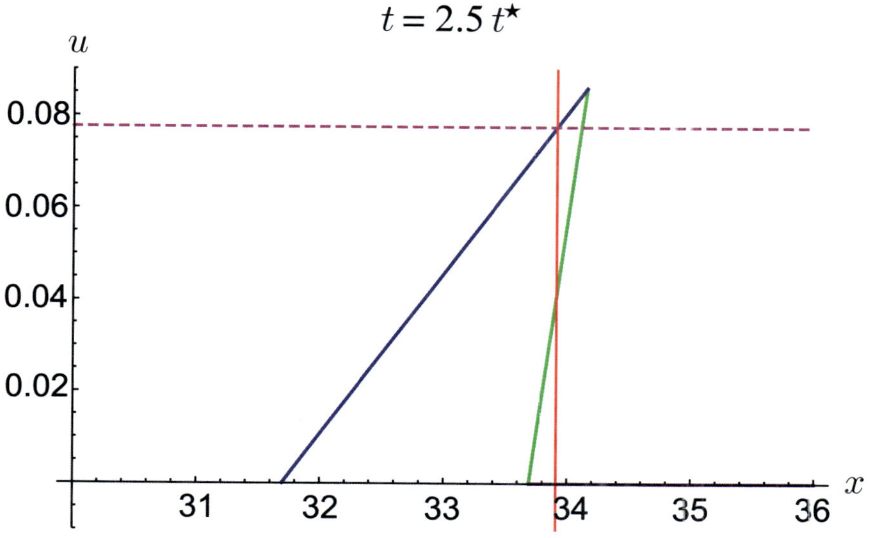

FIGURE 1. Plot of u vs. x based on the $t > t^\star$ case of Eq. (25), generated using $t = 2.5t^\star$, $\varkappa = 1.0$, $\epsilon = 0.1$, $\beta = 1.2$, and $\delta = 0.15$; here, $\alpha^\star = 0.625$ and $t^\star \approx 13.0777$. Black half-line: $x \leq t - 1/\varkappa$. Blue segment: $t-1/\varkappa < x \leq t+g(t)/\alpha^\star$. Green segment: $t+1/\varkappa < x < t + g(t)/\alpha^\star$. Purple half-line: $x \geq t + 1/\varkappa$. Red line: $x = \mathfrak{S}_{\rm sh}(2.5t^\star) \approx 33.9134$. Magenta dashed-line: $[\![u]\!]_{\rm sh}(2.5t^\star) \approx 0.0776$.

3.2. C^1-smooth case: $\cos^2$**-pulse.** Consider next the C^1-smooth IC

$$u_0(\xi) = \begin{cases} \cos^2(k\xi), & |\xi| < \pi/(2k), \\ 0, & |\xi| \geq \pi/(2k), \end{cases} \tag{35}$$

where $k(> 0)$ is a dimensionless wave number. In the case of this IC, finite-time gradient catastrophe occurs only when $\alpha^\star < k$, i.e., $t^\star \in \mathbb{R}^+$ only for $\alpha^\star < k$, where $t^\star$, the value of t at which gradient catastrophe occurs, is given by

$$t^\star = -2\delta^{-1}\ln(1 - \alpha^\star/k). \tag{36}$$

Here, as in Refs. [**1**, **9**], we regard k as a free parameter and, throughout this subsection, we shall assume that $\alpha^\star < k$ holds.

On substituting this IC into Sys. (8) we obtain

$$u(x,t) = \begin{cases} 0, & x \leq t - \pi/(2k), \\ 0, & x \geq t + \pi/(2k), \end{cases} \tag{37}$$

while, for the curved section defined over the support of our IC, Eq. (8b) yields the following variant of *Kepler's equation* [**19**, p. 552]:

(38)

$$2k\eta - (k/\alpha^\star)g(t) + \pi/2 = (2k\xi + \pi/2) + (k/\alpha^\star)g(t)\sin(2k\xi + \pi/2) \qquad (|\xi| < \tfrac{1}{2}\pi/k),$$

where we have made use of the trigonometric identity

$$\cos^2(k\xi) = \tfrac{1}{2}\left[1 + \sin(2k\xi + \pi/2)\right]. \tag{39}$$

Here, we have $2k\eta - (k/\alpha^\star)g(t) + \pi/2$ in the role of "mean anomaly" [**19**, p. 552], $2k\xi + \pi/2$ in the role of "eccentric anomaly" [**19**, p. 552], and $-(k/\alpha^\star)g(t)$ in the role of "negative eccentricity" [**20**, p. 144].

Now solving Eq. (38) for ξ with the aid of Ref. [**20**, Eq. (4.3.19)], and then making use of Eq. (8a), it is readily established, after incorporating Eq. (37), that the following is the exact, *explicit*, solution of IVP (1) with Eq. (35) as IC:

$$(40)\quad u(x,t) = \exp(-\delta t/2) \times \begin{cases} 0, & x \le t - \pi/(2k), \\ \dfrac{1}{2} - \dfrac{\alpha^\star}{k} \displaystyle\sum_{n=1}^{\infty} (-1)^n \frac{J_n\left[n(k/\alpha^\star)g(t)\right]}{g(t)} \\ \quad \times \dfrac{\sin\left[2nk(x-t) - (nk/\alpha^\star)g(t) + n\pi/2\right]}{n}, & t - \pi/(2k) < x < t + \pi/(2k), \\ 0, & x \ge t + \pi/(2k), \end{cases}$$

for $0 < t < t^\star$, where $J_n(\cdot)$ denotes the Bessel function of the first kind of order n. Here, the restriction $0 < t < t^\star$, which follows from the "$\epsilon < 1$" criterion noted by Watson [**19**, p. 553], is required to ensure convergence of our Bessel–Fubini series; i.e., imposing this restriction (and recalling our assumption $\alpha^\star < k$), ensures that $0 < (k/\alpha^\star)g(t) < 1$; see also the "cos-pulse" case in Ref. [**9**].

Now, the application of $\partial/\partial x$ to Eq. (40) yields

$$(41)\quad u_x(x,t) = -2\alpha^\star \left(\frac{\exp(-\delta t/2)}{g(t)}\right) \times \begin{cases} 0, & x \le t - \pi/(2k), \\ \displaystyle\sum_{n=1}^{\infty} (-1)^n J_n\left[n(k/\alpha^\star)g(t)\right] \\ \quad \times \cos\left[2nk(x-t) - (nk/\alpha^\star)g(t) + n\pi/2\right], & t - \pi/(2k) < x < t + \pi/(2k), \\ 0, & x \ge t + \pi/(2k), \end{cases}$$

for $0 < t < t^\star$. On setting $\mathfrak{F} = u_{xx}$ in Eq. (12), where applying $\partial/\partial x$ to Eq. (41) yields

$$(42)\quad u_{xx}(x,t) = 4k\alpha^\star \left(\frac{\exp(-\delta t/2)}{g(t)}\right) \times \begin{cases} 0, & x \le t - \pi/(2k), \\ \displaystyle\sum_{n=1}^{\infty} n(-1)^n J_n\left[n(k/\alpha^\star)g(t)\right] \\ \quad \times \sin\left[2nk(x-t) - (nk/\alpha^\star)g(t) + n\pi/2\right], & t - \pi/(2k) < x < t + \pi/(2k), \\ 0, & x \ge t + \pi/(2k), \end{cases}$$

for $0 < t < t^\star$, it is readily established that

$$(43)\quad [\![u_{xx}]\!]_{\mathrm{l,r}}(t) = \pm 4k\alpha^\star \left(\frac{\exp(-\delta t/2)}{g(t)}\right) \times \sum_{n=1}^{\infty} n(-1)^n J_n\left[n(k/\alpha^\star)g(t)\right] \sin\left[(nk/\alpha^\star)g(t) \mp n(2 \pm 1)\pi/2\right] \qquad (0 < t < t^\star),$$

respectively, and, moreover, that

$$\lim_{t \to t^{\star}\uparrow} [\![u_{xx}]\!]_{\mathrm{l,r}}(t) = \mp 2k^2, \tag{44}$$

respectively. In this subsection, " l " and " r " subscripts denote the amplitudes of the second order weaker waves (recall Section 2.4) that occur across the wavefronts $\mathfrak{S}_{\mathrm{l,r}}(t) = t \mp \pi/(2k)$, respectively, under the present IC.

From Eq. (42) we can show that $u_{xx}(x_{\mathrm{T}}(t), t) = 0$, where

$$x_{\mathrm{T}}(t) = t + \tfrac{1}{2}g(t)/\alpha^{\star} + \pi/(4k) \qquad (0 < t < t^{\star}). \tag{45}$$

Clearly, then, the point $(x_{\mathrm{T}}(t), \exp(-\delta t/2)/2)$, where we observe that $u(x_{\mathrm{T}}(t), t) = \exp(-\delta t/2)/2$, is the inflection point on the leading (i.e., right-hand) side of the u vs. x profile corresponding to Eq. (40). Using Eq. (41), it is a relatively simple matter to determine the equation of the tangent line through $(x_{\mathrm{T}}(t), \exp(-\delta t/2)/2)$; omitting the details, we opt to express this linear function of x as

$$u_{\mathrm{tang}}(x,t) = \exp(-\delta t/2)\{1/2 - k(x - x_{\mathrm{T}}(t))[1 - (k/\alpha^{\star})g(t)]^{-1}\} \quad (0 < t < t^{\star}). \tag{46}$$

Of greater interest, however, is the fact that

$$\lim_{t \to t^{\star}\uparrow} u_x(x_{\mathrm{T}}(t), t) = -\infty, \tag{47}$$

where we also note that

$$\lim_{t \to t^{\star}\uparrow} (x_{\mathrm{T}}(t),\, u(x_{\mathrm{T}}(t), t)) = \left(\frac{1}{2k} + \frac{\pi}{4k} - 2\delta^{-1}\ln(1 - \alpha^{\star}/k),\ \tfrac{1}{2}(1 - \alpha^{\star}/k)\right). \tag{48}$$

That is, the tangent line that passes through the leading-side inflection point of the solution profile generated by Eq. (40) is the one that suffers gradient catastrophe at time $t = t^{\star}$; see the sequence given in Fig. 2.

REMARK 4. Also noteworthy is the fact that

$$\max_{x \in \mathbb{R}}[u(x,t)] = u(\overline{x}(t), t) = \exp(-\delta t/2) \qquad (t > 0), \tag{49}$$

where $\overline{x}(t) = t + g(t)/\alpha^{\star}$ in the case of the present IC. With regard to $\overline{x}(t)$ we observe that $k \le k^{\bullet} \implies \mathfrak{S}_{\mathrm{l}}(t) < \overline{x}(t) < \mathfrak{S}_{\mathrm{r}}(t)$, for all $t > 0$, where we have set $k^{\bullet} := \alpha^{\star}\pi/2$; if $k > k^{\bullet}$, then $\overline{x}(t^{\bullet}) = \mathfrak{S}_{\mathrm{r}}(t^{\bullet})$, where $t^{\bullet} = -2\delta^{-1}\ln[1 - \alpha^{\star}\pi/(2/k)]$.

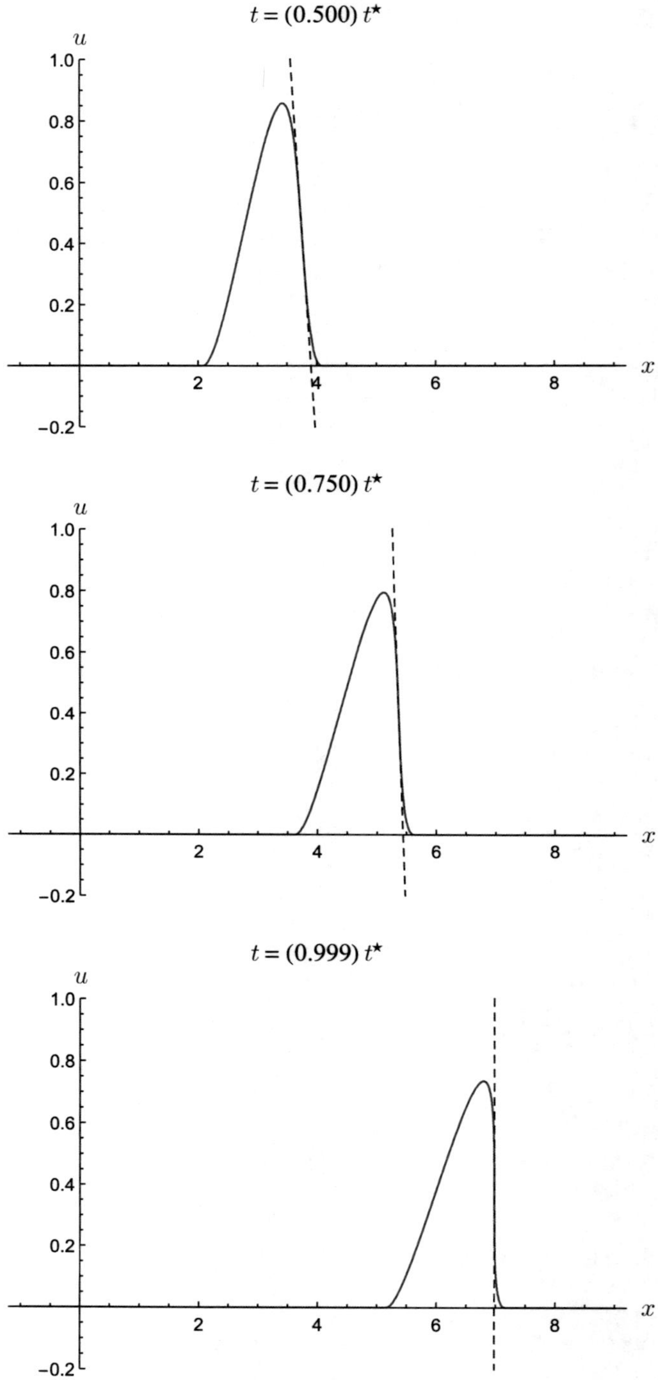

FIGURE 2. Three snapshots in the evolution of the u vs. x solution profile (blue curves) that were generated from Eq. (40) using $\beta = 1.2$, $\epsilon = \delta = 0.1$, and $k = \pi/2$; here, $\alpha^\star \approx 0.4167$ and $t^\star \approx 6.1647$. Black broken-lines: Plots of the tangent line equation Eq. (46) evaluated at indicated time-stamps.

3.3. Taylor-pulse IC. Now consider (what we term) the *Taylor-pulse* IC:

$$u_0(\xi) = \begin{cases} \frac{2}{1+\exp(-b\xi)}, & \xi < 0, \\ 1, & \xi = 0, \\ \frac{2}{1+\exp(b\xi)}, & \xi > 0, \end{cases} \tag{50}$$

where $b(>0)$ is another free parameter. For this IC, $\sup|u_0'(\xi)| = b/2$, where

$$u_0'(\xi) = \begin{cases} \frac{2b\exp(-b\xi)}{(1+\exp(-b\xi))^2}, & \xi < 0, \\ 0, & \xi = 0, \\ \frac{-2b\exp(b\xi)}{(1+\exp(b\xi))^2}, & \xi > 0. \end{cases} \tag{51}$$

Thus, it is not difficult to show that

$$t^\star = \begin{cases} -2\delta^{-1}\ln(1 - 2\alpha^\star/b), & \alpha^\star < b/2, \\ \infty, & \alpha^\star = b/2, \end{cases} \tag{52}$$

where we observe that $t^\star \in \mathbb{C}$ for $\alpha^\star > b/2$; i.e., in the case of our Taylor-pulse IC, finite-time gradient catastrophe *only* occurs when $\alpha^\star < b/2$.

On substituting Eq. (50) into Eq. (8b), the latter becomes, after some manipulation,

$$-1 = \begin{cases} \left[\dfrac{\mathfrak{z} + b(x-t)}{\mathfrak{z} + b(x - t - 2g(t)/\alpha^\star)}\right]\exp(\mathfrak{z}), & \xi < 0, \\ -1 + x - t - g(t)/\alpha^\star, & \xi = 0, \\ \left[\dfrac{\mathfrak{z} - b(x-t)}{\mathfrak{z} - b(x - t - 2g(t)/\alpha^\star)}\right]\exp(\mathfrak{z}), & \xi > 0, \end{cases} \tag{53}$$

where we have set $\xi := \mp\mathfrak{z}/b$ for the $\xi \lessgtr 0$ cases, respectively.

Making use of the generalized Lambert W-function, as defined by Mező and Keady [**21**, § 2.1], it is possible to solve the $\xi \neq 0$ cases of Eq. (53) for ξ in terms of x, t; doing so yields

$$\xi = b^{-1}\begin{cases} -W\left(\frac{-b(x-t)}{-b(x-t-2g(t)/\alpha^\star)}; -1\right), & \xi < 0, \\ W\left(\frac{b(x-t)}{b(x-t-2g(t)/\alpha^\star)}; -1\right), & \xi > 0, \end{cases} \tag{54}$$

where we have also employed the notation of Mező and Keady [**21**].

On substituting these two expressions, as well as the expression corresponding to $\xi = 0$ [see Eq. $(53)_2$], into Eq. (8a), the exact (explicit) solution of IVP (1), subject to our Taylor-pulse IC, is found to be

$$u(x,t) = \exp(-\delta t/2) \times \begin{cases} \dfrac{2}{1 + \exp\left[W\left(\frac{-b(x-t)}{-b(x-t-2g(t)/\alpha^\star)}; -1\right)\right]}, & x < t + g(t)/\alpha^\star, \\ 1, & x = t + g(t)/\alpha^\star, \quad 0 < t < t^\star. \\ \dfrac{2}{1 + \exp\left[W\left(\frac{b(x-t)}{b(x-t-2g(t)/\alpha^\star)}; -1\right)\right]}, & x > t + g(t)/\alpha^\star, \end{cases} \tag{55}$$

Here, due to the nascent nature of its development, we have opted to regard the generalized Lambert W-function as the *single-valued* inverse of the Taylor-pulse special case of Eq. (8b), hence the (temporal) range of validity stated.

According to Mező and Baricz [**22**, § 3.1], the Taylor series of the present special case of the generalized Lambert W-function about $\varsigma = 0$ is

$$W\left(\begin{matrix} X \\ Y \end{matrix}; \varsigma\right) = X - \mho \sum_{n=1}^{\infty} \varsigma^n L_n'(n\mho) \left(\frac{\exp(-nX)}{n}\right) \qquad (\mho \neq 0), \tag{56}$$

where $L_n'(\cdot)$ denotes the derivative of the nth Laguerre polynomial and $\mho := X - Y$. Unfortunately, however, the radius of convergence of this series for $\mho > 0$, which is the case corresponding to Eq. (55), has yet to be established; see Ref. [**22**, p. 7921].

Here, as in the previous subsection, it is a relatively simple matter to determine the equation of the tangent line that will become vertical at time $t = t^\star$; again omitting the details, this linear function of x can be written as

$$u_{\text{tang}}(x, t) = \frac{\exp(-\delta t/2)[2 - b(x - t)]}{2 - (b/\alpha^\star)g(t)} \qquad (0 < t < t^\star), \tag{57}$$

where we observe that $u_{\text{tang}}(x, t) \to 1 - bx/2$ as $t \to 0$.

From the sequence presented in Fig. 3, which it should be noted was *not* generated from Eq. (55), we see the steepening of the slope of the tangent line [see Eq. (57)] on the right-hand side of the u vs. x solution profile. The profile will exhibit a vertical tangent, which will pass through its vertex, at time $t = t^\star$; here, we observe that the x-coordinate of the vertex at said time is $x = \bar{x}(t^\star)$, where

$$\bar{x}(t^\star) = t^\star + 2/b. \tag{58}$$

Of course, $x = \bar{x}(t^\star)$ is also the equation of the (vertical) tangent line at time $t = t^\star$.

REMARK 5. According to Mező and Baricz [**22**, Thm 3], the present special case of the generalized Lambert W-function can also be expressed as

$$W\left(\begin{matrix} X \\ Y \end{matrix}; \varsigma\right) = X + W_{\mathfrak{r}}\left(\varsigma \mho \, \mathrm{e}^{-X}\right), \tag{59}$$

where, as asserted by Mező and Baricz, $\mathfrak{r} = -\varsigma \exp(-X)$. The function $W_{\mathfrak{r}}(\cdot)$, as noted in Ref. [**22**], is known as the $\mathfrak{r}$-Lambert function.

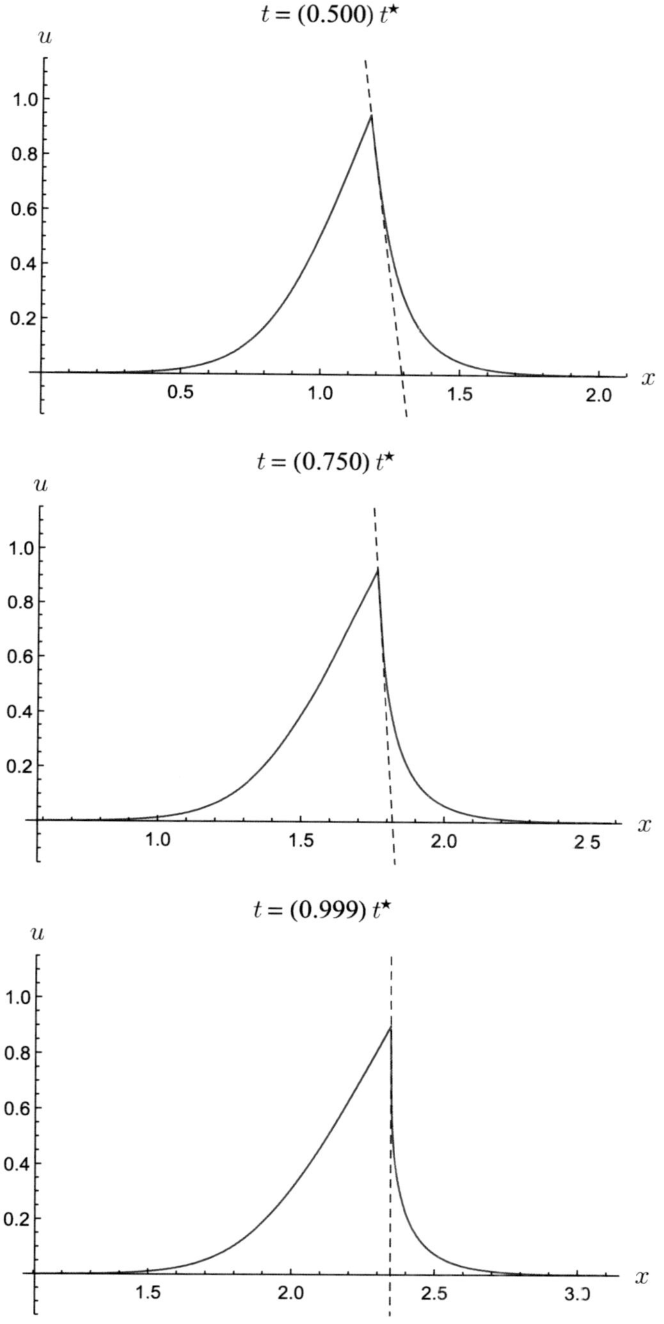

FIGURE 3. Three snapshots in the evolution of the u vs. x solution profile (blue curves) that were generated from Sys. (8) with $u_0(\xi)$ given by Eq. (50); here, we took $\epsilon = \delta = 0.1$ and $\beta = 1.2$, which gives $\alpha^\star \approx 0.41676$, and $b = 20\alpha^\star$, where for these parameter value $t^\star \approx 2.1072$. The black-broken tangent lines were generated using Eq. (57).

4. Numerical scheme construction

The main idea behind our numerical scheme for solving IVP (1) is to find all roots of Eq. (8b) for selected points in the xt-plane, in such a way as to balance the needs of reliability and efficiency.

4.1. Initial discretization. We choose a time step Δt, and define $t_n = n\Delta t$. For each time step from t_n to t_{n+1}, we will use a spatial grid $\{x_i^n\}_{i=1}^N$, with spacing $\Delta x_i^n = x_{i+1}^n - x_i^n$, for $i = 1, \dots, N-1$. For convenience, our initial grid $\{x_i^0\}_{i=1}^N$ will use uniform spacing. With each grid point x_i^n, we will associate an interval $[\xi_{i,L}^n, \xi_{i,R}^n]$, within which we will seek solutions of Eq. (8b) for $\eta = x_i^n - t_{n+1}$. In the case of $n = 0$, as Eq. (8b) has the solution $\xi = x$ for $t = 0$, we will set

$$\xi_{i,L}^0 = x_i^0 - \Delta x_i^0, \quad \xi_{i,R}^0 = x_i^0 + \Delta x_i^0, \quad i = 1, \dots, N-1,$$

and for the last grid point,

$$\xi_{N,L}^0 = x_N^0 - \Delta x_{N-1}^0, \quad \xi_{N,R}^0 = x_N^0 + \Delta x_{N-1}^0.$$

4.2. Finding roots. For each $i = 1, \dots, N$, we seek all roots of Eq. (8b) on the interval $I_i^n = [\xi_{i,L}^n, \xi_{i,R}^n]$, for $\eta = x_i^n - t_{n+1}$. To accomplish this, we divide I_i^n into M_i^n sub-intervals of maximum width $\Delta\xi_i^n$. We then define a sub-grid

$$\xi_{ij}^n = \xi_{i,L}^n + (j-1)\Delta\xi_i^n, \quad j = 0, 1, 2, \dots, M_i^n, \quad \Delta\xi_i^n = \frac{\xi_{i,R}^n - \xi_{i,L}^n}{M_i^n}.$$

Then, we define

$$F_i^n(\xi) = x_i^n - t_{n+1} - \xi - (\alpha^\star)^{-1} u_0(\xi) g(t_{n+1}) \tag{60}$$

and compute $y_{ij}^n = F_i^n(\xi_{ij}^n)$ for $j = 0, 1, \dots, M_i^n$.

For $j = 0, \dots, M_i^n - 1$, we check if $y_{ij}^n y_{i,j+1}^n < 0$. If so, then we use the secant method to find a root of $F_i^n(\xi)$ on $I_{ij}^n = [\xi_{ij}^n, \xi_{i,j+1}^n]$, with the endpoints of I_{ij}^n serving as the initial guesses. If the secant method generates an iterate that lies outside of I_{ij}^n, then one iteration of bisection is performed, and the secant method is applied on the half of I_{ij}^n that is guaranteed to contain a root, with the endpoints of that half serving as the initial guesses. This process is continued until a root is found in I_{ij}^n. All discovered roots are then substituted for ξ in Eq. (8a), with $x = x_i^n$ and $t = t^{n+1}$, to obtain values of the solution $u(x,t)$.

4.3. Updating intervals. If no root of Eq. (8b), with $\xi = x_i^n$ and $t = t_{n+1}$, is found in I_i^n, then we let $r_i = (\xi_{i,R}^n - \xi_{i,L}^n)/2$ and then update I_i^n by setting

$$\xi_{i,L}^n = \xi_{i,L}^n - r_i, \quad \xi_{i,R}^n = \xi_{i,R}^n + r_i. \tag{61}$$

We then apply the root-finding procedure described in the previous subsection to the updated I_i^n. This process is repeated until at least one root is found in I_i^n.

Let $\xi_{i,1}^*, \dots, \xi_{i,m_i}^*$ be the m_i distinct roots that are found in I_i^n. We then initialize $I_i^{n+1} = [\xi_{i,L}^{n+1}, \xi_{i,R}^{n+1}]$ as follows:

$$\xi_{i,L}^{n+1} = \min_{1\le k\le m_i} \xi_{i,k}^* - r_i/2, \qquad \xi_{i,R}^{n+1} = \max_{1\le k\le m_i} \xi_{i,k}^* + r_i/2, \tag{62}$$

where, as before, $r_i = (\xi_{i,R}^n - \xi_{i,L}^n)/2$. The rationale for this initialization is that roots in the next time step are likely to be found near the roots obtained during the current time step. The "padding" amount of $r_i/2$ is chosen to account for the shifting of roots over time; it is important to not pad by too much as that causes a loss of efficiency due to intervals I_i^n becoming unnecessarily large.

Next, to ensure that the onset of gradient catastrophe is properly resolved, we update I_i^{n+1} to include neighborhoods of solutions corresponding to nearby x-values. That is, we set

$$\xi_{i,L}^{n+1} = \min_{1\leq j\leq N, |i-j|\leq 2} \xi_{j,L}^{n+1}, \quad \xi_{i,R}^{n+1} = \max_{1\leq j\leq N, |i-j|\leq 2} \xi_{j,R}^{n+1}.$$

The choice of neighborhood size, in this case $\{j : |i-j| \leq 2\}$, is chosen to strike a balance between the need to make each neighborhood sufficiently large so that roots are not missed, and small enough to maintain efficiency. Future work will investigate setting the size of the neighborhood adaptively, based on the behavior of the solution.

4.4. Grid adaptivity. Near occurrences of gradient catastrophe, it is advisable to refine the spatial grid. This is based on the observation that when a uniform spatial grid is used, values of $u(x,t)$ obtained from computed roots of Eq. (8b) are spaced further apart within the xu-plane, thus yielding a low-resolution representation of the solution exactly where higher resolution is warranted.

Therefore, after the intervals I_i^{n+1}, $i = 1, \ldots, N$, have been constructed as described above, we adapt the spatial grid $\{x_i^n\}_{i=1}^N$ through the following procedure. As before, we denote by m_i the number of distinct roots of Eq. (8b) that have been obtained, where $\xi = x_i^n$ and $t = t_{n+1}$. We also let m and M, $M \geq m$, be thresholds for $|u_x(x,t)|$ (where u is single-valued) to guide adaptivity.

Initially, we define G_0 to consist of all sets of the form $\{x_i^0, [\xi_{i,L}^0, \xi_{i,R}^0]\}$, for $i = 1, \ldots, N$. Then, for $n = 0, 1, 2, \ldots$, G_n is updated to produce G_{n+1} by the following algorithm.

- Choose minimum and maximum grid spacings $\Delta x_{\min}$ and $\Delta x_{\max}$.
- Let $G_{n+1} = G_n$.
- For $i = 1, \ldots, N-1$, if $\Delta x_i = x_{i+1} - x_i$ satisfies $\frac{1}{2}\Delta x_i > \Delta x_{\min}$, refine the grid as follows:
  - If $m_i > 1$ or $m_{i+1} > 1$, or if

    $$\left| \frac{u(x_{i+1}^n, t_{n+1}) - u(x_i^n, t_{n+1})}{x_{i+1}^n - x_i^n} \right| > M,$$

    then we add the following points to G_{n+1}:
    * $\{(3x_i^n + x_{i+1}^n)/4, [\xi_{i,L}^n, \xi_{i+1,R}^n]\}$
    * $\{(x_i^n + x_{i+1}^n)/2, [\xi_{i,L}^n, \xi_{i+1,R}^n]\}$
    * $\{(x_i^n + 3x_{i+1}^n)/4, [\xi_{i,L}^n, \xi_{i+1,R}^n]\}$
    * $\{(x_{i-1}^n + x_i^n)/2, [\xi_{i-1,L}^n, \xi_{i,R}^n]\}$, if $i > 1$
    * $\{(x_{i+1}^n + x_{i+2}^n)/2, [\xi_{i+1,L}^n, \xi_{i+2,R}^n]\}$, if $i < N-1$
    * $\{(x_{i+2}^n + x_{i+3}^n)/2, [\xi_{i+2,L}^n, \xi_{i+3,R}^n]\}$, if $i < N-2$

    The asymmetry of the refinement is due to the rightward propagation of the waveform.
- For $i = 2, 4, 6, \ldots, N-1$, if Δx_i satisfies $\Delta x_i < \Delta x_{\max}$, coarsen the grid as follows: if $m_{i-1} = m_i = m_{i+1} = 1$, and if

  $$\left| \frac{u(x_{i+1}^n, t_{n+1}) - u(x_{i-1}^n, t_{n+1})}{x_{i+1}^n - x_{i-1}^n} \right| < m,$$

  then

  - For $j = i-1, i, i+1$, remove $\{x_j^n, [\xi_{j,L}^n, \xi_{j,R}^n]\}$ from G_{n+1}.
  - Add $\{x_{i-1}^n, [\xi_{i-1,L}^n, \xi_{i,R}^n]\}$ and $\{x_{i+1}^n, [\xi_{i,L}^n, \xi_{i+1,R}^n]\}$ to G_{n+1}.
- Let $N = |G_{n+1}|$.

Then, the new spatial grid $\{x_i^{n+1}\}_{i=1}^N$ is obtained from the first elements of the sets that comprise G_{n+1}, in sorted order, and the corresponding intervals $I_i^{n+1} = [\xi_{i,L}^{n+1}, \xi_{i,R}^{n+1}]$, for $i = 1, \dots, N$, are obtained from the second elements of these sets.

5. Numerical results

We now demonstrate the effectiveness of our numerical algorithm for several initial conditions. For each initial condition, we will solve IVP (1) on some interval $[a, b]$, with a spatial grid consisting of N points that are, initially, equally spaced in $[a, b]$, as discussed in Section 4.1. Unless otherwise stated, the following plots were generated using minimum and maximum grid spacings $\Delta x_{\min} = 0.01$ and $\Delta x_{\max} = 0.5$. We defined our adaptivity thresholds using $m = s(x_{i+1} - x_i)$ and $M = S/(x_{i+1} - x_i)$. Unless otherwise stated, the default values $s = 1$ and $S = 1/10$ were used to produce the plots throughout this section.

For computing the roots of Eq. (8b), we set the number of sub-grid points to be $M_i^n = 100(\xi_{i,R}^n - \xi_{i,L}^n)/\Delta x_i^0$.

5.1. Lorentzian IC. In Ref. [**1**], IVP (1) was first solved for the Lorentzian IC, namely,

$$u_0(\xi) = \frac{\varkappa^2}{\varkappa^2 + \xi^2}, \tag{63}$$

under which $t^\star$ is given by

$$t^\star = \begin{cases} -2\delta^{-1}\ln(1 - \varkappa/\varkappa^\star), & \varkappa < \varkappa^\star, \\ \infty, & \varkappa = \varkappa^\star, \end{cases} \tag{64}$$

and where $\varkappa^\star$ is defined in Ref. [**1**, Eq. (27)]. Substituting Eq. (63) into Sys. (8) gives

$$u(x,t) = \varkappa^2 \exp(-\delta t/2)/(\varkappa^2 + \xi^2), \tag{65a}$$

$$\xi^3 - \eta\xi^2 + \varkappa^2\xi + \varkappa^2[(\alpha^\star)^{-1}g(t) - \eta] = 0, \tag{65b}$$

and our task is to determine the real roots of this cubic.

In Figs. 4a, 4b; 5a, 5b; and 6a, 6b, our numerical solution[4] is compared with the analytical solution from Ref. [**1**, § 4.3]. For all three cases, we used $\beta = 1.2$ and $\epsilon = \delta = 0.1$, and the solution at time $t = 2t^\star$ is shown. The remaining parameter values taken are as follows:

Figures 4a and 4b: $[a, b] = [8.5, 10.5]$, $\varkappa = (0.20)\varkappa^\star$.
Figures 5a and 5b: $[a, b] = [50, 55]$, $\varkappa = (0.72)\varkappa^\star$.
Figures 6a and 6b: $[a, b] = [70, 72]$, $\varkappa = (0.82)\varkappa^\star$.

[4]It should be noted that a numerical investigation of the Lorentzian IC case of IVP (1) has also been performed by Montecinos [**23**, § 4.2], who, we observe, employs a finite-volume methodology.

In all of the figures in this section depicting *numerical solutions*, the blue circles represent the first root ξ of Eq. (8b) found for each (x,t), red circles represent the second root, and green circles represent any additional roots found.

The numerical solutions are consistent with the analytical solution from Ref. [**1**], up to limitations in spatial or temporal resolution. For example, in the case of $\varkappa = (0.20)\varkappa^\star$, the numerical solution, which is computed using $\Delta t = 0.04$, displays the first occurrence of gradient catastrophe, or multi-valuedness, at $(x,t) \approx (5.3631, 4.8)$, while the analytically determined time at which gradient catastrophe occurs is $t = t^\star \approx 4.4629$.

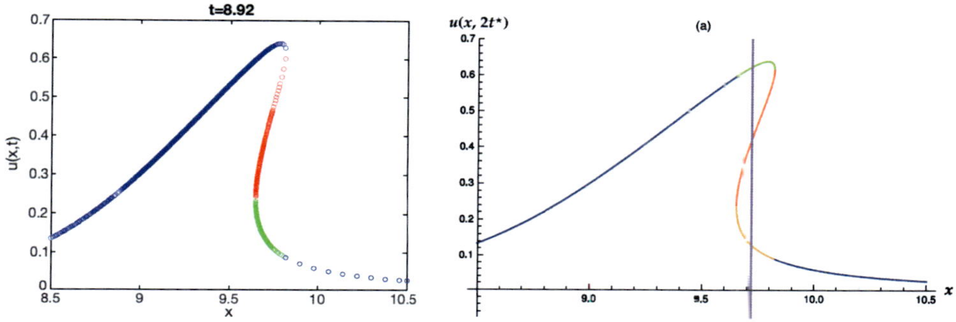

(a) Numerical solution using $N = 100$, $\Delta t = 0.04$, and $\Delta x_{\min} = 10^{-3}$.

(b) Analytical solution from Ref. [**1**, Fig. 1(a)].

FIGURE 4. Plots generated from numerical (plot (a)) and analytical (plot (b)) solutions of Eq. (65a), using $\varkappa = (0.20)\varkappa^\star$, at $t = 2t^\star \approx 8.9258$; here, $t^\star \approx 4.4629$.

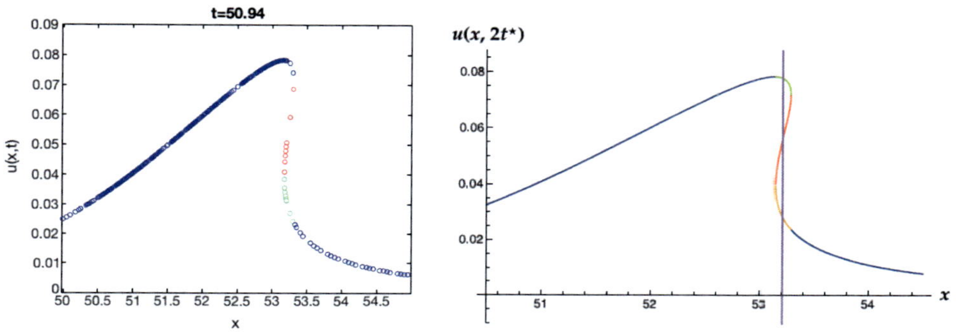

(a) Numerical solution using $N = 100$, $\Delta t = 0.1$, $\Delta x_{\min} = 10^{-4}$.

(b) Analytical solution from Ref. [**1**, Fig. 1(b)].

FIGURE 5. Plots generated from numerical (plot (a)) and analytical (plot (b)) solutions of Eq. (65a), using $\varkappa = (0.72)\varkappa^\star$, at $t = 2t^\star \approx 50.9186$; here, $t^\star \approx 25.4593$.

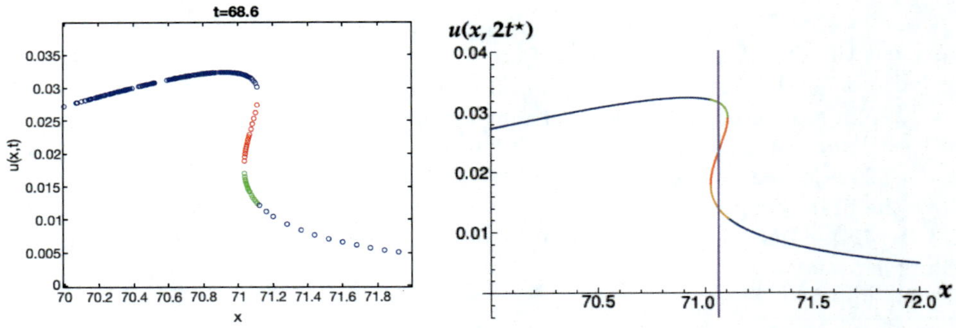

(a) Numerical solution with $N = 100$, $\Delta t = 0.04$, $\Delta x_{\text{min}} = 10^{-4}$.

(b) Analytical solution from Ref. [**1**, Fig. 1(c)].

FIGURE 6. Plots generated from numerical (plot (a)) and analytical (plot (b)) solutions of Eq. (65a), using $\varkappa = (0.82)\varkappa^\star$, at $t = 2t^\star \approx 68.5920$; here, $t^\star \approx 34.2960$. For plots of the numerical solution, $\Delta x_{\text{max}} = 0.1$.

5.2. Symmetric-exponential IC. Next, we consider the symmetric-exponential IC:

$$u_0(\xi) = \exp(-\varkappa|\xi|), \tag{66}$$

which was also analyzed in Ref. [**1**], and under which

$$t^\star = \begin{cases} -2\delta^{-1}\ln(1 - \alpha^\star/\varkappa), & \alpha^\star < \varkappa, \\ \infty, & \alpha^\star = \varkappa. \end{cases} \tag{67}$$

In Figs. 7a and 7b, our numerical solution is compared with the analytical solution from Ref. [**1**, § 5.1][5]. Using $N = 200$ grid points and time step $\Delta t = 0.1$, gradient catastrophe is not detected until $t = 20$, whereas with $\Delta t = 0.01$, the first instance occurs at $(x, t) \approx (21.2481, 19.2)$. As t increases, we see that the numerical method yields a solution that is consistent with the analytical one.

[5]See also Ref. [**9**, § 4, Item (iii)].

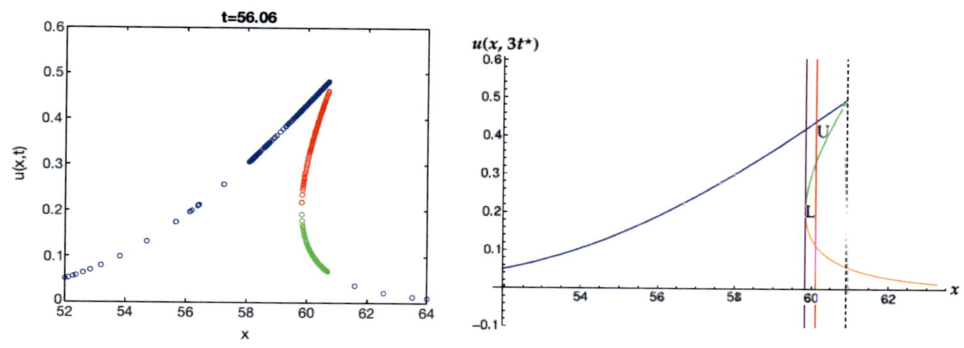

(a) Numerical solution with $N = 200$, $\Delta t = 0.1$, $\Delta x_{\max} = 1$, $s = 10$.

(b) Analytical solution from Ref. [**1**, Fig. 3].

FIGURE 7. Plots generated from numerical (plot (a)) and analytical (plot (b)) solutions of IVP (1) for the symmetric-exponential IC [see Eq. (66)] at time $t = 3t^\star \approx 56.0676$ using $\beta = 1.2$, $\epsilon = 0.1$, $\delta = 0.025$, and $\varkappa = 0.5$; here, $t^\star \approx 18.6892$.

5.3. Witch's hat IC. Next, we consider the piecewise linear IC given in Eq. (22). With $N = 200$ grid points and time step $\Delta t = 0.1$, the numerical solution displayed the first instance of gradient catastrophe at $(x, t) \approx (6.2027, 5.2)$. As shown in Figs. 8a and 8b, once again the numerical solution is consistent with the analytical one.

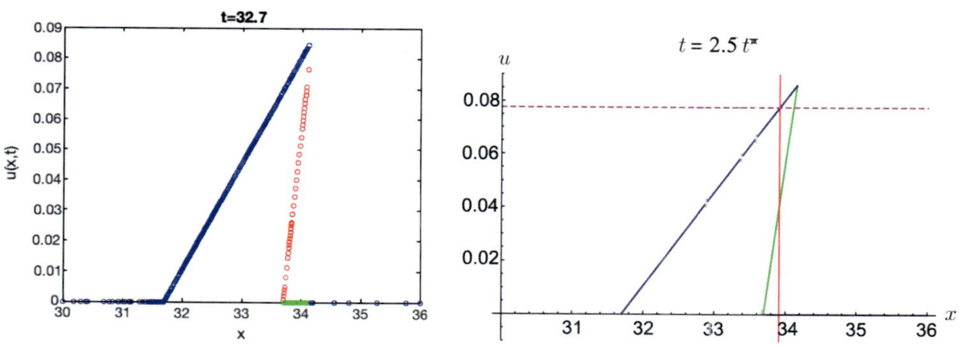

(a) Numerical solution with $N = 200$ and $\Delta t = 0.1$.

(b) Analytical solution from Eq. (25); representation of Fig. 1.

FIGURE 8. Plots generated from numerical (plot (a)) and analytical (plot (b)) solutions of IVP (1) for the IC given in Eq. (22) using $t = 2.5t^\star \approx 32.6943$, $\beta = 1.2$, $\epsilon = 0.1$, $\delta = 0.15$, and $\varkappa = 1.0$, where for these plots $t^\star \approx 13.0777$.

5.4. $\cos^2$-pulse IC. We next consider the $\cos^2$-pulse IC from Eq. (35). Focusing on the interval $[a, b] = [6, 8]$, the first instance of gradient catastrophe is detected at $(x, t) \approx (5.9643, 6.3)$. Applying the numerical method to Eq. (35), the following plots (see Figs. 9a and 9b) are produced—the first, shortly after the onset of gradient catastrophe; the second, much later in time, where there is significant overturning in the graph.

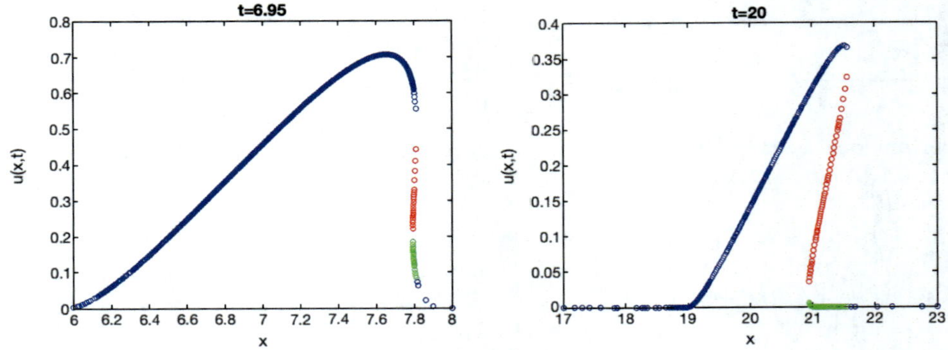

(a) Numerical solution with $N = 200$, $\Delta t = 0.01$, $\Delta x_{\min} = 10^{-6}$, and $\Delta x_{\max} = 0.01$.

(b) Numerical solution with $N = 200$, $\Delta t = 0.1$, and default values of other parameters.

FIGURE 9. Numerical solution of IVP (1) for the $\cos^2$-pulse IC given in Eq. (35), with parameter values $\delta = 0.1$, $\epsilon = 0.1$, $\beta = 1.2$, and $k = \pi/2$; here, $t^\star \approx 6.1647$. The discretization uses $N = 200$ grid points and time step $\Delta t = 0.01$. Here, $\Delta x_{\min} = 10^{-6}$ and $\Delta x_{\max} = 0.1$.

5.5. Cos-pulse IC. Similarly, we also consider the cos-pulse IC, namely,

$$u_0(\xi) = \begin{cases} \cos(k\xi), & |\xi| < \pi/(2k), \\ 0, & |\xi| \geq \pi/(2k), \end{cases} \tag{68}$$

which was treated via analytical means in Ref. [**9**, § 2.1].

The solution is shown in Figs. 10a–10c. The first instance of gradient catastrophe is detected at $(x, t) \approx (27.8874, 25.22)$, whereas Eq. (36) yields $t^\star \approx 24.5759$. It is worth noting that unlike the $\cos^2$-pulse case, the peak of the present waveform does not overtake the shock-front; this feature, in fact, stems from the different k-values taken, not the different (i.e., two vs. one) exponents involved; see Remark 4 (above) and the last paragraph of Ref. [**9**, § 2.1].

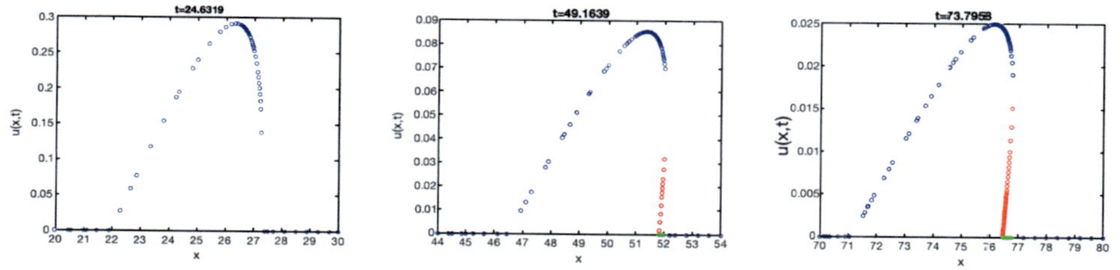

(a) Numerical solution at time $t = t^{\star}$.

(b) Numerical solution at time $t = 2t^{\star}$.

(c) Numerical solution at time $t = 3t^{\star}$.

FIGURE 10. Numerical solution of IVP (1) for the cos-pulse IC [i.e., Eq. (68)] using $N = 200$ and $\Delta t = 0.1$. The parameter values taken are $\beta = 1.2$, $\delta = 0.1$, $\epsilon = 0.1$, and $k = 0.9\alpha^{\star}\pi/2$; here, $t^{\star} \approx 24.5759$.

5.6. Taylor-pulse IC. We now apply our numerical method to the Taylor-pulse IC from Eq. (50), on the interval $[a, b] = [0, 10]$. We use the parameter values $\beta = 1.2$, $\epsilon = \delta = 0.1$, and $b = 20\alpha^{\star}$, which yields $t^{\star} \approx 2.1072$. Gradient catastrophe was first detected at $(x, t) \approx (2.4719, 2.22)$. Solutions are depicted in Figs. 11a and 11b.

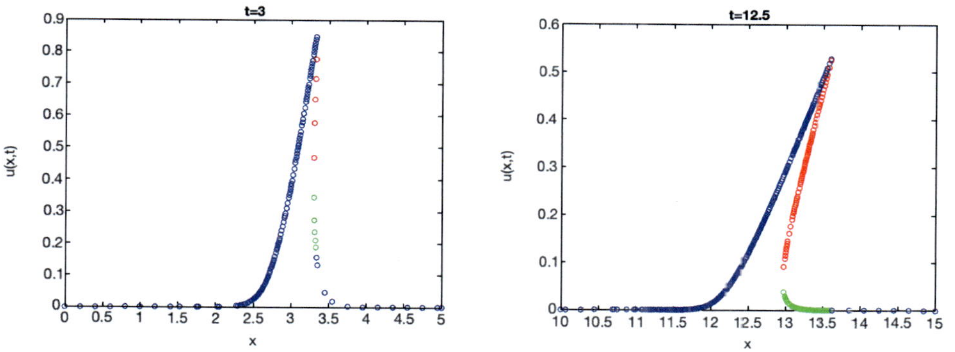

(a) Numerical solution computed shortly after the onset of gradient catastrophe, at $t = 3$

(b) Numerical solution computed well past the onset of gradient catastrophe, at $t = 12.5$.

FIGURE 11. Numerical solution of IVP (1) for the Taylor-pulse IC from Eq. (50), and parameter values $\delta = 0.1$, $\epsilon = 0.1$, $\beta = 1.2$, and $b = 20\alpha^{\star}$. Here, $\Delta x_{\max} = 0.1$. The discretization uses $N = 200$ grid points and time step $\Delta t = 0.01$.

5.7. Gaussian IC. Lastly, we consider the Gaussian IC

$$u_0(\xi) = \exp\left(-\frac{1}{2}\varkappa\xi^2\right), \tag{69}$$

under which gradient catastrophe occurs at time $t = t^\star_{\rm G}$, where [**1**, Eq. (106)]

$$t^\star_{\rm G} = \begin{cases} -2\delta^{-1}\ln(1-\sqrt{\varkappa^\star_{\rm G}/\varkappa}), & \varkappa > \varkappa^\star_{\rm G}, \\ \infty, & \varkappa = \varkappa^\star_{\rm G}, \end{cases} \tag{70}$$

and where $\varkappa^\star_{\rm G} = (\alpha^\star)^2 \exp(1)$.

As discussed in Ref. [**1**, § 6], an exact, explicit, solution is not possible under this IC. For our numerical treatment of this case we use the parameter values $\beta = 1.2$, $\epsilon = \delta = 0.1$, and $\varkappa = 20\varkappa^\star_{\rm G}$, from which we obtain $t^\star_{\rm G} \approx 5.0619$. When we compute a numerical solution on the interval $[a, b] = [0, 20]$, with $N = 200$ grid points and time step $\Delta t = 0.1$, multi-valuedness first occurs at $(x, t) \approx (5.9643, 5.3)$. The solution is displayed in Figs. 12a and 12b.

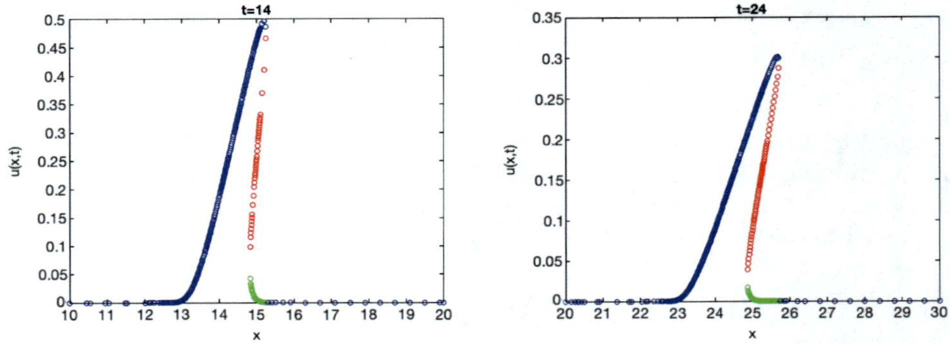

(a) Numerical solution at $t = 14$.

(b) Numerical solution at $t = 24$.

FIGURE 12. Numerical solution of IVP (1) for the Gaussian IC from Eq. (69), and parameter values $\delta = 0.1$, $\epsilon = 0.1$, $\beta = 1.2$, and $\varkappa = 20\varkappa^\star_{\rm G}$, where $\varkappa^\star_{\rm G} = (\alpha^\star)^2 \exp(1)$. The discretization uses $N = 200$ grid points and time step $\Delta t = 0.1$.

5.8. Comparison with WENO scheme. At the behest of Reviewer 3, we now compare our numerical approach with a particular case of a class of schemes known as *weighted essentially non-oscillatory* (WENO) schemes. WENO schemes for hyperbolic PDEs (see, e.g., Ref. [**24**] and those cited therein) are finite difference or finite volume [**14**] schemes in which an adaptive reconstruction procedure is used to prevent oscillations in the solution near discontinuities, while high-order formal accuracy is achieved in smooth regions.

In Figs. 13a, 13b, and 13c, a MATLAB implementation from Ref. [**25**] of a fully conservative, 5th-order, finite volume WENO scheme, in which time-stepping is performed using *strong stability-preserving 3rd-order Runge–Kutta* (SSP–RK3) [**26**], is applied to the Lorentzian, symmetric-exponential, and witch's hat ICs, respectively; the outputs are then compared with the corresponding solution profiles plotted from the (tri-valued) analytical solutions given in Ref. [**1**, § 4.3], Ref. [**1**, § 5.1], and Eq. (25), respectively. It can be seen that the WENO scheme yields a solution that, while nearly vertical as it crosses the plane of the shock wave, is also excessively smooth immediately on either side of it. Another aspect of the WENO solution profile, which can be seen in all three sub-figures, is the rather inaccurate, specifically, *under*, approximation of the shock amplitudes.

An advantage of WENO schemes is that they do not require tuning of any parameters, while the performance of our method can be impacted by the values of parameters such as those listed at the beginning of this section. That said, an advantage of our algorithm is that it is not necessary to commence time-stepping at $t = 0$; in the numerical experiments presented in this section, the initial time was set to $\lfloor \frac{3}{4}\frac{t_f}{\Delta t} \rfloor \Delta t$, where t_f is the time at which the solution is desired. Even though this results in poor initial guesses for finding roots of Eq. (8b), and thus expanding search intervals through Eq. (61), the updating of intervals through Eq. (62) has the effect of "catching up" so that the scheme can perform as efficiently, on a per-time step basis, as it would if time-stepping had begun at $t = 0$.

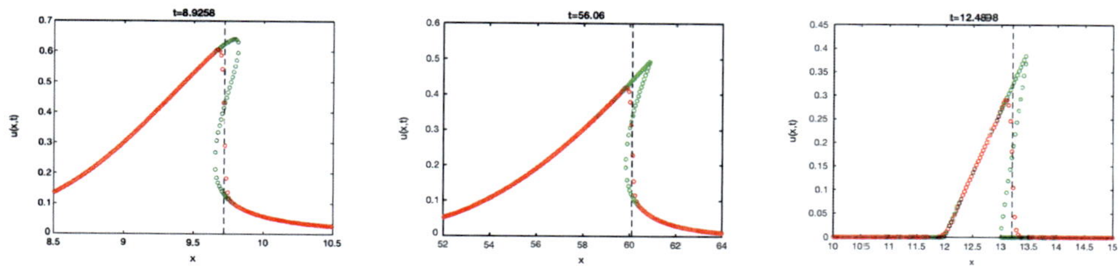

(a) Lorentzian IC $t_f \approx 8.926$, $\mathfrak{S}sh(t_f) \approx 9.719$

(b) Symmetric-exp. IC $t_f \approx 56.068$, $\mathfrak{S}sh(t_f) \approx 60.104$

(c) Lorentzian IC $t_f \approx 8.926$, $\mathfrak{S}sh(t_f) \approx 9.719$

FIGURE 13. Green profiles (a–c): Plotted from tri-valued analytical solutions given in Ref. [**1**, § 4.3], Ref. [**1**, § 5.1], and Eq. (25), respectively. Red profiles: Corresponding WENO-based numerical solutions. Broken vertical lines: $x = \mathfrak{S}_{\rm sh}(t_f)$, i.e., shock locations at times $t = t_f$. Fig. 13a: Same parameters values used for Ref. [**1**, Fig. 1(a)]. Fig. 13b: Same parameters values used for Ref. [**1**, Fig. 3]. Fig. 13c: Generated using $\epsilon = 0.1$, $\beta = 1.2$, $\delta = 0.15$, and $\varkappa = 2.0$, which yield $t^\star \approx 4.996$; here, $t^\star$ and $\mathfrak{S}_{\rm sh}(t_f)$ were computed based on Eqs. (23) and (28), respectively.

6. Closure

We have developed a numerical method that solves IVP (1), even when the solution is multi-valued. Through an analysis of the behavior of the solution over time, the spatial discretization is adapted so that all solution values at each point (x, t) in the space-time grid can be computed through targeted application of root-finding methods, and without any knowledge of an analytical solution. The effectiveness of the method—what might be termed the "Adaptive Multi-characteristic Method" (AMcM)—has been demonstrated using a variety of physically plausible ICs.

Future work on this topic will include the incorporation of adaptive time-stepping, and improvement of the algorithm for selecting and updating intervals for root-finding processes, to reduce gaps between points on the solution curve while also preserving efficiency and ensuring that no roots are overlooked. Furthermore, in its present form, the algorithm relies on certain parameters to guide adaptivity; a goal is to rely as little as possible on setting such parameters manually.

The onset of gradient catastrophe coincides with the equation $F_i^n(\xi) = 0$, where $F_i^n(\xi)$ is defined in Eq. (60), having a multiple root for some index i, $1 \le i \le N$, with $F_i^{n-1}(\xi)$ having only one root for $i = 1, \ldots, |G_{n-1}|$. If t^{n+1} is only slightly greater than $t^\star$, then these multiple roots may have spacing less than the sub-grid spacing $\Delta\xi_i^n$, and therefore may be missed by our hybrid root-finding procedure. Since $t^\star$ can be determined analytically, we can detect when roots have been missed, in which case we can modify our root-finding procedure as follows:

- Adaptively increase the number of sub-grid intervals, M_I^n, in regions where $F_i^n(\xi)$ is not monotone, and
- Search for roots of $F_i^n(\xi)/(F_i^n)'(\xi)$ instead of $F_i^n(\xi)$, to induce a sign change near closely-clustered roots so that they can be detected via bisection.

Future work will include the development of this approach so that all roots of Eq. (8b) can be reliably obtained for any $t > t^\star$, to the extent possible given limited precision in floating-point arithmetic.

Another avenue for future work is the estimation of the shock location through the EAR. After the nth time step is completed, let a and b be the minimum and maximum x-values, respectively, at which three roots of $F_i^n(\xi)$ have been obtained, and assume that for each i, the roots are sorted in increasing order by corresponding u-value. Then, we can use the trapezoidal rule and bisection to approximate a solution of the equation $\mathcal{F}[\mathfrak{S}_{\rm sh}(t)] = 0$; here, we have defined

$$\text{(71)}\quad \mathcal{F}[\varsigma] := \int_a^\varsigma [\hat{u}_2(x, t_{n+1}) - \hat{u}_1(x, t_{n+1})]\,\mathrm{d}x - \int_\varsigma^b [\hat{u}_3(x, t_{n+1}) - \hat{u}_2(x, t_{n+1})]\,\mathrm{d}x \qquad (t > t^\star),$$

where $u_j = \exp(-\delta t/2)\hat{u}_j$, and we recall that $x = \mathfrak{S}_{\rm sh}(t)$ denotes the location of the shock along the x-axis. For $j = 1, 2, 3$, $\hat{u}_j(x, t_{n+1})$ is obtained by substituting the root of Eq. (8b) with the jth smallest u-value for ξ into $u_0(\xi)$, with $t = t_{n+1}$. The approximate shock location obtained via the bisection method can be improved by refining the spatial grid as needed.

Acknowledgments

The authors thank the three anonymous reviewers for their instructive comments and helpful suggestions. The authors also thank Prof. Abba Gumel for his kind invitation to contribute to this special volume honoring Prof. Mickens. P.M.J. thanks Prof. Mickens himself, for many helpful discussions, and Prof. István Mező, for providing insight on, and the LaTeX source code to generate, the generalized Lambert W-function.

Figures 4b, 5b, 6b, and 7b are reprinted from Wave Motion, Volume 94, P. M. Jordan, *Poroacoustic solitary waves under the unidirectional Darcy-Jordan model*, 102498, Copyright(2020), with permission from Elsevier.

References

[1] P. M. Jordan, *Poroacoustic solitary waves under the unidirectional Darcy-Jordan model*, Wave Motion **94** (2020), 102498, 16, DOI 10.1016/j.wavemoti.2019.102498. MR4046352

[2] P. M. Jordan, *Growth and decay of acoustic acceleration waves in Darcy-type porous media*, Proc. R. Soc. Lond. Ser. A Math. Phys. Eng. Sci. **461** (2005), no. 2061, 2749–2766, DOI 10.1098/rspa.2005.1477. MR2165509

[3] M. Ciarletta and B. Straughan, *Poroacoustic acceleration waves*, Proc. R. Soc. Lond. Ser. A Math. Phys. Eng. Sci. **462** (2006), no. 2075, 3493–3499, DOI 10.1098/rspa.2006.1730. MR2269797

[4] Brian Straughan, *Stability and wave motion in porous media*, Applied Mathematical Sciences, vol. 165, Springer, New York, 2008. MR2433781

[5] R. T. Beyer, *The parameter B/A*, in: Nonlinear Acoustics, M. F. Hamilton and D. T. Blackstock (Eds.), Academic Press, 1998, pp. 25–39.

[6] A. D. Pierce, *Acoustics: An introduction to its physical principles and applications*, Acoustical Society of America, 1989.

[7] Ronald E. Mickens and P. M. Jordan, *A positivity-preserving nonstandard finite difference scheme for the damped wave equation*, Numer. Methods Partial Differential Equations **20** (2004), no. 5, 639–649, DOI 10.1002/num.20003. MR2076341

[8] G. B. Whitham, *Linear and nonlinear waves*, Pure and Applied Mathematics, Wiley-Interscience [John Wiley & Sons], New York-London-Sydney, 1974. MR0483954

[9] P. M. Jordan, *Addendum to "Poroacoustic solitary waves under the unidirectional Darcy-Jordan model" [Wave Motion 94 (2020) 102498]*, Wave Motion **119** (2023), Paper No. 103135, 6, DOI 10.1016/j.wavemoti.2023.103135. MR4568828

[10] Peter D. Lax, *The formation and decay of shock waves*, Amer. Math. Monthly **79** (1972), 227–241, DOI 10.2307/2316618. MR298252

[11] P. A. Thompson, *Compressible-fluid dynamics*, McGraw–Hill, 1972.

[12] D. G. Crighton, *Propagation of finite-amplitude waves in fluids*, in: Handbook of Acoustics, M. J. Crocker (Ed.), Wiley–Interscience, 1998, Chap. 17.

[13] J. David Logan, *An introduction to nonlinear partial differential equations*, 2nd ed., Pure and Applied Mathematics (Hoboken), Wiley-Interscience [John Wiley & Sons], Hoboken, NJ, 2008, DOI 10.1002/9780470287095. MR2397521

[14] Randall J. LeVeque, *Finite volume methods for hyperbolic problems*, Cambridge Texts in Applied Mathematics, Cambridge University Press, Cambridge, 2002, DOI 10.1017/CBO9780511791253. MR1925043

[15] C. Truesdell and R. Toupin, *The classical field theories*, Handbuch der Physik, Bd. III/1, Springer, Berlin. With an appendix on tensor fields by J. L. Ericksen. MR0118005

[16] D. R. Bland, *Wave theory and applications*, Oxford University Press, 1988.

[17] K. A. Lindsay and B. Straughan, *Acceleration waves and second sound in a perfect fluid*, Arch. Rational Mech. Anal. **68** (1978), no. 1, 53–87, DOI 10.1007/BF00276179. MR489302

[18] R. E. Mickens, *Mathematical modelling with differential equations*, CRC Press, 2022.

[19] G. N. Watson, *A Treatise on the Theory of Bessel Functions*, Cambridge University Press, Cambridge, England; The Macmillan Company, New York, 1944. §17.2 MR0010746

[20] Peter R. Holland, *The quantum theory of motion: An account of the de Broglie-Bohm causal interpretation of quantum mechanics*, Cambridge University Press, Cambridge, 1995. §4.3. MR1341368

[21] I. Mező and G. Keady, *Some physical applications of generalized Lambert functions*, Eur. J. Phys. **37** (2016), 065802.

[22] István Mező and Árpád Baricz, *On the generalization of the Lambert W function*, Trans. Amer. Math. Soc. **369** (2017), no. 11, 7917–7934, DOI 10.1090/tran/6911. MR3695849

[23] Gino I. Montecinos, *An iterative scaling function procedure for solving scalar non-linear hyperbolic balance laws*, Appl. Numer. Math. **162** (2021), 35–52, DOI 10.1016/j.apnum.2020.12.009. MR4190514

[24] Chi-Wang Shu, *High order weighted essentially nonoscillatory schemes for convection dominated problems*, SIAM Rev. **51** (2009), no. 1, 82–126, DOI 10.1137/070679065. MR2481112

[25] M. A. Diaz, *Weighted essentially non-oscillatory (WENO) scheme*[6], MATLAB Central File Exchange (2023). Retrieved: 22 January 2023.

[26] Sigal Gottlieb, Chi-Wang Shu, and Eitan Tadmor, *Strong stability-preserving high-order time discretization methods*, SIAM Rev. **43** (2001), no. 1, 89–112, DOI 10.1137/S003614450036757X. MR1854647

[6]Available at: `https://www.mathworks.com/matlabcentral/fileexchange/44639-weighted-essentially-non-oscillatory-weno-scheme`

School of Mathematics and Natural Sciences, The University of Southern Mississippi, Hattiesburg, Mississippi 39406

School of Mathematics and Natural Sciences, The University of Southern Mississippi, Hattiesburg, Mississippi 39406

Acoustics Division, U.S. Naval Research Laboratory, Stennis Space Center, Mississippi 39529
Email address: pedro.jordan@nrlssc.navy.mil

Contemporary Mathematics
Volume **793**, 2024
https://doi.org/10.1090/conm/793/15880

A fractional-order equation and its finite difference scheme for approximating a delay equation

Cui-Cui Ji and Weizhong Dai

This paper is dedicated to Professor Ronald E. Mickens for his 80th birthday.

ABSTRACT. This study aims at seeking a fractional-order equation that is a good approximation for a delay equation. To this end, we consider a delay equation with simple initial and boundary conditions and obtain a fractional-order equation and an associated numerical method for approximating the solution of the delay equation. In order to determine the fractional-order equation that is a better approximation of the Delay equation, the Levenberg-Marquardt iterative method is employed to estimate the optimal parameters in the fractional-order equation. This obtained fractional-order equation is then tested and compared its solution with the solution of the delay equation. Results show that the fractional method is indeed a good approximation for the Delay equation.

1. Introduction

In this chapter, we consider a delay equation (DE) as

$$\frac{\partial u(x,t+\kappa_0)}{\partial t} = \nabla^2 u(x,t) + Q(x,t), \tag{1.1}$$

which is related to the phase-lagging equation (PLE) [**1**, **2**], where κ_0 is the time lagging and $Q(x,t)$ is a source term. In general, finding the analytical or numerical solution of Equation (1.1) is tedious. Under some special cases, the analytical solution of the DE can be expressed as an expression of series in relation to the time variable (see [**3**, **4**]). To avoid this trouble, some researchers (see [**5**, **6**]) considered the first-order approximation of $\partial u(x,t+\kappa_0)/\partial t$, which yields the so-called damped wave equation (DWE)

$$\frac{\partial u(x,t)}{\partial t} + \kappa_0 \frac{\partial^2 u(x,t)}{\partial t^2} = \nabla^2 u(x,t) + Q(x,t). \tag{1.2}$$

2020 *Mathematics Subject Classification.* Primary 35A35, 65M06, 65M12, 65M32.

Key words and phrases. Phase-lagging equation, fractional-order heat equation, numerical scheme, parameter estimation.

The first author was supported in part by National Natural Science Foundation of China (Grant No. 12001307) and Natural Science Foundation of Shandong Province (Grant No. ZR2020QA033).

The second author is the corresponding author.

Dai and his co-workers [**3**,**4**] investigated the difference of the solutions between the DE and the DWE. It shows the DWE is a good approximation for the DE when κ_0 is small.

Recently, fractional calculus has become an emerging field in mathematics with deep applications in all related fields of science and engineering, such as in physics, biology, material science, etc; for example, see [**7**–**17**]. For instance, in biology there is a well-known Cole expression for the so-called membrane reaction, which is given in the frequency domain [**8**]. To fit the experimental data, the power of frequency in the expression was given to be non-integer (between 0 and 1). Thus, transforming from the frequency domain to the time domain leads to a fractional-order derivative with respect to time [**8**]. As a useful tool, some fractional models have been successfully applied to simulate the heat and thermal transfer in non-uniform porous medium, viscoelastic materials, dynamic electro-magnetic fields [**18**–**26**]. Since the convolution structure, fractional-order derivatives can accurately depict motions of different substances with memory and hereditary properties. In the 2020 American Mathematical Society (AMS) Joint Meeting, Professor R.E. Mickens discussed with the second author about the solution behavior of the DE. Since the solution of the DE shows a strong dependence on the time interval, which has the same behavior as the solution of fractional-order equation, it is possible that the DE can be replaced by a fractional-order equation. Note that in the DWE, Equation (1.2), involves a first-order derivative and a second-order derivative. If we replace the second-order derivative with respect to time in the DWE with a fractional-order (between 1 and 2) derivative with respect to time, will the obtained fractional-order equation provide a more accurate solution than the DWE does? Recently, we have obtained a fractional-order equation for approximating the DE and developed a numerical scheme for solving the fractional-order equation. Numerical simulations demonstrate that the fractional-order equation is indeed a good alternative for the DE. Results have been accepted for publication in International Journal of Numerical Analysis and Modeling (see [**27**]).

In this paper, we extend our study [**27**] to the delay equation with a source term and would like to take this special occasion in honor of Professor R.E. Mickens for his 80th birthday and for his great contribution to mathematical modeling and computation. For this purpose, we consider an extension of our obtained fractional-order equation in [**27**] to the following fractional-order equation with a source term as

$$\kappa_1 \cdot \frac{\partial u(x,t)}{\partial t} + \kappa_2 \cdot {}_0^C D_t^\alpha u(x,t) = \nabla^2 u(x,t) + Q(x,t), \tag{1.3}$$

where κ_1 and κ_2 are two constants to be determined, and ${}_0^C D_t^\alpha$ is the Caputo fractional derivative of order $\alpha \in (1,2)$. In general, let n be an integer, when $n-1 < \gamma < n$, the Caputo fractional derivative of order γ is defined by [**8**]

$${}_0^C D_t^\gamma f(t) = \int_0^t w_{n-\gamma}(t-s) f^{(n)}(s) ds, \quad \text{where } w_\mu(t) = t^{\mu-1}/\Gamma(\mu). \tag{1.4}$$

We then compare the difference of the solutions between the delay equation and the fractional-order equation (1.3). Based on research results of the solution of the factional partial differential equations (FPDEs), the analytical solutions of the most generalized FPDEs cannot be worked out. Up to now, robust numerical methods for solving the FPDEs have been investigated, such as finite difference methods

[28–33], finite element methods [34–36], and spectral methods [37–39], etc. Meanwhile, recent researches [31–33] have confirmed the weak regularity at the initial time $t = 0$ of the solution of the time FPDEs. Therefore, an efficient approximation to the Caputo derivative is desperately needed in constructing numerical methods for the fractional-order equation (1.3). In this study, to overcome initial time singularity, we adopt a time discretization technique based on the graded time mesh [31, 33]. And the unconditional stability and convergence of the present numerical method are proved.

The rest of this chapter is organized as follows. In Section 2, we consider the fractional-order equation with a source term and obtain an energy estimate for ensuring the mathematical model to be well-posed. In Section 3, based on the graded time mesh, we construct a difference scheme for solving the fractional-order equation. In Section 4, we analyze the unconditional stability and convergence of the scheme by the discrete energy method. In Section 5, we employ the Levenberg–Marquardt iterative method to estimate the parameters of the fractional-order equation. In Section 6, we compare the solutions between the fractional-order equation and the DE by investigating a testing problem. We summarize the major results of this work in Section 7.

2. Fractional-order equation

We consider the following dimensionless fractional-order equation with simple initial and boundary conditions, i.e.,

$$\kappa_1 \cdot u_t(x,t) + \kappa_2 \cdot {}_0^C D_t^\alpha u(x,t) = \frac{\partial^2 u(x,t)}{\partial x^2} + Q(x,t), \qquad (x,t) \in (0,1) \times (0,\infty), \tag{2.1}$$

$$u(0,t) = 0, \quad u(1,t) = 0, \quad t > 0, \tag{2.2}$$

$$u(x,0) = \psi(x), \quad \partial u(x,0)/\partial t = \phi(x), \quad x \in (0,1), \tag{2.3}$$

where the operator ${}_0^C D_t^\alpha$ denotes the Caputo fractional derivative of order $\alpha \in (1,2)$, i.e.,

$${}_0^C D_t^\alpha f(t) = \frac{1}{\Gamma(2-\alpha)} \int_0^t \frac{f''(s)}{(t-s)^{\alpha-1}} ds. \tag{2.4}$$

We next give the energy estimation of the fractional-order model in Equations (2.1)–(2.3). To this end, we present two useful lemmas with respect to the Caputo factional derivative, which will be used for estimating the energy of the governing model in Equations (2.1)–(2.3).

LEMMA 2.1 ([40]). *For any function $f(t) \in C^1([0,T])$, if $0 < \beta < 1$, it holds that*

$$f(t) \cdot {}_0^C D_t^\beta f(t) \geq \frac{1}{2} {}_0^C D_t^\beta f^2(t), \tag{2.5}$$

where the operator ${}_0^C D_t^\beta$ denotes the Caputo fractional derivative of order $\beta \in (0,1)$, i.e.,

$${}_0^C D_t^\beta f(t) = \frac{1}{\Gamma(1-\beta)} \int_0^t \frac{f'(s)}{(t-s)^\beta} ds. \tag{2.6}$$

LEMMA 2.2 ([**19**]). *Let the function $f(t) \in C^1([0,T])$. If $0 < \beta < 1$, then it holds that*

$$\int_0^t {}_0^C D_t^\beta f(\eta) d\eta = {}_0^{RL} D_t^{\beta-1} f(t) - \frac{t^{1-\beta}}{\Gamma(2-\beta)} f(0), \tag{2.7}$$

where the operator ${}_0^{RL} D_t^{-\beta}$ denotes the Riemann-Liouville (RL) fractional integral of order β, i.e.,

$$ {}_0^{RL} D_t^{-\beta} f(t) = \frac{1}{\Gamma(\beta)} \int_0^t \frac{f(s)ds}{(t-s)^{1-\beta}}, \quad t > 0, \quad \beta > 0. \tag{2.8}$$

THEOREM 2.3. *Let $u(x,t)$ be the solution of the fractional-order equation (2.1)–(2.3). And we assume that $u(x,t) \in C^{2,2}([0,1] \times [0,T])$. Then, it holds that, for $t > 0$,*

$$\begin{aligned} &\kappa_1 \cdot \int_0^t \int_0^1 u_s^2(x,s) dx ds + \kappa_2 \cdot {}_0^{RL} D_t^{\alpha-2} \left(\int_0^1 u_t^2(x,t) dx \right) + \int_0^1 u_x^2(x,t) dx \\ &\leq \frac{\kappa_2 \cdot t^{2-\alpha}}{\Gamma(3-\alpha)} \int_0^t u_t^2(x,0) dx + \int_0^1 u_x^2(x,0) dx + \frac{1}{\kappa_1} \int_0^t \int_0^1 Q^2(x,t) dx ds. \end{aligned}$$

PROOF. Taking an inner product of Equation (2.1) with $u_t(x,t)$, we have

$$\begin{aligned} &\kappa_1 \cdot \int_0^1 u_t^2(x,t) dx + \kappa_2 \cdot \int_0^1 {}_0^C D_t^\alpha u(x,t) \cdot u_t(x,t) dx \\ &= \int_0^1 u_{xx}(x,t) \cdot u_t(x,t) dx + \int_0^1 Q(x,t) \cdot u_t(x,t) dx. \end{aligned} \tag{2.9}$$

Noticing the relation ${}_0^C D_t^\alpha u(x,t) = {}_0^C D_t^{\alpha-1} u_t(x,t)$ and using Lemma 2.1, we obtain the following estimate as

$$\int_0^1 {}_0^C D_t^\alpha u(x,t) \cdot u_t(x,t) dx \geq \frac{1}{2} {}_0^C D_t^{\alpha-1} \left(\int_0^1 u_t^2(x,t) dx \right). \tag{2.10}$$

Applying the integration by parts and noticing the boundary condition in Equation (2.2), it holds that

$$\int_0^1 u_{xx}(x,t) \cdot u_t(x,t) dx = -\frac{1}{2} \frac{d}{dt} \int_0^1 u_x^2(x,t) dx. \tag{2.11}$$

Using the Cauchy-Schwarz inequality for the second term on the right-hand-side of Equation (2.9), we have

$$\int_0^1 Q(x,t) \cdot u_t(x,t) dx \leq \frac{\kappa_1}{2} \int_0^1 u_t^2(x,t) dx + \frac{1}{2\kappa_1} \int_0^1 Q^2(x,t) dx. \tag{2.12}$$

Inserting Equations (2.10)–(2.12) into Equation (2.9) and multiplying the result by 2 lead to

$$\begin{aligned} &\kappa_1 \cdot \int_0^1 u_t^2(x,t) dx + \kappa_2 \cdot {}_0^C D_t^{\alpha-1} \left(\int_0^1 u_t^2(x,t) dx \right) \\ &\leq -\frac{d}{dt} \int_0^1 u_x^2(x,t) dx + \frac{1}{\kappa_1} \int_0^1 Q^2(x,t) dx. \end{aligned} \tag{2.13}$$

Replacing the variable t with s on both sides of Equation (2.13), integrating the result with respect to the variable s from 0 to t, and then using Lemma 2.2, we arrive at the conclusion. □

REMARK 2.4. Recent studies in the time FPDEs have discovered that the solutions of the time FPDEs may have a weak singularity near the initial time $t = 0$ under certain circumstances. Based on the analysis in [**41**], the weak singularity may occur when the solution of the fractional-order equation (2.1) satisfies

$$\left|\frac{\partial^l u}{\partial t^l}(x,t)\right| \le C_u(1+t^{\alpha-l}), \quad l = 0,1,2,3. \tag{2.14}$$

In our proof for Theorem 2.3, we obtained an energy estimate based on Lemmas 2.1 and 2.2 by assuming $u(x,t) \in C^{2,2}([0,1]\times[0,T])$. However, in the case of weak singularity at $t = 0$, Lemmas 2.1 and 2.2 may not work for the energy estimate. Thus, how to estimate the energy of Equation (2.1) under the required regularity in Equation (2.14) needs to be further analyzed. We would like to point out that our focus in this study is to see if the delay equation (1.1) can be replaced with a fractional-order equation since finding the analytical or numerical solution of the DE is tedious in general.

3. Finite difference scheme for fractional-order equation

Due to the difficulty in finding the exact solution of the fractional-order equation with a source term in Equations (2.1)–(2.3), we next propose a numerical scheme for solving this fractional-order equation (2.1)–(2.3). In addition, since the weak regularity near the initial time $t = 0$ of the solution of the time FPDEs, we consider the numerical method on a graded time mesh $\{t_n \mid t_n = T(n/N)^r\}_{n=0}^N$, where $r(\ge 1)$ is a grading parameter and N is a given positive integer.

Before introducing an approximation formula for the Caputo fractional derivative ${}_0^C D_t^\alpha u(t)$ of order $\alpha \in (1,2)$, we denote $t_{n-\frac12} = \frac12(t_n + t_{n-1})$, $\tau_n = t_n - t_{n-1}$ for $1 \le n \le N$; $\bar\tau_0 = t_{\frac12} - t_0$, $\bar\tau_n = t_{n+\frac12} - t_{n-\frac12}$ for $1 \le n \le N-1$; and $\Omega_\tau = \{t_n \mid 0 \le n \le N\}$, $\hat\Omega_\tau = \{t_{n-\frac12} \mid 1 \le n \le N\}$. Let $u^k = u(t_k)$ for $k \ge 1$. And we denote the following two difference quotients

$$\begin{aligned} \delta_t u^{n-\frac12} &= \frac{1}{\tau_n}(u(t_n) - u(t_{n-1})), \quad 1 \le n \le N; \\ \delta_t^2 u^n &= \frac{1}{\bar\tau_n}(\delta_t u^{n+\frac12} - \delta_t u^{n-\frac12}), \quad 1 \le n \le N-1. \end{aligned}$$

We consider ${}_0^C D_t^\alpha u(t)$ at $t = t_{n-\frac12}$, which yields

$$\begin{aligned} {}_0^C D_t^\alpha u(t_{n-\frac12}) &= \frac{1}{\Gamma(2-\alpha)} \int_0^{t_{n-\frac12}} \frac{u''(s)}{(t_{n-\frac12} - s)^{\alpha-1}} ds \\ &= \frac{1}{\Gamma(2-\alpha)} \left[\int_0^{t_{\frac12}} \frac{u''(s)}{(t_{n-\frac12} - s)^{\alpha-1}} ds + \sum_{k=1}^{n-1} \int_{t_{k-\frac12}}^{t_{k+\frac12}} \frac{u''(s)}{(t_{n-\frac12} - s)^{\alpha-1}} ds \right]. \end{aligned}$$

Let $\beta = \alpha - 1 \in (0,1)$ and $v(t) = u'(t)$. Denote $v^n = v(t_n)$ for $n \ge 0$ and $v^{n-\frac12} = v(t_{n-\frac12})$ for $n \ge 1$. Next, we first consider a numerical approximation for the Caputo derivative ${}_0^C D_t^\beta v(t)$ of order β at $t = t_{n-\frac{1}{2}}$ as follows:

$$
\begin{aligned}
{}_0^C D_t^\beta v(t_{n-\frac{1}{2}}) &= \frac{1}{\Gamma(1-\beta)} \int_0^{t_{n-\frac{1}{2}}} \frac{v'(s)}{(t_{n-\frac{1}{2}}-s)^\beta} ds \\
&= \frac{1}{\Gamma(1-\beta)} \left[\int_0^{t_{\frac{1}{2}}} \frac{v'(s)}{(t_{n-\frac{1}{2}}-s)^\beta} ds + \sum_{k=1}^{n-1} \int_{t_{k-\frac{1}{2}}}^{t_{k+\frac{1}{2}}} \frac{v'(s)}{(t_{n-\frac{1}{2}}-s)^\beta} ds \right] \\
&\approx \frac{1}{\Gamma(1-\beta)} \left[\int_0^{t_{\frac{1}{2}}} (t_{n-\frac{1}{2}}-s)^{-\beta} \frac{v(t_{\frac{1}{2}}) - v(t_0))}{\bar{\tau}_0} ds \right. \\
&\quad \left. + \sum_{k=1}^{n-1} \int_{t_{k-\frac{1}{2}}}^{t_{k+\frac{1}{2}}} (t_{n-\frac{1}{2}}-s)^{-\beta} \frac{v(t_{k+\frac{1}{2}}) - v(t_{k-\frac{1}{2}})}{\bar{\tau}_k} ds \right]
\end{aligned}
$$

$$
\begin{aligned}
&\approx \frac{1}{\Gamma(1-\beta)} \left[\hat{a}_{n-1}^{(n,\beta)} (v^{\frac{1}{2}} - v^0) + \sum_{k=1}^{n-1} \hat{a}_{n-k-1}^{(n,\beta)} (v^{k+\frac{1}{2}} - v^{k-\frac{1}{2}}) \right] \\
&\stackrel{\text{def}}{=} \delta_t^\beta (v^{n-\frac{1}{2}}, v^0),
\end{aligned} \tag{3.1}
$$

where

$$
\begin{aligned}
\hat{a}_{n-1}^{(n,\beta)} &= \frac{1}{\bar{\tau}_0} \int_0^{t_{\frac{1}{2}}} (t_{n-\frac{1}{2}}-s)^{-\beta} ds, \\
\hat{a}_{n-k-1}^{(n,\beta)} &= \frac{1}{\bar{\tau}_k} \int_{t_{k-\frac{1}{2}}}^{t_{k+\frac{1}{2}}} (t_{n-\frac{1}{2}}-s)^{-\beta} ds, \quad 1 \le k \le n-1.
\end{aligned} \tag{3.2}
$$

We insert the following approximation

$$
v^{n-\frac{1}{2}} = u'(t_{n-\frac{1}{2}}) \approx \delta_t u^{n-\frac{1}{2}}, \quad 1 \le n \le N. \tag{3.3}
$$

into Equation (3.1), and notice the relation $\beta = \alpha - 1$, which gives a numerical approximation formula for the Caputo derivative ${}_0^C D_t^\alpha u(t)$ of order α at $t = t_{n-\frac{1}{2}}$ as

$$
\begin{aligned}
&{}_0^C D_t^\alpha u(t_{n-\frac{1}{2}}) \\
&= \frac{1}{\Gamma(2-\alpha)} \left[a_{n-1}^{(n,\alpha)} (\delta_t u^{\frac{1}{2}} - u'(t_0)) + \sum_{k=1}^{n-1} a_{n-k-1}^{(n,\alpha)} (\delta_t u^{k+\frac{1}{2}} - \delta_t u^{k-\frac{1}{2}}) \right] \\
&\stackrel{\text{def}}{=} \delta_t^\alpha (\delta_t u^{n-\frac{1}{2}}, u'(t_0)),
\end{aligned} \tag{3.4}
$$

where

$$
\begin{aligned}
a_{n-1}^{(n,\alpha)} &= \frac{1}{\bar{\tau}_0} \int_0^{t_{\frac{1}{2}}} (t_{n-\frac{1}{2}}-s)^{1-\alpha} ds, \\
a_{n-k-1}^{(n,\alpha)} &= \frac{1}{\bar{\tau}_k} \int_{t_{k-\frac{1}{2}}}^{t_{k+\frac{1}{2}}} (t_{n-\frac{1}{2}}-s)^{1-\alpha} ds, \quad 1 \le k \le n-1.
\end{aligned} \tag{3.5}
$$

We also rewrite the approximation operator in Equation (3.4) as

$$
\begin{aligned}
&\delta_t^{\alpha}(\delta_t u^{n-\frac{1}{2}}, u'(t_0)) \\
(3.6)\qquad &= \frac{1}{\Gamma(2-\alpha)}\Big[a_0^{(n,\alpha)}\delta_t u^{n-\frac{1}{2}} - \sum_{k=1}^{n-1}(a_{n-k-1}^{(n,\alpha)} - c_{n-k}^{(n,\alpha)})\delta_t u^{k-\frac{1}{2}} \\
&\quad - a_{n-1}^{(n,\alpha)}u'(t_0)\Big], \quad 1 \le n \le N.
\end{aligned}
$$

For the space domain, we divide the interval $[0,1]$ into M-subintervals with a step size $\Delta x = \frac{1}{M}$ and denote $x_i = i\Delta x,\ 0 \le i \le M$. Let $\mathcal{U}_h = \{u \mid u = (u_0, u_1, \cdots, u_M)\}$ be the grid function space defined on $\Omega_h = \{x_i \mid 0 \le i \le M\}$. For any $u, v \in \mathcal{U}_h$, we denote $\delta_x u_{i-\frac{1}{2}} = (u_i - u_{i-1})/\Delta x$. $\delta_x^2 u_i = (\delta_x u_{i+\frac{1}{2}} - \delta_x u_{i-\frac{1}{2}})/\Delta x,\ 1 \le i \le M-1$, and define two inner products as $(u,v) = \Delta x \sum_{i=1}^{M-1} u_i v_i$, $(\delta_x u, \delta_x v) = \Delta x \sum_{i=1}^{M}(\delta_x u_{i-\frac{1}{2}})\delta_x v_{i-\frac{1}{2}}$ and the corresponding norms as $\|u\| = \sqrt{(u,u)}$, $\|\delta_x u\| = \sqrt{(\delta_x u, \delta_x u)}$. Throughout this chapter, we use the notation $u \lesssim v$ to present that there is a generic constant C which is independent of the time steps τ_n, $\bar{\tau}_n$ and the space step Δx such that $u \le Cv$.

We now derive a difference scheme for the model in Equations (2.1)–(2.3). Assume that $u(x,t)$ is the solution of Equations (2.1)–(2.3). We define $U_i(t) = u(x_i,t)$ on Ω_h and $U_i^n = u(x_i,t_n)$ on grid points (x_i,t_n), $0 \le i \le M,\ 0 \le n \le N$. For the space discretization, considering Equation (2.1) at the grid point (x_i,t), we have

$$
\begin{aligned}
(3.7)\qquad &\kappa_1 \cdot u_t(x_i,t) + \kappa_2 \cdot {}_0^C D_t^{\alpha} u(x_i,t) = u_{xx}(x_i,t) + Q(x_i,t), \\
&\qquad\qquad 0 \le i \le M,\ t \in (0,T].
\end{aligned}
$$

Using the second-order difference quotient for $u_{xx}(x_i,t)$ in Equation (3.7), we obtain

$$
\begin{aligned}
(3.8)\qquad &\kappa_1 \frac{d}{dt}U_i(t) + \kappa_2 \cdot {}_0^C D_t^{\alpha} U_i(t) \\
&= \delta_x^2 U_i(t) + Q(x_i,t) + (\Delta x)^2 \cdot r_i(t), \quad 0 \le i \le M, \quad t \in (0,T].
\end{aligned}
$$

For the time discretization, considering Equation (3.8) at $t = t_{n-\frac{1}{2}}$, we have

$$
\begin{aligned}
(3.9)\quad &\kappa_1 \frac{d}{dt}U_i(t_{n-\frac{1}{2}}) + \kappa_2 \cdot {}_0^C D_t^{\alpha} U_i(t_{n-\frac{1}{2}}) = \delta_x^2 U_i(t_{n-\frac{1}{2}}) + Q(x_i, t_{n-\frac{1}{2}}) \\
&\qquad + (\Delta x)^2 \cdot r_i(t_{n-\frac{1}{2}}), \quad 0 \le i \le M, \quad t \in (0,T].
\end{aligned}
$$

We use the difference quotient $\delta_t U_i^{n-\frac{1}{2}}$ to approximate the first-order derivative, the formula $\delta_t^{\alpha}(\delta_t U_i^{n-\frac{1}{2}}, \phi(x_i))$ to approximate the Caputo derivative of order α, and the average denoted by $U_i^{n-\frac{1}{2}} = (U_i^n + U_i^{n-1})/2$ to approximate $U_i(t_{n-\frac{1}{2}})$ in Equation (3.9), respectively. That is,

$$
\begin{aligned}
(3.10)\qquad &\kappa_1 \cdot \delta_t U_i^{n-\frac{1}{2}} + \kappa_2 \cdot \delta_t^{\alpha}(\delta_t U_i^{n-\frac{1}{2}}, \phi(x_i)) = \delta_x^2 U_i^{n-\frac{1}{2}} + Q_i^{n-\frac{1}{2}} \\
&\quad + (R^x)_i^{n-\frac{1}{2}} + (R^t)_i^{n-\frac{1}{2}}, \quad 0 \le i \le M, \quad 1 \le n \le N,
\end{aligned}
$$

where the truncation errors in space satisfy

$$|(R^x)_i^{n-\frac{1}{2}}| = (\Delta x)^2 \left|\frac{r_i(t_n) + r_i(t_{n-1})}{2}\right| \lesssim (\Delta x)^2, \quad 0 \le i \le M, \tag{3.11}$$

for $n = 1, \cdots, N$. Noticing the boundary conditions in Equation (2.2) and the initial condition in Equation (2.3), i.e.,

$$U_0^n = 0, \quad U_M^n = 0, \ 1 \le n \le N, \tag{3.12}$$

$$U_i^0 = \psi(x_i), \quad 0 \le i \le M, \tag{3.13}$$

and dropping small terms in Equation (3.10), and replacing U_i^n with its numerical approximation u_i^n, we obtain a finite difference scheme for the problem in Equations (2.1)–(2.3) as follows

$$\kappa_1 \cdot \delta_t u_i^{n-\frac{1}{2}} + \kappa_2 \cdot \delta_t^\alpha (\delta_t u_i^{n-\frac{1}{2}}, \phi(x_i)) = \delta_x^2 u_i^{n-\frac{1}{2}} + Q_i^{n-\frac{1}{2}}, \quad 1 \le i \le M-1, \ 1 \le n \le N, \tag{3.14}$$

$$u_0^n = 0, \quad u_M^n = 0, \quad 1 \le n \le N, \tag{3.15}$$

$$u_i^0 = \psi(x_i), \quad 0 \le i \le M. \tag{3.16}$$

4. Stability and convergence of the finite difference scheme

In this section, we analyze the stability and convergence of the finite difference scheme (3.14)–(3.16) based on a graded time mesh. Since the nonuniform mesh in time, the theoretical analysis of the difference scheme (3.14)–(3.16) becomes more complicated than that in [**27**].

4.1. Stability analysis of the scheme. We first consider the stability of the difference scheme (3.14)–(3.16). Heretofore, we prove the following monotonicity of the discretization coefficients $\{a_k^{(n,\alpha)}\}$ in Equation (3.1), which will be used in the subsequent analysis.

LEMMA 4.1. *If $\{a_k^{(n,\alpha)} \mid 1 \le k \le n, \ 1 \le n \le N\}$ are defined by Equation (3.1), then it holds that*

$$0 < a_{n-1}^{(n,\alpha)} < a_{n-2}^{(n,\alpha)} < \cdots < a_1^{(n,\alpha)} < a_0^{(n,\alpha)}, \quad 1 \le n \le N. \tag{4.1}$$

PROOF. By the mean-value theorem, for $1 \le n \le N$, we have

$$a_{n-k-1}^{(n,\alpha)} = \frac{1}{\bar{\tau}_k} \int_{t_{k-\frac{1}{2}}}^{t_{k+\frac{1}{2}}} (t_{n-\frac{1}{2}} - s)^{1-\alpha} ds = (t_{n-\frac{1}{2}} - \xi_k)^{1-\alpha}, \ 1 \le k \le n-1, \tag{4.2}$$

where $\xi_k \in (t_{k-\frac{1}{2}}, t_{k+\frac{1}{2}})$. Similarly, we have

$$a_{n-1}^{(n,\alpha)} = \frac{1}{\bar{\tau}_0} \int_0^{t_{\frac{1}{2}}} (t_{n-\frac{1}{2}} - s)^{1-\alpha} ds = (t_{n-\frac{1}{2}} - \xi_0)^{1-\alpha}, \tag{4.3}$$

where $\xi_0 \in (0, t_{\frac{1}{2}})$. Hence, the conclusion in Equation (4.1) holds. □

Suppose that $\{w_i^n \mid 0 \le i \le M,\ 0 \le n \le N\}$ is the solution of

$$\kappa_1 \cdot \delta_t w_i^{n-\frac{1}{2}} + \kappa_2 \cdot \delta_t^\alpha(\delta_t w_i^{n-\frac{1}{2}}, \phi(x_i)) = \delta_x^2 w_i^{n-\frac{1}{2}} + Q_i^{n-\frac{1}{2}} + G_i^{n-\frac{1}{2}}, \quad 1 \le i \le M-1,\ 1 \le n \le N, \tag{4.4}$$

$$w_0^n = 0, \quad w_M^n = 0, \quad 1 \le n \le N, \tag{4.5}$$

$$w_i^0 = \psi(x_i) + \tilde{\psi}_i, \quad 0 \le i \le M, \tag{4.6}$$

where $\tilde{\psi}_i$ and $G_i^{n-\frac{1}{2}}$ are small perturbation functions. Denote the perturbation term $\eta_i^n = w_i^n - u_i^n$, $0 \le i \le M$, $0 \le n \le N$. Then, we have the perturbation system

$$\kappa_1 \cdot \delta_t \eta_i^{n-\frac{1}{2}} + \kappa_2 \cdot \delta_t^\alpha(\delta_t \eta_i^{n-\frac{1}{2}}, 0) = \delta_x^2 \eta_i^{n-\frac{1}{2}} + G_i^{n-\frac{1}{2}}, \quad 1 \le i \le M-1, \quad 1 \le n \le N, \tag{4.7}$$

$$\eta_0^n = 0, \quad \eta_M^n = 0, \quad 1 \le n \le N, \tag{4.8}$$

$$\eta_i^0 = \tilde{\psi}_i, \quad 0 \le i \le M. \tag{4.9}$$

THEOREM 4.2 (Stability). *Let $\{\eta_i^n \mid 0 \le i \le M, 0 \le n \le N\}$ be the solution of the perturbation system in Equations (4.7)–(4.9). It holds that the difference scheme (3.14)–(3.16) is stable with respect to the perturbations $\{G_i^{n-\frac{1}{2}}, \tilde{\psi}_i\}$, i.e.,*

$$\|\delta_x \eta^n\|^2 \lesssim \|\delta_x \tilde{\psi}\|^2 + \sum_{m=1}^{n} \tau_m t_m^{\alpha-1} \|G^{m-\frac{1}{2}}\|^2, \quad 1 \le n \le N. \tag{4.10}$$

PROOF. We take an inner product of Equation (4.7) with $\delta_t \eta^{n-\frac{1}{2}}$ and obtain

$$\begin{aligned}&\left(\kappa_1 \cdot \delta_t \eta^{n-\frac{1}{2}} + \kappa_2 \cdot \delta_t^\alpha(\delta_t \eta^{n-\frac{1}{2}}, 0)\right), \delta_t \eta^{n-\frac{1}{2}}\Big) \\ &= (\delta_x^2 \eta^{n-\frac{1}{2}}, \delta_t \eta^{n-\frac{1}{2}}) + (G^{n-\frac{1}{2}}, \delta_t \eta^{n-\frac{1}{2}}), \quad 1 \le n \le N.\end{aligned} \tag{4.11}$$

Using the expression of $\delta_t^\alpha(\delta_t \eta^{n-\frac{1}{2}}, 0)$ defined by Equation (3.4) and the Cauchy-Schwarz inequality, we have

$$\begin{aligned}\Big(\kappa_2 \cdot \delta_t^\alpha(\delta_t \eta^{n-\frac{1}{2}}, 0)\big), \delta_t \eta^{n-\frac{1}{2}}\Big) &= \frac{\kappa_2}{\Gamma(2-\alpha)} \Big[a_0^{(n,\alpha)} \|\delta_t \eta^{n-\frac{1}{2}}\|^2 \\ &- \sum_{k=1}^{n-1} (a_{n-k-1}^{(n,\alpha)} - a_{n-k}^{(n,\alpha)})(\delta_t \eta^{k-\frac{1}{2}}, \delta_t \eta^{n-\frac{1}{2}})\Big] \\ &\ge \frac{\kappa_2}{\Gamma(2-\alpha)} \Big[a_0^{(n,\alpha)} \|\delta_t \eta^{n-\frac{1}{2}}\|^2 - \sum_{k=1}^{n-1} (a_{n-k-1}^{(n,\alpha)} - a_{n-k}^{(n,\alpha)}) \\ &\cdot \frac{\|\delta_t \eta^{k-\frac{1}{2}}\|^2 + \|\delta_t \eta^{n-\frac{1}{2}}\|^2}{2}\Big].\end{aligned} \tag{4.12}$$

Using the summation by parts for the first term on the right-hand-side of Equation (4.11), we obtain

$$(\delta_x^2 \eta^{n-\frac{1}{2}}, \delta_t \eta^{n-\frac{1}{2}}) = -\frac{1}{2\tau_n}(\|\delta_x \eta^n\|^2 - \|\delta_x \eta^{n-1}\|^2). \tag{4.13}$$

Using the Cauchy-Schwarz inequality again for the second term on the right-hand-side of Equation (4.11), we have

$$\begin{aligned}(G^{n-\frac{1}{2}},\delta_t\eta^{n-\frac{1}{2}}) &\le \|\delta_t\eta^{n-\frac{1}{2}}\|\cdot\|G^{n-\frac{1}{2}}\|\\ &\le \frac{\kappa_2(1-\epsilon)}{2\Gamma(2-\alpha)}a_{n-1}^{(n,\alpha)}\|\delta_t\eta^{n-\frac{1}{2}}\|^2+\frac{\Gamma(2-\alpha)}{2\kappa_2(1-\epsilon)a_{n-1}^{(n,\alpha)}}\|G^{n-\frac{1}{2}}\|^2\end{aligned}\tag{4.14}$$

for any $\epsilon\in[0,1)$.

Inserting Equations (4.12)–(4.14) into Equation (4.11) yields

$$\begin{aligned}&\frac{1}{2\tau_n}(\|\delta_x\eta^n\|^2-\|\delta_x\eta^{n-1}\|^2)+\frac{\kappa_2\epsilon}{2\Gamma(2-\alpha)}a_{n-1}^{(n,\alpha)}\|\delta_t\eta^{n-\frac{1}{2}}\|^2\\ &+\frac{\kappa_2}{2\Gamma(2-\alpha)}\sum_{k=1}^{n}a_{n-k}^{(n,\alpha)}\|\delta_t\eta^{k-\frac{1}{2}}\|^2-\frac{\kappa_2}{2\Gamma(2-\alpha)}\sum_{k=1}^{n-1}a_{n-k-1}^{(n,\alpha)}\|\delta_t\eta^{k-\frac{1}{2}}\|^2\\ &\le\frac{\Gamma(2-\alpha)}{2\kappa_2(1-\epsilon)a_{n-1}^{(n,\alpha)}}\|G^{n-\frac{1}{2}}\|^2.\end{aligned}$$

Multiplying the above relation by $2\tau_n$, and using the following estimate

$$\frac{1}{a_{n-1}^{(n,\alpha)}}=(t_{n-\frac{1}{2}}-\xi_0)^{\alpha-1}<t_{n-\frac{1}{2}}^{\alpha-1}<t_n^{\alpha-1},\tag{4.15}$$

we have

$$\|\delta_x\eta^n\|^2-\|\delta_x\eta^{n-1}\|^2+\frac{\kappa_2A^n}{\Gamma(2-\alpha)}\le\frac{\Gamma(2-\alpha)t_n^{\alpha-1}}{\kappa_2(1-\epsilon)}\tau_n\|G^{n-\frac{1}{2}}\|^2,\tag{4.16}$$

where

$$\begin{aligned}A^n&=\epsilon\tau_na_{n-1}^{(n,\alpha)}\|\delta_t\eta^{n-\frac{1}{2}}\|^2\\ &+\tau_n\sum_{k=1}^{n}a_{n-k}^{(n,\alpha)}\|\delta_t\eta^{k-\frac{1}{2}}\|^2-\tau_n\sum_{k=1}^{n-1}a_{n-k-1}^{(n,\alpha)}\|\delta_t\eta^{k-\frac{1}{2}}\|^2.\end{aligned}\tag{4.17}$$

In Equation (4.16), replacing the index n by m, and then summing the result from $m=1$ to $m=n$, we obtain

$$\begin{aligned}&\|\delta_x\eta^n\|^2-\|\delta_x\eta^0\|^2+\frac{\kappa_2}{\Gamma(2-\alpha)}\sum_{m=1}^{n}A^m\\ &\le\frac{\Gamma(2-\alpha)}{\kappa_2(1-\epsilon)}\sum_{m=1}^{n}\tau_mt_m^{\alpha-1}\|G^{m-\frac{1}{2}}\|^2.\end{aligned}\tag{4.18}$$

For the third term on the left-hand-side of Equation (4.18), we exchange the summation order and then use assumption 4.1 in [**42**], yielding

$$\begin{aligned}\frac{\kappa_2}{\Gamma(2-\alpha)}\sum_{m=1}^{n}A^m&=\frac{\kappa_2}{\Gamma(2-\alpha)}\sum_{k=1}^{n}\Big(\epsilon\tau_ka_{k-1}^{(k,\alpha)}+\sum_{m=k}^{n}\tau_ma_{m-k}^{(m,\alpha)}\\ &-\sum_{m=k+1}^{n}\tau_ma_{m-k-1}^{(m,\alpha)}\Big)\|\delta_t\eta^{k-\frac{1}{2}}\|^2\ge0.\end{aligned}\tag{4.19}$$

Inserting Equation (4.19) into Equation (4.18) gives Equation (4.10) and hence the theorem holds. □

4.2. Convergence analysis of the scheme. Before analyzing the convergence of the difference scheme (3.14)–(3.16), we present the following results [**42**] of the local truncation errors on a graded time mesh $\{t_n \mid t_n = T(n/N)^r\}_{n=0}^N$, which will be used in estimating the optimal error of the present scheme. In this subsection, we denote $\alpha \in (1,2)$ and $\beta = \alpha - 1 \in (0,1)$.

LEMMA 4.3. *Suppose $f(t) \in C[0,T] \cap C^2(0,T]$, and $|f'(t)| + t|f''(t)| \lesssim t^{\beta-1}$ for $t \in (0,T]$. Then, it holds that*

$$\left|{}_0^C D_t^\beta f(t_{n-\frac{1}{2}}) - \delta_t^\beta(f^{n-\frac{1}{2}}, f^0)\right| \lesssim n^{-\min\{r(\beta+1),2-\beta\}}, \quad n = 1, 2, \cdots, N,$$

where $f^{n-\frac{1}{2}} = f(t_{n-\frac{1}{2}})$.

LEMMA 4.4. *Suppose $|f''(t)| \lesssim t^{\alpha-2}$ for $t \in (0,T]$. Then, it holds that*

$$\left|f(t_{n-\frac{1}{2}}) - \frac{f(t_n) + f(t_{n-1})}{2}\right| \lesssim \tau_n^2 t_n^{\alpha-2}, \quad n = 1, 2, \cdots, N.$$

LEMMA 4.5. *Suppose $|f'''(t)| \lesssim t^{\alpha-3}$ for $t \in (0,T]$. Then, it holds that*

$$\left|f'(t_{n-\frac{1}{2}}) - \frac{f(t_n) - f(t_{n-1})}{\tau_n}\right| \lesssim \tau_n^2 t_n^{\alpha-3}, \quad n = 1, 2, \cdots, N.$$

Since it is difficult to obtain directly the local truncation error of the numerical approximation $\delta_t^\alpha(\delta_t u^{n-\frac{1}{2}}, u'(t_0))$ to the Caputo derivative ${}_0^C D_t^\alpha u(t)$ of order α in Equation (3.4), we take a roundabout route to prove the convergence of the difference scheme (3.14)–(3.16). Denote $v_i^{n-\frac{1}{2}} = \delta_t u_i^{n-\frac{1}{2}}$, $0 \le i \le M$, $1 \le n \le N$. The difference scheme (3.14)–(3.16) is equivalent to the following scheme

$$\kappa_1 \cdot v_i^{n-\frac{1}{2}} + \kappa_2 \cdot \delta_t^\beta(v_i^{n-\frac{1}{2}}, \phi(x_i)) = \delta_x^2 u_i^{n-\frac{1}{2}} + Q_i^{n-\frac{1}{2}}, \quad 1 \le i \le M-1,\ 1 \le n \le N, \tag{4.20}$$

$$\delta_t u_i^{n-\frac{1}{2}} = v_i^{n-\frac{1}{2}}, \quad 1 \le i \le M-1,\ 1 \le n \le N, \tag{4.21}$$

$$u_0^n = 0, \quad u_M^n = 0, \quad 1 \le n \le N, \tag{4.22}$$

$$u_i^0 = \psi(x_i), \quad 0 \le i \le M. \tag{4.23}$$

Next, we consider the optimal error estimate of the equivalent form in Equations (4.20)–(4.23) of the scheme (3.14)–(3.16). Let $V_i^{n-\frac{1}{2}} = v(x_i, t_{n-\frac{1}{2}})$ for $0 \le i \le M$ and $1 \le n \le N$. Based on the derivation of the approximate formula in Equation (3.4) of the α-th Caputo derivative, we have

$$\kappa_1 \cdot V_i^{n-\frac{1}{2}} + \kappa_2 \cdot \delta_t^\beta(V_i^{n-\frac{1}{2}}, \phi(x_i)) = \delta_x^2 U_i^{n-\frac{1}{2}} + Q_i^{n-\frac{1}{2}} - (R^x)_i^{n-\frac{1}{2}} + (R_1^t)_i^{n-\frac{1}{2}}, \quad 1 \le i \le M-1,\ 1 \le n \le N, \tag{4.24}$$

$$\delta_t U_i^{n-\frac{1}{2}} = V_i^{n-\frac{1}{2}} + (R_2^t)_i^{n-\frac{1}{2}}, \quad 1 \le i \le M-1,\ 1 \le n \le N, \tag{4.25}$$

where the local truncation errors satisfy

$$|(R^x)_i^{n-\frac{1}{2}}| \lesssim h^2,\ 1 \le i \le M-1,\ 1 \le n \le N-1, \tag{4.26}$$

$$\begin{aligned}&|(R_1^t)_i^{n-\frac{1}{2}}| \lesssim n^{-\min\{r\alpha,3-\alpha\}} + \tau_n^2 t_n^{\alpha-2},\\ &\qquad 1 \le i \le M-1,\ 1 \le n \le N-1,\end{aligned} \tag{4.27}$$

$$|\delta_x(R_2^t)_i^{n-\frac{1}{2}}| \lesssim \tau_n^2 t_n^{\alpha-3} + h^2,\ 1 \le i \le M-1,\ 1 \le n \le N-1. \tag{4.28}$$

Denote

$$e_i^n = U_i^n - u_i^n, \quad 0 \le n \le N, \quad 0 \le i \le M, \tag{4.29}$$

$$\sigma_i^{n-\frac{1}{2}} = V_i^{n-\frac{1}{2}} - v_i^{n-\frac{1}{2}}, \quad 1 \le n \le N, \quad 0 \le i \le M. \tag{4.30}$$

We obtain the following error equation

$$\begin{aligned}&\kappa_1 \cdot \sigma_i^{n-\frac{1}{2}} + \kappa_2 \cdot \delta_t^\beta(\sigma_i^{n-\frac{1}{2}},0) = \delta_x^2 e_i^{n-\frac{1}{2}} + (R^x)_i^{n-\frac{1}{2}} + (R_1^t)_i^{n-\frac{1}{2}},\\ &\qquad\qquad 1 \le i \le M-1,\ 1 \le n \le N,\end{aligned} \tag{4.31}$$

$$\delta_t e_i^{n-\frac{1}{2}} = \sigma_i^{n-\frac{1}{2}} + (R_2^t)_i^{n-\frac{1}{2}}, \quad 1 \le i \le M-1,\ 1 \le n \le N, \tag{4.32}$$

with

$$e_0^n = e_M^n = 0, \quad \sigma_0^{n-\frac{1}{2}} = \sigma_M^{n-\frac{1}{2}} = 0, \quad 1 \le n \le N, \tag{4.33}$$

$$e_i^0 = 0, \quad 0 \le i \le M. \tag{4.34}$$

THEOREM 4.6. *Suppose that the solution $u(x,t)$ of the fractional-order equation (2.1)–(2.3) satisfies the regularity assumption in Equation (2.14). Let $\{u_i^n \mid 0 \le i \le M, 0 \le n \le N\}$ be the solution of the difference scheme (3.14)–(3.16) on the graded time meshes $t_n = T(n/N)^r$. Then, the following optimal error estimate holds*

$$\|U^n - u^n\|_\infty \lesssim h^2 + \begin{cases}(N^{-1})^{\frac{r\alpha}{2}} & \text{if } \frac{r\alpha}{2} < 3-\alpha,\\ (N^{-1})^{3-\alpha}\sqrt{\ln n} & \text{if } \frac{r\alpha}{2} = 3-\alpha,\\ (N^{-1})^{3-\alpha}(\frac{n}{N})^{\frac{r\alpha}{2}-(3-\alpha)} & \text{if } \frac{r\alpha}{2} > 3-\alpha.\end{cases} \tag{4.35}$$

PROOF. We take an inner product of (4.31) with $\sigma^{n-\frac{1}{2}}$ and obtain

$$\begin{aligned}&\Big(\kappa_1 \cdot \sigma^{n-\frac{1}{2}} + \kappa_2 \cdot \delta_t^\beta(\sigma^{n-\frac{1}{2}},0)), \sigma^{n-\frac{1}{2}}\Big) = (\delta_x^2 e^{n-\frac{1}{2}}, \sigma^{n-\frac{1}{2}})\\ &\quad +((R^x)^{n-\frac{1}{2}}, \sigma^{n-\frac{1}{2}}) + ((R_1^t)^{n-\frac{1}{2}}, \sigma^{n-\frac{1}{2}}), \quad 1 \le n \le N.\end{aligned} \tag{4.36}$$

For the left-hand-side of Equation (4.36), we use a similar argument for Equation (4.12) and use the Cauchy-Schwarz inequality and notice the relation $\beta = \alpha - 1$. This gives

$$\begin{aligned}&\Big(\kappa_1 \cdot \sigma^{n-\frac{1}{2}} + \kappa_2 \cdot \delta_t^\beta(\sigma^{n-\frac{1}{2}},0)), \sigma^{n-\frac{1}{2}}\Big)\\ &\ge \kappa_1\|\sigma^{n-\frac{1}{2}}\|^2 + \frac{\kappa_2}{\Gamma(2-\alpha)}\Big[a_0^{(n,\alpha)}\|\sigma^{n-\frac{1}{2}}\|^2\\ &\quad - \sum_{k=1}^{n-1}(a_{n-k-1}^{(n,\alpha)} - a_{n-k}^{(n,\alpha)})\frac{\|\sigma^{k-\frac{1}{2}}\|^2 + \|\sigma^{n-\frac{1}{2}}\|^2}{2}\Big].\end{aligned} \tag{4.37}$$

For the first term on the right-hand-side of Equation (4.36), we use Equation (4.32) to obtain

$$\begin{aligned}\left(\delta_x^2 e^{n-\frac{1}{2}}, \sigma^{n-\frac{1}{2}}\right) &= \left(\delta_x^2 e^{n-\frac{1}{2}}, \delta_t e^{n-\frac{1}{2}} - (R_2^t)^{n-\frac{1}{2}}\right)\\ &= \left(\delta_x^2 e^{n-\frac{1}{2}}, \delta_t e^{n-\frac{1}{2}}\right) - \left(\delta_x^2 e^{n-\frac{1}{2}}, (R_2^t)^{n-\frac{1}{2}}\right).\end{aligned} \tag{4.38}$$

Applying the summation by parts for both terms on the right-hand-side of Equation (4.38), and then applying the Cauchy-Schwarz inequality to the result of the second term, we have

$$\left(\delta_x^2 e^{n-\frac{1}{2}}, \delta_t e^{n-\frac{1}{2}}\right) = -\frac{\|\delta_x e^n\|^2 - \|\delta_x e^{n-1}\|^2}{2\tau_n}, \tag{4.39}$$

and

$$\begin{aligned}-\left(\delta_x^2 e^{n-\frac{1}{2}}, (R_2^t)^{n-\frac{1}{2}}\right) &= \left(\delta_x e^{n-\frac{1}{2}}, \delta_x (R_2^t)^{n-\frac{1}{2}}\right)\\ &\le \frac{1}{2}\left(\|\delta_x e^n\|^2 + \|\delta_x e^{n-1}\|^2\right) + \frac{1}{4}\|\delta_x (R_2^t)^{n-\frac{1}{2}}\|^2.\end{aligned} \tag{4.40}$$

Inserting Equation (4.39) and Equation (4.40) into Equation (4.38) yields

$$\begin{aligned}\left(\delta_x^2 e^{n-\frac{1}{2}}, \sigma^{n-\frac{1}{2}}\right) &\le -\frac{\|\delta_x e^n\|^2 - \|\delta_x e^{n-1}\|^2}{2\tau_n}\\ &+ \frac{1}{2}\left(\|\delta_x e^n\|^2 + \|\delta_x e^{n-1}\|^2\right) + \frac{1}{4}\|\delta_x (R_2^t)^{n-\frac{1}{2}}\|^2.\end{aligned} \tag{4.41}$$

Using the Cauchy-Schwarz inequality again for the last two terms on the right-hand-side of Equation (4.36), we have

$$((R^x)^{n-\frac{1}{2}}, \sigma^{n-\frac{1}{2}}) \le \kappa_1 \|\sigma^{n-\frac{1}{2}}\|^2 + \frac{1}{4\kappa_1}\|(R^x)^{n-\frac{1}{2}}\|^2 \tag{4.42}$$

and

$$\begin{aligned}&((R_1^t)^{n-\frac{1}{2}}, \sigma^{n-\frac{1}{2}})\\ &\le \frac{\kappa_2(1-\epsilon)}{2\Gamma(2-\alpha)} a_{n-1}^{(n,\alpha)} \|\sigma^{n-\frac{1}{2}}\|^2 + \frac{\Gamma(2-\alpha)}{2\kappa_2(1-\epsilon)a_{n-1}^{(n,\alpha)}}\|(R_1^t)^{n-\frac{1}{2}}\|^2.\end{aligned} \tag{4.43}$$

Let

$$\begin{aligned}B^n &= \epsilon \tau_n a_{n-1}^{(n,\alpha)} \|\sigma^{n-\frac{1}{2}}\|^2\\ &+ \tau_n \sum_{k=1}^{n} a_{n-k}^{(n,\alpha)} \|\sigma^{k-\frac{1}{2}}\|^2 - \tau_n \sum_{k=1}^{n-1} a_{n-k-1}^{(n,\alpha)} \|\sigma^{k-\frac{1}{2}}\|^2.\end{aligned} \tag{4.44}$$

Inserting Equation (4.37) and Equations (4.41)–(4.43) into Equation (4.36) and then multiplying the result by $2\tau_n$, we have

$$\begin{aligned}&\|\delta_x e^n\|^2 - \|\delta_x e^{n-1}\|^2 + \frac{\kappa_2 B^n}{\Gamma(2-\alpha)}\\ &\le \tau_n\left(\|\delta_x e^n\|^2 + \|\delta_x e^{n-1}\|^2\right) + \frac{\Gamma(2-\alpha)\tau_n}{\kappa_2(1-\epsilon)a_{n-1}^{(n,\alpha)}}\|(R_1^t)^{n-\frac{1}{2}}\|^2\\ &+ \frac{\tau_n}{2\kappa_1}\|(R^x)^{n-\frac{1}{2}}\|^2 + \frac{\tau_n}{2}\|\delta_x (R_2^t)^{n-\frac{1}{2}}\|^2.\end{aligned} \tag{4.45}$$

In Equation (4.45), replacing the index n by m, and then summing the result from $m = 1$ to $m = n$ and noticing the result in Equation (4.15) and Equation (4.19), we obtain

$$
\begin{aligned}
\|\delta_x e^n\|^2 \lesssim & \sum_{m=1}^{n} \tau_m \left(\|\delta_x e^m\|^2 + \|\delta_x e^{m-1}\|^2 \right) + \sum_{m=1}^{n} \tau_m \|(R_1^t)^{m-\frac{1}{2}}\|^2 \\
& + \sum_{m=1}^{n} \tau_m \|(R^x)^{m-\frac{1}{2}}\|^2 + \sum_{m=1}^{n} \tau_m \|\delta_x (R_2^t)^{m-\frac{1}{2}}\|^2.
\end{aligned} \tag{4.46}
$$

Using the Gronwall inequality and the local truncation errors in Equations (4.26)–(4.28) for Equation (4.46), we have

(4.47)

$$
\begin{aligned}
\|\delta_x e^n\|^2 & \lesssim \sum_{m=1}^{n} \tau_m \|(R_1^t)^{m-\frac{1}{2}}\|^2 + \sum_{m=1}^{n} \tau_m \|(R^x)^{m-\frac{1}{2}}\|^2 + \sum_{m=1}^{n} \tau_m \|\delta_x (R_2^t)^{m-\frac{1}{2}}\|^2 \\
& \lesssim \sum_{m=1}^{n} \tau_m t_n^{\alpha-1} [m^{-2\min\{r\alpha, 3-\alpha\}} + (\tau_m^2 t_m^{\alpha-2})^2] \\
& \quad + \sum_{m=1}^{n} \tau_m (\tau_m^2 t_m^{\alpha-3})^2 + \sum_{m=1}^{n} \tau_m h^4.
\end{aligned}
$$

Following the direct result in [**42**] for the right-hand-side of Equation (4.47), and using the estimate $\|e^n\|_\infty \leq \frac{1}{2}\|\delta_x e^n\|$, we obtain the convergence conclusion. □

5. Parameter estimation for the fractional-order equation using the Levenberg-Marquardt algorithm

In this section, we regard the parameter estimation as an optimization problem and use the Levenberg-Marquardt algorithm to estimate the parameters α, κ_1, κ_2. Let $\mathbf{P}$ be a vector of the unknown parameters α, κ_1 and κ_2, i.e., $\mathbf{P} = [\alpha, \kappa_1, \kappa_2]$. Then we need find $\mathbf{P}$ to minimize the objective function $S(\mathbf{P})$, i.e.

$$
S(\mathbf{P}) = [\mathbf{Y} - \mathbf{U}(\mathbf{P})]^{\mathrm{T}} [\mathbf{Y} - \mathbf{U}(\mathbf{P})], \tag{5.1}
$$

where $\mathbf{Y} = [y_1, y_2, \cdots, y_\Omega]^{\mathrm{T}}$ is a measured heat flux vector obtained from the solution of the direct problem at the measurement location, $\mathbf{U}(\mathbf{P}) = [U_1(\mathbf{P}), \cdots, U_\Omega(\mathbf{P})]^{\mathrm{T}}$ is an estimated heat flux vector calculated from the direct problem by the difference method in Equations (3.14)–(3.16).

The Levenberg-Marquardt algorithm is an optimization method for solving nonlinear least-squares problem of parameter estimation by iteration. This iterative algorithm is given by [**43**]

$$
\mathbf{P}^{(k+1)} = \mathbf{P}^{(k)} + [(\mathbf{J}^{(k)})^{\mathrm{T}} \mathbf{J}^{(k)} + \mu_k \mathbf{D}^{(k)}]^{-1} (\mathbf{J}^{(k)})^{\mathrm{T}} [\mathbf{Y} - \mathbf{U}(\mathbf{P}^{(k)})], \tag{5.2}
$$

where the superscript (k) is the number of iterations, μ_k is a damping parameter, and the Jacobian matrix $\mathbf{J}$ is the sensitivity coefficient matrix defined by $\mathbf{J} \equiv \partial \mathbf{U} / \partial \mathbf{P}$. The matrix $\mathbf{D}^{(k)} = \mathrm{diag}((\mathbf{J}^{(k)})^{\mathrm{T}} \mathbf{J}^{(k)})$ is a diagonal matrix. In each iteration, the matrix term $\mu_k \mathbf{D}^{(k)}$ can be viewed as a regulator to make the matrix $(\mathbf{J}^{(k)})^{\mathrm{T}} \mathbf{J}^{(k)}$ reversible if necessary.

The Levenberg-Marquardt algorithm is summarized as follows. Suppose that the measurement value $\mathbf{Y}$ is available, the initial guess $\mathbf{P}^0$ is given for the unknown parameter $\mathbf{P}$, choose $\nu = 2$, $\eta = 1$, $\mu_0 = \epsilon \cdot \max\{\text{diag}((\mathbf{J}^{(k)})^{\mathrm{T}}\mathbf{J}^{(k)})\}$, where ϵ is chosen by the user, for example, $\epsilon = 10^{-1}$, and set $k = 0$.

(1) Solve the direct problem with the difference scheme (3.14)–(3.16) by using the estimated value $\mathbf{P}^{(k)} = \{\alpha, \kappa_1, \kappa_2\}^{\mathrm{T}}$ at the k-th iteration and compute $\mathbf{U}(\mathbf{P}^{(k)})$;

(2) Compute $S(\mathbf{P}^{(k)})$;

(3) Solve the direct problem with the difference scheme (3.14)–(3.16) three more times with each time perturbing only one of the parameters by a small amount and compute

$$\mathbf{U}^{(k)}(\alpha + \Delta\alpha, \kappa_1, \kappa_2; t)$$
$$\mathbf{U}^{(k)}(\alpha, \kappa_1 + \Delta\kappa_1, \kappa_2; t)$$
$$\mathbf{U}^{(k)}(\alpha, \kappa_1, \kappa_2 + \Delta\kappa_2; t)$$

For each case $U_i^{(k)} \equiv T(t_i)$, $i = 1, \cdots, \Omega$ are readily available.

(4) Compute the sensitivity coefficients in the Jacobian matrix $\mathbf{J}$ for each parameter $\alpha, \kappa_1, \kappa_2$. For example, with respect to α, we compute

$$\frac{\partial U_i^{(k)}}{\partial \alpha} = \frac{\mathbf{U}^{(k)}(\alpha + \Delta\alpha, \kappa_1, \kappa_2; t_i) - \mathbf{U}^{(k)}(\alpha, \kappa_1, \kappa_2; t_i)}{\Delta\alpha},$$

for $i = 1, \cdots, \Omega$, then determine the Jacobian matrix $\mathbf{J}^{(k)}$;

(5) Compute $\mathbf{g}^{(k)} = (\mathbf{J}^{(k)})^{\mathrm{T}}[\mathbf{Y} - \mathbf{U}(\mathbf{P}^{(k)})]$ and $\mathbf{d}^{(k)} = [(\mathbf{J}^{(k)})^{\mathrm{T}}\mathbf{J}^{(k)} + \mu_k\mathbf{D}^{(k)}]^{-1}\mathbf{g}^{(k)}$;

(6) Compute step size η by linear search method, and compute $\mathbf{h}^{(k)} = \eta\mathbf{d}^{(k)}$ and $\mathbf{P}^{(k+1)} = \mathbf{P}^{(k)} + \mathbf{d}^{(k)}$;

(7) Solve the direct problem with the difference scheme (3.14)–(3.16) by using the new estimated value $\mathbf{P}^{(k+1)}$ and compute $\mathbf{U}(\mathbf{P}^{(k+1)})$, $S(\mathbf{P}^{(k+1)})$, and compute

$$\rho = \frac{S(\mathbf{P}^{(k)}) - S(\mathbf{P}^{(k+1)})}{(\mathbf{h}^{(k)})^{\mathrm{T}}(\mu_k\mathbf{h}^{(k)} - g^{(k)})};$$

(8) If $\rho > 0$, replace μ_k by $\mu_k \max\{\frac{1}{3}, 1 - (2\rho - 1)^2\}$, and $\nu = 2$;

(9) If $\rho \leq 0$, replace μ_k by $\nu * \mu_k$, and $\nu = 2 * \mu$;

(10) Check the stopping criteria. If satisfied, stop the iteration; otherwise, replace k by $k + 1$, and return to Step (3).

6. Numerical results and testing

In this section, we compare the difference of the solutions between the fractional-order equation (1.3) and the delay equation (1.1). Meanwhile, we will show the advantage of the fractional-order equation for approximating the DE over the damped wave equation (DWE) and the traditional diffusion equation.

We consider the DE with a time delay κ_0 and a source term, coupled with simple initial and boundary conditions as follows:

$$\frac{\partial u(x, t + \kappa_0)}{\partial t} = \frac{\partial^2 u(x, t)}{\partial x^2} + e^{-(t+\kappa_0)}\sin[\pi x], \ (x, t) \in (0, 1) \times (0, \infty), \tag{6.1}$$

$$u(0, t) = 0, \quad u(1, t) = 0, \quad t > 0, \tag{6.2}$$

$$u(x, t) = \sin[\pi x], \quad (x, t) \in (0, 1) \times [-\kappa_0, 0]. \tag{6.3}$$

Using the Laplace transform coupled with the separation of variables method, the exact solution to the governing equations (6.1)–(6.3) can be found to be

$$\begin{aligned} u(x,t) = & \left\{ 1 - \sum_{n=0}^{\lfloor t/\kappa_0 \rfloor} (-1)^n \pi^{2(n+1)} \frac{(t-\kappa_0 n)^{n+1}}{(n+1)!} \right. \\ & \left. + \sum_{n=0}^{\lfloor t/\kappa_0 \rfloor} \frac{(-1)^n \pi^{2n} A(n)}{n!} \right\} \sin[\pi x], \end{aligned} \tag{6.4}$$

where $A(n)$ is calculated by the formula

$$\begin{aligned} A(n) = & (t-\kappa_0 n)^n - (-\kappa_0 n)^n e^{-t} \\ & + \sum_{m=1}^{n} \left\{ (-1)^{n-m+1} \left(\prod_{j=m}^{n} j \right) \left[(t-\kappa_0 n)^{m-1} - (-\kappa_0 n)^{m-1} e^{-t} \right] \right\}, \end{aligned} \tag{6.5}$$

and $\lfloor \cdot \rfloor$ denotes the greatest integer (or floor) function, i.e., $\lfloor p \rfloor$ denotes the greatest integer no larger than the real number $\lfloor p \rfloor$.

The fractional-order approximation for Equations (6.1)–(6.3) can be expressed as

$$\kappa_1 \cdot u_t(x,t) + \kappa_2 \cdot {}_0^C D_t^{\alpha} u(x,t) = \frac{\partial^2 u(x,t)}{\partial x^2} + e^{-(t+\kappa_0)} \sin[\pi x], \quad (x,t) \in (0,1) \times (0,\infty), \tag{6.6}$$

$$u(0,t) = 0, \quad u(1,t) = 0, \quad t > 0, \tag{6.7}$$

$$u(x,0) = \sin[\pi x], \quad \partial u(x,0)/\partial t = 0, \quad x \in (0,1). \tag{6.8}$$

The solution of the fractional-order equation (6.6)–(6.8) is approximated by using the finite difference scheme (3.14)–(3.16).

The damped wave equation (DWE) is presented as follows [**4**]:

$$\frac{\partial u(x,t)}{\partial t} + \kappa_0 \frac{\partial^2 u(x,t)}{\partial t^2} = \frac{\partial^2 u(x,t)}{\partial x^2} + e^{-(t+\kappa_0)} \sin[\pi x], \quad (x,t) \in (0,1) \times (0,\infty), \tag{6.9}$$

$$u(0,t) = 0, \quad u(1,t) = 0, \quad t > 0, \tag{6.10}$$

$$u(x,0) = \sin[\pi x], \quad \partial u(x,0)/\partial t = 0, \quad x \in (0,1). \tag{6.11}$$

The exact solution to the above initial-boundary condition problem can be obtained using the separation of variables method as follows:

$$u(x,t) = \sin[\pi x] \cdot \begin{cases} e^{-\frac{t}{2\kappa_0}}\left[\cosh(wt) + \frac{\sinh(wt)}{\sqrt{|\Delta|}}\right] \\ -\frac{2\kappa_0^2 e^{-\kappa_0}}{\sqrt{|\Delta|}(1-\sqrt{|\Delta|}-2\kappa_0)} e^{\frac{-1+\sqrt{|\Delta|}}{2\kappa_0}t} \\ +\frac{2\kappa_0^2 e^{-\kappa_0}}{\sqrt{|\Delta|}(1+\sqrt{|\Delta|}-2\kappa_0)} e^{\frac{-1-\sqrt{|\Delta|}}{2\kappa_0}t} \\ +\frac{4\kappa_0^2}{(1-2\kappa_0)^2-\Delta} e^{-(t+\kappa_0)}, \quad \kappa_0 < \kappa_c, \\ e^{-\frac{t}{2\kappa_0}}\left[1 + \frac{t}{2\kappa_0} - \left(\frac{2\kappa_0}{1-2\kappa_0}\right)^2 e^{-\kappa_0} - \frac{2\kappa_0 e^{-\kappa_0}}{1-2\kappa_0}t\right] \\ +\left(\frac{2\kappa_0}{1-2\kappa_0}\right)^2 e^{-(t+\kappa_0)}, \quad \kappa_0 = \kappa_c, \\ e^{\gamma t}\left[\cos(wt) + \frac{\sin(wt)}{\sqrt{|\Delta|}}\right] \\ -e^{\gamma t-\kappa_0} \cdot \frac{1}{(\gamma+1)^2+w^2}\left[\cos(wt) - \frac{\gamma+1}{w}\sin(wt)\right] \\ +e^{-(t+\kappa_0)} \cdot \frac{1}{(\gamma+1)^2+w^2}, \quad \kappa_0 > \kappa_c, \end{cases} \tag{6.12}$$

where $\kappa_c = (2\pi)^{-2}$ is a critical value of the lag time, $\gamma = \frac{-1}{2\kappa_0}$, $w = (2\kappa_0)^{-1}\sqrt{|\Delta|}$, and $\Delta = 1 - 4\pi^2\kappa_0$. However, we must reject the case $\kappa_0 > \kappa_c$ as it allows u to assume negative values, in opposition to the fact that u denotes an absolute quantity [**6**]. Furthermore, by letting $\kappa_0 \to 0$ in Equation (6.12), the solution of the traditional diffusion equation is recovered, i.e.,

$$u(x,t) = \left(\frac{e^{-t}}{\pi^2 - 1} + \frac{\pi^2 - 2}{\pi^2 - 1}e^{-\pi^2 t}\right)\sin[\pi x]. \tag{6.13}$$

6.1. Convergence test of the difference scheme. We first check the efficiency and convergence orders of the finite difference scheme (3.14)–(3.16). In numerical simulations, we computed the time domain $t \in [0,1]$, and chose the parameters $\kappa_0 = 1$, $\kappa_1 = 1$, $\kappa_2 = 1$ and the fractional order $\alpha = 1.3, 1.5, 1.7$, respectively, in Equations (6.6)–(6.8). Since the exact solution is not available in general, we use

$$\text{Error}_1(\Delta t) = \max_{0\le i\le M}\left|u_i^N(\Delta x, \Delta t) - u_i^{2N}\left(\Delta x, \frac{\Delta t}{2}\right)\right|$$

$$\text{Error}_2(\Delta x) = \max_{0\le i\le M}\left|u_i^N(\Delta x, \Delta t) - u_{2i}^{N}\left(\frac{\Delta x}{2}, \Delta t\right)\right|$$

to measure the numerical errors in time and in space, respectively, where $u_i^N(\Delta x, \Delta t)$ denotes the numerical solution at grids (x_i, t_N). The corresponding temporal and spatial convergence orders are defined by

$$\text{Order}_t = \log_2\frac{\text{Error}_1(2\Delta t)}{\text{Error}_1(\Delta t)}, \quad \text{Order}_x = \log_2\frac{\text{Error}_2(2\Delta x)}{\text{Error}_2(\Delta x)}.$$

Next, we reported the maximum norm errors of numerical solution of the fractional-order equation (6.6)–(6.8) and the corresponding convergence accuracy, which was solved via the difference scheme (3.14)–(3.16). In order to test the temporal convergence order, we fixed the space step $\Delta x = 1/500$, which ensures that the errors in time predominate. From Table 1, one can observe that the temporal convergence accuracy reaches only first-order when taking an uniform time mesh (i.e., $r = 1$), and achieves the optimal $(3-\alpha)$-order accuracy when taking the grading parameter $r = 3$. This indicates that the larger value of the grading parameter r will concentrate more mesh points near $t = 0$ to conquer the weak regularity at the initial time of the solution of the fractional-order equation. Similarly, in order to test the spatial convergence order, we fixed the number of the temporal grid points $N = 1000$, which ensures that the errors in space predominate. From Table 2, numerical results show the second-order convergence in space of the difference scheme (3.14)–(3.16) for both $r = 1$ and $r = 3$.

TABLE 1. Maximum errors and temporal convergence orders ($M = 500$).

		$r=1$		$r=3$	
α	N	$\text{Error}_1(\Delta t)$	Order_t	$\text{Error}_1(\Delta t)$	Order_t
1.3	50	3.622E–02		8.836E–04	
	100	1.947E–02	0.895	2.694E–04	1.714
	200	1.011E–02	0.946	8.197E–05	1.717
	400	5.152E–03	0.972	2.495E–05	1.716
	800	2.601E–03	0.986	7.625E–06	1.710
1.5	50	2.846E–02		4.234E–03	
	100	1.548E–02	0.878	1.489E–03	1.507
	200	8.103E–03	0.934	5.243E–04	1.506
	400	4.153E–03	0.964	1.848E–04	1.504
	800	2.105E–03	0.980	6.522E–05	1.503
1.7	50	1.344E–02		1.610E–02	
	100	7.728E–03	0.798	6.687E–03	1.268
	200	4.232E–03	0.869	2.743E–03	1.286
	400	2.250E–03	0.912	1.119E–03	1.294
	800	1.174E–03	0.938	4.553E–04	1.297

6.2. Fractional-order approximation for the DE with a source term. In the following computation, we determine the optimal parameters α, κ_1 and κ_2 in the fractional-order equation (6.6)–(6.8) with various values of κ_0. In our computation, we computed $T = 0.8$, and chose the space step (Denoted by Δx) to be 0.01, and took a graded time mesh $\{t_n \mid t_n = T(n/N)^r\}_{n=0}^N$ with the grading parameter $r = 3$ and $N = 1000$. We chose different initial guess $\mathbf{P}^0 = [\alpha^0, \kappa_1^0, \kappa_2^0]$, the tolerance $\epsilon = 10^{-13}$, the measurement value $u_{\text{measurement}} = u_{\text{exact}}$ from the analytical solution of the direct problem (6.1)–(6.3) on the graded time mesh Ω_τ. Table 3 lists the estimated values of the parameter $\mathbf{P}^{\text{inv}} = [\alpha, \kappa_1, \kappa_2]^{\text{inv}}$ and the number of iterations for different initial guess with various values of $\kappa_0 = 0.25\kappa_c, 0.5\kappa_c, \kappa_c$. From the numerical results in Tables 3, one can see that, after several iterations, the optimal parameter $\mathbf{P}^{\text{inv}} = [\alpha, \kappa_1, \kappa_2]^{\text{inv}}$ can be obtained, and the different initial guesses have no affect on the final estimation. This indicates that the present difference scheme with the Levenberg-Marquardt algorithm is effective to obtain the estimation of the parameters α, κ_1, κ_2.

TABLE 2. Maximum errors and spatial convergence orders ($N = 1000$).

α	M	$r=1$		$r=3$	
		$\text{Error}_2(\Delta x)$	Order_x	$\text{Error}_2(\Delta x)$	Order_x
1.3	6	2.302E–03		2.242E–03	
	12	5.924E–04	1.958	5.773E–04	1.957
	24	1.491E–04	1.990	1.453E–04	1.990
	48	3.734E–05	1.998	3.639E–05	1.998
1.5	6	8.700E–03		8.679E–03	
	12	2.199E–03	1.984	2.194E–03	1.984
	24	5.511E–04	1.996	5.498E–04	1.996
	48	1.379E–04	1.999	1.375E–04	1.999
1.7	6	2.049E–02		2.054E–02	
	12	5.190E–03	1.981	5.202E–03	1.981
	24	1.301E–03	1.996	1.305E–03	1.996
	48	3.256E–04	1.999	3.264E–04	1.999

TABLE 3. Optimal estimation of the parameter $\mathbf{P} = [\alpha, \kappa_1, \kappa_2]$.

Value κ_0	Initial guess $\mathbf{P}^0$	Number of iterations	Estimated value $\mathbf{P}^{\text{inv}}$
$0.25\kappa_c$	$(1.30, 0.90, \kappa_0)$	15	$(1.237876, 0.935840, 0.000392)$
	$(1.20, 1.00, \kappa_0)$	9	$(1.237876, 0.935840, 0.000392)$
	$(1.25, 1.00, \kappa_0/10)$	12	$(1.237877, 0.935840, 0.000392)$
	$(1.10, 0.90, \kappa_0/10)$	19	$(1.237877, 0.935840, 0.000392)$
$0.5\kappa_c$	$(1.30, 0.90, \kappa_0)$	32	$(1.234992, 0.866506, 0.000764)$
	$(1.20, 1.00, \kappa_0)$	33	$(1.234991, 0.866506, 0.000764)$
	$(1.25, 1.00, \kappa_0/10)$	36	$(1.234990, 0.866506, 0.000764)$
	$(1.10, 0.90, \kappa_0/10)$	37	$(1.234991, 0.866506, 0.000764)$
κ_c	$(1.30, 0.90, \kappa_0)$	33	$(1.227515, 0.701276, 0.001407)$
	$(1.20, 1.00, \kappa_0)$	22	$(1.227516, 0.701276, 0.001407)$
	$(1.25, 1.00, \kappa_0/10)$	39	$(1.227517, 0.701276, 0.001407)$
	$(1.10, 0.90, \kappa_0/10)$	41	$(1.227515, 0.701276, 0.001407)$

Finally, we computed the solution of the fractional-order equation (6.6)–(6.8) based on the difference scheme (3.14)–(3.16), the analytical solution in Equation (6.4) of the DE (6.1)–(6.3), the analytical solution in Equation (6.12) of the DWE (6.9)–(6.11), and the analytical solution in Equation (6.13) of the diffusion equation. The values of the parameters α, κ_1, κ_2 in Equation (6.6) were chosen based on the corresponding estimated values in Table 3. In Fig. 1–Fig. 3, we plotted the temporal evolution of the temperature vs. x profile for $\kappa_0 = 0.25\kappa_c, 0.5\kappa_c, \kappa_c$, at the time $t = t_5$, t_{50}, t_{100}, t_{500}, t_{800}, t_{1000}, respectively. From Fig. 1–Fig. 3, one may see that the solution of the fractional-order equation agrees very well with the solution of the DE (6.1)–(6.3). Fig. 4 shows 3D comparison between the numerical solution of the difference scheme (3.14)–(3.16) and the analytical solution of the DE (6.1)–(6.3) at three different values of κ_0. Results indicate that the fractional-order equation (6.6)–(6.8) is a good approximation for the DE (6.1)–(6.3).

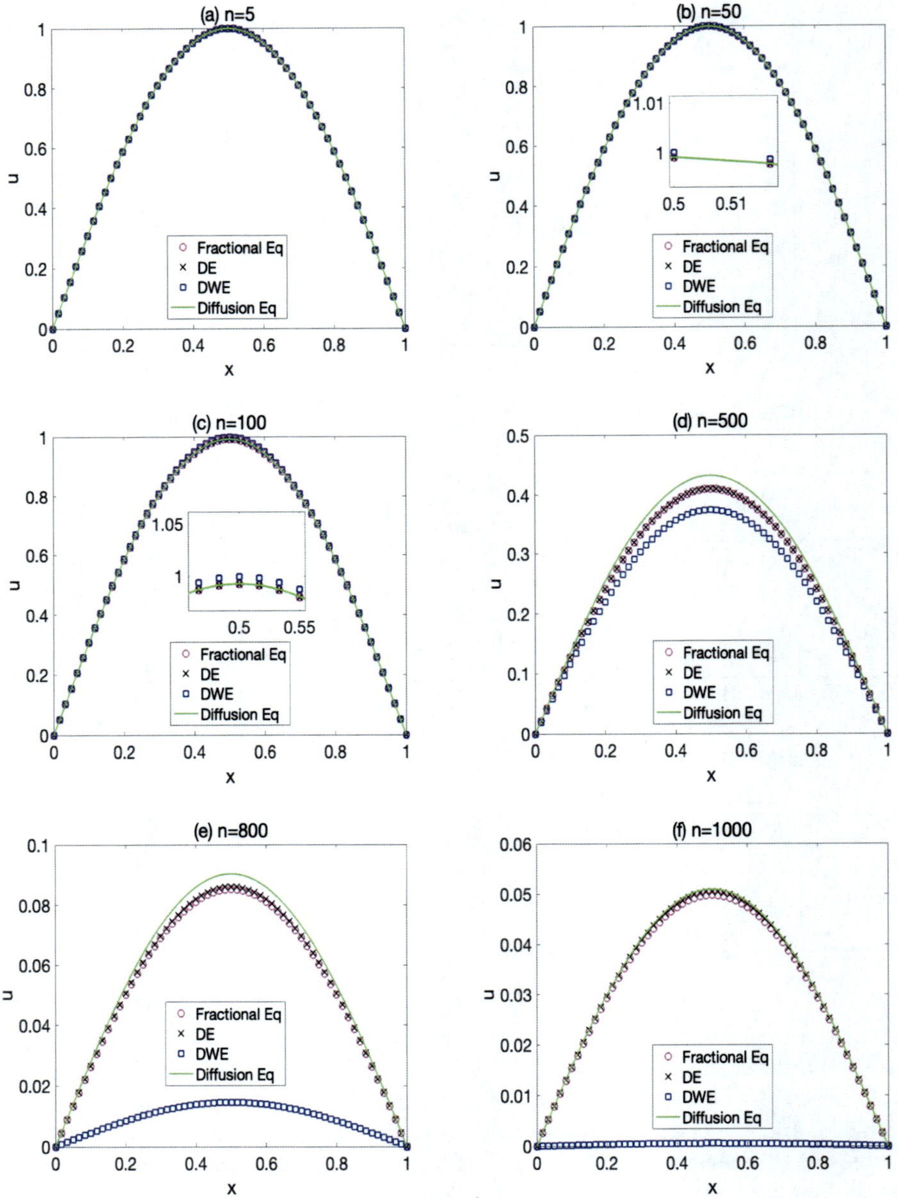

FIGURE 1. u vs. x for $\kappa_0 = 0.25\kappa_c$ at different times $t_n = T(n/N)^r$ with $r = 3$ and $n = 5, 50, 100, 500, 800, 1000$, respectively.

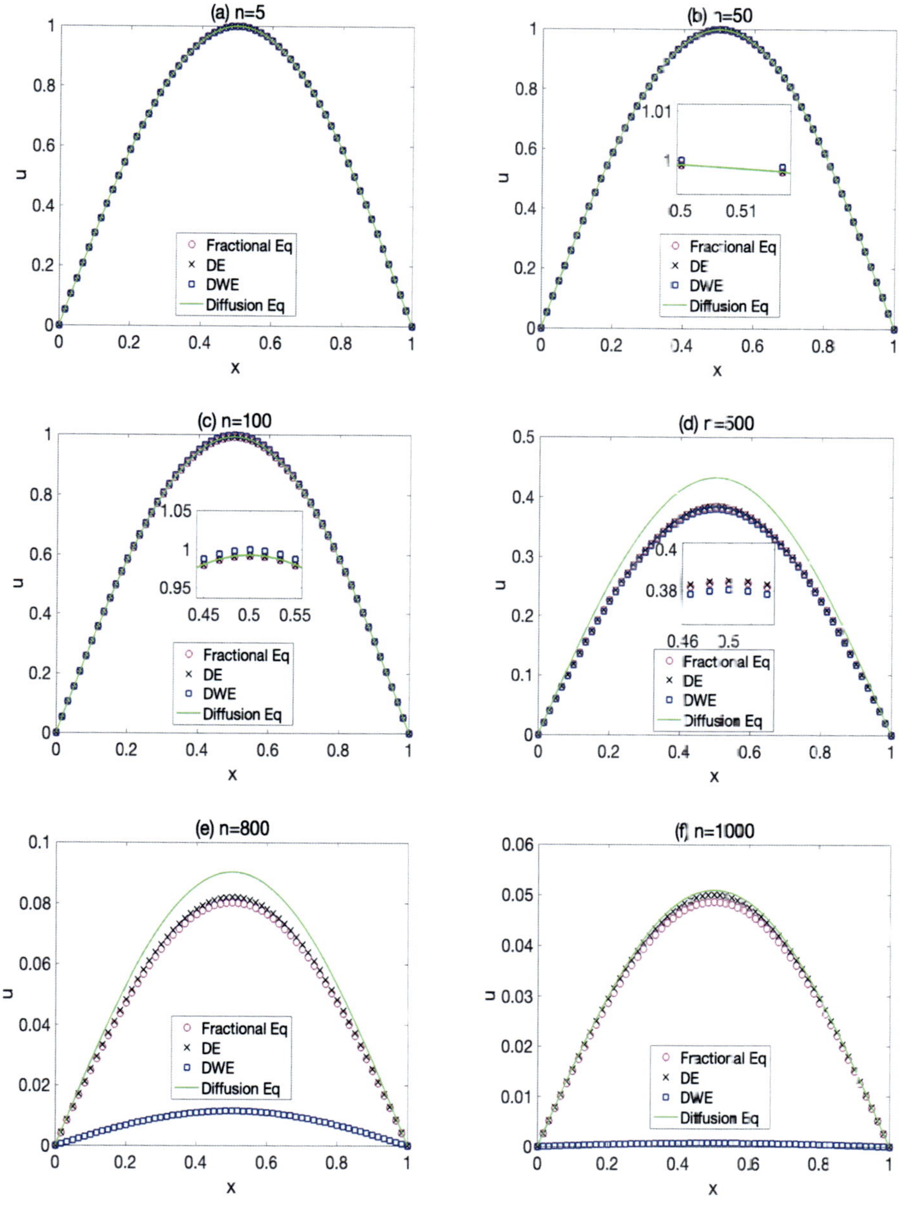

FIGURE 2. u vs. x for $\kappa_0 = 0.5\kappa_c$ at different times $t_n = T(n/N)^r$ with $r = 3$ and $n = 5, 50, 100, 500, 800, 1000$, respectively.

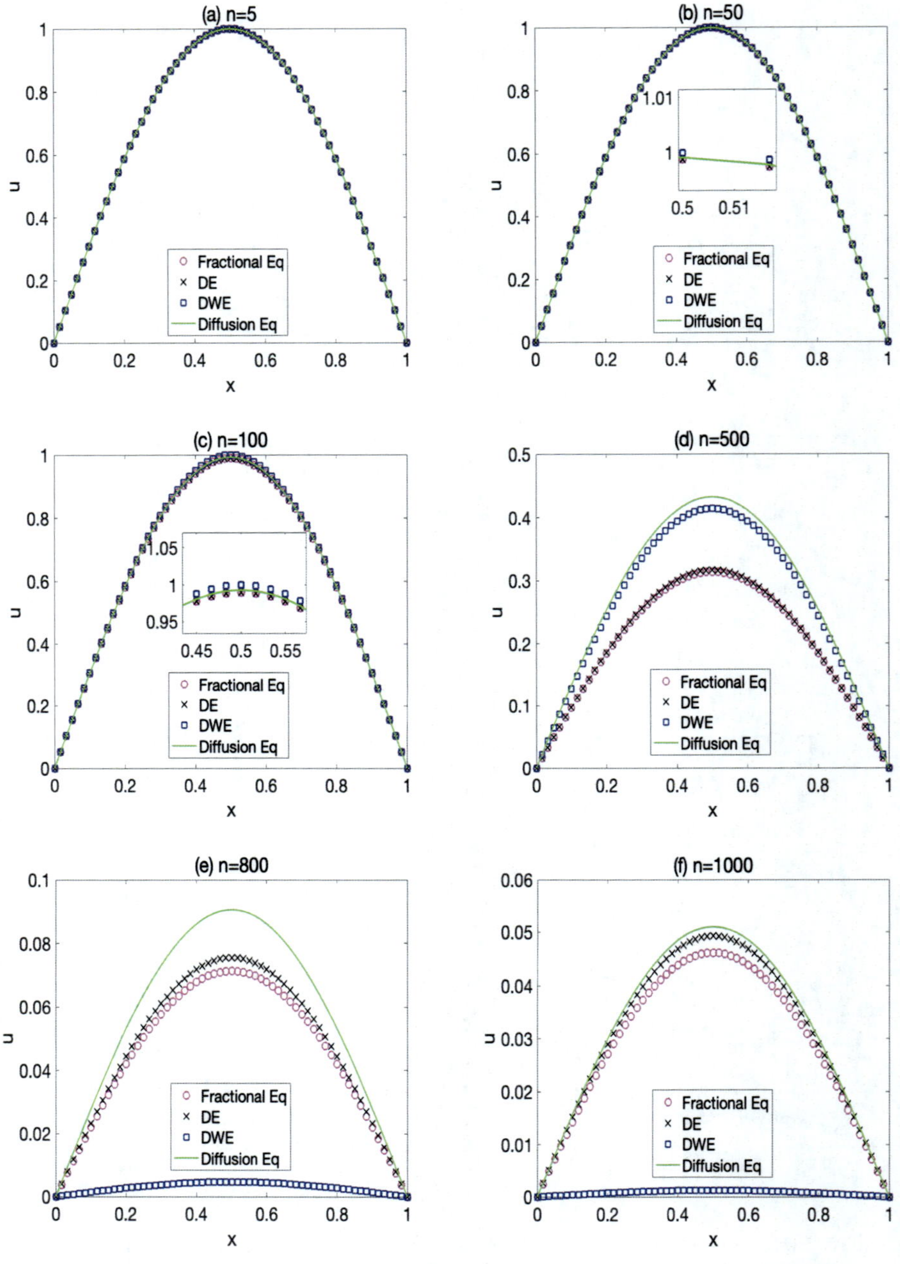

FIGURE 3. u vs. x for $\kappa_0 = \kappa_c$ at different times $t_n = T(n/N)^r$ with $r = 3$ and $n = 5, 50, 100, 500, 800, 1000$, respectively.

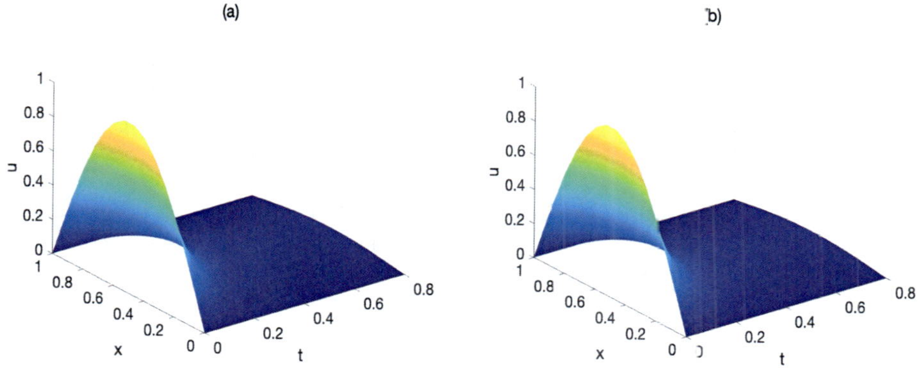

Case 1. $\kappa_0 = 0.25\kappa_c$

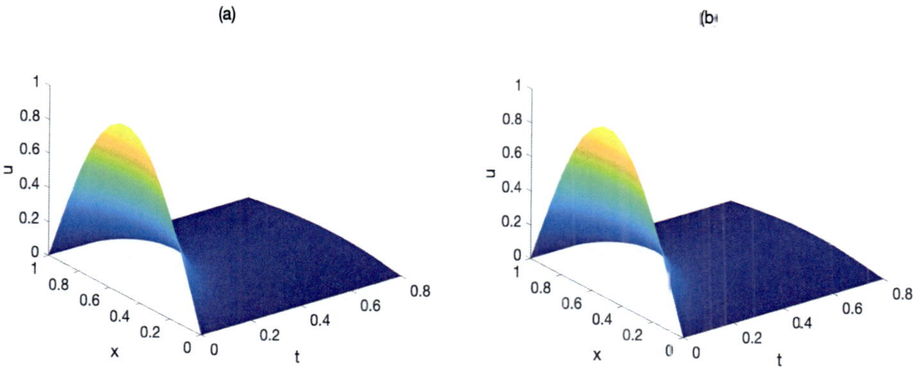

Case 2. $\kappa_0 = 0.5\kappa_c$

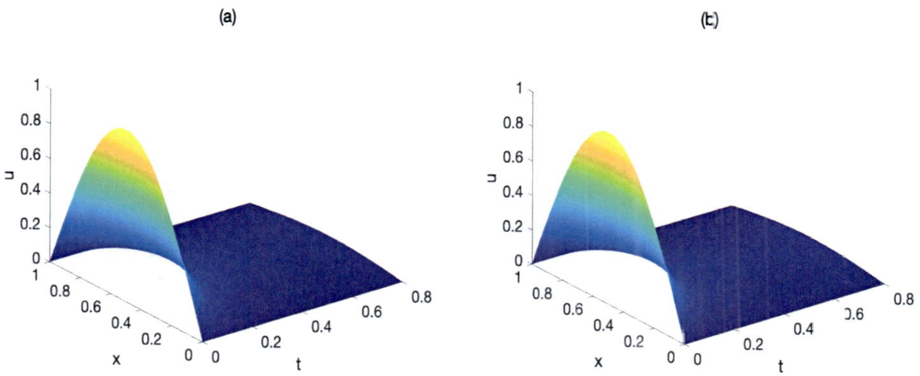

Case 3. $\kappa_0 = \kappa_c$

FIGURE 4. The numerical solution (left) of the fractional-order equation (6.6)–(6.8) and the exact solution (right) of the DE (6.1)–(6.3) at $\kappa_0 = 0.25\kappa_c, 0.5\kappa_c, \kappa_c$, respectively.

7. Conclusion

In this paper, we have proposed a fractional-order equation for approximating a delay equation with a source term. An energy estimation of the obtained fractional-order equation is proved. To obtain the solution of the fractional-order equation, we have further developed an unconditionally stable difference scheme with a graded time mesh. By optimizing the parameters α, κ_1, κ_2 in the fractional-order equation based on the time lagging κ_c, we have obtained an optimal fractional-order equation and an associated numerical scheme whose solutions we take as approximations to the corresponding solutions of the DE with a source term.

References

[1] Carlo Cattaneo, *Sur une forme de l'équation de la chaleur éliminant le paradoxe d'une propagation instantanée* (French), C. R. Acad. Sci. Paris **247** (1958), 431–433. MR95680

[2] Pierre Vernotte, *Les paradoxes de la théorie continue de l'équation de la chaleur* (French), C. R. Acad. Sci. Paris **246** (1958), 3154–3155. MR95679

[3] S. Su, W. Dai, P.M. Jordan, R.E. Mickens, *Comparison of the solutions of a phase-lagging heat transport equation and damped wave equation*, Int. J. Heat Mass Transf. **48** (2005), 2233–2241.

[4] S. Su, W. Dai, *Comparison of the solutions of a phase-lagging heat transport equation and damped wave equation with a heat source*, Int. J. Heat Mass Transf. **49** (2006), 2793–2801.

[5] J.K. Chen, J.E. Beraun, C.L. Tham, *Investigation of thermal response caused by pulse laser heating*, Numer. Heat Transfer A 44 (2003) 705–722.

[6] Ronald E. Mickens and P. M. Jordan, *A positivity-preserving nonstandard finite difference scheme for the damped wave equation*, Numer. Methods Partial Differential Equations **20** (2004), no. 5, 639–649, DOI 10.1002/num.20003. MR2076341

[7] A. Carpinteri and F. Mainardi, *Fractals and fractional calculus in continuum mechanics*, Springer-Verlag Wien, 1997.

[8] Igor Podlubny, *Fractional differential equations: An introduction to fractional derivatives, fractional differential equations, to methods of their solution and some of their applications*, Mathematics in Science and Engineering, vol. 198, Academic Press, Inc., San Diego, CA, 1999. MR1658022

[9] E. Barkai, R. Metzler, and J. Klafter, *From continuous time random walks to the fractional Fokker-Planck equation*, Phys. Rev. E (3) **61** (2000), no. 1, 132–138, DOI 10.1103/PhysRevE.61.132. MR1736459

[10] Anatoly A. Kilbas, Hari M. Srivastava, and Juan J. Trujillo, *Theory and applications of fractional differential equations*, North-Holland Mathematics Studies, vol. 204, Elsevier Science B.V., Amsterdam, 2006. MR2218073

[11] R. Klages, G. Radons, I.M. Sokolov, *Anomalous transport: Foundations and applications*, Wiley-VCH, 2008.

[12] V. E. Tarasov, *Fractional dynamics: applications of fractional calculus to dynamics of particles*, fields and media. Springer, 2011.

[13] Antoine Mellet, Stéphane Mischler, and Clément Mouhot, *Fractional diffusion limit for collisional kinetic equations*, Arch. Ration. Mech. Anal. **199** (2011), no. 2, 493–525, DOI 10.1007/s00205-010-0354-2. MR2763032

[14] Dumitru Baleanu, Kai Diethelm, Enrico Scalas, and Juan J. Trujillo, *Fractional calculus: Models and numerical methods*, Series on Complexity, Nonlinearity and Chaos, vol. 3, World Scientific Publishing Co. Pte. Ltd., Hackensack, NJ, 2012, DOI 10.1142/9789814355216. MR2894576

[15] V. Uchaikin, R. Sibatov, *Fractional kinetics in solids: Anomalous charge transport in semiconductors, dielectrics and nanosystems*, World Scientific Publishing Company, 2013.

[16] Yuriy Povstenko, *Fractional thermoelasticity*, Solid Mechanics and its Applications, vol. 219, Springer, Cham, 2015, DOI 10.1007/978-3-319-15335-3. MR3287225

[17] H.G. Sun, Y. Zhang, D. Baleanu, W. Chen, Y.Q. Chen, *A new collection of real world applications of fractional calculus in science and engineering*, Commun. Nonlinear. Sci. Numer. Simulat. **64** (2018) 213–231.

[18] Cui-Cui Ji, Weizhong Dai, and Zhi-zhong Sun, *Numerical method for solving the time-fractional dual-phase-lagging heat conduction equation with the temperature-jump boundary condition*, J. Sci. Comput. **75** (2018), no. 3, 1307–1336, DOI 10.1007/s10915-017-0588-3. MR3798103

[19] Cui-Cui Ji, Weizhong Dai, and Zhi-zhong Sun, *Numerical schemes for solving the time-fractional dual-phase-lagging heat conduction model in a double-layered nanoscale thin film*, J. Sci. Comput. **81** (2019), no. 3, 1767–1800, DOI 10.1007/s10915-019-01062-6. MR4036969

[20] Cui-Cui Ji and Weizhong Dai, *Numerical algorithm with fourth-order spatial accuracy for solving the time-fractional dual-phase-lagging nanoscale heat conduction equation*, Numer. Math. Theory Methods Appl. **16** (2023), no. 2, 511–540. MR4573912

[21] Wenzhi Yang and Zengtao Chen, *Fractional single-phase lag heat conduction and transient thermal fracture in cracked viscoelastic materials*, Acta Mech. **230** (2019), no. 10, 3723–3740, DOI 10.1007/s00707-019-02474-z. MR4017672

[22] Bo Yu, Xiaoyun Jiang, and Haitao Qi, *Numerical method for the estimation of the fractional parameters in the fractional mobile/immobile advection-diffusion model*, Int. J. Comput. Math. **95** (2018), no. 6-7, 1131–1150, DOI 10.1080/00207160.2017.1378811. MR3783174

[23] Xiaoqing Chi, Bo Yu, and Xiaoyun Jiang, *Parameter estimation for the time fractional heat conduction model based on experimental heat flux data*, Appl. Math. Lett. **102** (2020), 106094, 8, DOI 10.1016/j.aml.2019.106094. MR4022044

[24] Weilin Yang, Mahsa Nourazar, Zengtao Chen, Keqiang Hu, and Xueyang Zhang, *Dynamic response of a cracked thermopiezoelectric strip under thermoelectric loading using fractional heat conduction*, Appl. Math. Model. **103** (2022), 580–603, DOI 10.1016/j.apm.2021.11.005. MR4344957

[25] A.E. Abouelregal, H. Ersoy, O. Civalek, *A new heat conduction model for viscoelastic micro beams considering the magnetic field and thermal effects*, Wave. Random. Complex. 2021, DOI:10.1080/17455030.2021.2009152.

[26] X.P. Wang, H.T. Qi, X. Yang, H.Y. Xu, *Analysis of the time-space fractional bioheat transfer equation for biological tissues during laser irradiation*, Int. J. Heat Mass Transf. **177** (2021), 121555.

[27] Cui-Cui Ji, Weizhong Dai, and Ronald E. Mickens, *A fractional-order alternative for phase-lagging equation*, Int. J. Numer. Anal. Model. **20** (2023), no. 3, 391–406, DOI 10.4208/ijnam2023-1016. MR4594672

[28] Cui-Cui Ji and Zhi-zhong Sun, *A high-order compact finite difference scheme for the fractional sub-diffusion equation*, J. Sci. Comput. **64** (2015), no. 3, 959–985, DOI 10.1007/s10915-014-9956-4. MR3377846

[29] Cui-Cui Ji, Zhi-zhong Sun, and Zhao-peng Hao, *Numerical algorithms with high spatial accuracy for the fourth-order fractional sub-diffusion equations with the first Dirichlet boundary conditions*, J. Sci. Comput. **66** (2016), no. 3, 1148–1174, DOI 10.1007/s10915-015-0059-7. MR3456968

[30] Anatoly A. Alikhanov, *A new difference scheme for the time fractional diffusion equation*, J. Comput. Phys. **280** (2015), 424–438, DOI 10.1016/j.jcp.2014.09.031. MR3273144

[31] Martin Stynes, Eugene O'Riordan, and José Luis Gracia, *Error analysis of a finite difference method on graded meshes for a time-fractional diffusion equation*, SIAM J. Numer. Anal. **55** (2017), no. 2, 1057–1079, DOI 10.1137/16M1082329. MR3639581

[32] Bangti Jin, Buyang Li, and Zhi Zhou, *Correction of high-order BDF convolution quadrature for fractional evolution equations*, SIAM J. Sci. Comput. **39** (2017), no. 6, A3129–A3152, DOI 10.1137/17M1118816. MR3738850

[33] Hong-lin Liao, Dongfang Li, and Jiwei Zhang, *Sharp error estimate of the nonuniform L1 formula for linear reaction-subdiffusion equations*, SIAM J. Numer. Anal. **56** (2018), no. 2, 1112–1133, DOI 10.1137/17M1131829. MR3790081

[34] Hong Wang and Danping Yang, *Wellposedness of variable-coefficient conservative fractional elliptic differential equations*, SIAM J. Numer. Anal. **51** (2013), no. 2, 1088–1107, DOI 10.1137/120892295. MR3036999

[35] Fanhai Zeng, Changpin Li, Fawang Liu, and Ian Turner, *Numerical algorithms for time-fractional subdiffusion equation with second-order accuracy*, SIAM J. Sci. Comput. **37** (2015), no. 1, A55–A78, DOI 10.1137/14096390X. MR3296626

[36] Yirang Yuan, Changfeng Li, and Qing Yang, *Mixed finite element-second order upwind fractional step difference scheme of Darcy-Forchheimer miscible displacement and its numerical

analysis, J. Sci. Comput. **86** (2021), no. 2, Paper No. 24, 19, DOI 10.1007/s10915-020-01393-9. MR4197404

[37] Yumin Lin and Chuanju Xu, *Finite difference/spectral approximations for the time-fractional diffusion equation*, J. Comput. Phys. **225** (2007), no. 2, 1533–1552, DOI 10.1016/j.jcp.2007.02.001. MR2349193

[38] Mohsen Zayernouri and George Em Karniadakis, *Discontinuous spectral element methods for time- and space-fractional advection equations*, SIAM J. Sci. Comput. **36** (2014), no. 4, B684–B707, DOI 10.1137/130940967. MR3240858

[39] Muhammad Usman, Muhammad Hamid, and Moubin Liu, *Novel operational matrices-based finite difference/spectral algorithm for a class of time-fractional Burger equation in multidimensions*, Chaos Solitons Fractals **144** (2021), Paper No. 110701, 20, DOI 10.1016/j.chaos.2021.110701. MR4209683

[40] A. A. Alikhanov, *A priori estimates for solutions of boundary value problems for equations of fractional order* (Russian, with Russian summary), Differ. Uravn. **46** (2010), no. 5, 658–664, DOI 10.1134/S0012266110050058; English transl., Differ. Equ. **46** (2010), no. 5, 660–666. MR2797545

[41] Xiangcheng Zheng and Hong Wang, *Wellposedness and regularity of a nonlinear variable-order fractional wave equation*, Appl. Math. Lett. **95** (2019), 29–35, DOI 10.1016/j.aml.2019.03.015. MR3928677

[42] Jinye Shen, Martin Stynes, and Zhi-Zhong Sun, *Two finite difference schemes for multi-dimensional fractional wave equations with weakly singular solutions*, Comput. Methods Appl. Math. **21** (2021), no. 4, 913–928, DOI 10.1515/cmam-2020-0101. MR4319052

[43] M.N. Ozisik, *Heat conduction*, 2nd ed., Wiley, New York, 1993.

School of Mathematics and Statistics, Qingdao University, Qingdao 266071, People's Republic of China
Email address: `cuicuiahuan@163.com`

Mathematics and Statistics, Louisiana Tech University, Ruston, Louisiana 71272
Email address: `dai@coes.latech.edu`

Contemporary Mathematics
Volume **793**, 2024
https://doi.org/10.1090/conm/793/15902

Multistability in a discrete-time SI epidemic model with Ricker growth: Infection-induced changes in population dynamics

Laura F. Strube and Lauren M. Childs

ABSTRACT. One-dimensional discrete-time population models, such as those with logistic or Ricker growth, may exhibit periodic or chaotic dynamics depending on the parameter values. Adding epidemiological interactions into a population model increases its dimension and the resulting complexity of its dynamics. Previous work showed that a discrete susceptible-infectious-recovered (SIR) model with Ricker growth and density-dependent, non-fatal infection exhibits qualitatively similar total population dynamics in the presence and absence of disease. In contrast, a more complicated three-class susceptible-infectious-virus (SIV) system that includes disease-induced mortality does not. Instead, infection in the SIV system shifts the periodic behavior in a manner that distinguishes it from the corresponding disease-free system. Here, we examine a two-class susceptible-infectious (SI) model with Ricker population growth, density-dependent infection, and parameters that tune disease-induced mortality and the capacity of infected individuals to reproduce. We use numerical bifurcation analysis to determine the influence of infection on the qualitative structure of the long-time behavior. We show that when disease is allowed to alter reproduction or disease-induced mortality, infection produces distinctly different bifurcation structures than that of the underlying disease-free system. In particular, it shifts both the location of period-doubling bifurcations and the onset of chaos. Additionally, we show that disease-induced mortality introduces multistability into the system such that a given set of model parameters can produce multiple distinct qualitative behaviors depending upon initial conditions. This work demonstrates that the infection-induced changes in dynamics observed by previous authors do not require the presence of infecting virus particles in the environment. In doing so, our work also advances understanding of the conditions under which discrete epidemiological models exhibit multistability.

2020 *Mathematics Subject Classification.* Primary 92B05, 39A60; Secondary 39A33, 92D40.
Key words and phrases. Discrete-time model, epidemiology, population dynamics, period-doubling, chaos, density-dependence, disease-induced mortality.
This work was supported by NSF Grant # 1853495. LMC and LFS thank Ryan Hughes for initial ideas. LFS thanks Shelby M. Scott and Jessica Crawshaw for helpful feedback on an early draft.

1. Introduction

Discrete-time population models have a long history of application to biological systems dating back to the early 1900s when they were used to describe fish populations [**7**, **22**]. These models are valuable for their utility in describing ecological and epidemiological systems in which population dynamics are governed by biological events that occur at regular time intervals such as seasonal birth or death. Discrete-time models have been used to describe other systems as well, such as insect populations and infectious disease, e.g., [**4**, **26**]. One distinct feature of discrete dynamical systems is that the are able to produce complex dynamics at low dimension [**16**, **17**, **19**]. For example, while systems of continuous ordinary differential equations require a minimum of two equations to produce periodic dynamics, a discrete difference equation of the form

$$S_{t+1} = rS_te^{-bS_t} + (1-d)S_t, \tag{1.1}$$

which models a single population, can produce periodic behavior without interactions from any other population classes. It can also produce chaotic behavior in which the long-time population dynamics move between multiple population sizes without pattern instead of settling into either a fixed point or cyclic behavior [**3**, **17**, **19**]. The qualitative type of dynamics this equation, or any discrete-time population model, exhibits (fixed point, cyclic, chaotic, etc.) depends on the model parameters.

The growth term in Equation (1.1), rXe^{-bX}, was proposed by William Ricker for the study of fishery populations [**22**]. It is one of a number of functional forms including logistic that are classified as overcompensatory, and describe a type of reproduction in which the maximum level of growth occurs at intermediate population sizes and is reduced for population sizes above or below this level [**2**, **6**, **13**, **20**]. This form is used to describe scenarios in which competition for resources at large population size drives per-capita resource levels below the threshold needed for reproduction, also known as scramble competition [**6**]. In the Ricker term, e^{-bX} describes the population-size-dependent probability of reproduction in a given time step, and r describes the intrinsic per-capita reproduction count for the population in a given time step. That is, r is the maximum level of reproduction and e^{-bX} is a scale factor in $[0, 1]$ that tunes reproduction based on population size.

Recently, van den Driessche and Yakubu [**23**] used discrete-time population models to examine the influence of infection in two distinct disease systems in which the underlying population exhibits Ricker growth and each time step is considered a generation. They began by considering a three-dimensional Susceptible-Infectious-Recovered (SIR) model in which disease is transmitted via direct contact with infected individuals and may be eventually cleared. In this SIR model (Figure 1), recovery of infectious individuals occurs with constant probability and transmission is assumed to occur via a density-dependent Poisson process. This infection form is analogous to a mass-action infection process in the continuous ordinary differential equation framework.

They also considered a three-dimensional Susceptible-Infectious-Virus (SIV) model inspired by the Infectious Salmon Anemia Virus (ISAv). In this SIV model (Figure 1), infection is permanent such that infectious individuals do not recover and do not contribute to reproduction. Disease may be transmitted via direct contact with infectious individuals or via contact with virus particles. In each generation, infected individuals shed virus particles which decay over time.

A key difference between these two models is that infection in the SIR model serves merely to subdivide the total population into classes. It does not influence either fecundity (the quantity of offspring successfully produced by an individual) or mortality. In contrast, infection in the SIV model increases mortality and completely disrupts the fecundity of the infectious individuals so that they do not contribute to reproduction. van den Driessche and Yakubu found that in the SIV model the presence of infection shifted the behavior of the system such that specific pairs of model parameters, related to natural death and reproduction, produced different qualitative behaviors in the presence and absence of disease. In contrast the SIR model produced the same qualitative dynamic behaviors (for each pair of parameter values) regardless of the presence of infection.

The capacity of discrete systems to produce complicated dynamics, even under simplifying assumptions regarding demography and infection, has been demonstrated in an array of contexts [**1**, **2**, **9**–**11**, **13**, **16**, **28**, **29**] and references therein. In particular, significant work has been done to consider a two-dimensional SIS system where infection is frequency-dependent, described by $\phi = e^{-\beta I_t/N_t}$, and a fraction recover and are once again fully susceptible [**9**–**11**, **14**, **27**]. The combination of Ricker growth and feedback from the infectious class to the susceptible class results in a complex set of dynamics including equilibria, cycles, and chaotic behavior, but typically the population dynamics drive the disease dynamics provided the disease persists [**5**, **9**–**11**, **14**]. Exceptions to this include scenarios with multiple patches [**11**] or when infectivity is a function of infection, e.g. $\phi = e^{\beta(I)I/N}$ [**10**, **11**]. This latter case produces a backward bifurcation and bistability between the disease free and endemic equilibria [**10**, **11**]. Importantly, the addition of disease-induced mortality to the SIS system alters the bifurcation structure of the system, introduces additional dynamics, and produces multistability with fractal basins of attraction [**14**, **27**]. Recent work provided analytical evidence for a sequence of bifurcations in this system including transcritical, period doubling (referred to as flip), and Neimark-Sacker. It also provided a numerical exploration of the fractal basins of attraction [**27**].

The work discussed above focused on SIS models and showed that with simplifying assumptions, such as constant infectivity and no disease-induced mortality, the population dynamics drive the disease dynamics [**9**–**12**, **14**]. More recently, van den Driessche and Yakubu [**23**–**25**] revisited epidemic models with Ricker growth and elucidated the global stability of the disease-free equilibrium under simplifying assumptions in an SEIR model [**24**]. They also derived the basic reproductive number with [**25**] and without [**23**] age structure in the SIR and SIV models discussed above. In their examination, they noted the open question of why infection alters dynamics in the SIV model but not the SIR model [**23**].

In this chapter, we use a simplified SI model to address this open question. Our model allows for disruption of fecundity in infected individuals and incorporates disease-induced mortality via tunable parameters. Using numerical bifurcation analysis we show that the long-time dynamics of this system and its qualitative behavior can be altered via infection. Furthermore, we find that infection-induced mortality introduces multistability such that the stable long-time behavior of the population depends sensitively on initial conditions. In doing so, our work extends understanding of the capacity of infection to alter the periodic behavior of populations exhibiting Ricker growth.

2. Model

To examine the influence of infection on a population subject to Ricker growth, we consider a simple two-dimensional model consisting of susceptible (S) and infectious (I) individuals described by

$$S_{t+1} = \underbrace{r \cdot g(S_t + wI_t)}_{\text{Ricker growth}} + \underbrace{(1-d)}_{\text{survival}} \underbrace{S_t\phi(I_t)}_{\substack{\text{avoidance}\\ \text{of infection}}}, \tag{2.1}$$

$$I_{t+1} = \underbrace{(1-d)}_{\text{survival}} \Big(\underbrace{S_t\,(1-\phi(I_t))}_{\substack{\text{newly}\\ \text{infected}}} + \underbrace{(1-\mu)I_t}_{\substack{\text{avoidance of}\\ \text{disease-induced}\\ \text{mortality}}} \Big).$$

In this system, $g(X) = Xe^{-bX}$ is the Ricker growth term and is scaled by the intrinsic per-capita reproduction count, r, of the population. The form of the term indicates that all of the susceptible and a fraction, w, of the infectious individuals contribute to reproduction in a given generation, and that the production of reproducing individuals decreases exponentially as the size of the reproducing population, $S + wI$, increases. As in the SIV model described above, infection is assumed be density-dependent and occur via a Poisson process such that susceptible individuals have a probability $\phi(I_t) = e^{-\beta I_t}$ of evading disease and a probability $(1 - \phi(I_t))$ of contracting the disease in a given generation. The per-capita probability of natural death is given by d and the probability of disease-induced death is given by μ. This system is a simplification of two models analyzed by van den Driessche and Yakubu [**23**], and was also considered by Jaramillo et al. [**15**].

In addition to structural changes, i.e., only two classes instead of three, our model makes additional assumptions about the impact of the disease which distinguish it from the SIR and SIV models presented by van den Driessche and Yakubu (Figure 1). First, like the SIV model, but distinct from the SIR model, we assume that disease may be fatal and increases the per-capita probability of mortality by μ. Second, we allow for a partial reduction in fecundity due to infection, via the parameter w, and require neither the complete loss of fecundity (as in the SIV model) nor complete reproductive capacity of all individuals (as in the SIR model). Thus, our SI model reduces the dimension of the system while retaining key biological processes (Figure 1). We show that the SI model retains the capacity for infection-induced changes in system dynamics as observed in the SIV model. We use the parameters w and μ to examine these changes in detail.

Order of events have been shown to play an important role in determining the phenomena exhibited by discrete systems [**8**], thus before beginning the analysis of this model, it is important to point out that its structure implies an ordered sequence of four events within each generation. First, a generation begins with the individuals present at the end of the previous generation reproducing into the susceptible class. Second, disease transmission occurs between infectious individuals and those susceptible individuals present at the end of the previous generation. Third, infectious individuals die due to infection. Finally, at the end of each generation, individuals from all population classes die due to natural causes. Two

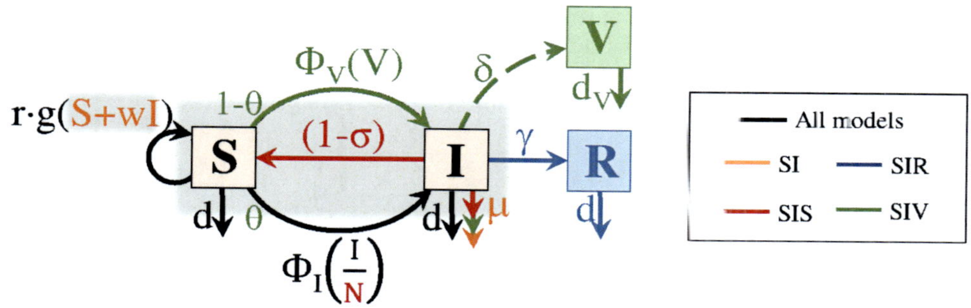

FIGURE 1. **Model schematic.** SI model in the context of previous SIR, SIV, and SIS models [**23**, **27**]. Model processes include Ricker growth; Poisson, density-dependent, disease transmission; and constant per-capita mortality. Newly born individuals are protected from natural mortality and disease until the following generation. Newly infected individuals are protected from disease-induced mortality and are not infectious until the following generation. SI, SIR, and SIV models use density-dependent infection; the SIS models use frequency-dependent infection. $N = S+I+R$. The SIR and SIV parameters: γ, θ, $(1-\theta)$, Φ_V, δ, and d_v are defined by [**23**] and describe the probability of recovery, fraction of susceptible individuals contacting infectious individuals, the fraction of susceptible individuals contacting viral particles, the probability of being infected by virus particles, the per-capita production of virus particles, and the probability of virus decay respectively. The parameter Φ_I=Φ represents the probability of being infected by infectious individuals. Grey shading indicates the reproducing populations of the SI model. In the SIR and SIS models infection does not alter reproduction. In the SIV model only susceptible individuals reproduce.

important comments on these events: (i) susceptible individuals are protected from disease and natural mortality in the generation of their birth and (ii) the disease has an incubation period such that infected individuals do not transmit disease or die due to infection until the generation following contraction of the disease.

3. Parameters and analysis

We analyze the SI model using numerical bifurcation analysis and focus primarily on natural mortality (d) and intrinsic reproduction (r), and the influence of disease-induced mortality (μ) and infection-reductions in fecundity (w) on the dynamics produced by specific (d,r) pairs. Parameters for the model are chosen to align with that of recent work of van den Driessche and Yakubu [**23**]. Deviations from these values are described as relative to the "baseline" parameter set (Table 1). Initial conditions are set to (S_0,I_0)=(150,20) unless otherwise noted.

Our figures fall into two broad categories: one-parameter diagrams and two-parameter diagrams. For the one-parameter diagrams, we use a discrete analogue of the one-parameter bifurcation analysis used in continuous ODE frameworks. Here,

TABLE 1. **Description of parameters.** Values listed are baseline values referenced, while ranges are the standard range of values varied across in our simulations. Baseline values for r, b, d, β from [**23**]. Here, indiv. refers to individuals.

Symbols	Description	Value	Range
r	Intrinsic per-capita reproduction per generation	$e^4 \approx 54.6$	[0,320]
b	Scaling parameter for density-dependent growth	0.1	–
d	Probability of death due to natural causes	0.5	[0,1]
β	Poisson constant for infection transmission	0.056	–
w	Relative reproduction capacity of infected indiv.	1	[0,1]
μ	Probability of death due to infection	0	[0,1]

all parameters but one are fixed and the steady state of a model variable is depicted with respect to the value of the remaining "bifurcation" parameter. Unlike their continuous counterparts which are produced via continuation methods, these discrete diagrams are produced via simulation. Initial conditions are selected, the system is simulated for long time using a single value of the bifurcation parameter, transient data is discarded, and the remaining trajectory points, here, population sizes, are plotted versus the bifurcation parameter. When more complicated long-term behaviors occur, this is represented by multiple points appearing on the plot for the same bifurcation parameter. The process is then repeated at increments along the chosen range of the bifurcation parameter using the same set of initial conditions throughout, typically $(S_0,I_0) = (150,20)$. We refer to this type of diagram as a "fixed-r" slice when d is used as the bifurcation parameter (Figure 2a, center) and a "fixed-d" slice when r is used as the bifurcation parameter in reference to the relationship to the two-parameter diagrams in which two parameters are varied simultaneously.

Two-parameter diagrams, which we also refer to as heat maps, can be understood as discrete analogues of the two-parameter bifurcation diagrams of ODE systems. They are constructed using the same simulation technique used for the discrete one-parameter diagrams except that two bifurcation parameters are incremented along their chosen ranges and a summary of each simulation is reported on a grid. For our two-parameter diagrams, we report the quantity of distinct values in the long-time trajectory of a simulation (after removal of transient behavior) and encode this count by color at the corresponding (d,r)-grid point. The result is a heat map in which boundaries between color regions indicate bifurcation points across which the qualitative behavior of the system changes. Regions reported as a single distinct point (dark blue) represent non-oscillatory fixed-point dynamics at steady state and regions reported as multiple points (2-medium blue, 3-light blue, etc.) represent periodic behavior (2-cycles, 3-cycles, etc.) at steady state. When the dynamics are more complicated, such as long-time trajectories with greater than 16 distinct points, these are all reported together in a "17+" category (colored black). Thus, while we observe apparently chaotic behavior within the 17+ regions for all cases where we examine fixed-r or fixed-d slices that traverse these regions, our two-parameter diagrams do not distinguish between higher order limit cycles and chaos.

Two important notes to keep in mind when reading these diagrams. First, our numerical bifurcation analysis method cannot detect unbounded trajectories, although these are not observed in our system. This is consistent with the proof by van den Driessche and Yakubu [**23**] showing that solutions of the SIR system are bounded. Second, unlike their continuous counterparts, which are generated via continuation methods, the simulation-based methods underlying our diagrams are specific to an individual set of initial conditions and do not directly include any information regarding the potential multistability.

To get around this second limitation, we construct a series of two-parameter diagrams generated using a distribution of randomly selected initial conditions. We combine the resulting data to trace out the maximal parameter domain for each qualitative dynamic classification and locate regions of overlap between them to identify parameter regimes where we observe multistability.

A precise description of all methods and the associated numerics are provided in the Appendix A.1. Code was developed in MATLAB 2021b, and simulations were performed in MATLAB 2021b and 2020a. Codes and data are available at `https://github.com/laurenchilds/SIdiscrete`.

4. Results

4.1. Sufficiently high natural mortality drives the disease to extinction. To examine the role of infection on populations with Ricker growth, we extend the one-dimensional disease-free system by the addition of an infectious class, which does not reduce fecundity, w=1, or increase mortality, μ=0 (System 2.1). In this case the total population, $N_t = S_t + I_t$, satisfies the difference equation $N_{t+1} = rg(N_t) + (1 - d)N_t$ and has the same form as the disease-free system $S_{t+1} = rg(S_t) + (1 - d)S_t$. Thus, the structure of the total population dynamics as natural reproduction (r) and mortality (d) are varied is identical to that of the disease-free system. Like the disease-free system, the full system exhibits fixed points, periodic behavior, and chaotic behavior depending on the values of natural mortality and fecundity (Figures 2a, left and A.1, left) Additionally, the qualitative long-time dynamics of the susceptible population mirrors that of the disease-free system, as the parameter combinations that produce specific qualitative dynamics are identical (Figure A.1, left and center). However, in contrast, the parameter regions corresponding to distinct long-time behaviors in the infectious population show a non-linear threshold depending on d and r, such that for d values above that threshold, all periodic behavior in the infectious population disappears (Figure A.1, right; Figure 2a, right). Further examination of this threshold for regions of parameter space exhibiting fixed-point or n-cycles ($n < 17$) in the absence of disease show that it represents parameter values at which the system transitions from $\mathcal{R}_0 > 1$ to $\mathcal{R}_0 < 1$ (See Appendix A.2 for $\mathcal{R}_0$ derivation), and that this value does not depend on w. Thus, sufficiently large natural mortality leads to elimination of infection from the population, and the relationship between intrinsic reproduction, r, and the natural mortality, d, determines the threshold at which this occurs.

4.2. Reducing the fecundity of infected individuals distorts the bifurcation structure of the with-disease system. For the SI system (System 2.1), we consider that a fraction of infectious individuals produce viable offspring despite infection, through the use of the parameter w. This was motivated by work by van den Driessche and Yakubu [**23**], who showed that the introduction of infection shifted the period of the long-time population dynamics in the SIV model at $(d,r) = (0.5,e^4)$, while it did not in the SIR model. A key feature of the SIV model was a complete loss of successful reproduction from infected individuals, which contrasts with the lack of influence on fecundity by infection in their SIR model. By using w, our SI model can range between these two scenarios of no loss of fecundity to complete loss of fecundity of infectious individuals.

As seen in Figure 2a, when there is no effect on fecundity, $w = 1$, (and no disease-induced mortality, $\mu = 0$) there is no shift in the long-time population dynamics. That is, the parameter values at the total population bifurcation points are identical to those in the disease-free system despite the higher dimensionality that results from splitting the population into susceptible and infectious classes. This is because the form of the total population equation in the case $(w,\mu)=(1,0)$ is identical to that of the disease-free system.

Decreasing w reduces the contribution of the infectious individuals to the population's fecundity and shifts the boundaries between regions in parameter space that produce distinct long-time behaviors. The range of d values that lead to fixed point behavior becomes narrower, and the regions to the left of the infection extinction threshold are stretched in both the d and r directions (Figure 2). Indeed, as w is decreased and infection persists in the population, the locations of period-doubling bifurcations occur at lower d values, and the overall population sizes are larger. While altering fecundity does not shift the d-values at which disease-extinction occurs, the long-time dynamics of the infectious population for values of d below the extinction threshold ($d \approx 0.761$) are altered by w. This shows that the simple two-dimensional SI model retains the capacity for infection to shift the long-term total population dynamics as observed in the SIV system. However, the complete loss of reproductive capacity due to infection, as occurs in the SIV model, is not necessary for this shift.

4.3. The inclusion of disease-induced mortality introduces additional dynamic structures. To consider the influence of disease-induced mortality on the long-time system behavior, we fix w=1 so that infection does not influence fecundity, and vary μ. We find that while w appeared to stretch and distort existing dynamic regions (Figure 2), μ introduces additional dynamic structures (Figure 3), which appear to "interrupt" those observed in the disease-free system. Additionally, increases in μ cause a non-uniform stretching of the dynamic regions so that the once-concentric regions overlap. For example, when μ=0.25 the region in intermediate-d and high-r parameter space characterized by a 3-cycle (light blue), 6-cycle (green), 12-cycle (red) and 17+ distinct points (black) stretches horizontally and vertically so that the surrounding 17+ region narrows. And even more erratically, when μ=0.50 a new region, characterized by the presence of a 4 cycle (very light blue) and an 8-cycle (yellow) emerges for small-d/large-r pairs. Furthermore, when μ=0.75, a similar region with 4- and 8-cycles appears to thread through the structures of the disease-free system and a new region characterized by a 5-cycle (forest green) and a 10-cycle (dark orange) appears for small-d/large-r pairs.

4.4. With-disease system dynamics *jump* at the boundaries between structural regions. To examine the mechanism underlying the mortality-induced

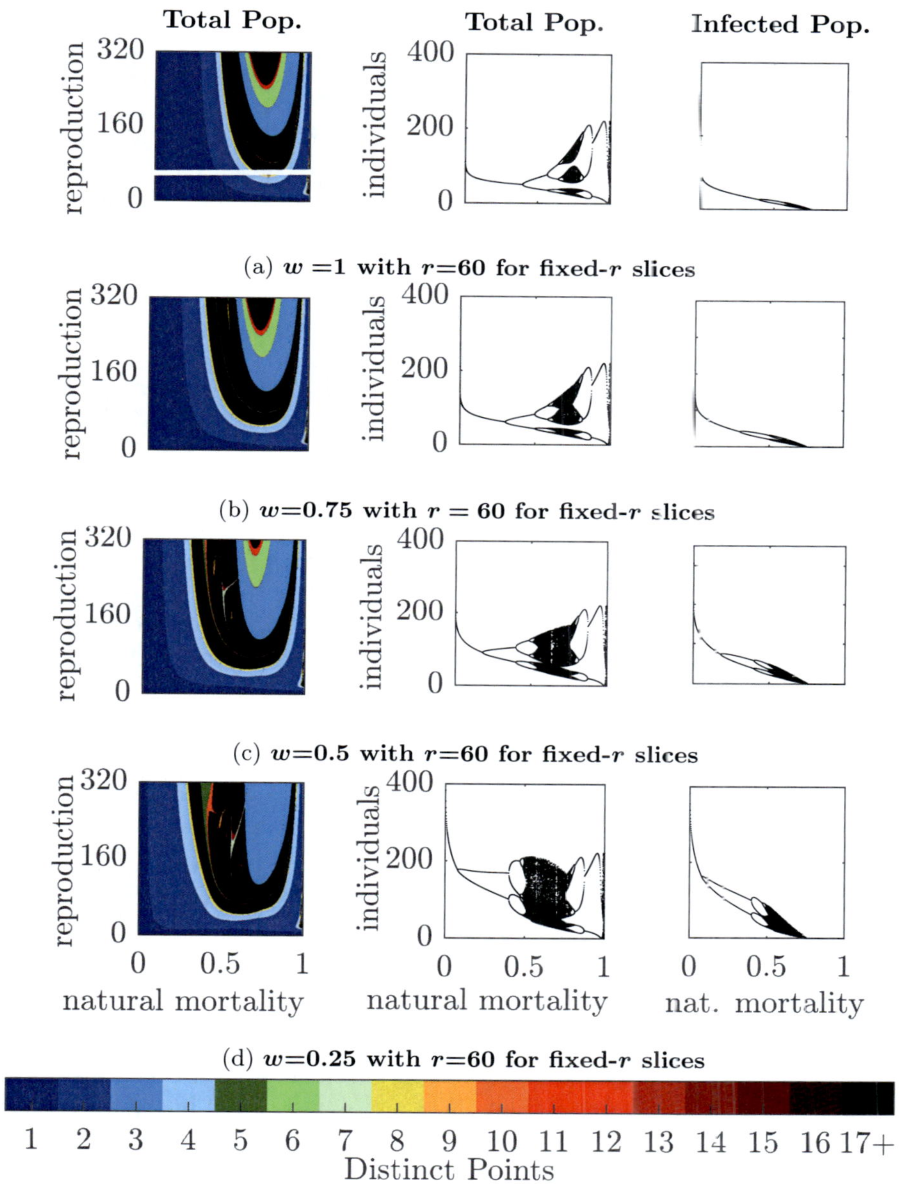

FIGURE 2. **Shifts in long-time population dynamics due to reductions in fecundity.** (a)-(d) correspond to w=1.00, 0.75, 0.5, 0.25, respectively, with (left) qualitative long-time total population behavior, (middle) long-time total population size when r=60, and (right) long-time infectious population size when r=60. White line in (a, left) indicates the r=60 "slice". Disease extinction occurs at $d \approx 0.761$ for all w in the with-disease system. The disease-free system exhibits a bifurcation structure identical to the total population in the w=1 case.

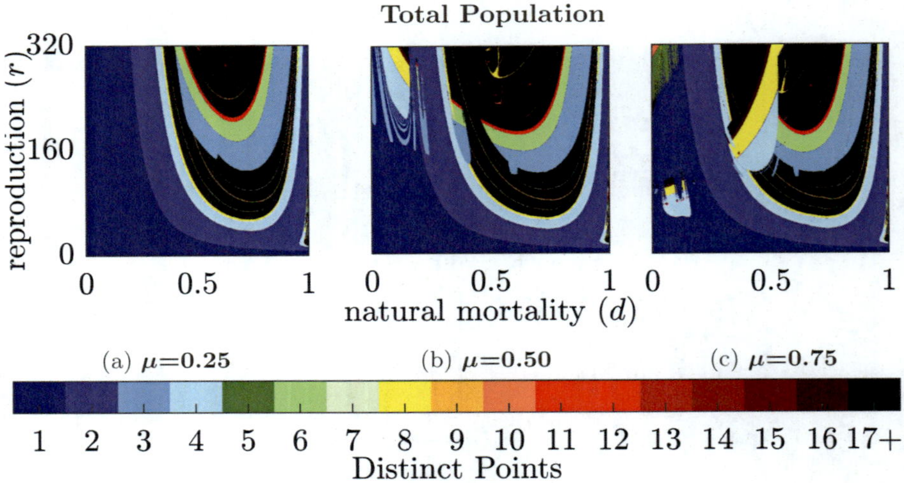

FIGURE 3. **Appearance of additional dynamical regions for increases in disease-induced mortality.** (a)-(c) Qualitative long-time total population behavior corresponding to μ=0.25, 0.50, 0.75, respectively. For the μ=0 case, see Figure 2a. Baseline parameters, including no effect on fecundity (w=1), except as indicated.

introduction of dynamic regions, we focus on $(\mu,w)=(0.75,1)$. For these parameter values and $(S_0,I_0)=(150,20)$ the appearance of additional structures is very clear for small-d/large-r and for low to intermediate d, as described above (Figures 3c and 4a) Similar regions are also found using a second set of initial conditions $(S_0,I_0)=(125,45)$ (Figure 4b). A fixed-d slice at d=0.475 when $(S_0,I_0)=(150,20)$ shows a transition between structures at r=130.5, where the long-time trajectories jump from a chaotic regime into a 4-cycle (Figure 4a,c). In contrast to the chaotic-to-cyclic transitions we saw previously in both the disease-free system and with-disease system lacking impacts on fecundity and disease-induced mortality (Figure 2a, left), this transition is non-smooth and appears to be between two unrelated long-time dynamic trajectories. Taking a similar fixed-d slice for d=0.475 when $(S_0,I_0)=(125,45)$ reveals a similar chaotic-to-4 cycle jump but now occurring at a lower r value, r=115.5 (Figure 4d).

4.5. Long-time dynamics depend on initial conditions. The apparent jump between distinct dynamic structures in Figure 4 and the shift in its location with a change in initial conditions suggests that the qualitative long-time dynamic structure of the SI system depends on the initial conditions of the population. To examine the role of initial conditions, we consider the case when $(\mu,w)=(0.75,1)$ and choose two (d,r) pairs: $(d_1,r_1)=(0.475,120)$ and $(d_2,r_2)=(0.475,140)$ based on the behavior they produced for $(S_0,I_0)=(150,20)$ and $(S_0,I_0)=(125,45)$ (Figure 4). At these chosen (d,r) pairs, we vary the initial conditions, S_0 and I_0, between 0 and 1000 with steps of size one to produce initial condition heat maps (Figure 5).

As expected, in the case $(d,r)=(0.475,120)$, our standard initial conditions of $(S_0,I_0)=(150,20)$ lead to long-time apparently chaotic behavior (Figure 5a, black)

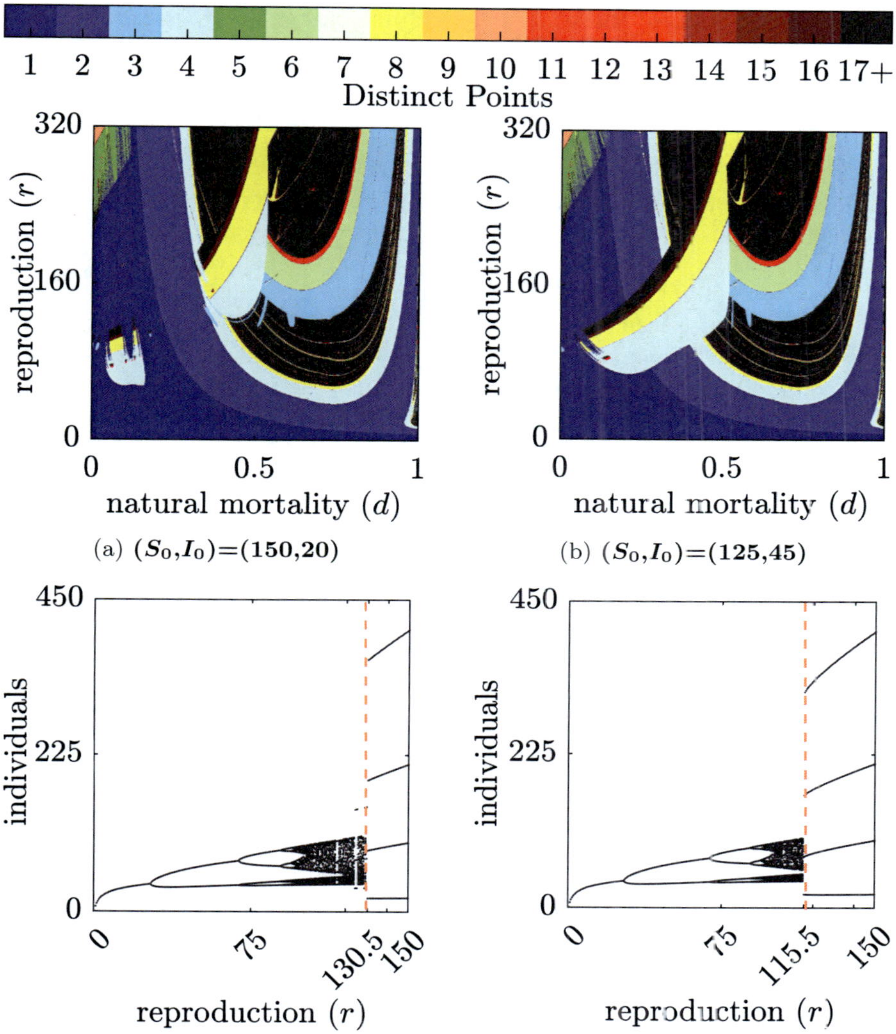

(a) (S_0,I_0)=(150,20) (b) (S_0,I_0)=(125,45)

(c) (S_0,I_0)=(150,20) for d=0.475 (d) (S_0,I_0)=(125,45) for d=0.475

FIGURE 4. **High disease-induced mortality leads to jumps between long-time stable behaviors.** (a),(b) Qualitative long-time behavior of the total population. (c),(d) Total population in a fixed-d slice (d=0.475). For (c),(d), unlabeled ticks are r=120 and r=140. (a),(c) Standard initial conditions of (S_0,I_0)=(150,20). (b),(d) Modified initial conditions of (S_0,I_0)=(125,45). Baseline parameters, except as indicated and disease-induced mortality of μ=0.75.

while (S_0,I_0)=(125,45) results in a 4-cycle (Figure 5b, light blue). In contrast, for (d,r)=(0.475,140), both sets of initial conditions lead to a 4-cycle (Figure 5a and b, light blue). Furthermore, for the wide range of initial conditions considered at the chosen parameter values, the system can exhibit one of three distinct qualitative behaviors depending on the initial conditions (Figure 5). These behaviors include a 3-cycle, a 4-cycle, and 17+ distinct points (likely chaos). Shifting r from 120 to 140 causes the regions of initial conditions with stable cyclic behavior (Figure 5, both shades of blue) to enlarge and become closer together.

4.6. Consideration of multiple initial conditions reveals that dynamic regions span wide ranges of parameter space. To examine the full range of qualitatively distinct long-time behaviors displayed by this system for $r \in$[0,320] and $d \in$ [0,1], we use 100 randomly-generated, unique pairs of initial conditions (Figure A.2) to construct a series of (d,r)-heat maps (not shown). From these, we extract all (d,r) parameter combinations leading to a given qualitative long-time behavior (2-cycle, 3-cycle, etc.) (Figure 6). Each type of long-time behavior occurs for a wide range of d and r values and, for some portions of parameter space, multiple long-time behaviors are possible, depending on the initial conditions. Additionally, the parameter regions appear to belong to three distinct structures, (1) a u-shaped region resulting from repeated period doubling out of a 2-cycle (2-, 4-, 8-, 16-cycles along with the 17+ region) for intermediate to high d-values (Figure 6 a-d and k), (2) a band-shaped region resulting from repeated period doubling out of a 4-cycle (4-, 8-, 16-cycles, and the 17+ region) for low to intermediate d values (Figure 6 b-d and k), and (3) a region resulting from a single period doubling out of a 5-cycle (5-, 10-cycles) for low d and high r (Figure 6,h-i).

4.7. Multistability is present in the SI system. If we consider the quantity of distinct qualitative behaviors at each (d,r) parameter pair using the data generated above, we see several regions of multi-stability (Figures 6 and 7). The first occurs for small d and large r where the system either exhibits a simple fixed point or behaviors from the 5- or 10-cycle regions (Figure 7). The second occurs for a band of (d,r) stretching from low to intermediate d. For small d, r the band lies roughly between 45 and 110, and for intermediate d, the band reaches the maximum considered value of r=320. This band is the arising dynamical region, which occurs with the presence of disease-induced morality (non-zero μ), and is overlaid across the dynamic regions exhibited by the disease-free system (Figures 3 and 6). This results in several different patterns of bistability: (i) one of two stable cyclic behaviors, (ii) one of two distinct, apparently aperiodic chaotic behaviors, or (iii) either a cyclic behavior or apparent chaos. Outside of these multistability regions, the dynamical structures from the disease-free system remain. Infection still alters the total population size, but not the qualitative long-time behavior.

It is important to note that while our analysis does detect multistability arising out of fundamental changes in system dynamics it does not distinguish between structurally distinct dynamics with identical cycle lengths. That is, it is possible for the system to jump between two distinct n-cycles with identical n. This is not detected in our two-parameter bifurcation diagrams, though it is suggested by the crossing of regions. An example of such a jump between distinct 4-cycles can be found in Figure A.3, located at the parameters indicated by the white dot in Figure 7. Additionally, a similar numerical analysis of the disease-free system illustrates the presence of numerical error at the boundaries of dynamic regions (Figure A.4). Namely this two-parameter diagram shows multistability which does not actually exist. This is due entirely to numerical precision of trajectories that take a long

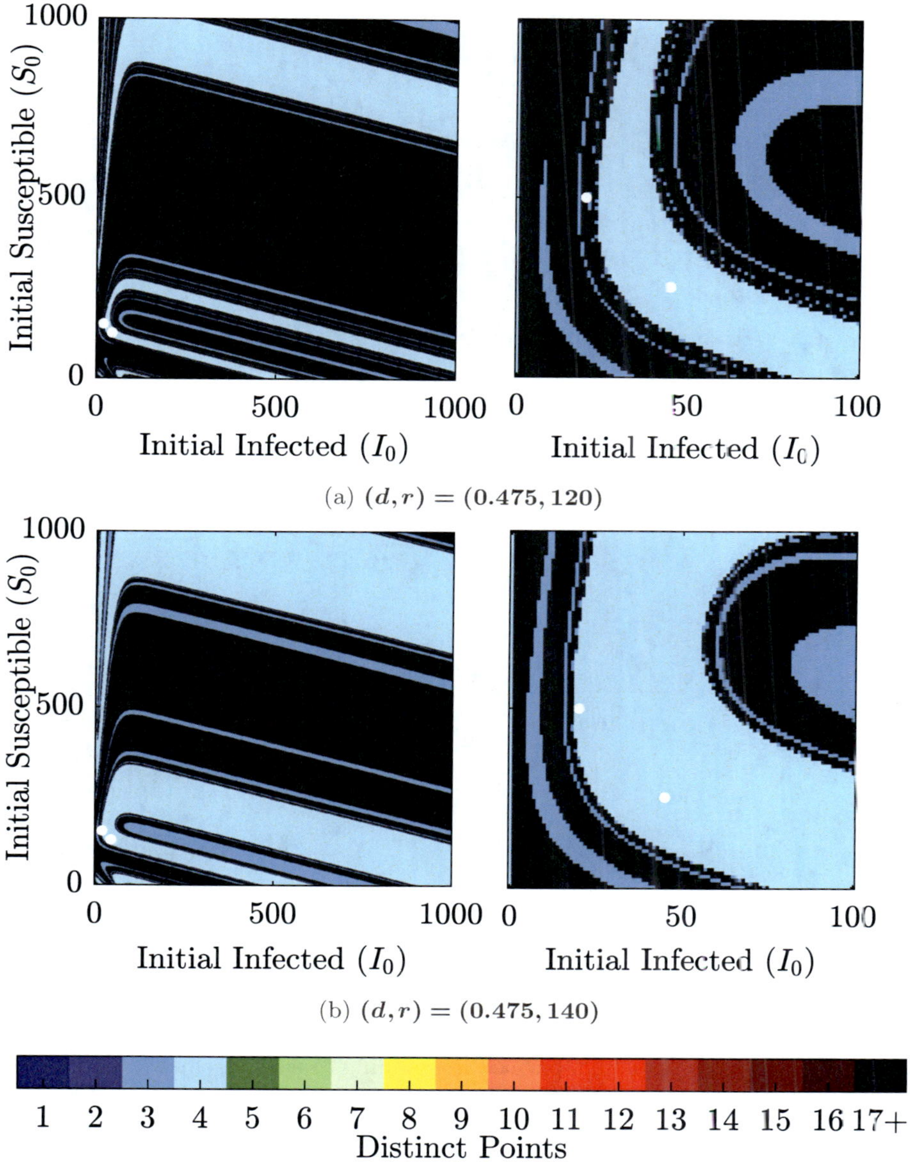

(a) $(d, r) = (0.475, 120)$

(b) $(d, r) = (0.475, 140)$

FIGURE 5. **Sensitivity to initial conditions for select pairs of natural mortality and reproduction.** Natural mortality of d=0.475, and reproduction of (a) r=120 and (b) r=140. Initial conditions (left) $S_0 \in[0,1000]$, $I_0 \in[0,1000]$ and (right) $S_0 \in [0, 100]$, $I_0 \in[100,200]$. Step size is 1. White dots indicate the location of specific initial conditions (S_0,I_0)=(150,20) and (S_0,I_0)=(125,45) from Figure 4. Regions observed are 3-cycles (blue), 4-cycles (light blue), and 17+ distinct points (black).

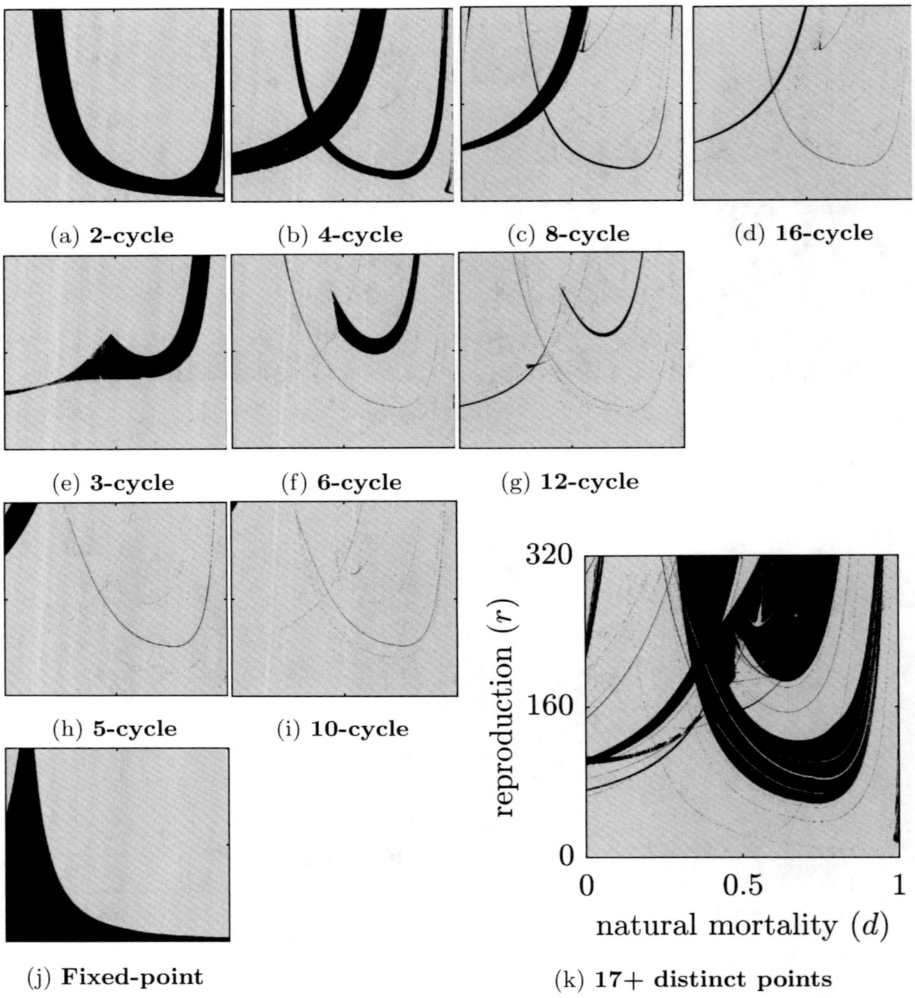

FIGURE 6. **Compilation of distinct dynamic regions.** Regions of (d,r) parameter space corresponding to each distinct total-population dynamic in the with-disease system. Compilation of data from 100 heat maps, each generated by simulation from one of 100 distinct and randomly chosen initial conditions (Figure A.2). Baseline parameters, except μ=0.75. Not shown: 7-, 9-, 11-, 13-, 14-, and 15- cycles. Parameter regions producing these behaviors are imperceptibly narrow for the parameter ranges sampled.

time to settle. Traces of similar issues with numerical precision at the boundaries of dynamic regions can be seen in Figure 7.

5. Discussion

The study of discrete-time dynamical systems has a rich history both due to the complexity of dynamical behaviors even simple models can produce [**17**, **19**]

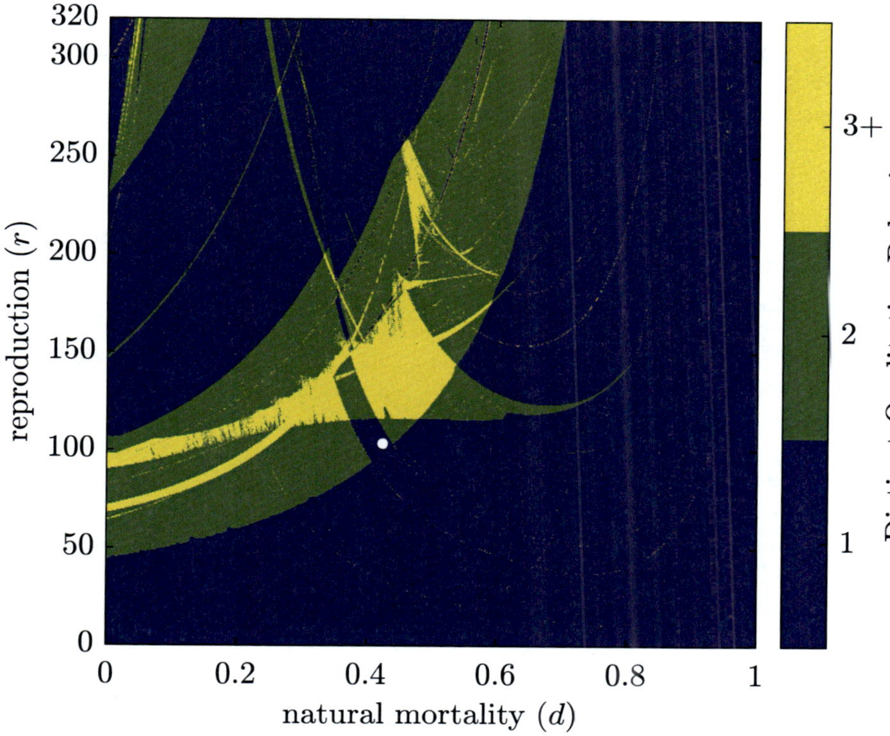

FIGURE 7. **Regions of multistability for the with-disease system.** Blue regions indicate a single type of qualitative long-time behavior across all initial conditions sampled. Green and yellow regions indicate two and three or more stable qualitative long-time behaviors, respectively, when different initial conditions are considered. Compilation of data from 100 heat maps, each generated by simulation from one of 100 distinct and randomly chosen initial conditions (Figure A.2). White dot indicates the location of the jump in Figure A.3.

and in its application to populations such as fisheries and insects [**7**, **21**, **22**]. A single population that exhibits overcompensatory reproduction (i.e., reproduction which is maximal at intermediate population sizes) can produce fixed-point, cyclic, or chaotic behavior depending on model parameter values [**3**, **13**, **17**, **19**, **20**]. This is a direct result of an interplay between the overcompensatory growth form and mortality [**18**].

While the complicated dynamics arising out of overcompensatory growth are well-studied, the influence of infection on such populations is not as well understood. Work on SIS systems showed that infection-induced multistability can arise in a number of contexts including non-constant infection and influence from patch dispersal mechanisms [**11**, **12**]. However, analysis of simple SIS models with Ricker-growth and frequency-dependent infection demonstrated that in the absence of disease-induced mortality and alterations in reproduction, the system dynamics are determined by that of the underlying demographic equation [**10, 14**]. Furthermore, disease-induced mortality has the capacity to fundamentally alter the bifurcation structure of the system and introduce multistability [**14, 27**]. These studies left open the question of how recovery and the mode of infection, i.e. frequency- vs. density-dependence, influences these observations and did not consider alterations of fecundity in their analysis.

Recent work by van den Driessche and Yakubu analyzed two discrete-time epidemiological models in which the underlying population exhibits Ricker growth However, in contrast to the SIS models referenced above, they considered a scenario in which infection is density-dependent instead of frequency-dependent. They showed that, in the context of an SIV model, disease that influences mortality and prevents reproduction of infectious individuals has the capacity to shift the long-time qualitative behavior of a system. That is, pairs of parameters describing natural mortality and reproduction produce qualitatively distinct cyclic behaviors in the disease-free and with-disease systems [**23**]. Similar phenomena were not observed in their analysis of an SIR model in which infection serves only to compartmentalize the model and does not influence either fecundity or mortality.

Here, we sought to further analyze the capacity of infection to alter the qualitative behavior of discrete dynamical systems with Ricker growth by examining a two-dimensional SI model with density-dependent infection and parameters that tune the influence of infection on reproduction and mortality. Using numerical bifurcation analysis, we found that reductions in fecundity serve to stretch and distort the parameter regions that produce cyclic and chaotic behavior while infection-induced increases in mortality introduce dynamic regions that interrupt those established by the disease-free system.

Closer examination of these "interruptions" revealed locations of parameter space in which small changes in parameters produce dramatic jumps between distinct long-time dynamic structures. Furthermore, the location of these jumps can be altered by a change in initial conditions. Investigation of multiple initial conditions for the parameter space around these jumps revealed that the system is multistable in these regions and produces 3-cycles, 4-cycles, or 17+ distinct point long-time behavior depending on initial conditions. Compilation of the possible long-time behavior, via random sampling of initial conditions, revealed that the "interrupting" regions introduced by infection-induced increases in mortality actually span wide swaths of parameter space. Thus, the abrupt transitions in dynamics are the result of parameter-induced changes in the basins of attraction for the dynamic behaviors.

Our results mirror those of previous work [**14, 27**] in that our model, like theirs, shows that infection-induced mortality is capable of introducing multistability into a population that exhibits Ricker growth. The SIS disease system with disease-induced mortality exhibits a Neimark-Sacker bifurcation and resulting fractal basins of attraction [**14, 27**]. However, the pattern of dependence on initial conditions in our system suggests that the type of dynamic behavior produced from a particular set of initial conditions depends on the ratio of susceptible to infected individuals. For example, in Figure 5, when I_0 is small, variation in initial conditions that maintains a ratio greater than one preserves the qualitative nature of the dynamics. Similarly, when I_0 is large, variation in initial conditions that maintains a ratio less than one preserves the qualitative nature of the dynamics. It remains an open question what processes determine these ratios and the (S_C, I_0)-dependent threshold between them.

While our work does not exhaustively examine all of parameter space, it serves to expand understanding of the rich breadth of behavior overcompensatory growth can produce when coupled with infection in deceptively simple systems. In addition to the potential for infection-induced multistability, we also observe an infection-induced increase in total population size via a reduction in fecundity (Figure 2). In previous work on the SI system, Jaramillo et al. referred to this phenomenon as an "infection-induced hydra effect" and demonstrated the capacity of infection to produce this effect in regions of parameter space where the population reaches an equilibrium [**15**]. Via alterations in reproduction, we show an infection-induced hydra effect in regions of parameter space that exhibit population cycles (Figure 2). It remains to be determined whether or not infection-induced mortality can also elicit a hydra-effect in cyclic regions of parameter space. We anticipate that additional analysis that considers the joint influence of infection-induced alterations in fecundity and mortality will reveal even greater complexities in dynamics. This will likely involve further characterization of the hydra effect and multistable regions.

While it is clear that Ricker growth can produce complex dynamics in simple systems, it remains an open question exactly how overcompensatory growth combined with infection produces these complex behaviors. Since the intersection of the Ricker growth curve and the natural mortality curve allows for oscillations in the disease-free system, it is reasonable to assume that dynamical changes produced by alterations in w and μ are the direct result of how they reduce the net reproductive capacity of the population. Recent work by Xiang et al. begins to address this through the derivation of analytical conditions for a variety of bifurcations in an SIS model with disease-induced mortality [**27**]. However, the interplay of infection-induced mortality, reduction in fecundity, and transmission requires further examination.

The SI model studied here was motivated by an epidemiological model of Infectious Salmon Anemia Virus described in work from van den Driessche and Yakubu [**23**]. Our work shows that interventions that eliminate disease transmission via virus particles in the environment are not sufficient to prevent the qualitative alterations in population dynamics observed in the SIV system. That is, infection-induced changes in the bifurcation structure relative to the disease-free system are observed in our simple SI model which does not include virus dynamics. Further, our work demonstrates that the multistability observed by Franke et al. and Yakubu et al. does not depend on recovery from disease or on frequency-dependent infection [**14, 29**]. Thus, we advance understanding of the complicated dynamics and the presence multistability arising from infection-induced changes to demographic processes in a system exhibiting overcompensatory growth and in doing so provides new insights into the interplay of population and epidemic dynamics.

A. Appendix

A.1. Detailed methods. Our results are depicted with two primary figure types: one-parameter diagrams that depict the long-time behavior along a one-dimensional "slice" of parameter values incremented in a single dimension and two-parameter diagrams which summarize the qualitative behaviors of a system as parameters are incremented along two dimensions. These diagrams are analogous to the one- and two-parameter bifurcation diagrams in continuous ODE systems, with one caveat. While continuous bifurcation diagrams typically use continuation methods to determine the behavior of a system across all possible initial conditions, construction of our diagrams depend on simulation. Thus they depict the system behavior emanating from a single set of initial conditions across a range of parameter values. Our diagrams depict the long-time behavior that results from a single set of "baseline" initial conditions, $(S_0,I_0)=(150,20)$. When r is used as a bifurcation parameter it is incremented in steps of size 0.5 in [0,320] and when d is used as a bifurcation parameter it is incremented in steps of size 0.001 in [0,1]. Deviations from these values are indicated clearly in the text.

A.1.1. *One-parameter diagrams: fixed parameter slices.* Fixed parameter "slices" (sometimes called orbit maps) in this chapter depict the long-time behavior of System (2.1) with respect to a chosen parameter. Each diagram is constructed via a series of simulations in which the value of the chosen bifurcation parameter is varied while other parameters and the initial conditions are held constant. For each simulation, the system is updated for 10,048 generations and the first 10,000 generations are discarded to remove transient behavior. The population counts during the remaining 48 generations, assumed to be representative of the long-time behavior of the system, are plotted against the chosen value of the bifurcation parameter. The choice to retain the behavior of 48 generations was to encompass a minimum of three realizations for each cycle size up to a 16-cycle.

A.1.2. *Two-parameter diagrams: qualitative heat maps.* To summarize the long-time behavior of the model as we vary two parameters, we create a series of heat maps in which each cell represents the quantity of distinct points in the long-time behavior of a simulation trajectory. We categorize these behaviors into 17 groups; values between 1 and 16 refer to the quantity of distinct points in a cycle. The 17+ category contains higher-order cycles as well as apparent aperiodic or chaotic behavior. The quantity of distinct points is determined by running long-time simulations, as described in Section A.1.1, rounding population counts to seven decimal places, and then identifying the quantity of unique points using the `unique` function in MATLAB. Seven decimal places was chosen as the rounding cut-off to minimize the numerical error when comparing the disease-free system and with-disease system for $(\mu,w)=(0,1)$, in which the total population sizes are mathematically equivalent.

A.1.3. *Initial condition heat maps.* While many of the fixed parameter slices and heat maps use the baseline initial conditions, $(S_0,I_0)=(150,20)$, and vary the parameters d and r, we also examine two specific scenarios in which d and r are fixed and the initial conditions allowed to vary between 0 and 1000. These initial condition heat maps were constructed using the method described in Section A.1.2, but instead of fixed initial conditions, the d and r values are fixed and all other parameters are fixed, and S_0 and I_0 serve as the bifurcation parameters.

A.1.4. *Structure compilation diagrams.* To explore the presence of multiple stable long-time structures, we construct a set of heat maps that combines information from multiple sets of initial conditions. To produce each diagram, we perform the simulations necessary to create 100 heat maps (not shown) each with all (d,r) combinations (as described in Section A.1.2) and using one pair from a set of 100 distinct initial conditions. We then combine the data from these 100 heatmaps to find all (d,r) combinations where a particular type of long-time behavior (e.g., fixed point, 2-cycle, 17+, etc.) occurs. To efficiently sample the initial condition space, our 100 initial condition pairs are chosen via Latin Hypercube Sampling from $S_0 \in [0,1000]$ and $I_0 \in [0,1000]$ with equally spaced bins. Chosen sets of initial conditions are shown in Figure A.2. The multistability plots are constructed by counting the quantity of distinct dynamic structures present at each (d,r) pair.

A.2. Calculation of $\mathcal{R}_0$. We follow the methodology presented by van den Driessche and Yakubu [**23**] that allows for calculation of $\mathcal{R}_0$ from k-cycles in addition to fixed points. Using this implementation of the next generation method, we obtain

$$\mathcal{R}_0 = \frac{(1-d)^k \left(\Pi_{j=1}^k\big(\beta z_{j\infty} + (1-\mu)\big) - (1-\mu)^k\right)}{1-\big((1-d)(1-\mu)\big)^k},$$

with the k-cycle given by $\{z_{1\infty}, z_{2\infty}, \cdots, z_{k\infty}\}$. Note that $\mathcal{R}_0$ never depends on w, the reduction in fecundity of infected individuals.

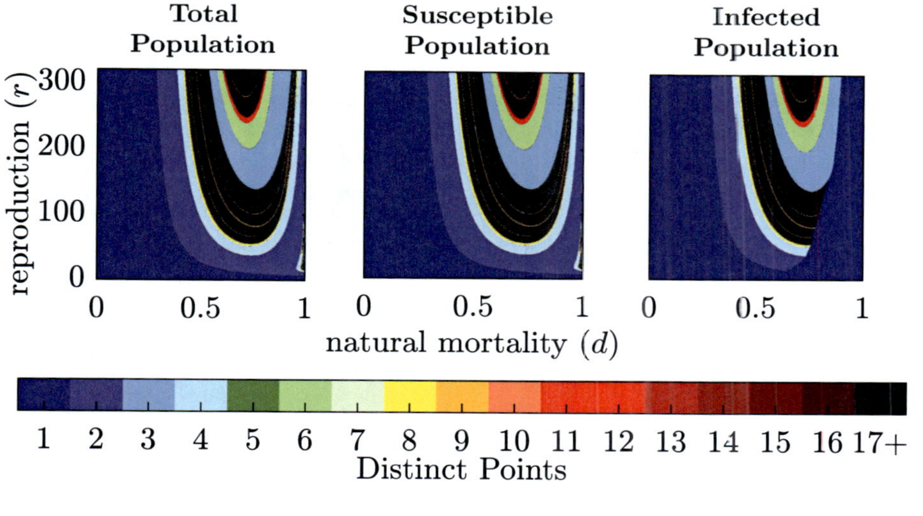

FIGURE A.1. **Long-time behavior of with-disease system in the absence of infection-influence on reproduction or fecundity.** Number of distinct points in the total population (left), susceptible population (middle) and infectious population (right) for pairs of (d,r). Baseline parameters including w=1 and $\mu = 0$. The disease-free system exhibits a bifurcation structure identical to the total population in the (w,μ)=(1,0) case.

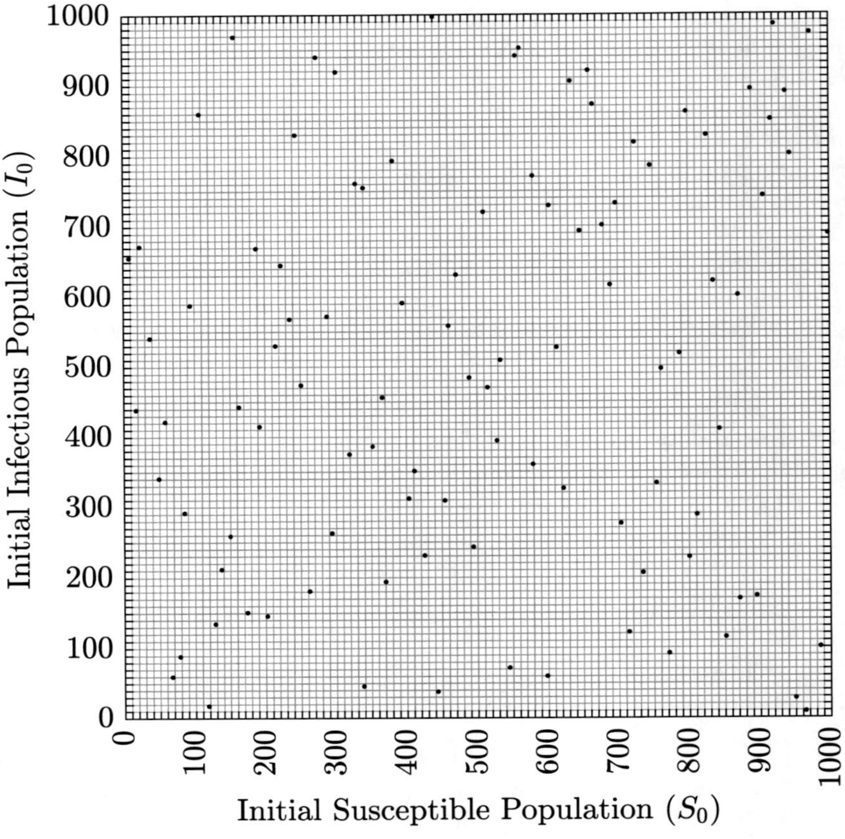

FIGURE A.2. **Distinct initial conditions pairs chosen via Latin hypercube sampling.** Sampling of 100 pairs was done via a uniform distribution of $S_0 \in [0,1000]$ and $I_0 \in [0, 1000]$. Points represent location of chosen initial conditions.

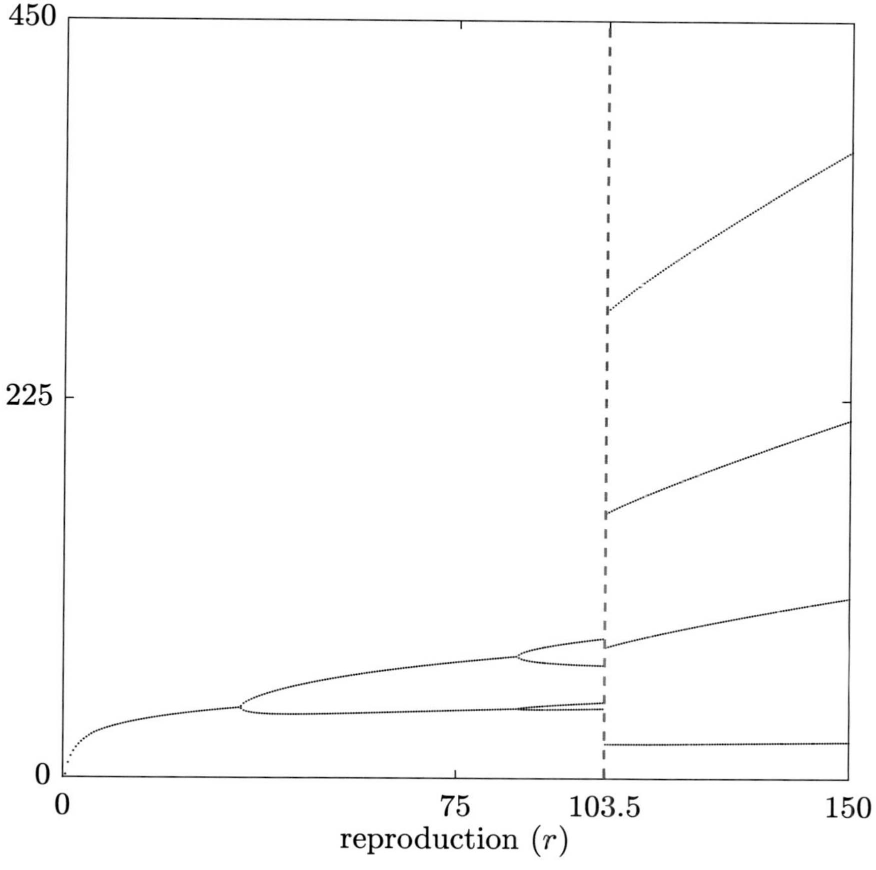

FIGURE A.3. **Jumps within qualitatively similar dynamic regions.** Fixed-d slice for d=0.424 and (S_0,I_0)=(150,20) exhibits a 4-cycle to 4-cycle "jump" at r=103.5. Baseline parameters, except as indicated.

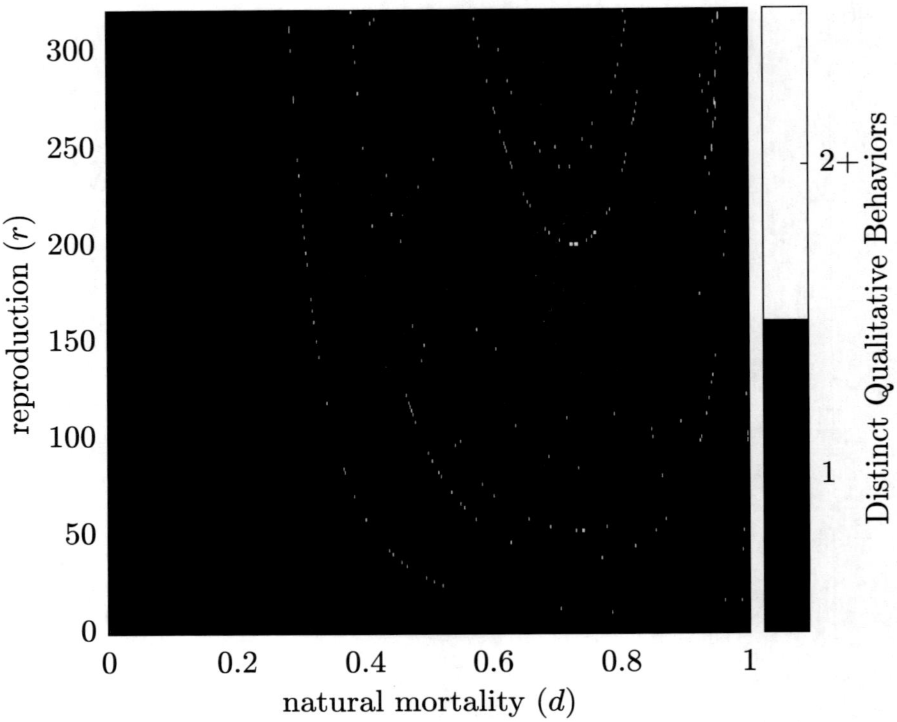

FIGURE A.4. **Multistability diagram for the disease-free system.** Compilation of data from 100 heat maps, each generated by simulation from one of 100 distinct and randomly chosen initial conditions (Figure A.2). Black indicates regions in which our methods identify a single stable qualitative behavior. White indicates regions in which our methods identify more than one stable qualitative behavior. As there is no multistability in the disease-free system, regions of parameter space suggesting multistability are the result of numerical error arising at the boundary between qualitative dynamic regions. Here the time necessary to settle into the true long term behavior is longer than the 10,048 time steps utilized by our methods For visual clarity, the parameters r and d are incremented in coarse steps of size 1 and .004 respectively.

References

[1] Linda J. S. Allen, *Some discrete-time SI, SIR, and SIS epidemic models*, Math. Biosci. **124** (1994), 83–105.

[2] Linda J. S. Allen, *An introduction to mathematical biology*, Pearson/Prentice Hall, 2007.

[3] Roy M. Anderson and Robert M. May, *Population biology of infectious diseases: Part I*, Nature **280** (1979), 361–367.

[4] Roy M. Anderson and Robert M. May, *The logic of vaccination*, The New Scientist **96** (1982), 410–415.

[5] Jaime H. Barrera, Ariel Cintron-Arias, Nicolas Davidenko, Lisa Denogean, and Saul R. Franco-Gonzalez, *Dynamics of a two-dimensional discrete-time SIS model*, Biometrics Unit Technical Reports **BU-1518-M** (1999), 1–23.

[6] T. S. Bellows Jr., *The descriptive properties of some models for density dependence*, J. Animal Ecology **50** (1981), no. 1, 139–156, DOI 10.2307/4037. MR644721

[7] Raymond J. H. Beverton and Sidney J. Holt, *On the dynamics of exploited fish populations*, Fish and Fisheries, vol. 11, Springer-Science + Business Media, 1957.

[8] Erin N. Bodine, Louis J. Gross, and Suzanne Lenhart, *Order of events matter: comparing discrete models for optimal control of species augmentation*, J. Biol. Dyn. **6** (2012), no. suppl. 2, 31–49, DOI 10.1080/17513758.2012.697197. MR2994278

[9] Carlos Castillo-Chavez and Abdul-Aziz Yakubu, *Discrete-time S-I-S models with simple and complex population dynamics*, Mathematical approaches for emerging and reemerging infectious diseases: an introduction (Minneapolis, MN, 1999), IMA Vol. Math. Appl., vol. 125, Springer, New York, 2002, pp. 153–163, DOI 10.1007/978-1-4757-3667-0_9. MR1938884

[10] Carlos Castillo-Chavez and Abdul-Aziz Yakubu, *Discrete-time S-I-S models with complex dynamics*, Proceedings of the Third World Congress of Nonlinear Analysts, Part 7 (Catania, 2000), Nonlinear Anal. **47** (2001), no. 7, 4753–4762, DOI 10.1016/S0362-546X(01)00587-9. MR1975868

[11] Carlos Castillo-Chavez and Abdul-Aziz Yakubu, *Dispersal, disease and life-history evolution*, Math. Biosci. **173** (2001), no. 1, 35–53, DOI 10.1016/S0025-5564(01)00065-7. MR1860421

[12] Carlos Castillo-Chavez and Abdul-Aziz Yakubu, *Intraspecific competition, dispersal and disease dynamics in discrete-time patchy environments*, IMA Vol. Math. Appl. pp. 165–181, Springer-Verlag, 2002.

[13] Leah Edelstein-Keshet, *Mathematical models in biology*, Classics in Applied Mathematics, vol. 46, Society for Industrial and Applied Mathematics (SIAM), Philadelphia, PA, 2005. Reprint of the 1988 original, DOI 10.1137/1.9780898719147. MR2131632

[14] John E. Franke and Abdul-Aziz Yakubu, *Disease-induced mortality in density-dependent discrete-time S-I-S epidemic models*, J. Math. Biol. **57** (2008), no. 6, 755–790, DOI 10.1007/s00285-008-0188-9. MR2439672

[15] J. M. Jaramillo, Junling Ma, P. van den Driessche, and Abdul-Aziz Yakubu, *Disease-induced hydra effect with overcompensatory recruitment*, Bull. Math. Biol. **84** (2022), no. 1, Paper No. 17, 15, DOI 10.1007/s11538-021-00975-4. MR4354914

[16] Maia Martcheva, *An introduction to mathematical epidemiology*, Texts in Applied Mathematics, vol. 61, Springer, New York, 2015, DOI 10.1007/978-1-4899-7612-3. MR3409181

[17] Robert M. May, *Biological populations with nonoverlapping generations: Stable points, stable cycles, and chaos*, Science **186** (1974), 645–647.

[18] Robert M. May, *Simple mathematical models with very complicated dynamics*, Nature **261** (1976), 459–467.

[19] Robert M. May, *Chaos and the dynamics of biological populations*, Proc. Roy. Soc. London Ser. A **413** (1987), no. 1844, 27–44. MR909266

[20] J. D. Murray, *Mathematical biology. I: An introduction*, 3rd ed., Interdisciplinary Applied Mathematics, vol. 17, Springer-Verlag, New York, 2002. MR1908418

[21] A. J. Nicholson and V. A. Bailey, *The balance of animal populations - part I*, Proceedings of the Zoological Society of London **105** (1935), 551–598.

[22] W. E. Ricker, *Stock and recruitment*, Journal of the Fisheries Board of Canada **11** (1954), 559–623.

[23] P. van den Driessche and Abdul-Aziz Yakubu, *Demographic population cycles and $\mathcal{R}_0$ in discrete-time epidemic models*, J. Biol. Dyn. **12** (2018), no. 1, 961–982, DOI 10.1080/17513758.2018.1537449. MR3870233

[24] P. van den Driessche and Abdul-Aziz Yakubu, *Disease extinction versus persistence in discrete-time epidemic models*, Bull. Math. Biol. **81** (2019), no. 11, 4412–4446, DOI 10.1007/s11538-018-0426-2. MR4034832

[25] P. van den Driessche and Abdul-Aziz Yakubu, *Age structured discrete-time disease models with demographic population cycles*, J. Biol. Dyn. **14** (2020), no. 1, 308–331, DOI 10.1080/17513758.2020.1743885. MR4085845

[26] G. F. Webb, *The prime number periodical cicada problem*, Discrete Contin. Dyn. Syst. Ser. B **1** (2001), no. 3, 387–399, DOI 10.3934/dcdsb.2001.1.387. MR1849824

[27] Lei Xiang, Yuyue Zhang, Jicai Huang, and Shigui Ruan, *Complex dynamics in a discrete sis epidemic model with ricker-type recruitment and disease-induced death*, Nonlinear Dynamics **104** (2021), 4635–4654.

[28] Abdul-Aziz Yakubu, *Allee effects in a discrete-time SIS epidemic model with infected newborns*, J. Difference Equ. Appl. **13** (2007), no. 4, 341–356, DOI 10.1080/10236190601079076. MR2311065

[29] Abdul-Aziz Yakubu, *Demographic population cycles in infectious salmon anemia models*, Notices Amer. Math. Soc. **67** (2020), no. 2, 170–180. MR3971640

Department of Mathematics, Virginia Tech, Blacksburg, Virginia, 24061

Current address: Department of Immunology, Department of Computational and Systems Biology, and Center for Systems Immunology, University of Pittsburgh, Pittsburgh, Pennsylvania, 15260

Email address: `strube.laura@pitt.edu`

Department of Mathematics, Virginia Tech, Blacksburg, Virginia, 24061

Email address: `lchilds@vt.edu`

Contemporary Mathematics
Volume **793**, 2024
https://doi.org/10.1090/conm/793/15908

A genetic-epidemiology modeling framework for malaria mosquitoes and disease

Jemal Mohammed-Awel and Abba B. Gumel

This chapter is dedicated in celebration of Professor Ronald E. Mickens' 80th birthday.

ABSTRACT. The large-scale and widespread use of insecticide-based mosquito control has resulted in a dramatic reduction of malaria burden in endemic areas over the last two decades, prompting a renewed concerted global effort to eradicate malaria. Such a widespread and large-scale use of insecticides has, however, resulted in the emergence of insecticide resistance in the endemic areas. This study presents a genetic-epidemiology mathematical modeling framework for assessing the impacts of insecticide resistance on the population abundance of malaria mosquitoes and disease. In particular, two mathematical models are presented. In the first model, insecticide resistance is determined by a single gene with one allele (monoploid), and in the second resistance is determined by a single gene with two alleles (diploid). The models, which take the form of deterministic system of nonlinear differential equations, are rigorously analysed to gain insight into the asymptotic stability properties of their associated non-trivial disease-free equilibria. These analyses revealed that, for each of the two models, the associated generalized non-trivial co-existent disease-free equilibrium is globally-asymptotically stable for a special case (with negligible disease-induced mortality in the human host population) if the corresponding reproduction number of the model is less than unity (the parameters related to the fitness costs of insecticide resistance play a major role in bringing, and maintaining, the value of the reproduction numbers below one). Using numerical simulations, we identified two scenarios, in parameter space, where malaria can be eliminated or persist in the population even when all mosquitoes are fully resistant at steady-state. The study shows that the prospect for malaria elimination is promising using existing insecticide-based mosquito control interventions. It further emphasizes the need to generate the genotype-specific laboratory and field mosquito data needed for more realistic estimation of the parameters related to the fitness costs of insecticide resistance in malaria mosquitoes.

Key words and phrases. insecticide resistance; LLINs; mosquito genotype; allele frequency; reproduction number.

The second author acknowledges the support, in part, of the National Science Foundation (DMS-2052363). The first author acknowledges the support, in part, of the National Science Foundation (DMS-2052355 (transferred to DMS-2221794)).

Corresponding author: Jemal Mohammed-Awel (email: jemal.mohammed-awel@morgan.edu).

1. Introduction

Malaria, caused by the protozoan parasite *Plasmodium* and transmitted to humans *via* the bite of an infectious adult female *Anopheles* mosquito, is the most important mosquito-borne disease of mankind. The disease is endemic in tropical and subtropical regions of the world, representing at least half the human population. Data from the World Health Organization (WHO) showed that, in the year 2021 the disease accounted for 247 million cases and 619,000 deaths globally [**40**]. Adult female mosquitoes bite humans to take a bloodmeal, which is needed for the process of the development and maturation of eggs [**11, 40**] (susceptible humans acquire malaria infection if the mosquito that bites them is infectious, and susceptible mosquitoes can acquire infection if they bite an infectious human). There are five species of malaria parasites, namely *P. falciparum*, *P. malariae*, *P. vivax*, *P. ovale* and *P. knowlesi*. Malaria caused by the *P. falciparum* parasite, if untreated, may result in life-threatening infection including kidney and liver failure [**22, 40**]. *Falciparum* malaria is common in endemic areas in sub-Saharan Africa.

A dramatic reduction in global malaria burden was recorded over the last two decades, largely due to the expansion of access to malaria prevention tools, the widespread and large-scale implementation of insecticide-based vector control methods (notably the use of insecticide-treated bednets (ITNs), later replaced by long-lasting insecticidal nets (LLINs); indoor residual spraying (IRS)) and the use of anti-malarial drugs (particularly *artemisinin* combination therapy) [**11, 29–31, 40**]. Specifically, while nearly all malaria morbidity and mortality is still concentrated in children in sub-Saharan Africa, malaria burden has fallen greatly in this region owing to the heavy and widespread use of the aforementioned insecticide-based vector control measures (especially the use of LLINs, which is believed to account for nearly 50% reduction in sub-Saharan Africa) [**11, 29–31, 40**]. Five major chemical compounds (insecticides) are used in malaria vector control efforts, namely *pyrethroids, organochlorines, organophosphates, neonicotinoids* and *carbamates.* While all the five are used in IRS, only the *pyrethroids*, owing to their low mammalian toxicity and irritant effect on the mosquitoes, are approved (by the WHO) for use in LLINs [**26**].

Unfortunately, the aforementioned large-scale and heavy use of insecticide-based vector control methods has resulted in *Anopheles* resistance to all the five chemical insecticides used in vector control (see [**11, 29–31, 40**] and some of the references therein). Mosquitoes are said to be *resistant* to an insecticide if the ability of the chemical insecticide to kill (or repel) them upon contact with the insecticide is greatly reduced [**11, 24, 31**]. There are four categories of mosquito insecticide resistance mechanisms, namely target-site resistance, metabolic resistance, cuticular resistance, and behavioral resistance [**24, 33**]. While target-site resistance occurs when target site stop the insecticide from effectively binding there [**24, 33**], metabolic resistant mosquitoes detoxify the toxin of the chemical insecticide faster than the insecticide-sensitive mosquitoes using their internal enzyme systems [**24, 33**]. Cuticular resistance occurs when mosquitoes develop cuticular barriers by evolving a thicker cuticle that enables them to absorb the toxin slower than susceptible mosquitoes [**4, 24, 33**]. In behavioral resistance, resistant mosquitoes may detect or recognize, and avoid, the toxin embedded in the chemical insecticide [**24, 33**].

DDT (*Dichlorodiphenyltrichloroethane*) was the first insecticide used for mosquito control (introduced in 1946) [**15**]. The first case of DDT resistance was reported in

An. sacharovi in Greece in 1953 [**15**]. Since then, DDT resistance appeared in several *Anopheles* mosquito species and in several countries [**15**]. Consequently, DDT was disbanded in many countries in the 1980s, and was replaced by other chemical insecticides, such as *organophosphate* (OP) and *pyrethroids* [**15**]. However, mosquito insecticide resistance continued to emerge in all these chemical insecticides, including in *pyrethroids* [**11, 15, 31, 40**]. To mitigate the potential adverse impacts of mosquito insecticide resistance on malaria disease, effective insecticide resistance management (IRM) strategies need to be implemented. Some of the IRM strategies encouraged by the WHO include the use of insecticides in mixtures, sequences, mosaic, rotations and the incorporation of *refugia* [**4, 24, 33**]. The insecticide management strategies reduce the selective pressure of resistance against one particular chemical insecticide by avoiding the use of the same class of chemical for longer periods of time [**24**].

As large-scale field and laboratory trials require years of testing, mathematical models continue to be the most efficient tools for use to realistically assess the effectiveness of different insecticide resistance management strategies and to explore the impact of insecticide resistance in *Anopheles* mosquitoes on malaria transmission dynamics and control. Specifically, mathematical models can be used to address or answer the *key question* in the malaria ecology literature on whether or not insecticide resistance increases malaria transmission [**11, 31**]. Although numerous genetic mathematical models have been developed and used in the literature to study the evolution and spread of insecticide resistance in vector populations [**5, 6, 9, 10, 28, 39**], very few mathematical models that incorporate both the population genetics of malaria mosquitoes and malaria epidemiology have been developed [**27, 29–31**]. One of these few studies included the deterministic model developed by Mohammed-Awel and Gumel [**29**], which stratifies the total mosquito population in terms of type (wild or resistant to the chemical insecticides used in vector control) and feeding preferences (indoor or outdoor). An extended genetic-epidemiology malaria model, which couples the epidemiology of malaria disease with vector population genetics, was developed in [**30**]. The model in [**30**] was further extended in [**31**] by explicitly accounting for the dynamics of mosquitoes in the aquatic stage of the vector lifecycle and adult mosquito sex-structure. A genetic-ecology model was recently developed by Brozak *et al.* [**11**] to assess the combined impacts of insecticide resistance, local temperature variability, and insecticide-based interventions on the population abundance and control of malaria mosquitoes by genotype.

This chapter is focused on using mathematical modeling, together with rigorous analysis and computation, to study the impact of insecticide resistance on the population abundance of malaria mosquitoes and disease in a malaria-endemic area. Two genetic-epidemiology models for malaria transmission dynamics will be presented. The first is developed based on the simplifying assumption that insecticide resistance in the *Anopheles* mosquitoes is determined by a single gene of one allele (*monoploid*), whereas in the second model, a single gene of two alleles (*diploid*) determines insecticide resistance in the mosquitoes. In both models, several parameters associated with the fitness costs of insecticide resistance are accounted for. The chapter is organized as follows. The basic (monoploid) genetic-epidemiology malaria transmission model is presented, analysed and simulated in Section 2. The extended (diploid) model is formulated and studied in Section 3. Concluding remarks are given in Section 4.

2. Basic epidemiology-genetic model for malaria transmission dynamics

In this section, a basic epidemiology-genetic mathematical model for the transmission dynamics of malaria in a population is presented. The total human population at time t, denoted by $N_h(t)$, is sub-divided into the mutually-exclusive compartments of susceptible ($S_h(t)$), exposed ($E_h(t)$), infectious ($I_h(t)$), and recovered ($R_h(t)$) humans. Hence,

$$N_h(t) = S_h(t) + E_h(t) + I_h(t) + R_h(t).$$

In the formulation of this basic model, it is assumed, for simplicity, that insecticide resistance in mosquitoes is determined by a single gene of one allele, and mosquito offsprings inherit the allele from their parents [**23, 27, 29–31**]. That is, the local mosquitoes are either fully insecticide sensitive (i.e., they carry insecticide sensitive alleles only) or fully resistant (i.e., they carry insecticide resistant alleles only), and no mosquitoes have both sensitive and resistant alleles. In this study, mosquitoes that are sensitive to the chemical insecticides used in vector control are referred to as *wild-type* mosquitoes, while those that are resistant to the insecticides are referred to as *resistant-type*. Furthermore, for simplicity, it is assumed that wild-type adult mosquitoes mate with wild-type adult mosquitoes only, and resistant-type adult mosquitoes mate with resistant-type only (i.e., mating is not assumed to be random; this simplifying assumption was also made in [**29**]). The two assumptions will be relaxed in the extended model to be developed in Section 3.

The total adult female *Anopheles* mosquito population at time t, denoted by $N_v(t)$, is stratified into the total wild-type ($N_{vw}(t)$) and resistant-type ($N_{vr}(t)$) mosquitoes. Furthermore, the total population of the adult female insecticide-sensitive mosquitoes is sub-divided into the sub-populations of susceptible ($S_{vw}(t)$), exposed ($E_{vw}(t)$) and infectious ($I_{vw}(t)$) sensitive adult female mosquitoes. Similarly, the total population of adult female insecticide-resistant mosquitoes is sub-divided into the sub-populations of susceptible ($S_{vr}(t)$), exposed ($E_{vr}(t)$) and infectious ($I_{vr}(t)$) adult female resistant-type mosquitoes, so that

$$N_{vw}(t) = S_{vw}(t) + E_{vw}(t) + I_{vw}(t),\ N_{vr}(t) = S_{vr}(t) + E_{vr}(t) + I_{vr}(t)$$
$$\text{and } N_v(t) = N_{vw}(t) + N_{vr}(t).$$

The basic epidemiology-genetic model for the transmission dynamics of malaria is given by the following deterministic system of non-linear differential equations (Figure 1 depicts a schematic description of the basic model. The basic model (2.1) is a simplified version of the model developed in [**29**]):

$$
\begin{aligned}
\frac{dS_h}{dt} &= \Lambda_h + \rho_h R_h - \lambda_{vh} S_h - \mu_h S_h,\\
\frac{dE_h}{dt} &= \lambda_{vh} S_h - (\sigma_h + \mu_h) E_h,\\
\frac{dI_h}{dt} &= \sigma_h E_h - (\gamma_h + \mu_h + \delta_h) I_h,\\
\frac{dR_h}{dt} &= \gamma_h I_h - (\rho_h + \mu_h) R_h,
\end{aligned}
\tag{2.1}
$$

$$
\begin{aligned}
\frac{dS_{vw}}{dt} &= r_{vw}\left(1-\frac{N_v}{K_v}\right) N_{vw} - \lambda_{hv} S_{vw} - (\kappa_{wr} + \delta_{iw} + \mu_v) S_{vw},\\
\frac{dE_{vw}}{dt} &= \lambda_{hv} S_{vw} - (\sigma_{vw} + \kappa_{wr} + \delta_{iw} + \mu_v) E_{vw},\\
\frac{dI_{vw}}{dt} &= \sigma_{vw} E_{vw} - (\kappa_{wr} + \delta_{iw} + \mu_v) I_{vw},\\
\frac{dS_{vr}}{dt} &= r_{vr}\left(1-\frac{N_v}{K_v}\right) N_{vr} + \kappa_{wr} S_{vw} - \lambda_{hv} S_{vr} - [(1-u)\delta_{iw} + \mu_v]\, S_{vr},\\
\frac{dE_{vr}}{dt} &= \lambda_{hv} S_{vr} + \kappa_{wr} E_{vw} - [\sigma_{vr} + (1-u)\delta_{iw} + \mu_v]\, E_{vr}\\
\frac{dI_{vr}}{dt} &= \sigma_{vr} E_{vr} + \kappa_{wr} I_{vw} - [(1-u)\delta_{iw} + \mu_v]\, I_{vr},
\end{aligned}
$$

where, b is the *per capita* biting rate, λ_{hv} is the transmission rate from infectious humans to susceptible mosquitoes, λ_{vh} is the transmission rate from infectious mosquitoes to susceptible humans, and are given, respectively, by [**31**]:

$$
b = \frac{(1-\varepsilon_B C_B) b_{max}}{1+\varepsilon_B C_B b_{max}}, \quad \lambda_{hv} = (\beta_v b)\left(\frac{I_h}{N_h}\right) \text{ and } \lambda_{vh} = (\beta_h b)\left(\frac{I_{vw}+I_{vr}}{N_h}\right), \tag{2.2}
$$

where ε_B is efficacy of LLINs to protect humans from contact with the mosquito, C_B is the proportion of humans who are protected by the LLINs, b_{max} is the maximum mosquito biting rate and β_h (β_v) is transmission probability from infectious mosquitoes (humans) to susceptible humans (mosquitoes).

In the basic model (2.1), the parameter Λ_h represents the recruitment rate of humans (due to immigration or birth). Furthermore, ρ_h is the rate of loss of infection-acquired (natural) immunity. Susceptible humans acquire malaria infection following an effective bite by an infectious adult female *Anopheles* mosquito (at a rate λ_{vh}). It is assumed that humans in all epidemiological classes suffer natural death at a rate μ_h. Exposed humans develop clinical symptoms of malaria (and become infectious) at a rate of σ_h. Infectious humans recover from clinical malaria at a rate γ_h. Infected humans suffer disease-induced mortality at a rate δ_h. The parameters $r_{vw} > 0$ and $r_{vr} > 0$ represent the production (birth) rates of new adult insecticide-sensitive and insecticide-resistant adult female mosquitoes, respectively. Furthermore, $K_v > 0$ is the environmental carrying capacity of adult female mosquitoes (it is assumed that $N_v(t) \leq K_v$ for all $t \geq 0$, so that the logistic term,

$1 - N_v/K_v$, is non-negative). Susceptible adult female mosquitoes (S_{vw} and S_{vr}) acquire malaria infection at the rate λ_{hv}. It is assumed that wild-type mosquitoes develop resistance to the chemical insecticide at a rate κ_{wr} (where κ_{wr} can be also be considered to be the mutation rate). Adult female *Anopheles* mosquitoes can die when they come in contact with LLINs [**29**]. For this reason, it is assumed that wild-type mosquitoes suffer additional death due to exposure to the chemical insecticides embedded in LLINs at a rate δ_{wi}. Similarly, resistant-type mosquitoes also suffer additional mortality at a rate of $(1-u)\delta_{iw}$, with $0 \leq u \leq 1$. The parameter $0 \leq u \leq 1$ is a modification parameter accounting for the fact that the insecticide-induced death rate of resistant-type mosquito is lower than that of wild-type mosquitoes [**29–31**]. Mosquitoes of wild-type (resistant-type) are assumed to suffer natural death at a rate μ_v, and exposed mosquitoes of wild-type (resistant-type) develop clinical symptoms of malaria (and become infectious) at a rate of σ_{vw} (σ_{vr}). The state variables and parameters of the basic model (2.1) are described in Tables 1 and 2, respectively.

2.1. Qualitative properties of the basic model. Since the state variables of the basic model (2.1) represent the temporal dynamics of human and adult female mosquito populations, they must all be non-negative for all time t [**29**]. It is convenient to define the following quantities: $k_1 = \sigma_h + \mu_h$, $k_2 = \gamma_h + \delta_h + \mu_h$, $k_3 = \rho_h + \mu_h$, $k_4 = \kappa_{wr} + \delta_{iw} + \mu_v$ $k_5 = \sigma_{vw} + \kappa_{wr} + \delta_{iw} + \mu_v$, $k_6 = (1-u)\delta_{iw} + \mu_v$ and $k_7 = \sigma_{vr} + (1-u)\delta_{iw} + \mu_v$. Further, let $r_v = \max\{r_{vw}, r_{vr}\}$, $\mathcal{R}_v = \frac{r_v}{k_6}$ and

$$(N_h)_{\min} = \min\left\{N_h(0), \frac{\Lambda_h}{\mu_h + \delta_h}\right\}. \tag{2.3}$$

It can be shown that $(N_h)_{\min} \leq N_h(t) \leq \Lambda_h/\mu_h$ and $N_v(t) \leq K_v(\mathcal{R}_v - 1)/\mathcal{R}_v$ for all $t \geq 0$ (see, for instance, [**29, 32**]). It should be observed that the upper bound of $N_v(t)$, given by $K_v(\mathcal{R}_v - 1)/\mathcal{R}_v$, is positive if $\mathcal{R}_v > 1$. That is, the quantity $K_v(\mathcal{R}_v - 1)/\mathcal{R}_v$ is positive if the maximum of the growth rates of the wild-type and resistant-type mosquitoes (r_v) exceeds the sum of the death rates of resistant-type mosquitoes due to natural causes (μ_v) and due to the insecticides ($(1-u)\delta_{iw}$; note that the sum equals k_6 in the expression for $\mathcal{R}_v$ above). It follows from the expression for $\mathcal{R}_v$ that, in a closed environment (where there is no inflow or outflow of mosquitoes from, and into, external sources), if the total death rate of mosquitoes (due to natural causes and the chemical insecticides used in vector control), denoted by k_6, exceeds the growth (birth) rate of the mosquitoes (r_v), then $\mathcal{R}_v < 1$ (consequently, the mosquito population dies out asymptotically). To ensure the presence of mosquitoes in the environment, it is assumed, for the rest of this section, that $\mathcal{R}_v > 1$. Moreover, it is assumed that the mutation rate (κ_{wr}) is positive (i.e., $\kappa_{wr} > 0$).

Define the following feasible region for the basic model (2.1):

$$\Omega = \Big\{(S_h, E_h, I_h, R_h, S_{vw}, E_{vw}, I_{vw}, S_{vr}, E_{vr}, I_{vr}) \in \mathbb{R}^4_+ \cup \mathbb{R}^6_+ : 0 < (N_h)_{min} \leq N_h(t) \leq \frac{\Lambda_h}{\mu_h}, \tag{2.4}$$
$$\text{and } 0 \leq N_v \leq K_v(\mathcal{R}_v - 1)/\mathcal{R}_v\Big\}.$$

The following result can be established for the basic model (the proof is standard; see, for instance, [**29, 32**]).

LEMMA 1. *Consider the basic model (2.1) with $\mathcal{R}_v > 1$ and $N_h(0) > 0$. The region Ω is positively-invariant and attracts all solutions of the basic model (2.1) in $\mathbb{R}^4_+ \cup \mathbb{R}^6_+$.*

2.2. Existence and asymptotic stability of equilibria of the basic model. In this section, the qualitative properties of the basic model (2.1), with respect to the existence and asymptotic stability of its equilibria, will be rigorously analysed.

2.2.1. *Existence of disease-free equilibria.* The basic model (2.1) has three disease-free equilibria, namely a trivial disease-free equilibrium (denoted by $\mathcal{E}_T$), a non-trivial resistant-only disease-free boundary equilibrium (denoted by $\mathcal{E}_R$) and a non-trivial co-existence disease-free equilibrium (denoted by $\mathcal{E}_C$) given below:

$$
\begin{aligned}
\mathcal{E}_T &= \left(S_h^{*T}, E_h^{*T}, I_h^{*T}, R_h^{*T}, S_{vw}^{*T}, E_{vw}^{*T}, I_{vw}^{*}, S_{vr}^{*T}, E_{vr}^{*T}, I_{vr}^{*T}\right)\\
&= \left(S_h^{*T}, 0,0,0,0,0,0,0,0,0,0,0,0,0,0,0\right),\\
\mathcal{E}_R &= \left(S_h^{*R}, E_h^{*R}, I_h^{*R}, R_h^{*R}, S_{vw}^{*R}, E_{vw}^{*R}, I_{vw}^{*R}, S_{vr}^{*R}, E_{vr}^{*R}, I_{vr}^{*R}\right)\\
&= \left(S_h^{*R}, 0,0,0,0,0,0, S_{vr}^{*R}, 0,0\right),\\
\mathcal{E}_C &= \left(S_h^{*C}, E_h^{*C}, I_h^{*C}, R_h^{*C}, S_{vw}^{*C}, E_{vw}^{C*}, I_{vw}^{*C}, S_{vr}^{*C}, E_{vr}^{*C}, I_{vro}^{*C}\right)\\
&= \left(S_h^{*C}, 0,0,0, S_{vw}^{*C}, 0,0, S_{vr}^{*C}, 0,0\right),
\end{aligned}
\tag{2.5}
$$

where,

$$
\begin{aligned}
&S_h^{*T} = S_h^{*R} = S_h^{*C} = \frac{\Lambda_h}{\mu_h},\ S_{vr}^{*R} = \frac{K_v(\mathcal{R}_r - 1)}{\mathcal{R}_r},\\
&S_{vw}^{*C} = \frac{K_v r_{vr}(\mathcal{R}_w - 1)(\mathcal{R}_w - \mathcal{R}_r)}{\mathcal{R}_w\left[\kappa_{wr}\mathcal{R}_w\mathcal{R}_r + r_{vr}(\mathcal{R}_w - \mathcal{R}_r)\right]} \text{ and } S_{vr}^{*C} = \frac{K_v \kappa_{wr}\mathcal{R}_r(\mathcal{R}_w - 1)}{\kappa_{wr}\mathcal{R}_w\mathcal{R}_r + r_{vr}(\mathcal{R}_w - \mathcal{R}_r)},
\end{aligned}
\tag{2.6}
$$

with,

$$
\mathcal{R}_w = \frac{r_{vw}}{k_4} \text{ and } \mathcal{R}_r = \frac{r_{vr}}{k_6}. \tag{2.7}
$$

It is worth stating that since the basic model (2.1) allows for the development of resistance by wild-type adult female mosquitoes (due to the mutation of the wild-type adult female mosquitoes to resistant-type adult female mosquitoes) at the rate $\kappa_{wr} > 0$, the basic model cannot have a boundary disease-free equilibrium with only the wild-type mosquitoes (i.e., the basic model (2.1) does not have a wild-type disease-free boundary equilibrium due to the fact that $\kappa_{wr} > 0$). It follows from Equations (2.5)–(2.7) that:

(i) the resistant-only boundary disease-free equilibrium ($\mathcal{E}_R$) exists if and only if $\mathcal{R}_r > 1$,

(ii) the co-existence disease-free equilibrium ($\mathcal{E}_C$) exists if and only if $\mathcal{R}_w > \max\{\mathcal{R}_r, 1\}$.

It is convenient to define, for $\mathcal{R}_w > 1$, the following general non-trivial disease-free equilibrium of the basic model (2.1):

$$
\begin{aligned}
\mathcal{E}_{df} &= (S_h, E_h, I_h, R_h, S_{vw}, E_{vw}, I_{vw}, S_{vr}, E_{vr}, I_{vr})\\
&= \left(\frac{\Lambda_h}{\mu_h}, 0,0,0, S_{vw}^{\diamond}, 0,0, S_{vr}^{\diamond}, 0,0\right).
\end{aligned}
\tag{2.8}
$$

The following relations hold (by using (2.8) in (2.5)):

(i) If $\mathcal{E}_{df} = \mathcal{E}_R$, then $S_{vw}^{\diamond} = 0$ and $S_{vr}^{\diamond} = S_{vr}^{*R}$.
(ii) If $\mathcal{E}_{df} = \mathcal{E}_C$, then $S_{vw}^{\diamond} = S_{vw}^{*C}$ and $S_{vr}^{\diamond} = S_{vr}^{*C}$.

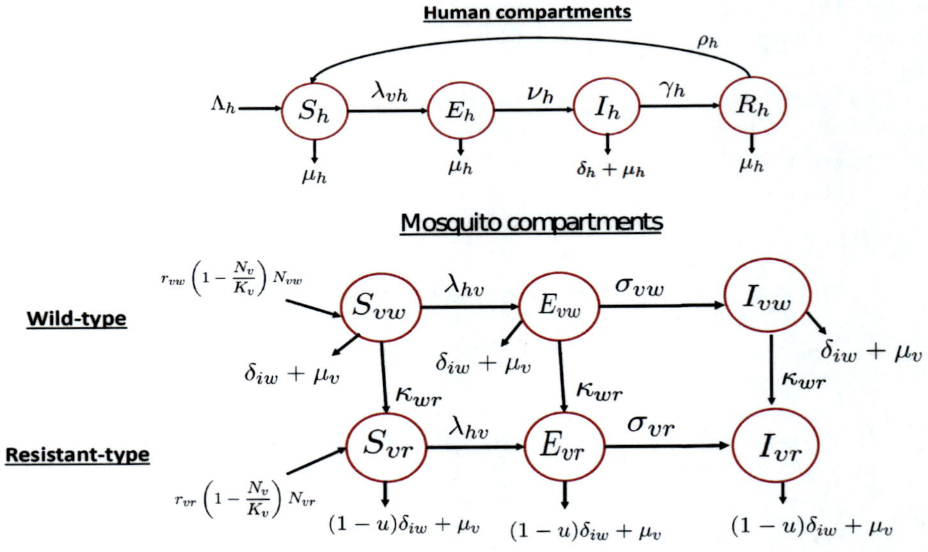

FIGURE 1. Flow diagram of the basic model (2.1).

TABLE 1. Description of the state variables of the basic model (2.1)

Human classes	Description
S_h	Population of susceptible humans
E_h	Population of exposed (infected but not symptomatic) humans
I_h	Population of infectious (symptomatic) humans
R_h	Population of recovered humans
N_h	Total human population
Mosquito classes	**Description**
S_{vw}	Population of susceptible wild-type adult female mosquitoes
E_{vw}	Population of exposed wild-type adult female mosquitoes
I_{vw}	Population of infectious wild-type adult female mosquitoes
S_{vr}	Population of susceptible resistant-type adult female mosquitoes
E_{vr}	Population of exposed resistant-type adult female mosquitoes
I_{vr}	Population of infectious resistant-type adult female mosquitoes
N_{vw}	Total number of wild-type adult female mosquitoes
N_{vr}	Total number of resistant-type adult female mosquitoes
N_v	Total number of adult female mosquitoes

2.2.2. *Asymptotic stability of generalized non-trivial disease-free equilibrium.* In this section, the asymptotic stability of the generalized non-trivial disease-free equilibrium ($\mathcal{E}_{df}$) will be explored.

Local asymptotic stability. : the *next generation operator* method [**17**, **18**] can be used to analyse the linear stability property of the generalized non-trivial disease-free equilibrium ($\mathcal{E}_{df}$) of the basic model (2.1). Specifically, it can be shown, using the notation in [**18**], that the associated next generation matrices F and V, for the new infection terms and the remaining transfer terms, respectively (obtained from the equations of the infected classes of the model (2.1), using the ordering (E_h, I_h, E_{vw}, I_{vw}, E_{vr}, I_{vr})) are given, respectively, by:

$$F = \begin{pmatrix} 0 & 0 & 0 & \beta_h b & 0 & \beta_h b \\ 0 & 0 & 0 & 0 & 0 & 0 \\ 0 & \dfrac{\beta_v b \mu_h S^{\diamond}_{vw}}{\Lambda_h} & 0 & 0 & 0 & 0 \\ 0 & 0 & 0 & 0 & 0 & 0 \\ 0 & \dfrac{\beta_v b \mu_h S^{\diamond}_{vr}}{\Lambda_h} & 0 & 0 & 0 & 0 \\ 0 & 0 & 0 & 0 & 0 & 0 \end{pmatrix} \tag{2.9}$$

$$\text{and } V = \begin{pmatrix} k_1 & 0 & 0 & 0 & 0 & 0 \\ -\sigma_h & k_2 & 0 & 0 & 0 & 0 \\ 0 & 0 & k_5 & 0 & 0 & 0 \\ 0 & 0 & -\sigma_{vw} & k_4 & 0 & 0 \\ 0 & 0 & -\kappa_{wr} & 0 & k_7 & 0 \\ 0 & 0 & 0 & -\kappa_{wr} & -\sigma_{vr} & k_6 \end{pmatrix}.$$

It is convenient to define the following quantity [**17**, **18**]:

$$\mathcal{R}_0 = \rho(FV^{-1}) = \sqrt{\mathcal{R}_{HM}\mathcal{R}_{MH}}, \tag{2.10}$$

where,

$$\mathcal{R}_{HM} = \frac{b\beta_h \mu_h \sigma_h}{\Lambda_h k_1 k_2},$$

$$\text{and } \mathcal{R}_{MH} = b\beta_v \left(\frac{\sigma_{vw} S^{\diamond}_{vw}}{k_4 k_5} + \frac{\kappa_{wr}\sigma_{vw} S^{\diamond}_{vw}}{k_4 k_5 k_6} + \frac{\kappa_{wr}\sigma_{vr} S^{\diamond}_{vw}}{k_5 k_6 k_7} + \frac{\sigma_{vr} S^{\diamond}_{vr}}{k_6 k_7} \right).$$

Furthermore, let

(2.11)

$$\mathcal{R}_0^{0R} = \mathcal{R}_0|_{\mathcal{E}_{df}=\mathcal{E}_R} = \sqrt{\frac{b\beta_h \mu_h \sigma_h}{\Lambda_h k_1 k_2} \left(\frac{b\beta_v \sigma_{vr} S^{*R}_{vr}}{k_6 k_7} \right)}, \text{ and}$$

$$\mathcal{R}_0^{0C} = \mathcal{R}_0|_{\mathcal{E}_{df}=\mathcal{E}_C} = \sqrt{\mathcal{R}_{HM} \left[b\beta_v \left(\frac{\sigma_{vw} S^{*C}_{vw}}{k_4 k_5} + \frac{\kappa_{wr}\sigma_{vw} S^{*C}_{vw}}{k_4 k_5 k_6} + \frac{\kappa_{wr}\sigma_{vr} S^{*C}_{vw}}{k_5 k_6 k_7} + \frac{\sigma_{vr} S^{*C}_{vr}}{k_6 k_7} \right) \right]}.$$

Hence, using the above analyses and Theorem 2 in [**18**], the result below follows (see also [**29–31**]):

THEOREM 2.1. *Consider the basic model* (2.1).

(a). *If $\mathcal{R}_r > 1$, then the resistant-type disease-free equilibrium ($\mathcal{E}_R$) is locally-asymptotically stable if $\mathcal{R}_0^{0R} < 1$, and unstable if $\mathcal{R}_0^{0R} > 1$.*

(b). *If $\mathcal{R}_w > \max\{R_r, 1\}$, then the co-existence disease-free equilibrium ($\mathcal{E}_C$) is locally-asymptotically stable if $\mathcal{R}_0^{0C} < 1$, and unstable if $\mathcal{R}_0^{0C} > 1$.*

The control reproduction number $\mathcal{R}_0$, defined in (2.10), represents the average number of new infections (in humans or mosquitoes) generated by a typical infectious individual (or mosquito) introduced into a community where insecticide-based vector control measures are implemented [**17, 18**]. It is the geometric mean of the constituent reproduction numbers for human-to-mosquito ($\mathcal{R}_{HM}$; given in (2.10)) and the mosquito-to-human ($\mathcal{R}_{MH}$) transmission of malaria. Specifically, $\mathcal{R}_{HM}$ ($\mathcal{R}_{MH}$) represents the average number of secondary malaria infections generated by a single infectious human (mosquito) if introduced in a population of mosquitoes (humans) where mosquito control interventions are implemented. The epidemiological implication of Theorem 2.1 is that malaria can be effectively controlled in the community, when $\mathcal{R}_0^{OC} < 1$ ($\mathcal{R}_0^{OR} < 1$), if the initial sizes of the sub-populations of the basic model (2.1) are in the basin of attraction of the co-existence disease-free equilibrium ($\mathcal{E}_C$) [resistant-only disease-free equilibrium ($\mathcal{E}_R$)]. In order for such effective control to be independent of the initial sizes of the sub-populations of the model, a global asymptotic stability result must be established for the generalized non-trivial disease-free equilibrium of the basic model (2.1). This is explored below, for a special case.

TABLE 2. Descriptions and baseline values of the parameters of the model (2.1)

Parameters	Description	Baseline Value (per day)	Source
Λ_h	Human recruitment rate	2.19	Estimated from [16]
μ_h	Natural death rate for humans	4.405×10^{-5}	Estimated from [14]
δ_h	Disease-induced death rate for humans	9.0×10^{-5}	[35]
σ_h	Rate at which exposed humans become infectious	(1/14)	[29, 34]
γ_h	Recovery rate of humans	6.58×10^{-3}	[36]
ρ_h	Rate of loss of natural immunity	5.6×10^{-3}	[35]
K_v	Environmental carrying capacity of mosquitoes	10^5 (dimensionless)	[29]
r_{vw}	Production (birth) rates of new adult wild-type female mosquitoes	77.4/365	[29]
r_{vr}	Production (birth) rates of new adult resistant-type female mosquitoes	76.3/365	[29]
μ_v	Natural death rate of mosquitoes	(1/14)	[35]
σ_{vw}	Rate at which exposed wild-type adult resistant-type become infectious	36.5/365	[29]
σ_{vr}	Rate at which exposed resistant-type mosquitoes become infectious	43.6/365	[29]
δ_{iw}	Insecticide induced death rate of wild-type mosquitoes due to the use of LLINs	40.6/365	[29]
u	Modification parameter accounting for the assumed decrease in the mortality rate of adult mosquitoes of resistant-type due to the use of insecticides, in comparison to insecticide -induced mortality rate of wild-type (those exposed to insecticide) due to the use of LLINs	0.9 (dimensionless)	[30]
κ_{wr}	Rate of development of resistance by wild-type mosquitoes due to the widespread use of LLINs ("mutation rate")	10^{-3}	[29]
b_{max}	Maximum mosquito biting rate	Varies	[29–31]
b	Average mosquito biting rate	Expressed in terms of other parameters	[31]
β_h	Transmission probability from infectious mosquitoes to susceptible humans	$22/10^3$ (dimensionless)	[34]
β_v	Transmission probability from infectious humans to susceptible mosquitoes	$48/10^2$ (dimensionless)	[34]
C_B	LLINs coverage (or proportion of LLINs usage)	Varies (dimensionless)	[30, 31]
ε_B	Efficacy of LLINs ($0 \le \varepsilon_B \le 1$) to protect humans	0.85 (dimensionless)	Estimated [3, 31]

Global asymptotic stability: special case. in this section, the global asymptotic stability property of the generalized non-trivial disease-free equilibrium ($\mathcal{E}_{df}$) of the basic model (2.1) will be explored for the special case with negligible disease-induced mortality in the human population (i.e., we will consider the model (2.1) with $\delta_h = 0$). This assumption is made for mathematical tractability. It is convenient, first of all, to define the following feasible regions for the basic model:

$$\Omega_C = \{(S_h, E_h, I_h, R_h, S_{vw}, E_{vw}, I_{vw}, S_{vr}, E_{vr}, I_{vr}) \in \Omega : \tag{2.12}$$
$$N_h(t) \le \frac{\Lambda_h}{\mu_h},\ S_{vw}(t) \le S_{vw}^{*C},\ S_{vr}(t) \le S_{vr}^{*C} \Big\},$$

$$\Omega_R = \{(S_h, E_h, I_h, R_h, S_{vw}, E_{vw}, I_{vw}, S_{vr}, E_{vr}, I_{vr}) \in \Omega : \tag{2.13}$$
$$N_h(t) \le \frac{\Lambda_h}{\mu_h},\ S_{vw}(t) = 0, S_{vr}(t) \le S_{vr}^{*R} \Big\}.$$

LEMMA 2. *Consider the special case of the model (2.1) with $\delta_h = 0$.*

(a). *If $\mathcal{R}_w > \max\{\mathcal{R}_r, 1\}$, then the region Ω_C is positively-invariant for the basic model (2.1) provided that*

$$\min\left\{\frac{r_{vr}\left(\mathcal{R}_w-1\right)\left(\mathcal{R}_w-\mathcal{R}_r\right)}{\mathcal{R}_w^2\left[\kappa_{vw}\mathcal{R}_w\mathcal{R}_r+r_{rv}(\mathcal{R}_w-\mathcal{R}_r)\right]}, \frac{\kappa_{wr}(\mathcal{R}_w-1)}{\kappa_{vw}\mathcal{R}_w\mathcal{R}_r+r_{vr}(\mathcal{R}_w-\mathcal{R}_r)}\right\}\geq\frac{1}{4}.$$

(b). *If $\mathcal{R}_r > 1$ and $\dfrac{\mathcal{R}_r-1}{\mathcal{R}_r^2} \geq \frac{1}{4}$, then the region Ω_R is positively-invariant for the basic model (2.1).*

PROOF. (a). Let $\mathcal{R}_w > \max\{\mathcal{R}_r, 1\}$. Adding all the equations for the human components of the model (2.1), and noting that $\delta_h = 0$, gives

$$\frac{dN_h}{dt} = \Lambda_h - \mu_h N_h = \mu_h\left(\frac{\Lambda_h}{\mu_h} - N_h\right).$$

Hence, $N_h(t) = N_h(0)e^{-\mu_h t} + \frac{\Lambda_h}{\mu_h}(1-e^{-\mu_h t})$. Thus, if $N_h(0) \leq \frac{\Lambda_h}{\mu_h}$, then $N_h(t) \leq \frac{\Lambda_h}{\mu_h}$ for all $t \geq 0$.
Similarly, it follows from the equation of the model (2.1) that

$$\begin{aligned}\frac{dS_{vw}}{dt} &= r_{vw}N_{vw}\left(1-\frac{N_v}{K_v}\right) - \lambda_{hv}S_{vw} - k_4 S_{vw}\\ &\leq r_{vw}N_{vw}\left(1-\frac{N_v}{K_v}\right) - k_4S_{vw} \leq r_{vw}N_v\left(1-\frac{N_v}{K_v}\right) - k_4S_{vw}.\end{aligned}$$

Observe that the quadratic polynomial (in N_v) in the inequality above has a global maximum value $\frac{K_v}{4}$ (attained when $N_v = \frac{K_v}{2}$). Thus, the above inequality can be simplified as follows:

$$\begin{aligned}\frac{dS_{vw}}{dt} &\leq r_{vw}N_v\left(1-\frac{N_v}{K_v}\right) - k_4S_{vw} \leq r_{vw}\frac{K_v}{4} - k_4S_{vw}\\ &\leq r_{vw}\frac{K_v r_{vr}\left(\mathcal{R}_w-1\right)\left(\mathcal{R}_w-\mathcal{R}_r\right)}{\mathcal{R}_w^2\left[\kappa_{vw}\mathcal{R}_w\mathcal{R}_r+r_{rv}(\mathcal{R}_w-\mathcal{R}_r)\right]} - k_4S_{vw}\\ &= k_4\left[S_{vw}^{*C} - S_{vw}(t)\right] \text{ (since } \mathcal{R}_w > \max\{\mathcal{R}_r,1\}).\end{aligned}$$

Hence, $S_{vw}(t) \leq S_{vw}(0)e^{-k_4t} + S_{vw}^{*C}\left(1-e^{-k_4t}\right)$. Thus, if $S_{vw}(0) \leq S_{vw}^{*C}$, then $S_{vw}(t) \leq S_{vw}^{*C}$ for all $t \geq 0$. Using similar arguments, it can be shown that $S_{vr}(t) \leq S_{vr}^{*C}$ for all $t \geq 0$ if the relevant initial conditions are in Ω_C. Thus, the region Ω_C is positively-invariant for the special case of the model (2.1) provided that $\mathcal{R}_w > \max\{\mathcal{R}_r, 1\}$ and

$$\min\left\{\frac{r_{vr}\left(\mathcal{R}_w-1\right)\left(\mathcal{R}_w-\mathcal{R}_r\right)}{\mathcal{R}_w^2\left[\kappa_{vw}\mathcal{R}_w\mathcal{R}_r+r_{rv}(\mathcal{R}_w-\mathcal{R}_r)\right]}, \frac{\kappa_{wr}(\mathcal{R}_w-1)}{\kappa_{vw}\mathcal{R}_w\mathcal{R}_r+r_{vr}(\mathcal{R}_w-\mathcal{R}_r)}\right\}\geq\frac{1}{4}.$$

Item (b) can be proved similarly (hence, not repeated here). ☐

It is convenient to define $\overline{\mathcal{R}}_0^{0R} = \mathcal{R}_0^{0R}|_{\delta_h=0}$ and $\overline{\mathcal{R}}_0^{0C} = \mathcal{R}_0^{0C}|_{\delta_h=0}$.

THEOREM 2.2. *Consider the special case of the model (2.1) with $\delta_h = 0$.*

(a). *If $\mathcal{R}_w > \max\{\mathcal{R}_r, 1\}$, then the co-existence disease-free equilibrium ($\mathcal{E}_C$) is globally-asymptotically stable in $\Omega_C \setminus \{\mathcal{E}_T, \mathcal{E}_R\}$ whenever $\overline{\mathcal{R}}_0^{0C} < 1$ provided*

$$\min\left\{\frac{r_{vr}\left(\mathcal{R}_w-1\right)\left(\mathcal{R}_w-\mathcal{R}_r\right)}{\mathcal{R}_w^2\left[\kappa_{vw}\mathcal{R}_w\mathcal{R}_r+r_{rv}(\mathcal{R}_w-\mathcal{R}_r)\right]}, \frac{\kappa_{wr}(\mathcal{R}_w-1)}{\kappa_{vw}\mathcal{R}_w\mathcal{R}_r+r_{vr}(\mathcal{R}_w-\mathcal{R}_r)}\right\}\geq\frac{1}{4}.$$

(b). *If* $\mathcal{R}_r > 1$ *and* $\dfrac{(\mathcal{R}_r - 1)}{\mathcal{R}_r^2} \geq \frac{1}{4}$, *then the resistant-type disease-free equilibrium* ($\mathcal{E}_R$) *is globally-asymptotically stable in* $\Omega_R \setminus \{\mathcal{E}_T, \mathcal{E}_C\}$ *whenever* $\overline{\mathcal{R}}_0^{0R} < 1$.

The proof of Theorem 2.2, based on using a comparison theorem [**38**], is given in Appendix A. The epidemiological implication of Theorem 2.2 is that malaria can be effectively-controlled if the threshold quantity $\overline{\mathcal{R}}_0^{0C}$ (or $\overline{\mathcal{R}}_0^{0R}$) can be reduced to (and maintained at) a value less than unity. For example, for the special case of the model (2.1) with $\delta_h = 0$, if $\mathcal{R}_w > \max\{\mathcal{R}_r, 1\}$ and

$$\min\left\{\frac{r_{vr}\,(\mathcal{R}_w - 1)\,(\mathcal{R}_w - \mathcal{R}_r)}{\mathcal{R}_w^2\,[\kappa_{vw}\mathcal{R}_w\mathcal{R}_r + r_{rv}(\mathcal{R}_w - \mathcal{R}_r)]}, \frac{\kappa_{wr}(\mathcal{R}_w - 1)}{\kappa_{vw}\mathcal{R}_w\mathcal{R}_r + r_{vr}(\mathcal{R}_w - \mathcal{R}_r)}\right\} \geq \frac{1}{4},$$

then the classical epidemiological requirement of $\overline{\mathcal{R}}_0^{0C} < 1$ is necessary and sufficient for the effective control (or elimination) of malaria from the community. It should be mentioned that if the assumption for negligible disease-induced mortality in humans (i.e., $\delta_h \approx 0$) is relaxed, the basic model (2.1) may exhibit the dynamic phenomenon of a *backward bifurcation*, which is characterized by the co-existence of a stable non-trivial disease-free equilibrium ($\mathcal{E}_C$ or $\mathcal{E}_R$) and a stable endemic equilibrium when the associated control reproduction number of the basic model ($\overline{\mathcal{R}}_0^{0C} < 1$ or $\overline{\mathcal{R}}_0^{0R} < 1$) is less than one (see, for instance, [**7, 25, 32**]). The presence of backward bifurcation in the transmission dynamics of a disease makes its effective control more difficult (since, in a backward bifurcation situation, effective disease control when the associated reproduction number of the model is less than one depends on the initial sizes of the sub-populations of the model). The absence of backward bifurcation in the basic model when disease-induced mortality in the host population is negligible and the associated reproduction number less than unity (guaranteed by Theorem 2.2) shows that disease-induced mortality in the human population (at non-trivial level) is a sufficient condition for the presence of backward bifurcation in the basic model (2.1). The results of Theorem 2.2 are numerically illustrated by simulating the basic model (2.1) using the baseline parameter values given in Table 2, with $\delta_h = 0$, $r_{vr} = 58.3/365$, and $b_{max} = 0.1$ (so that $\mathcal{R}_r = 2$, $(\mathcal{R}_r - 1)/\mathcal{R}_r^2 = 0.25$ and $\overline{\mathcal{R}}_0^{0R} = 0.347$). Since, for this set of parameter values, the hypotheses of Theorem 2.2 (b) are satisfied, all solution trajectories of the basic model will converge to the globally-asymptotically stable resistant-type disease-free equilibrium ($\mathcal{E}_R$). This is shown in Figure 2. Specifically, the total number of infectious humans (Figure 2 (a)) and infectious mosquitoes (Figure 2 (b)) converge to zero (confirming malaria elimination for this set of parameter values). Furthermore, while the total number of wild-type adult female mosquitoes die out at equilibrium (Figure 2 (c)), all mosquitoes in the community are resistant-type (Figure 2 (d)), in line with Theorem 2.2 (b). It is worth emphasizing that Figure 2 shows that malaria can be eliminated even when all mosquitoes are resistant-type (Figure 2(d)). In other words, this study shows that, for this scenario, insecticide resistance does not increase malaria burden (hence, it does not inhibit malaria elimination).

The basic model is further simulated for the case when the reproduction threshold $\overline{\mathcal{R}}_0^{0R}$ exceeds unity. The results obtained, depicted in Figure 3, show the persistence of the disease in the human (Figure 3 (a)) and mosquito (Figure 3 (b)) populations.

Here, too, the population of wild-type mosquitoes die out at equilibrium (Figure 3 (c)), while that of resistant-type mosquitoes persist at equilibrium (Figure 3 (d)). It should be noted that, in this case with $\overline{\mathcal{R}}_0^{0R} > 1$, the resistant-type disease-free equilibrium ($\mathcal{E}_R$) is unstable (Theorem 2.1 (a)). In conclusion, the rigorous analyses and simulations of the basic model (2.1) conducted in this section show that malaria can persist in, or be eliminated from, a community even when all the malaria mosquitoes in the community (at steady-state) are resistant to the insecticides used in the community. In other words, insecticide resistance does not affect malaria elimination or persistence as long as the ecological and epidemiological conditions for achieving such (which depend on the values of the aforementioned associated reproduction thresholds) are satisfied.

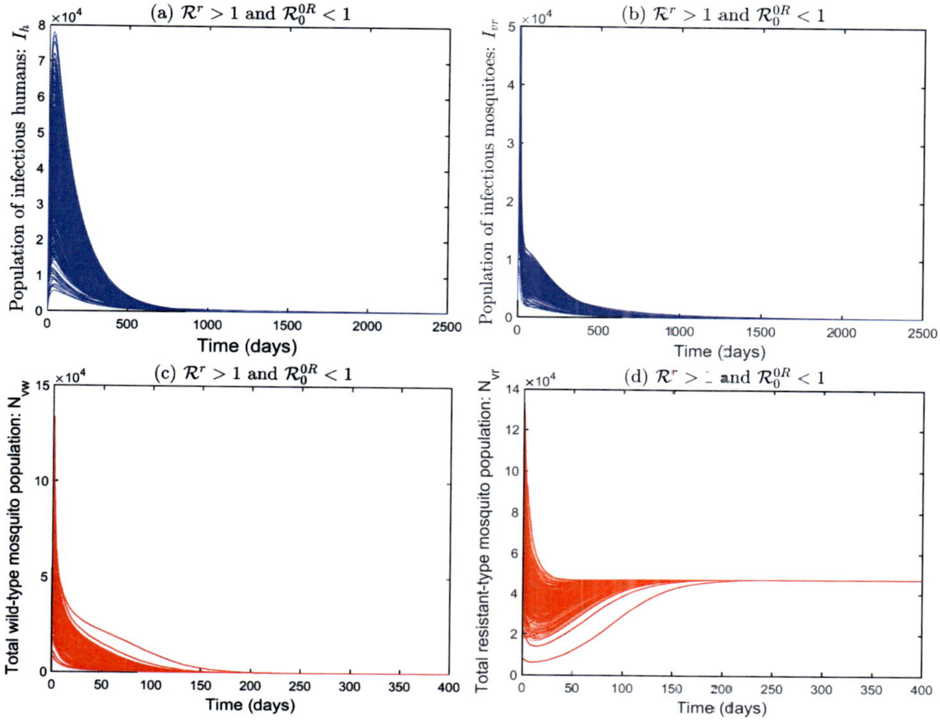

FIGURE 2. Numerical simulations of the model (2.1) showing convergence of initial solutions to the resistant-only non-trivial disease-free boundary equilibrium when $\mathcal{R}_r > 1$ and $\overline{\mathcal{R}}_0^{0R} < 1$. (a) population of infectious humans (I_h) as a function of time for various initial conditions, (b) population of infectious resistant-type mosquitoes (I_{vr}), (c) total population of wild-type mosquitoes (N_{vw}) as a function of time for various initial conditions, and (d) total population of resistant-type mosquitoes (N_{vr}) as a function of time for various initial conditions. Parameter values used are as given in Table 2 with $\delta_h = 0$, $C_B = 0$, $r_{vr} = 58.3/365$ and $b_{max} = 0.1$ (so that $\mathcal{R}_r = 2 > 1, \frac{(\mathcal{R}_r - 1)}{\mathcal{R}_r^2} = 0.25 \geq \frac{1}{4}$ and $\overline{\mathcal{R}}_0^{0R} = 0.347 < 1$; hence, by Theorem 2.2 (b), the resistant-only disease-free boundary equilibrium $(\mathcal{E}_R)$ is globally-asymptotically stable in $\Omega_R \setminus \{\mathcal{E}_T, \mathcal{E}_C\}$.

3. Malaria model with genetic structure in the mosquito population

It was assumed, in the formulation of the basic model (2.1), that mosquito insecticide resistance is determined by a single gene with one allele in a locus. This assumption is relaxed in this section, and replaced by the more realistic assumption that insecticide resistance is determined by two alleles at a particular locus (see Figure 4), as in [**11**, **27**, **30**, **31**]. That is, in the extended model to be developed in this section, insecticide resistance will be determined by a single gene of two alleles (namely, resistant (R) and susceptible (S) alleles), resulting in three different mosquito genotypes, namely homozygous resistant (RR), homozygous sensitive (SS), and heterozygous (RS). The three mosquito genotypes have different levels of sensitivity to anti-malaria chemical insecticides [**8**]. The state variables of the extended genetic-epidemiology malaria model to be developed are defined below.

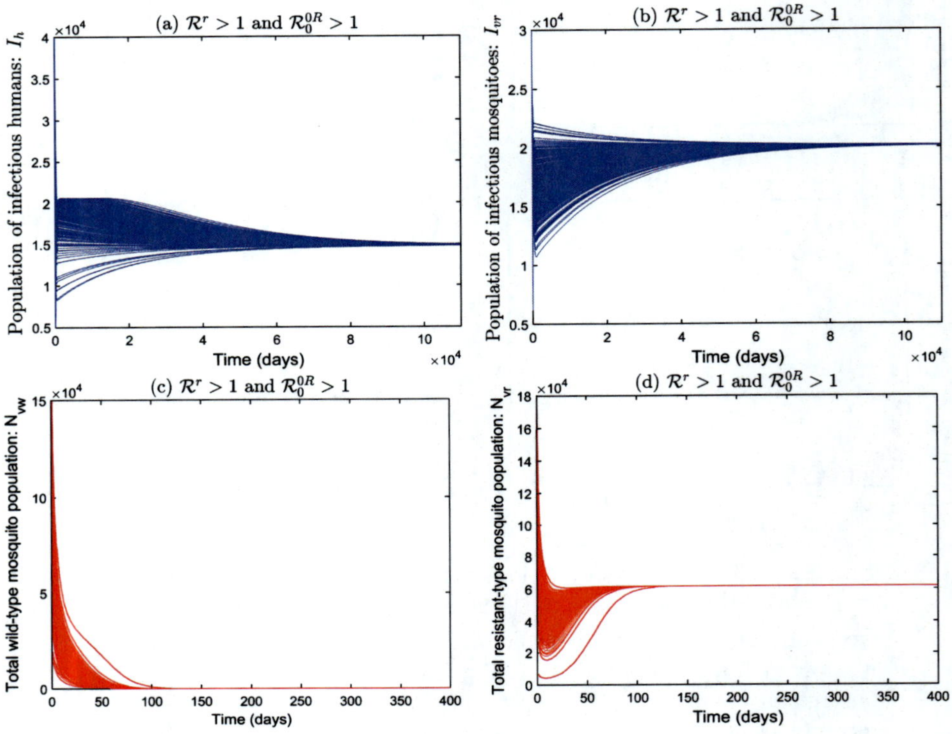

FIGURE 3. Numerical simulations of the model (2.1) convergence of initial solutions to the resistant-only boundary equilibrium when $\mathcal{R}_r > 1$ and $\overline{\mathcal{R}}_0^{0R} > 1$. (a) population of infectious humans (I_h) as a function of time for various initial conditions, (b) population of infectious resistant-type mosquitoes (I_{vr}), (c) total population of wild-type mosquitoes (N_{vw}) as a function of time for various initial conditions, and (d) total population of resistant-type mosquitoes (N_{vr}) as a function of time for various initial conditions. Parameter values used are as given in Table 2 with $\delta_h = 0$, $C_B = 0$, $r_{vr} = 58.3/365$ and $b_{max} = 0.67$ (so that $\mathcal{R}_r = 2 > 1, \frac{(\mathcal{R}_r - 1)}{\mathcal{R}_r^2} = 0.25 \geq \frac{1}{4}$ and $\overline{\mathcal{R}}_0^{0R} = 2.3251 > 1$).

State variables of the extended model: To formulate the extended genetic-epidemiology malaria model, the total population of adult female *Anopheles* mosquitoes in the community at time t, denoted by $N_M(t)$, is sub-divided into the sub-populations of homozygous resistant ($N_{RR}^M(t)$), homozygous sensitive ($N_{SS}^M(t)$) and heterozygous ($N_{RS}^M(t)$) mosquitoes. Furthermore, the total population of the homozygous resistant mosquitoes is stratified in terms of susceptible ($S_{RR}^M(t)$), exposed ($E_{RR}^M(t)$), and infectious ($I_{RR}^M(t)$) homozygous resistant mosquitoes, while the total population of the homozygous sensitive mosquitoes is subdivided into the sub-populations of susceptible ($S_{SS}^M(t)$), exposed ($E_{SS}^M(t)$), and infectious ($I_{SS}^M(t)$) homozygous sensitive mosquitoes. Finally, the total population of the heterozygous mosquitoes is split into the sub-populations of susceptible ($S_{RS}^M(t)$), exposed ($E_{RS}^M(t)$), and infectious ($I_{RS}^M(t)$) heterozygous mosquitoes. Thus,

$$N_{SS}^M(t) = S_{SS}^M(t) + E_{SS}^M(t) + I_{SS}^M(t),\ N_{RS}^M(t) = S_{RS}^M(t) + E_{RS}^M(t) + I_{RS}^M(t),$$
$$N_{RR}^M(t) = S_{RR}^M(t) + E_{RR}^M(t) + I_{RR}^M(t) \text{ and } N_M(t) = N_{RR}^M(t) + N_{RS}^M(t) + N_{SS}^M(t).$$

The total human population at time t, denoted by $N_H(t)$, is split into the mutually-exclusive compartments of susceptible ($S_H(t)$), exposed ($E_H(t)$), infectious ($I_H(t)$), and recovered ($R_H(t)$) humans, so that:

$$N_H(t) = S_H(t) + E_H(t) + I_H(t) + R_H(t).$$

Random mating and allele frequency: In the extended model to be formulated in this section, it is assumed that at each generation, there is a 50:50 sex ratio in mosquito population (that is, it is assumed that there are equal number of male and female mosquitoes in the community) [**20, 37**]. Furthermore, it is assumed that mating between adult mosquitoes (of opposite sex) is random [**11, 27, 30, 31**]. That is, all adult mosquitoes have equal chance of reproducing, and have the same probability of mating with any other adult mosquito (of opposite sex) in the environment. Unlike in the case of the basic model (2.1), where monoploid genetic structure was used, the diploid genetic structure in the extended model (3.4) enables for the inclusion of the mosquito random mating process (mating with all genotypes mixed) in the formulation of the mosquito birth functions by genotype.

Let $q(t)$ and $p(t)$ represent the frequencies of S and R alleles at time t, respectively. It should, first of all, be recalled that each gene of a homozygous SS (RR) genotype contains two S (R) alleles (i.e., $2S$ ($2R$) copies). Similarly, each gene of the heterozygous RS genotype contains one R allele and one S allele. Hence, the proportion (or frequency) of the sensitive alleles in the female mosquito population is given by:

$$q(t) = \frac{2N_{SS}^M(t) + N_{RS}^M(t)}{2N_M(t)} = \frac{N_{SS}^M(t) + \frac{1}{2}N_{RS}^M(t)}{N_M(t)}. \tag{3.1}$$

Similarly, the frequency of the resistant alleles in the community is given by:

$$p(t) = \frac{N_{RR}^M(t) + \frac{1}{2}N_{RS}^M(t)}{N_M(t)}. \tag{3.2}$$

Hardy-Weinberg condition at a fixed time t: The derivations of the expressions for the allele frequencies ($q(t)$ and $p(t)$, given by (3.1) and (3.2), respectively, for a fixed time t) were based on the assumptions for Hardy-Weinberg equilibrium holding [**19, 30**]. In other words, the expressions work in an ecological scenario where there is (a) random mating of mosquitoes, (b) no natural selection, (c) no mutation, (d) no migration or importation of alleles from external sources, and (e) large mosquito population size [**19**]. Owing to the 50:50 sex ratio assumption (i.e., there is equal proportions of adult male and female mosquitoes in the environment), the probability that the outcome of the mating between an adult female and an adult male mosquito will lead to the formation of eggs of SS genotype is $q(t) \times q(t)$. Similarly, the probability that such mating produces eggs of RS genotype is $p(t) \times q(t) + q(t) \times p(t)$ (i.e., $p(t) \times q(t)$ accounts for an R allele from the female and an S allele from the male partner) added to $q(t) \times p(t)$ (which accounts for an R allele from the male and an S allele from the female). Finally, the probability that such mating produces eggs of RR genotype is $p(t) \times p(t)$. Consequently, the proportion of the total eggs produced by adult female mosquitoes in the community with the SS, RS and RR genotype in the next generation (i.e., at time $t+\Delta t$, where $\Delta t > 0$ is an increment in time) is given by the quantities $q^2(t)$, $2p(t)q(t)$ and $p^2(t)$, respectively [**30, 31**]. It should be observed that $q(t) + p(t) = 1$, consequently, $(q(t))^2 + 2p(t)q(t) + (p(t))^2 = 1$ (which is the same as $(p(t) + q(t))^2 = 1$). It follows, by Hardy-Weinberg equilibrium, that, at a fixed time t, the genotype distribution (frequency) of the total number of eggs laid by adult female mosquitoes of SS, RS and RR genotype are $(q(t))^2$, $2p(t)q(t)$ and $(p(t))^2$, respectively [**19, 23, 30, 31**].

Mosquito birth functions: The allele frequencies ($p(t)$ and $q(t)$, given in (3.1) and (3.2), respectively) and the genotype distributions (frequencies) of eggs (derived above) will now be used to formulate the birth functions of mosquitoes by genotype. Let B_i (with $i = \{SS, RS, RR\}$) represent the number of newly-produced adult female mosquitoes of genotype i. Let $r_i > 0$ be the rate at which mosquito eggs genotype i develop to become adult female mosquitoes of genotype i. Based on the computations above, the genotype distribution (frequency) of the total number of eggs laid by adult female mosquitoes bearing the SS, RS and RR genotype are given, respectively, by $(q(t))^2$, $2p(t)q(t)$, and $(p(t))^2$. Consequently, at time (generation) t, the number of new adult female mosquitoes of genotype SS, RS and RR are $q^2(t)r_{SS}N_M(t)\left(1 - \dfrac{N_M(t)}{K_M}\right)$, $2p(t)q(t)r_{RS}N_M(t)\left(1 - \frac{N_M(t)}{K_M}\right)$, and $p^2(t)r_{RR}N_M(t)\left(1 - \frac{N_M(t)}{K_M}\right)$, respectively. It follows then that the total number of new adult mosquitoes of genotype i produced (due to birth) is given by:

$$
\begin{aligned}
B_{SS}(t) &= q^2(t)r_{SS}\left(1 - \frac{N_M(t)}{K_M}\right)N_M(t), \\
B_{RS}(t) &= 2q(t)p(t)r_{RS}\left(1 - \frac{N_M(t)}{K_M}\right)N_M(t) \text{ and} \\
B_{RR}(t) &= p^2(t)r_{RR}\left(1 - \frac{N_M(t)}{K_M}\right)N_M(t),
\end{aligned}
\tag{3.3}
$$

where $K_M > 0$ is the environmental carrying capacity of adult female mosquitoes (it is assumed that $N_M(t) \leq K_M$ for all $t \geq 0$, so that the logistic term, $1 - \frac{N_M(t)}{K_M}$, is always non-negative). The Verhulst-Pearl logistic function is chosen to capture the mosquito density-dependent competition for resources [**29–31**].

FIGURE 4. A schematic of a single gene with two alleles in a locus

3.1. Equations for the extended model. Based on the derivations and assumptions in the preceding sections, the genetic-epidemiology model for the transmission dynamics of malaria (in the presence of LLINs intervention) is given by the following deterministic system of non-linear differential equations (a schematic diagram of the model is depicted in Figure 5):

$$
\begin{aligned}
\frac{dS_H}{dt} &= \Pi_H + \rho_H R_H - \left(\lambda_{VH}^{SS} + \lambda_{VH}^{RS} + \lambda_{VH}^{RR}\right) S_H - \mu_H S_H,\\
\frac{dE_H}{dt} &= \left(\lambda_{VH}^{SS} + \lambda_{VH}^{RS} + \lambda_{VH}^{RR}\right) S_H - (\sigma_H + \mu_H) E_H,\\
\frac{dI_H}{dt} &= \sigma_H E_H - (\gamma_H + \delta_H + \mu_H) I_H,\\
\frac{dR_H}{dt} &= \gamma_H I_H - (\rho_H + \mu_H) R_H,\\
\frac{dS_{SS}^M}{dt} &= B_{SS} - \lambda_{HV}^{SS} S_{SS}^M - \left[C_B \delta_{(ins)} + \mu_M\right] S_{SS}^M,\\
\frac{dE_{SS}^M}{dt} &= \lambda_{HV}^{SS} S_{SS}^M - \left[\sigma_M + C_B \delta_{(ins)} + \mu_M\right] E_{SS}^M,\\
\frac{dI_{SS}^M}{dt} &= \sigma_M E_{SS}^M - \left[C_B \delta_{(ins)} + \mu_M\right] I_{SS}^M,\\
\frac{dS_{RS}^M}{dt} &= B_{RS} - \lambda_{HV}^{RS} S_{RS}^M - \left[C_B (1 - h\rho_i) \delta_{(ins)} + \alpha_{RS}\mu_M\right] S_{RS}^M,\\
\frac{dE_{RS}^M}{dt} &= \lambda_{HV}^{RS} S_{RS}^M - \left[\theta_{RS}\sigma_M + C_B (1 - h\rho_i) \delta_{(ins)} + \alpha_{RS}\mu_M\right] E_{RS}^M,\\
\frac{dI_{RS}^M}{dt} &= \theta_{RS}\sigma_M E_{RS}^M - \left[C_B (1 - h\rho_i) \delta_{(ins)} + \alpha_{RS}\mu_M\right] I_{RS}^M,\\
\frac{dS_{RR}^M}{dt} &= B_{RR} - \lambda_{HV}^{RR} S_{RR}^M - \left[C_B (1 - \rho_i) \delta_{(ins)} + \alpha_{RR}\mu_M\right] S_{RR}^M,\\
\frac{dE_{RR}^M}{dt} &= \lambda_{HV}^{RR} S_{RR}^M - \left[\theta_{RR}\sigma_M + C_B (1 - \rho_i) \delta_{(ins)} + \alpha_{RR}\mu_M\right] E_{RR}^M,\\
\frac{dI_{RR}^V}{dt} &= \theta_{RR}\sigma_M E_{RR}^M - \left[C_B (1 - \rho_i) \delta_{(ins)} \alpha_{RR} + \alpha_{RR}\mu_M\right] I_{RR}^M,
\end{aligned}
\tag{3.4}
$$

where b_i (with $i = SS,\ RS,\ RR$) is the average biting rate for mosquitoes of genotype i, given by:

$$
\begin{aligned}
b_i &= \frac{(1 - \varepsilon_B^i \mathcal{C}_B) b_{max}}{1 + \varepsilon_B^i \mathcal{C}_B b_{max}} \quad \text{and} \quad \lambda_{HV}^{SS} = \frac{\beta_M b_{SS} I_H}{N_H},\\
\lambda_{HV}^{RS} &= \frac{\beta_M b_{RS} I_H}{N_H}\ \lambda_{HV}^{RR} = \frac{\beta_M b_{RR} I_H}{N_H},\\
\lambda_{VH}^{SS} &= \frac{\beta_H b_{SS} I_{SS}^M}{N_H},\ \lambda_{VH}^{RS} = \frac{\beta_H b_{RS} I_{RS}^M}{N_H},\ \text{and}\ \lambda_{VH}^{RR} = \frac{\beta_H b_{RR} I_{RR}^M}{N_H}.
\end{aligned}
\tag{3.5}
$$

In the extended model (3.4), $0 \le \mathcal{C}_B \le 1$ is the proportion of individuals in the community who consistently sleep under a bednet (i.e., C_B represents LLINs coverage in the community), $0 \le \varepsilon_B^i \le 1$ (with $i = SS,\ RS,\ RR$) is the efficacy of the LLINs to protect humans from being bitten by adult female mosquitoes of genotype i and b_{max} is maximum mosquito biting rate) [**31**]. The expression λ_{HV}^i (λ_{VH}^i) represents the transmission rate from infectious humans (mosquitoes of genotype i) to

susceptible mosquitoes of genotype i (humans) and β_H (β_V) is the probability of malaria transmission from infectious mosquitoes of genotype i (humans) to susceptible humans (mosquitoes of genotype i). Furthermore, Π_H is the recruitment rate of humans (by immigration or birth) and ρ_H is the rate of loss of natural immunity (following recovery from the disease). Susceptible humans acquire malaria infection, following effective contact with an infectious adult female *Anopheles* mosquito of genotype i, at a rate $\left(\lambda_{VH}^{SS} + \lambda_{VH}^{RS} + \lambda_{VH}^{RR}\right)$. Humans in all epidemiological compartments are assumed to suffer natural death at a rate μ_H. Exposed humans develop clinical symptoms of malaria at a rate of σ_h. Infectious humans recover (naturally) at a rate γ_H, and infectious humans suffer disease-induced death at a rate δ_H.
Susceptible mosquitoes of genotype i (S_i^M) acquire malaria infection at the rate λ_{HV}^i (with $\{i = SS,\ RS,\ RR\}$). Adult female mosquitoes of SS genotype (i.e., S_{SS}^M,E_{SS}^M, and I_{SS}^M) suffer natural mortality at a rate of μ_M. Furthermore, adult female mosquitoes of SS genotype suffer additional mortality due to the encounter with the insecticide, at a rate $C_B\ \delta_{(ins)}$. Similarly, adult female mosquitoes of RS and RR genotype suffer insecticide-induced death at a rate $(1 - h\rho_i)C_B\delta_{(ins)}$ and $(1 - \rho_i)C_B\delta_{(ins)}$, respectively, where $0 \leq h \leq 1$ is a modification parameter accounting for the measure of the dominance of the resistant allele over the sensitive allele (i.e., $h = 1$ models the case where the resistant allele is dominant, and $h = 0$ represents the case when it is recessive) and $0 \leq \rho_i \leq 1$ is a modification parameter accounting for the assumed decrease in the mortality rate of the RR-genotype mosquitoes due to the use of insecticides, in comparison to the insecticides-induced death rate of adult mosquitoes of SS-genotype. Exposed adult female mosquitoes of SS genotype (E_{SS}^M) become infectious at a rate of σ_M. In the formulation of the model (3.4), we assumed a fitness cost for homozygous resistant and heterozygous adult female mosquitoes with respect to natural death rate (due to insecticide resistance) [**1**, **2**, **5**]. Specifically, it is assumed that mosquitoes of the RS and RR genotypes suffer natural mortality at a rate $\alpha_i\mu_M$ (for $\{i = RS,\ RR\}$), where $\alpha_i \geq 1$ accounts for the assumed increase in natural mortality rate for mosquitoes of the RS and RR genotypes, in comparison to the natural mortality rate for mosquitoes of the SS genotype. It should be mentioned that $\alpha_{RS} = \alpha_{RR}$ if the resistant allele is dominant, and $\alpha_{RS} = 1$ if the resistant allele is recessive). Similarly, exposed mosquitoes of genotypes RS and RR become infectious at a rate $\theta_i\sigma_M$ (for $\{i = RS,\ RR\}$), with $1 \leq \theta_{RS} \leq \theta_{RR}$ accounting for the assumed increase in disease progression rate of mosquitoes of genotypes RS and RR, in comparison to exposed mosquitoes of genotype SS (due to fitness cost of insecticide resistance). It is assumed that insecticide resistance increases the progression rate of mosquitoes from the exposed to the infectious class (i.e., it is assumed that $1 \leq \theta_{RS} \leq \theta_{RR}$ [**30**, **31**]).
Here, $\theta_{RS} = \theta_{RR}$ if the resistant allele is dominant, and $\theta_{RS} = 1$ if it is recessive. The state variables and parameters of the model (3.4) are described in Table 3 and Tables 4–6, respectively. It is worth mentioning that a non-linear genotype-specific mosquito biting rate function (b_i), which depends on LLINs coverage (C_B) and genotype-specific LLINs efficacy (ε_B^i, where $i = SS, RS, RR$) is used in the extended model (3.4) (linear biting rate functions, that do not depend on genotype, were used in [**29**, **30**]).

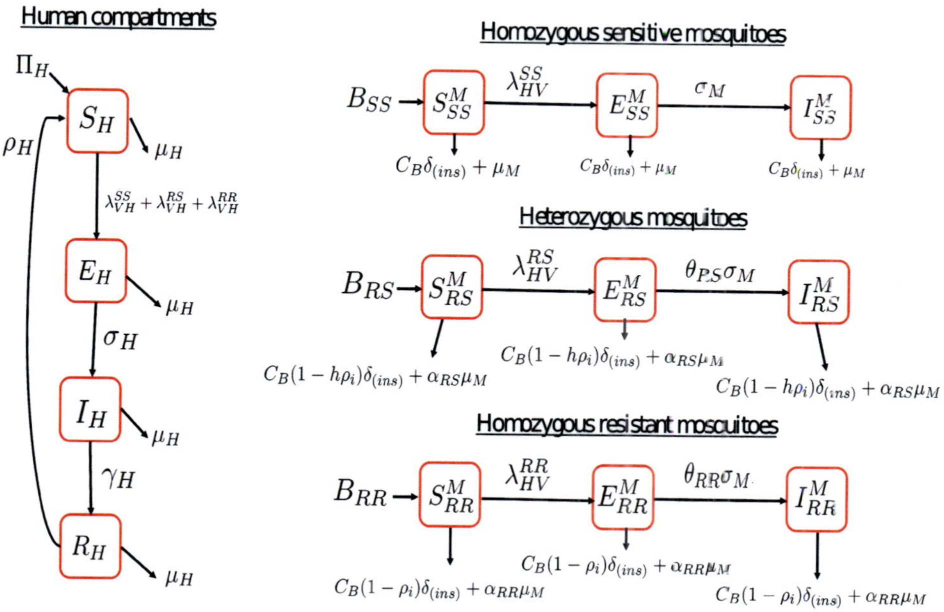

FIGURE 5. Schematic diagram of the extended genetic-epidemiology malaria model (3.4)

3.1.1. *Basic qualitative properties of the extended model.* All the state variables of the extended model (3.4) are non-negative (since they represent the temporal dynamics of human and adult female mosquito populations). We define, for mathematical convenience, the following quantities: $g_1 = \sigma_H + \mu_H$, $g_2 = \gamma_H + \delta_H + \mu_H$, $g_3 = \rho_H + \mu_H$, $g_4 = C_B\delta_{(ins)} + \mu_M$, $g_5 = \sigma_M + C_B\delta_{(ins)} + \mu_M$, $g_6 = C_B(1-h\rho_i)\delta_{(ins)} + \alpha_{RS}\mu_M$, $g_7 = \theta_{RS}\sigma_M + C_B(1-h\rho_i)\delta_{(ins)} + \alpha_{RS}\mu_M$, $g_8 = C_B(1-\rho_i)\delta_{(ins)} + \alpha_{RR}\mu_M$, and $g_9 = \theta_{RR}\sigma_M + C_B(1-\rho_i)\delta_{(ins)} + \alpha_{RR}\mu_M$. Define the following feasible region for the extended model (3.4):

$$\Omega^{ext} = \Big\{(S_H, E_H, I_H, R_H, S_{SS}^M, E_{SS}^M, I_{SS}^M, S_{RS}^M, E_{RS}^M, I_{RS}^M, S_{RR}^M, E_{RR}^M, I_{RR}^M) \in \mathbb{R}_+^4 \cup \mathbb{R}_+^9 : \tag{3.6}$$

$$0 < (N_H)_{min} \le N_H(t) \le \frac{\Pi_H}{\mu_H}, \text{ and } 0 \le N_M \le K_M(\mathcal{R}_M - 1)/\mathcal{R}_M\Big\},$$

where,

$$(N_H)_{min} = min\left(N_H(0), \frac{\Pi_H}{\mu_H + \delta_H}\right),$$

$$\omega_M = \min\{g_4, g_6, g_8\} \text{ and } \mathcal{R}_M = \frac{r_{SS} + r_{RS} + r_{RR}}{\omega_M}.$$

TABLE 3. Description of the state variables of the extended model (3.4)

Human classes	Description
S_H	Population of susceptible humans
E_H	Population of exposed humans
I_H	Population of infectious humans
R_H	Population of recovered humans
Mosquito classes	**Description**
S^M_{SS}	Population of adult female mosquitoes of SS genotype that are susceptible to malaria
E^M_{SS}	Population of adult female mosquitoes of SS genotype that are exposed to malaria
I^M_{SS}	Population of infectious adult female mosquitoes of SS genotype
S^M_{RS}	Population of adult female mosquitoes of RS genotype that are susceptible to malaria
E^M_{RS}	Population of adult female mosquitoes of RS genotype that are exposed to malaria
I^M_{RS}	Population of infectious adult female mosquitoes of RS genotype
S^M_{RR}	Population of adult female mosquitoes of RR genotype that are susceptible to malaria
E^M_{RR}	Population of adult female mosquitoes of RR genotype that are exposed to malaria
I^M_{RR}	Population of infectious adult female mosquitoes of RR genotype

TABLE 4. Descriptions and baseline values of the human population parameters of the extended model (3.4)

Parameters	Description	Baseline Value (day^{-1})	Source
Π_H	Human recruitment rate (due to birth or immigration)	2.19	[16]
μ_H	Natural death rate for humans	4.405×10^{-5}	[14]
δ_H	Disease-induced death rate for humans	9.0×10^{-5}	[35]
σ_H	Rate at which exposed humans become infectious	1/14	[34]
γ_H	Recovery rate of humans	6.58×10^{-3}	[36]
ρ_H	Rate of loss of natural immunity for humans	5.6×10^{-3}	[35]

TABLE 5. Descriptions and baseline values of the mosquito population parameters of the extended model (3.4)

Parameters	Description	Baseline Value (day^{-1})	Source
K_M	Environmental carrying capacity of mosquitoes	10^5 (dimensionless)	[27, 32]
r_{SS}	Development rate of eggs of SS-genotype to adult female mosquitoes	6.353	[27]
r_{RS}	Development rate of eggs of RS-genotype to adult female mosquitoes	$\frac{3}{4}r_{SS}$	[30, 31]
r_{RR}	Development rate of eggs of RR-genotype to adult female mosquitoes	$\frac{1}{2}r_{SS}$	[1, 31]
μ_M	Natural death rate of mosquitoes of SS-genotype	1/14	[35]
σ_M	Rate at which exposed mosquitoes of SS-genotype become infectious	0.1	[34]
β_{max}	Maximum mosquito biting rate	0.67	[36]
b_i	Average biting rate of mosquitoes of i genotype ($i = \{SS, RS.RR\}$	Expressed in terms of other parameters	[31] [31]
β_H	Transmission probability from infectious mosquitoes to susceptible humans	0.14 (dimensionless)	[13]
β_M	Transmission probability from infectious humans to susceptible mosquitoes	0.48 (dimensionless)	[13]

Here, too, the upper bound of $N_M(t)$, given by $K_M(\mathcal{R}_M - 1)/\mathcal{R}_M$ in (3.6), is positive if $\mathcal{R}_M > 1$. It should be observe that the quantity $\mathcal{R}_M$ is greater than unity if the sum of the mosquito growth rates of the three genotypes is higher than the minimum of the sum of the death rates due to natural and insecticide-induced death rates of all mosquito genotypes. In a closed environment, without immigration and migration, if mosquito death rate is higher than the birth rate, then the mosquito population will eventually die out. In this extended model, mosquito immigration and migration are not (since mosquitoes do not fly far from their breeding sites [**12**]). Thus, the mosquito population eventually dies out if $\mathcal{R}_M < 1$. For the rest of this section, we assume that $\mathcal{R}_M > 1$ (so that mosquitoes, of all three genotypes, always exist in the environment).

The following result can be established using the approach in Section 2 (see also [**30**, **31**]):

THEOREM 3.1. *Consider the extended model* (3.4) *with* $N_H(0) > 0$. *The biologically-feasible region* Ω^{ext} *is positively-invariant and attracts all solutions of the extended model* (3.4) *in* $\mathbb{R}^4_+ \cup \mathbb{R}^9_+$.

3.2. Existence and asymptotic stability of disease-free equilibria. In this section, conditions for the existence and asymptotic stability of the various disease-free equilibria of the extended model (3.4) will be derived.

3.2.1. *Existence.* Let $((S_H)^*, (R_H)^*, (S^M_{SS})^*, (S^M_{RS})^*, (S^M_{RR})^*)$ represent the components of $S_H(t), R_H(t), S^M_{SS}(t), S^M_{RS}(t)$ and $S^M_{RR}(t)$ at the disease-free steady-state, respectively. It can be shown, by setting the infected components components of the extended model to zero and solving for the non-infected state variables, that:

TABLE 6. Descriptions and baseline values of the fitness cost and insecticide parameters of the extended model (3.4)

Parameters	Description	Baseline Value (day^{-1})	Source
α_{RS}	Modification parameter for the increase in natural death rate of mosquitoes of RS-genotype due to fitness cost, compare to SS-genotype ($\alpha_{RS} \geq 1$)	1.5 (dimensionless)	[**1**, **2**]
α_{RR}	Modification parameter for the increase in natural death rate of mosquitoes of RR-genotype due to fitness cost, compare to SS-genotype ($\alpha_{RR} \geq 1$)	2.0 (dimensionless)	[**1**, **2**]
$\delta_{(ins)}$	Death rate of mosquitoes of SS-genotype due to IRS and LLINs	0.84	[**6**]
h	Modification parameter accounting for the measure of the dominance of the resistant allele over the sensitive allele ($0 \leq h \leq 1$)	0.25 (dimensionless)	[**6**]
ρ_i	Modification parameter for the decrease in mortality, due to insecticide, of vectors of RR-genotype, compare to SS-genotype ($0 \leq \rho_i \leq 1$)	0.95 (dimensionless)	[**6**]
θ_{RS}	Modification parameter for the increase in progression rate of exposed vectors of RS-genotype due to fitness cost, in comparison to exposed mosquitoes of SS-genotype ($\theta_{RS} \geq 1$)	1.5 (dimensionless)	[**1**, **2**]
θ_{RR}	Modification parameter for the increase in progression rate of exposed vectors of RR-genotype due to fitness cost, in comparison to exposed exposed mosquitoes of SS-genotype ($\theta_{RR} \geq 1$)	2.0 (dimensionless)	[**1**, **2**]
ε^i_B	The efficacy of the bednet (LLINs) to protect humans from bites by adult female mosquitoes of i-genotype ($0 \leq \varepsilon^{SS}_B \leq 1$) for $i = \{SS, RS, RR\}$	$\varepsilon^{SS}_B = 0.85$, $\varepsilon^{RS}_B = \frac{1}{1.25} \times \varepsilon^{SS}_B$, $\varepsilon^{RR}_B = \frac{1}{1.75} \times \varepsilon^{SS}_B$ (dimensionless)	[**30**, **31**]
C_B	Proportion of humans who protected by LLINs ($0 \leq C_B \leq 1$)	varies (dimensionless)	varies

(3.7)

$$(S_H)^* = \frac{\Pi_H}{\mu_H}, (R_H)^* = 0, q^* = \frac{(S^M_{SS})^* + \frac{1}{2}(S^M_{RS})^*}{N^*_M}, \text{and } p^* = \frac{(S^M_{RR})^* + \frac{1}{2}(S^M_{RS})^*}{N^*_M},$$

with,

$$N^*_M = (S^M_{SS})^* + (S^M_{RS})^* + (S^M_{RR})^*. \tag{3.8}$$

It can be shown from (3.7) with (3.8) that, at the disease-free steady-state, q^* satisfies:

$$q^*(q^* - 1)\left\{[(\mathcal{R}_{SS} - \mathcal{R}_{RS}) + (\mathcal{R}_{RR} - \mathcal{R}_{RS})]\, q^* - (\mathcal{R}_{RR} - \mathcal{R}_{RS})\right\} = 0, \tag{3.9}$$

where,

$$\mathcal{R}_{SS} = \frac{r_{SS}}{g_4}, \ \mathcal{R}_{RS} = \frac{r_{RS}}{g_6}, \text{ and } \mathcal{R}_{RR} = \frac{r_{RR}}{g_8}. \tag{3.10}$$

Hence, it follows from Equation (3.9), at disease-free equilibrium, that q^* and p^* can take values from any one of the following four cases (this follows from the solution of Equation (3.9)) (as also shown in [**30**]):

(i) $q^* = 0$ and $p^* = 1$, or
(ii) $q^* = 1$ and $p^* = 0$, or
(iii) $q^* = \dfrac{\mathcal{R}_{RR} - \mathcal{R}_{RS}}{(\mathcal{R}_{SS} - \mathcal{R}_{RS}) + (\mathcal{R}_{RR} - \mathcal{R}_{RS})}$ and $p^* = \dfrac{\mathcal{R}_{SS} - \mathcal{R}_{RS}}{(\mathcal{R}_{SS} - \mathcal{R}_{RS}) + (\mathcal{R}_{RR} - \mathcal{R}_{RS})}$, provided that $(\mathcal{R}_{SS} - \mathcal{R}_{RS})(\mathcal{R}_{RR} - \mathcal{R}_{RS}) > 0$, or
(iv) any value q^*, p^* in $[0, 1]$ with $p^* + q^* = 1$, if $\mathcal{R}_{SS} = \mathcal{R}_{RS} = \mathcal{R}_{RR}$.

It should be noted that q^* and p^* must satisfy the equality $p^* + q^* = 1$ in all cases. Therefore, the model (3.4) has four disease-free equilibria, namely a trivial disease-free equilibrium (denoted by $\mathcal{E}_T^{ext}$), a non-trivial sensitive-only disease-free boundary equilibrium (denoted by $\mathcal{E}_S^{ext}$, when $q^* = 1$ and $p^* = 0$), a non-trivial resistant-only disease-free boundary equilibrium (denoted by $\mathcal{E}_R^{ext}$, when $q^* = 0$ and $p^* = 1$) and a non-trivial co-existence disease-free equilibrium (denoted by $\mathcal{E}_C^{ext}$, when $q^* > 0$ and $p^* > 0$), as given below:

$$
\begin{aligned}
\mathcal{E}_T^{ext} &= \big((S_H)^{*T}, (E_H)^{*T}, (I_H)^{*T}, (R_H)^{*T}, (S_{SS}^M)^{*T}, (I_{SS}^M)^{*T}, (S_{RS}^M)^{*T}, (E_{RS}^M)^{*T},\\
&\quad (I_{RS}^M)^{*T}, (E_{SS}^M)^{*T}, (S_{RR}^M)^{*T}, (E_{RR}^M)^{*T}, (I_{RR}^M)^{*T}\big),\\
&= \big((S_H)^{*T}, 0, 0, 0, 0, 0, 0, 0, 0, 0, 0, 0, 0\big),\\
\mathcal{E}_S^{ext} &= \big((S_H)^{*S}, (E_H)^{*S}, (I_H)^{*S}, (R_H)^{*S}, (S_{SS}^M)^{*S}, (E_{SS}^M)^{*S}, (I_{SS}^M)^{*S}, (S_{RS}^M)^{*S},\\
&\quad (E_{RS}^M)^{*S}, (I_{RS}^M)^{*S}, (S_{RR}^M)^{*S}, (E_{RR}^M)^{*S}, (I_{RR}^M)^{*S}\big),\\
&= \big((S_H)^{*S}, 0, 0, 0, (S_{SS}^M)^{*S}, 0, 0, 0, 0, 0, 0, 0, 0\big),\\
\mathcal{E}_R^{ext} &= \big((S_H)^{*R}, (E_H)^{*R}, (I_H)^{*R}, (R_H)^{*R}, (S_{SS}^M)^{*R}, (E_{SS}^M)^{*R}, (I_{SS}^M)^{*R}, (S_{RS}^M)^{*R},\\
&\quad (E_{RS}^M)^{*R}, (I_{RS}^M)^{*R}, (S_{RR}^M)^{*R}, (E_{RR}^M)^{*R}, (I_{RR}^M)^{*R}\big),\\
&= \big((S_H)^{*R}, 0, 0, 0, 0, 0, 0, 0, 0, 0, (S_{RR}^M)^{*R}, 0, 0\big),\\
\mathcal{E}_C^{ext} &= \big((S_H)^{*C}, (E_H)^{*C}, (I_H)^{*C}, (R_H)^{*C}, (S_{SS}^M)^{*C}, (E_{SS}^M)^{*C}, (I_{SS}^M)^{*C}, (S_{RS}^M)^{*C},\\
&\quad (E_{RS}^M)^{*C}, (I_{RS}^M)^{*C}, (S_{RR}^M)^{*C}, (E_{RR}^M)^{*C}, (I_{RR}^M)^{*C}\big),\\
&= \big((S_H)^{*C}, 0, 0, 0, (S_{SS}^M)^{*C}, 0, 0, (S_{RS}^M)^{*C}, 0, 0, (S_{RR}^M)^{*C}, 0, 0\big),
\end{aligned}
$$

where,

(3.11)

$$
\begin{aligned}
&(S_H)^{*T} = (S_H)^{*S} = (S_H)^{*R} = (S_H)^{*C} = \frac{\Pi_H}{\mu_H},\ (S_{SS}^M)^{*S} = \frac{K_M(\mathcal{R}_{SS} - 1)}{\mathcal{R}_{SS}},\\
&(S_{RR}^M)^{*R} = \frac{K_M(\mathcal{R}_{RR} - 1)}{\mathcal{R}_{RR}},\ (S_{SS}^M)^{*C} = \frac{(q^*)^2\mathcal{R}_{SS}K_M(\mathcal{R}_C - 1)}{\mathcal{R}_C^2},\\
&(S_{RS}^M)^{*C} = \frac{2p^*q^*\mathcal{R}_{RS}K_M(\mathcal{R}_C - 1)}{\mathcal{R}_C^2},\ \text{and } (S_{RR}^M)^{*C} = \frac{(p^*)^2\mathcal{R}_{RR}K_M(\mathcal{R}_C - 1)}{\mathcal{R}_C^2},
\end{aligned}
$$

with, q^* and p^* are one of the possible solutions of Equation (3.9) described above, and

$$
\mathcal{R}_C = (q^*)^2\mathcal{R}_{SS} + 2p^*q^*\mathcal{R}_{RS} + (p^*)^2\mathcal{R}_{RR}. \tag{3.12}
$$

It follows from Equation (3.11) that:

(i) the non-trivial sensitive-only equilibrium ($\mathcal{E}_S^{ext}$) exists if and only if $\mathcal{R}_{SS} > 1$,
(ii) the non-trivial resistant-only equilibrium ($\mathcal{E}_R^{ext}$) exists if and only if $\mathcal{R}_{RR} > 1$, and
(iii) the non-trivial co-existence equilibrium ($\mathcal{E}_C^{ext}$) exists if and only if $(\mathcal{R}_{SS} - \mathcal{R}_{RS})(\mathcal{R}_{RR} - \mathcal{R}_{RS}) > 0$ and $\mathcal{R}_C > 1$, or if and only if $\mathcal{R}_{SS} = \mathcal{R}_{RS} = \mathcal{R}_{RR}$ and $\mathcal{R}_C > 1$.

It is worth mentioning that, unlike in the case of the basic model (2.1), a sensitive-only non-trivial disease-free equilibrium ($\mathcal{E}_S^{ext}$) exists for the extended model (3.4). This is due to the fact that, in the extended model, insecticide-sensitive do not transition (or mutate) to the insecticide-resistant classes (such mutation was allowed in the basic model). Although mutation is the original source of genetic variation in populations, mutation, in the context of insecticide resistance, is quite rare [**21**]. Furthermore, in comparison to selective pressure due to the use of insecticides, the impact of mutation on the frequency of alleles from one generation to the next is not very significant [**21**]. Based on these, insecticide resistance mutation is not included in the extended model (3.4). Furthermore, it is worth stating that the trivial disease-free equilibrium ($\mathcal{E}_0^{ext}$) of the extended model is not ecologically realistic (since it entails having no mosquitoes in the population, consequently, the asymptotic stability of this equilibrium not provided here). The local asymptotic stability of the other three non-trivial disease-free equilibria of the extended model will be explored in Section 3.2.2.
It is convenient to define, for $\mathcal{R}_M > 1$, the following general non-trivial disease-free equilibrium of the model (3.4), denoted by $\mathcal{E}_{df}^{ext}$:

$$\begin{aligned}(3.13)\quad \mathcal{E}_{df}^{ext} &= (S_H, E_H, I_H, R_H, S_{SS}^M, E_{SS}^M, I_{SS}^M, S_{RS}^M, E_{RS}^M, I_{RS}^M, S_{RR}^M, E_{RR}^M, I_{RR}^M),\\ &= \left(\frac{\Pi_H}{\mu_H}, 0, 0, 0, (S_{SS}^M)^\diamond, 0, 0, (S_{RS}^M)^\diamond, 0, 0, (S_{RR}^M)^\diamond, 0, 0\right).\end{aligned}$$

That is,

(a) If $\mathcal{E}_{df}^{ext} = \mathcal{E}_S^{ext}$, then $(S_{SS}^M)^\diamond = (S_{SS}^M)^{*S}$, $(S_{RS}^M)^\diamond = 0$, and $(S_{RR}^M)^\diamond = 0$.
(b) If $\mathcal{E}_{df}^{ext} = \mathcal{E}_R^{ext}$, then $(S_{SS}^M)^\diamond = 0$, $(S_{RS}^M)^\diamond = 0$, and $(S_{RR}^M)^\diamond = (S_{RR}^M)^{*R}$.
(c) If $\mathcal{E}_{df}^{ext} = \mathcal{E}_C^{ext}$, then $(S_{SS}^V)^\diamond = (S_{SS}^M)^{*C}$, $(S_{RS}^M)^\diamond = (S_{RS}^M)^{*C}$, and $(S_{RR}^M)^\diamond = (S_{RR}^M)^{*C}$.

3.2.2. *Local asymptotic stability of the generalized non-trivial disease-free equilibrium.* Here, too, the next generation operator method will be used to analyse the asymptotic stability property of the generalized non-trivial disease-free equilibrium ($\mathcal{E}_{df}^{ext}$) of the extended model (3.4). It can be seen that the next generation matrices [**17, 18**] , of the new infection terms (denoted by F_{ext}) and linear transition terms (denoted by V_{ext}), associated with the extended model are given, respectively, by:

$$F_{ext} = \begin{pmatrix} 0 & 0 & 0 & \beta_H b_{SS} & 0 & \beta_H b_{RS} & 0 & \beta_H b_{RR} \\ 0 & 0 & 0 & 0 & 0 & 0 & 0 & 0 \\ 0 & \frac{\beta_M b_{SS} \mu_H (S_{SS}^M)^\diamond}{\Pi_H} & 0 & 0 & 0 & 0 & 0 & 0 \\ 0 & 0 & 0 & 0 & 0 & 0 & 0 & 0 \\ 0 & \frac{\beta_M b_{RS} \mu_H (S_{RS}^M)^\diamond}{\Pi_H} & 0 & 0 & 0 & 0 & 0 & 0 \\ 0 & 0 & 0 & 0 & 0 & 0 & 0 & 0 \\ 0 & \frac{\beta_M b_{RR} \mu_H (S_{RR}^M)^\diamond}{\Pi_H} & 0 & 0 & 0 & 0 & 0 & 0 \\ 0 & 0 & 0 & 0 & 0 & 0 & 0 & 0 \end{pmatrix}, \tag{3.14}$$

and,

$$V_{ext} = \begin{pmatrix} g_1 & 0 & 0 & 0 & 0 & 0 & 0 & 0 \\ -\sigma_H & g_2 & 0 & 0 & 0 & 0 & 0 & 0 \\ 0 & 0 & g_5 & 0 & 0 & 0 & 0 & 0 \\ 0 & 0 & -\sigma_M & g_4 & 0 & 0 & 0 & 0 \\ 0 & 0 & 0 & 0 & g_7 & 0 & 0 & 0 \\ 0 & 0 & 0 & 0 & -\theta_{RS}\sigma_M & g_6 & 0 & 0 \\ 0 & 0 & 0 & 0 & 0 & 0 & g_9 & 0 \\ 0 & 0 & 0 & 0 & 0 & 0 & -\theta_{RR}\sigma_M & g_8 \end{pmatrix}, \tag{3.15}$$

Hence, the *control reproduction number* of the extended model (3.4), is given by [**17**, **18**]:

$$\mathcal{R}_0^{ext} = \rho\left[F_{ext}\,(V_{ext})^{-1}\right] = \sqrt{\mathcal{R}_{0_{HM}}^{SS}\mathcal{R}_{0_{MH}}^{SS} + \mathcal{R}_{0_{HM}}^{RS}\mathcal{R}_{0_{MH}}^{RS} + \mathcal{R}_{0_{HM}}^{RR}\mathcal{R}_{0_{MH}}^{RR}}, \tag{3.16}$$

where,

$$\begin{aligned} \mathcal{R}_{0_{HM}}^{SS} &= \frac{\beta_H\mu_H\sigma_H b_{SS}}{g_1g_2\Pi_H},\ \mathcal{R}_{0_{HM}}^{RS} = \frac{\beta_H\mu_H\sigma_H b_{RS}}{g_1g_2\Pi_H},\ \mathcal{R}_{0_{HM}}^{RR} = \frac{\beta_H\mu_H\sigma_H b_{RR}}{g_1g_2\Pi_H}, \\ \mathcal{R}_{0_{MH}}^{SS} &= \frac{\beta_M\sigma_M b_{SS}(S_{SS}^M)^\diamond}{g_4g_5}\ \mathcal{R}_{0_{MH}}^{RS} = \frac{\beta_M\theta_{RS}\sigma_M b_{RS}(S_{RS}^M)^\diamond}{g_6g_7} \\ \text{and } \mathcal{R}_{0_{MH}}^{RR} &= \frac{\beta_M\theta_{RR}\sigma_M b_{RR}(S_{RR}^M)^\diamond}{g_8g_9}. \end{aligned} \tag{3.17}$$

It is convenient to define the following threshold quantities:

$$\mathcal{R}_{0_S}^{ext} = \mathcal{R}_0^{ext}|_{\mathcal{E}_{df}^{ext}=\mathcal{E}_S^{ext}} = \sqrt{\frac{\beta_H\mu_H\sigma_H b_{SS}}{g_1g_2\Pi_H}\left(\frac{\beta_M\sigma_M b_{SS}(S_{SS}^M)^{*S}}{g_4g_5}\right)}, \tag{3.18}$$

$$\mathcal{R}_{0_R}^{ext} = \mathcal{R}_0^{ext}|_{\mathcal{E}_{df}^{ext}=\mathcal{E}_R^{ext}} = \sqrt{\frac{\beta_H\mu_H\sigma_H b_{RR}}{g_1g_2\Pi_H}\left(\frac{\beta_M\sigma_M b_{RR}(S_{RR}^M)^{*R}}{g_6g_7}\right)}, \tag{3.19}$$

$$\mathcal{R}_{0_C}^{ext} = \mathcal{R}_0^{ext}|_{\mathcal{E}_{df}^{ext}=\mathcal{E}_C^{ext}} = \sqrt{\mathcal{R}_{0_{HM}}^{SS}\mathcal{R}_{0_{SS}}^{0C} + \mathcal{R}_{0_{HM}}^{RS}\mathcal{R}_{0_{RS}}^{0C} + \mathcal{R}_{0_{HM}}^{RR}\mathcal{R}_{0_{RR}}^{0C}}, \tag{3.20}$$

with,

$$\mathcal{R}_{0_{SS}}^{0C} = \frac{\beta_M\sigma_M b_{SS}(S_{SS}^M)^{*C}}{g_4g_5},\ \mathcal{R}_{0_{RS}}^{0C} = \frac{\beta_M\theta_{RS}\sigma_M b_{RS}(S_{RS}^M)^{*C}}{g_6g_7}$$

$$\text{and } \mathcal{R}_{0_{RR}}^{0C} = \frac{\beta_M\sigma_M b_{RR}(S_{RR}^M)^{*C}}{g_8g_9}.$$

The result below follows from Theorem 2 in [**18**]:

THEOREM 3.2. *Consider the extended model* (3.4).

(a) *If* $\mathcal{R}_{SS} > 1$, *then* $\mathcal{E}_S^{ext}$ *is locally-asymptotically stable if* $\mathcal{R}_{0_S}^{ext} < 1$, *and unstable if* $\mathcal{R}_{0_S}^{ext} > 1$.

(b) *If* $\mathcal{R}_{RR} > 1$, *then* $\mathcal{E}_R^{ext}$ *is locally-asymptotically stable if* $\mathcal{R}_{0_R}^{ext} < 1$, *and unstable if* $\mathcal{R}_{0_R}^{ext} > 1$.

(c) *If* $(\mathcal{R}_{SS} - \mathcal{R}_{RS})(\mathcal{R}_{RR} - \mathcal{R}_{RS}) > 0$ *or* $\mathcal{R}_{SS} = \mathcal{R}_{RS} = \mathcal{R}_{RR}$, *then the non-trivial co-existence disease-free equilibrium* ($\mathcal{E}_C^{ext}$) *is locally-asymptotically stable if* $\mathcal{R}_{0_C}^{ext} < 1$, *and unstable if* $\mathcal{R}_{0_C}^{ext} > 1$ *provided that* $\mathcal{R}_C > 1$.

The epidemiological significance of Theorem 3.2 is that the disease can be effectively-controlled. Initial conditions of the extended model (3.4) will converge to the various disease-free equilibria ($\mathcal{E}_i^{ext}$, with $i = \{S, R, C\}$), when their associated reproduction numbers ($\mathcal{R}_{0_i}^{ext}$; $i = \{S, R, C\}$) are less than one, if they lie in the basin of attraction of the respective disease-free equilibria. In other words, the effective control of the disease when the respective reproduction numbers are less than one depends on the initial sizes of the sub-populations of the model. For such control to be independent on the initial sizes of the sub-populations, a global asymptotic stability result must be established for the non-trivial disease-free equilibria ($\mathcal{E}_{0i}$, with $i = \{S, R, C\}$). This is established for a special case below.

3.2.3. *Global asymptotic stability of the non-trivial disease-free equilibrium: special case.* Consider the special case of the extended model (3.4) with negligible disease-induced mortality in the human population (that is, $\delta_H \approx 0$). Furthermore, define the following feasible regions for the extended model:

$$\Omega_C^{ext} = \Big\{(S_H, E_H, I_H, R_H, S_{SS}^M, E_{SS}^M, I_{SS}^M, S_{RS}^M, E_{RS}^M, I_{RS}^M, S_{RR}^M, E_{RR}^M, I_{RR}^M) \in \Omega^{ext} :$$
(3.21)
$$N_H(t) \le \frac{\Pi_H}{\mu_H},\ S_{SS}^M(t) \le (S_{SS}^M)^{*C},\ S_{RS}^M(t) \le (S_{RS}^M)^{*C},\ S_{RR}^M(t) \le (S_{RR}^M)^{*C}\Big\},$$

$$\Omega_{SS}^{ext} = \Big\{(S_H, E_H, I_H, R_H, S_{SS}^M, E_{SS}^M, I_{SS}^M, S_{RS}^M, E_{RS}^M, I_{RS}^M, S_{RR}^M, E_{RR}^M, I_{RR}^M) \in \Omega^{ext} :$$
(3.22)
$$N_H(t) \le \frac{\Pi_H}{\mu_H},\ S_{SS}^M(t) \le (S_{SS}^M)^{*S},\ S_{RS}^M(t) = 0,\ S_{RR}^M(t) = 0\Big\},$$

and,

$$\Omega_{RR}^{ext} = \Big\{(S_H, E_H, I_H, R_H, S_{SS}^M, E_{SS}^M, I_{SS}^M, S_{RS}^M, E_{RS}^M, I_{RS}^M, S_{RR}^M, E_{RR}^M, I_{RR}^M) \in \Omega^{ext} :$$
(3.23)
$$N_H(t) \le \frac{\Pi_H}{\mu_H},\ S_{SS}^M(t) = 0,\ S_{RS}^M(t) = 0,\ S_{RR}^M(t) \le (S_{RR}^M)^{*R}\Big\},$$

where, $N_H(t) = S_H(t) + E_H(t) + I_H(t) + R_H(t)$. The results below can be shown using the approach in Section 2 (see also [**30, 31**]):

LEMMA 3. *Consider the special case of the extended model* (3.4) *with* $\delta_H = 0$.

(a) *If* $\mathcal{R}_i > 1$ *and* $\frac{\mathcal{R}_i - 1}{\mathcal{R}_i^2} \ge \frac{1}{4}$, *then the region* Ω_i^{ext} (*with* $i = SS, RR$) *is positively-invariant with respect to the extended model* (3.4).

(b) *If* $(\mathcal{R}_{SS} - \mathcal{R}_{RS})(\mathcal{R}_{RR} - \mathcal{R}_{RS}) > 0$ *or* $\mathcal{R}_{SS} = \mathcal{R}_{RS} = \mathcal{R}_{RR}$, *then the region* Ω_C^{ext} *is positively-invariant with respect to the extended model* (3.4) *provided that* $\mathcal{R}_C > 1$ *and* $\frac{(q^*)^2(\mathcal{R}_C - 1)}{\mathcal{R}_C^2} \geq \frac{1}{4}$.

It is convenient to define the quantities: $\overline{\mathcal{R}}_{0_S}^{ext} = \mathcal{R}_{0_S}^{ext}|_{\delta_H=0}$, $\overline{\mathcal{R}}_{0_R}^{ext} = \mathcal{R}_{0_R}^{ext}|_{\delta_H=0}$ and $\overline{\mathcal{R}}_{0_C}^{ext} = \mathcal{R}_{0_C}^{ext}|_{\delta_H=0}$. Furthermore, recall that the frequency of the sensitive alleles at the co-existence disease-free equilibrium $\mathcal{E}_C^{ext}$ (q^*) is given by:

$$q^* = \frac{(S_{SS}^M)^{*C} + 1/2(S_{RS}^M)^{*C}}{(S_{SS}^M)^{*C} + (S_{RS}^M)^{*C} + (S_{RR}^M)^{*C}}.$$

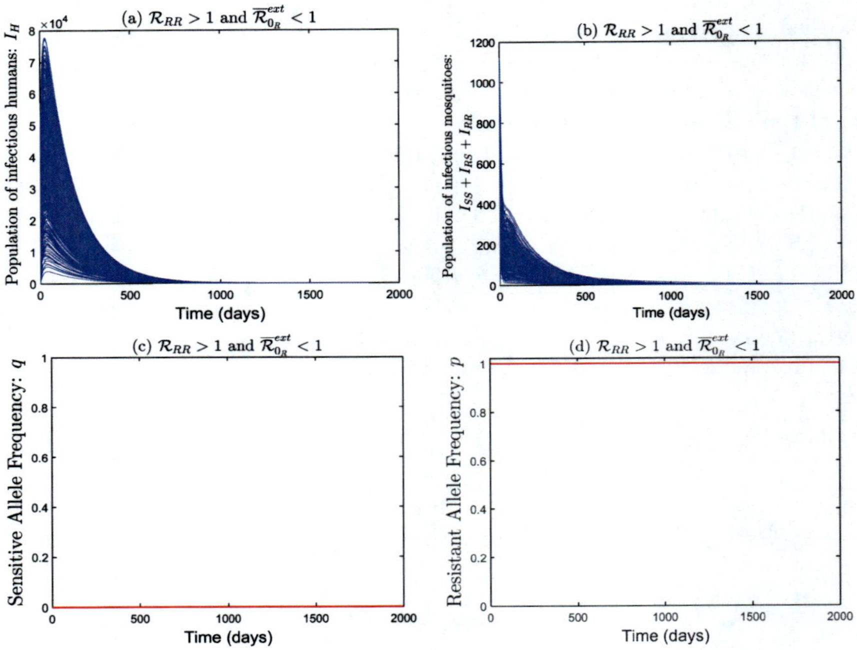

FIGURE 6. Numerical simulations of the model (3.4) showing convergence of initial solutions to the resistant-only non-trivial disease-free boundary equilibrium when $\mathcal{R}_{RR} > 1$ and $\overline{\mathcal{R}}_{0_R}^{ext} < 1$. (a) population of infectious humans (I_H) as a function of time for various initial conditions, (b) population of infectious mosquitoes $(I_{SS}+I_{RS}+I_{RR})$, (c) frequency of sensitive allele (q) as a function of time for various initial conditions, and (d) frequency of resistant allele (p) as a function of time for various initial conditions. Parameter values used are as given in Tables 4–6 with $\delta_H = 0$, $C_B = 0$, $r_{RR} = 0.045 r_{SS}$ and $\beta_{max} = 0.08$ (so that, $\dfrac{\mathcal{R}_{RR} - 1}{\mathcal{R}_{RR}^2} = 0.25 \geq \frac{1}{4}$, $\mathcal{R}_{RR} = 2 > 1$ and $\overline{\mathcal{R}}_{0_R}^{ext} = 0.4868 < 1$; hence, by Theorem 3.3 (b), the resistant-only non-trivial disease-free equilibrium $(\mathcal{E}_R^{ext})$ is globally-asymptotically stable in $\Omega_{RR}^{ext} \setminus \{\mathcal{E}_T^{ext}, \mathcal{E}_S^{ext}, \mathcal{E}_C^{ext}\}$.)

We claim the following results:

THEOREM 3.3. *Consider the special case of the extended model* (3.4) *with* $\delta_H = 0$.

(a). *If* $\mathcal{R}_{SS} > 1$ *and* $\dfrac{(\mathcal{R}_{SS}-1)}{\mathcal{R}_{SS}^2} \geq \frac{1}{4}$, *then the sensitive-only non-trivial disease-free equilibrium,* $\mathcal{E}_S^{ext}$, *is globally-asymptotically stable in* $\Omega_{SS}^{ext} \setminus \{\mathcal{E}_T^{ext}, \mathcal{E}_R^{ext}, \mathcal{E}_C^{ext}\}$ *whenever* $\overline{\mathcal{R}}_{0_S}^{ext} < 1$.

(b). *If* $\mathcal{R}_{RR} > 1$ *and* $\dfrac{(\mathcal{R}_{RR}-1)}{\mathcal{R}_{RR}^2} \geq \frac{1}{4}$, *then the resistant-only non-trivial disease-free equilibrium,* $\mathcal{E}_R^{ext}$, *is globally-asymptotically stable in* $\Omega_{RR}^{ext} \setminus \{\mathcal{E}_T^{ext}, \mathcal{E}_S^{ext}, \mathcal{E}_C^{ext}\}$ *whenever* $\overline{\mathcal{R}}_{0_R}^{ext} < 1$.

Theorem 3.3 can be proved using the approach in Section 2 or in [**30, 31**] (hence, the proof is not given here).

The theoretical result given in Theorem 3.3 (b) is numerically-illustrated by simulating the extended model (3.4), using the parameter values in Tables 4–6 such that the hypotheses of the theorem are satisfied, with multiple initial conditions. The simulation results obtained, depicted in Figure 6, show the convergence of all initial solutions to the resistant-only non-trivial disease-free equilibrium whenever $\mathcal{R}_{RR} > 1$ and $\overline{\mathcal{R}}_{0_R}^{ext} < 1$. Specifically, the total populations of infectious humans (Figure 6 (a)) and mosquitoes (Figure 6 (b)) converge to zero. Similarly, while the frequency of the sensitive alleles (Figure 6 (c)) reduced to zero, that of the resistant alleles reaches 100% (Figure 6 (d)).

We ran additional numerical simulations of the extended model (3.4) for the case where $\mathcal{R}_{RR} > 1$ but with $\overline{\mathcal{R}}_{0_R}^{ext} > 1$ (in this case, Theorem 3.3 (b) shows that the resistant-only non-trivial disease-free equilibrium is unstable). The simulation results obtained, depicted in Figure 7, show convergence of initial conditions to an endemic equilibrium, where the total population of infectious humans (Figure 7 (a)) reach a positive steady-state and all mosquitoes in the environment at steady-state are resistant (Figure 7 (b)). Here, too, the frequency of the sensitive alleles is zero (Figure 7 (c)) while that of the resistant alleles is 100% (Figure 7 (d)). Thus, it follows from the simulations in Figures 6 and 7 that, despite malaria mosquitoes in the community been fully-resistant at steady-state, malaria can be eliminated (Figures 6 (a) and (b), where $\overline{\mathcal{R}}_{0_R}^{ext} < 1$) or persist in the community (Figures 7 (a) and (b), where $\overline{\mathcal{R}}_{0_R}^{ext} > 1$). Thus, from these figures, effect of insecticide resistance on malaria elimination or persistence depends on the value of the reproduction threshold $\overline{\mathcal{R}}_{0_R}^{ext}$. Consequently, it is important that entomological data is collected to allow for the realistic estimation of the associated fitness costs and other factor/parameters of the model that affect the value of $\overline{\mathcal{R}}_{0_R}^{ext}$.

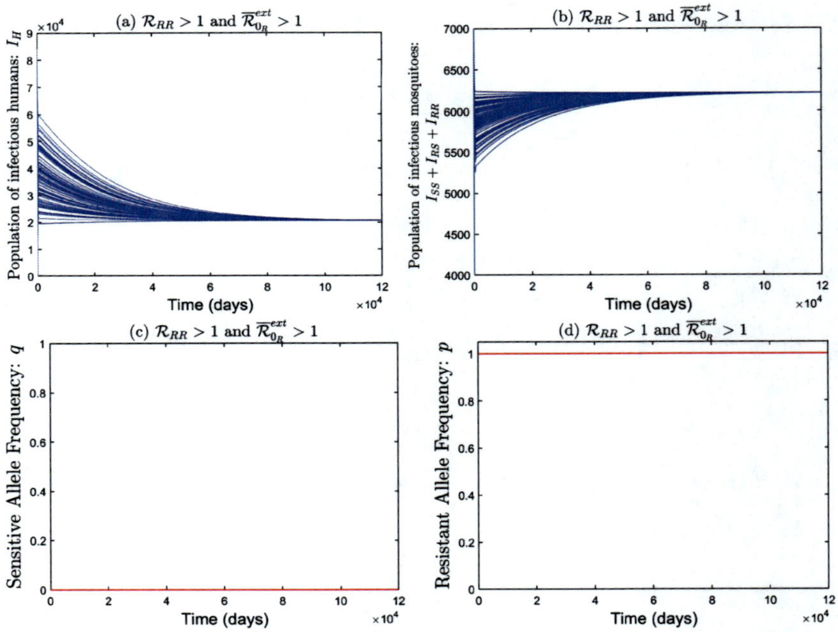

FIGURE 7. Numerical simulations of the model (3.4) showing convergence of initial solutions to the resistant-only boundary equilibrium when $\mathcal{R}_{RR} > 1$ and $\overline{\mathcal{R}}_{0_R}^{ext} > 1$. (a) population of infectious humans (I_H) as a function of time for various initial conditions, (b) population of infectious mosquitoes ($I_{SS} + I_{RS} + I_{RR}$), (c) frequency of sensitive allele (q) as a function of time for various initial conditions, and (d) frequency of resistant allele (p) as a function of time for various initial conditions. Parameter values used are as given in Tables 4–6 with $\delta_H = 0$, $C_B = 0$, $r_{RR} = 0.045 r_{SS}$ and $\beta_{max} = 0.67$ (so that, $\frac{\mathcal{R}_{RR}-1}{\mathcal{R}_{RR}^2} = 0.25 \geq \frac{1}{4}$, $\mathcal{R}_{RR} = 2 > 1$ and $\overline{\mathcal{R}}_{0_R}^{ext} = 4.077 > 1$).

THEOREM 3.4. *Consider the special case of the extended model (3.4) with $\delta_H = 0$. If $(\mathcal{R}_{SS} - \mathcal{R}_{RS})(\mathcal{R}_{RR} - \mathcal{R}_{RS}) > 0$ or $\mathcal{R}_{SS} = \mathcal{R}_{RS} = \mathcal{R}_{RR}$, then the non-trivial co-existence disease-free equilibrium, $\mathcal{E}_C^{ext}$, is globally-asymptotically stable in $\Omega_C^{ext} \setminus \{\mathcal{E}_T^{ext}, \mathcal{E}_S^{ext}, \mathcal{E}_R^{ext}\}$ whenever $\overline{\mathcal{R}}_{0_C}^{ext} < 1$, $\mathcal{R}_C > 1$ and $\dfrac{(q^*)^2(\mathcal{R}_C - 1)}{\mathcal{R}_C^2} \geq \frac{1}{4}$.*

The proof of Theorem 3.4, based on using a comparison theorem, is given in Appendix B. The epidemiological implication of Theorem 3.4 is that, if the hypotheses of the theorem are satisfied, malaria will be eliminated despite the persistence of the mosquitoes (of all three genotypes) in the community. In other words, this study shows that, for the special case of the extended model (3.4), malaria will be eliminated whenever $\overline{\mathcal{R}}_{0_C}^{ext} < 1$, $\mathcal{R}_C > 1$ and $\dfrac{(q^*)^2(\mathcal{R}_C - 1)}{\mathcal{R}_C^2} \geq \frac{1}{4}$, regardless of the persistence of the resistant mosquitoes in the community (i.e., bringing, and maintaining, the reproduction threshold, $\overline{\mathcal{R}}_{0_C}^{ext}$, to a value less than one is necessary and sufficient for malaria elimination in the community).

4. Concluding remarks

Malaria, a mosquito-borne parasitic disease spread to humans by infectious adult female *Anopheles* mosquitoes, is one of the deadliest diseases affecting over half the world's population. Significant progress has been made, owing to the large-scale and widespread use of insecticide-based mosquito control interventions (notably the use of long-lasting insecticidal nets and indoor residual spraying in malaria-endemic areas), over the last two decades, prompting a renewed concerted global effort for even greater reduction in malaria burden and its possible eradication [**11**, **29**–**31**, **40**]. For instance, the World Health Organization's Global Technical Strategy for Malaria 2016–2030 sets a series of ambitious global targets for combating malaria, such as reducing malaria case incidence and malaria mortality rates and by at least 90% by 2030 as well as eliminating malaria in at least 35 countries by 2030 [**40**]. There is also the *Zero X 2040* initiative, spearheaded by the world's leading agricultural companies, aimed at eradicating malaria by 2040 [**41**]. Unfortunately, the widespread and large-scale use of the aforementioned insecticide-based interventions resulted in widespread *Anopheles* resistance to all the five chemical insecticides currently used in vector control. This poses the key question on whether such insecticide resistance could potentially inhibit the current malaria eradication efforts.

This malaria modeling studies presented in this chapter are based on the design and analysis of two epidemiology-genetic models for malaria transmission dynamics that include insecticide resistance gene into the mosquito population. In the first model (named the *basic model*), insecticide resistance was determined by a single gene of one allele, and mosquito offsprings inherit the allele (insecticide-sensitive or insecticide-resistant) from their parents. In the second model (named the *extended model*), insecticide resistance was determined by a single gene with two alleles in a locus (one allele for insecticide-sensitive and and the other for insecticide resistant). In the extended model, the total mosquito population was stratified in terms of genotypes (SS-genotype for homozygous sensitive, RS-genotype for heterozygous and RR-genotype for homozygous resistant). In both models, mosquito population of each genotype is further divided into the sub-populations that are susceptible to malaria (S), exposed (E) or infectious (I) with malaria. Furthermore, in each of the models, the total human host population is sub-divided into the sub-populations of susceptible (S_H), exposed (E_H), infectious (I_H) and recovered (R_H) humans.

Each of the two models was rigorously analysed to gain insight into its qualitative properties, particularly with respect to the existence and asymptotic stability (both local and global, for special cases) of its associated non-trivial disease-free equilibria. Furthermore, numerical simulations were carried out to show two scenarios, in parameter space, where (i) malaria can be eliminated in a community where all mosquitoes carry the resistant allele at steady-state (as shown in Figures 2 and 6) and (ii) malaria persists in the community and all mosquitoes carry the resistant allele (as shown in Figures 3 and 7). These simulations show that, depending on the values of some of the parameters (particularly those related to the fitness cost and other factors that affect insecticide resistance) used in the simulations, insecticide resistance may or may not increase malaria transmission in a community. This finding is in agreement with the conclusions drawn in related genetic-epidemiology malaria modeling studies reported in [**11**, **29**–**31**].

Mathematical models for malaria transmission dynamics that adopt a genetic-epidemiology modeling framework (such as the two models presented in this study, as well those in [**11**, **29**–**31**]) can be used to gain deeper qualitative insight into (and quantify) the potential impacts of insecticide resistance on mosquito population abundance by genotype as well as malaria transmission dynamics and control. In particular, these models can be used to provide realistic answers to the important question on whether *Anopheles* resistance to all the currently-available insecticides used for its control increases malaria transmission or not. Mutations or genes conferring resistance are usually associated with a fitness cost. These costs manifest in numerous ways and forms. For instance, insecticide-resistant mosquitoes, in comparison to insecticide-sensitive mosquitoes, may have lower mating success, lower fecundity and fertility, higher developmental time and lower longevity. Thus, it is imperative that parameters related to these fitness cost processes are realistically estimated so that the true impact of these costs are accounted for in the simulations and predictions of the malaria transmission models (i.e., models for malaria dynamics and control that stratify the mosquito population by genotype).

This study shows that the generalized non-trivial co-existent disease-free equilibrium for each of the two models presented is globally-asymptotically stable, for the special case where the disease-induced mortality in the human host population is negligible, whenever the associated reproduction number of the model is less than unity. In other words, this study shows that, for the two models presented in this study, bringing (and maintaining) the reproduction numbers below one is necessary and sufficient for the effective control or elimination of malaria in the population. Parameters related to the fitness costs of insecticide resistance play a major role in determining the values of the reproduction numbers. As a consequence, this study emphasize the urgent need to carry out laboratory and field experiments to generate the genotype-specific mosquito data needed for the realistic estimation of the values of the fitness costs parameters in the models. This result is consistent with the results reported in [**11**, **29**–**31**] (where also the need for laboratory and field experiments were emphasized). Furthermore, it is important to incorporate (and realistically estimate in these models) other important factors that affect malaria transmission and the evolution of insecticide resistance in the *Anopheles* population, such as the effect of fluctuations in local climatic variables, insecticides deployment strategies (i.e., resistance management strategies implemented in the community), differences in demographic, landscape and socio-economic settings (for example, urban vs. rural, low-elevation *versus* high-elevation, irrigated *versus* rainfall agriculture, poor *versus* affluent community), genetic factors (such as, mutation, genetic-drift..etc) and other factors discussed in [**11**, **29**–**31**] (and the references therein). Finally, models for the transmission dynamics of vector-borne diseases, such as malaria, are known to undergo a backward bifurcation (a dynamic phenomenon characterized by the co-existence of multiple stable attractors when the associated reproduction number of the model is less than one) [**7**, **25**]. Although this phenomenon has not been explored in details in this study, the authors hope to address this problem (particularly whether or not insecticide resistance makes the occurrence of a backward bifurcation more or less likely in the genetic-epidemiology modeling framework for malaria dynamics) in a future study. Overall, in summary, this study shows that the prospect of malaria elimination is promising, even in the case where majority of (or all) mosquitoes in the population are fully- or partially-resistant to the insecticides being used for mosquito control, if the insecticide-based interventions are implemented at optimal levels.

Appendix A. Proof of Theorem 2.2

PROOF. The proof is based on using comparison theorem. Item (a). Let $\mathcal{R}_w > \max\{\mathcal{R}_r, 1\}$ $(\mathcal{E}_{df} = \mathcal{E}_C)$, $\delta_h = 0$, and

$$\min\left\{\frac{r_{vr}\,(\mathcal{R}_w - 1)\,(\mathcal{R}_w - \mathcal{R}_r)}{\mathcal{R}_w^2\,[\kappa_{vw}\mathcal{R}_w\mathcal{R}_r + r_{rv}(\mathcal{R}_w - \mathcal{R}_r)]}, \frac{\kappa_{wr}(\mathcal{R}_w - 1)}{\kappa_{vw}\mathcal{R}_w\mathcal{R}_r + r_{vr}(\mathcal{R}_w - \mathcal{R}_r)}\right\} \geq \frac{1}{4}.$$

Setting $\delta_h = 0$ in the model (2.1) shows that $\dfrac{dN_h}{dt} = \Lambda_h - \mu_h N_h$. Thus, $N_h(t) \to \frac{\Lambda_h}{\mu_h}$, as $t \to \infty$. Hence, from now on, the limiting system with $N_h(t) = N_h^* = \frac{\Lambda_h}{\mu_h}$ will be used. The equations for the rate of change of the infected components of the basic model (2.1) can then be written in the form:

$$\text{(A.1)}\qquad \frac{d}{dt}\begin{pmatrix} E_h(t)\\ I_h(t)\\ E_{vw}(t)\\ I_{vw}(t)\\ E_{vr}(t)\\ I_{vr}(t)\end{pmatrix} = (F - V - S)\begin{pmatrix} E_h(t)\\ I_h(t)\\ E_{vw}(t)\\ I_{vw}(t)\\ E_{vr}(t)\\ I_{vr}(t)\end{pmatrix},$$

where the next generation matrices F and V are as given in Section 2.2.2, and the matrix S is given by

$$S = \begin{pmatrix} 0 & 0 & 0 & s_{14} & 0 & s_{16}\\ 0 & s_{22} & 0 & 0 & 0 & 0\\ 0 & 0 & 0 & 0 & 0 & 0\\ 0 & s_{24} & 0 & 0 & 0 & 0\\ 0 & 0 & 0 & 0 & 0 & 0\\ 0 & 0 & 0 & 0 & 0 & 0\end{pmatrix},$$

where $S_h^* = \frac{\Lambda_h}{\mu_h}$, $s_{14} = s_{16} = \frac{\beta_h b(S_h^* - S_h(t))}{N_h^*}$, $s_{22} = \dfrac{\beta_v b(S_{vw}^{*C} - S_{vw}(t))}{N_h^*}$ and $s_{24} = \dfrac{\beta_v b(S_{vr}^{*C} - S_{vr}(t))}{N_h^*}$. Since $S_h(t) \leq S_h^*$, $S_{vw}(t) \leq S_{vw}^{*C}$ and $S_{vr}(t) \leq S_{vr}^{*C}$, in $\Omega_C \setminus \{\mathcal{E}_T, \mathcal{E}_R\}$ (which follows from Lemma 2), it follows that the matrix S is non-negative. Hence, Equation (A.1) can be expressed in terms of the following linear inequality:

$$\text{(A.2)}\qquad \frac{d}{dt}\begin{pmatrix} E_h(t)\\ I_h(t)\\ E_{vw}(t)\\ I_{vw}(t)\\ E_{vr}(t)\\ I_{vr}(t)\end{pmatrix} \leq (F - V)\begin{pmatrix} E_h(t)\\ I_h(t)\\ E_{vw}(t)\\ I_{vw}(t)\\ E_{vr}(t)\\ I_{vr}(t)\end{pmatrix}.$$

Since all the eigenvalues of $F - V$ have negative real parts for $\overline{\mathcal{R}}_0^{0C} < 1$ (from the local stability result in Theorem 2.1), it follows that the linearized differential inequality system (A.2) is stable if $\overline{\mathcal{R}}_0^{0C} < 1$. Hence,

$$(E_h(t), I_h(t), E_{vw}(t), I_{vw}(t), E_{vr}(t), I_{vr}(t)) \to (0,0,0,0,0,0), \text{ as } t \to \infty,$$

for this linear ODE system. Substituting $(E_h(t), I_h(t), E_{vw}(t), I_{vw}(t), E_{vr}(t), I_{vr}(t)) = (0,0,0,0,0,0)$ into the equations for the non-infected components of the basic model (2.1) show that:

$$(S_h(t), R_h(t), S_{vw}(t), S_{vr}(t)) \to (S_h^*, 0, S_{vw}^{*C}, S_{vr}^{*C}) \text{ as } t \to \infty.$$

Hence, the non-trivial co-existence disease-free equilibrium of the special case of the basic model (2.1), with $\delta_H = 0$, is globally-asymptotically stable in $\Omega_C \setminus \{\mathcal{E}_T, \mathcal{E}_R\}$ whenever $\overline{\mathcal{R}}_0^{0C} < 1$, with $\mathcal{R}_w > \max\{\mathcal{R}_r, 1\}$ and

$$\min\left\{\frac{r_{vr}(\mathcal{R}_w - 1)(\mathcal{R}_w - \mathcal{R}_r)}{\mathcal{R}_w^2[\kappa_{vw}\mathcal{R}_w\mathcal{R}_r + r_{rv}(\mathcal{R}_w - \mathcal{R}_r)]}, \frac{\kappa_{wr}(\mathcal{R}_w - 1)}{\kappa_{vw}\mathcal{R}_w\mathcal{R}_r + r_{vr}(\mathcal{R}_w - \mathcal{R}_r)}\right\} \geq \frac{1}{4}.$$

Item (b) of Theorem 2.2 can also be proved using this approach (hence, not repeated here). □

Appendix B. Proof of Theorem 3.4

PROOF. The proof of Theorem 3.4 is also based on using a comparison theorem. Let $\mathcal{R}_C > 1$ ($\mathcal{E}_{df} = \mathcal{E}_C$), $\delta_H = 0$ and $\frac{(q^*)^2(\mathcal{R}_C - 1)}{\mathcal{R}_C^2} \geq \frac{1}{4}$. Setting $\delta_H = 0$ in the model (3.4) shows that $\frac{dN_H}{dt} = \Pi_H - \mu_H N_H$. Thus, $N_H(t) \to \frac{\Pi_H}{\mu_H}$, as $t \to \infty$. Hence, in the rest of the proof, we replace $N_H(t)$ with $N_H^* = \Pi_H/\mu_H$. The resulting equations for the infected components of the extended model (3.4) can be re-written as:

$$\text{(B.1)} \qquad \frac{d}{dt}\begin{pmatrix} E_H(t) \\ I_H(t) \\ E_{SS}^M(t) \\ I_{SS}^M(t) \\ E_{RS}^M(t) \\ I_{RS}^M(t) \\ E_{RR}^M(t) \\ I_{RR}^M(t) \end{pmatrix} = (F_{ext} - V_{ext} - S_{ext})\begin{pmatrix} E_H(t) \\ I_H(t) \\ E_{SS}^M(t) \\ I_{SS}^M(t) \\ E_{RS}^M(t) \\ I_{RS}^M(t) \\ E_{RR}^M(t) \\ I_{RR}^M(t) \end{pmatrix},$$

where the next generation matrices F_{ext} and V_{ext} are as given by equations (3.14) and (3.15) (in Section 3.2.2), respectively, and the matrix S_{ext} is given by:

$$S_{ext} = \begin{pmatrix} 0 & 0 & 0 & s_{14}^{ext} & 0 & s_{16}^{ext} & 0 & s_{18}^{ext} \\ 0 & 0 & 0 & 0 & 0 & 0 & 0 & 0 \\ 0 & s_{32}^{ext} & 0 & 0 & 0 & 0 & 0 & 0 \\ 0 & 0 & 0 & 0 & 0 & 0 & 0 & 0 \\ 0 & s_{52}^{ext} & 0 & 0 & 0 & 0 & 0 & 0 \\ 0 & 0 & 0 & 0 & 0 & 0 & 0 & 0 \\ 0 & s_{72}^{ext} & 0 & 0 & 0 & 0 & 0 & 0 \\ 0 & 0 & 0 & 0 & 0 & 0 & 0 & 0 \end{pmatrix},$$

where $S_H^* = \frac{\Pi_H}{\mu_H}$, $s_{14}^{ext} = \frac{\beta_H b_{SS}(S_H^* - S_H(t))}{N_H^*}$, $s_{16}^{ext} = \frac{\beta_H b_{RS}(S_H^* - S_H(t))}{N_H^*}$, $s_{18}^{ext} = \frac{\beta_h b_{RR}(S_H^* - S_H(t))}{N_H^*}$, $s_{32}^{ext} = \dfrac{\beta_M b_{SS}((S_{SS}^M)^{*C} - S_{SS}^M(t))}{N_H^*}$, $s_{52}^{ext} = \dfrac{\beta_M b_{RS}((S_{RS}^M)^{*C} - S_{RS}^M(t))}{N_H^*}$, and $s_{72}^{ext} = \dfrac{\beta_M b_{RR}((S_{RR}^M)^{*C} - S_{RR}^M(t))}{N_H^*}$. Since $S_H(t) \leq (S_H)^* = \frac{\mu}{\Pi_H}$, $S_{SS}^M(t) \leq (S_{SS}^M)^{*C}$, $S_{RS}^M(t) \leq (S_{RS}^M)^{*C}$, $S_{RR}M(t) \leq (S_{RR}^M)^{*C}$, in $\Omega_C^{ext} \setminus \{\mathcal{E}_T^{ext}, \mathcal{E}_S^{ext}, \mathcal{E}_R^{ext}\}$ (which follows from Lemma 3), it follows that the matrix S_{ext} is non-negative. Hence, Equation (B.1) can be expressed in terms of the following linear inequality:

$$\text{(B.2)} \qquad \frac{d}{dt}\begin{pmatrix} E_H(t) \\ I_H(t) \\ E_{SS}^M(t) \\ I_{SS}^M(t) \\ E_{RS}^M(t) \\ I_{RS}^M(t) \\ E_{RR}^M(t) \\ I_{RR}^M(t) \end{pmatrix} \leq (F_{ext} - V_{ext}) \begin{pmatrix} E_H(t) \\ I_H(t) \\ E_{SS}^M(t) \\ I_{SS}^M(t) \\ E_{RS}^M(t) \\ I_{RS}^M(t) \\ E_{RR}^M(t) \\ I_{RR}^M(t) \end{pmatrix}.$$

The proof is concluded using the same approach in the proof of Theorem 3.3. Hence, the non-trivial co-existence disease-free equilibrium ($\mathcal{E}_C^{ext}$) of the extended model (3.4), with $\mathcal{R}_C > 1$, $(\mathcal{R}_{SS} - \mathcal{R}_{RS})(\mathcal{R}_{RR} - \mathcal{R}_{RS}) > 0$, $\frac{(q^*)^2(\mathcal{R}_C - 1)}{\mathcal{R}_C^2} \geq \frac{1}{4}$ and $\delta_H = 0$, is globally-asymptotically stable in $\Omega_C^{ext} \setminus \{\mathcal{E}_T^{ext}, \mathcal{E}_S^{ext}, \mathcal{E}_R^{ext}\}$ wherever $\overline{\mathcal{R}}_{0_C}^{ext} < 1$. □

Acknowledgments

The authors are grateful to the three anonymous reviewers and the handling editor for their very constructive comments.

References

[1] Alout H, Dabire RK, Djogbénou LS, Abate L, Corbel V, Chandre F, Cohuet A. 2016. *Interactive cost of Plasmodium infection and insecticide resistance in the malaria vector Anopheles gambiae*, Scientific Reports 6 , 1-11. Article number:29755.

[2] R. A Anderson, B. G. Knols, J. C. Koella, *Plasmodium falciparum sporozoites increase feeding-associated mortality of their mosquito hosts Anopheles gambiae s.I*, Parasitology 120 (4) (2000) 329–333.

[3] Saminu Bala and Bello Gimba, *Global sensitivity analysis to study the impacts of bed-nets, drug treatment, and their efficacies on a two-strain malaria model*, Math. Comput. Appl. **24** (2019), no. 1, Paper No. 32, 28, DOI 10.3390/mca24010032. MR3939221

[4] V. Balabanidou, L. Grigoraki, J. Vontas, *Insect cuticle: a critical determinant of insecticide resistance*, Curr Opin Insect Sci. 2018 Jun 27:68-74.

[5] S. Barbosa, I. M. Hastings, 2012. *The importance of modeling the spread of insecticide resistance in a heterogeneous environment: the example of adding synergists to bednets*, Malaria Journal 11:258.

[6] P. L. G. Birget, J. C. Koella, *A genetic model of the effects of insecticide-treated bed nets on the evolution of insecticide-resistance*, Evolution, Medicine, and Public Health 2015(1) (2015) 205–215.

[7] Kbenesh W. Blayneh, Abba B. Gumel, Suzanne Lenhart, and Tim Clayton, *Backward bifurcation and optimal control in transmission dynamics of West Nile virus*, Bull. Math. Biol. **72** (2010), no. 4, 1006–1028, DOI 10.1007/s11538-009-9480-0. MR2609598

[8] D. Bourguet, A. Genissel, M. Raymond, *Insecticide Resistance and Dominance Levels*, J. Econ. Entomol. 93 (6) (2000) 1588–1595.

[9] Briet OJT, Penny MA, Hardy D, Awolola TS, Bortel WB, Corbel V, Dabire RK, Etang J, Koudou BG, Tungu PK, Chitnis N. 2013. *Effects of pyrethroid resistance on the cost effectiveness of a mass distribution of longlasting insecticidal nets: a modelling study.* Malaria Journal. 12: 77. http://www.malariajournal.com/content/12/1/77

[10] Z. S. Brown, K. L. Dickinson, R. A. Kramer, *Insecticide Resistance and Malaria Vector Control: The Importance of Fitness Cost Mechanisms in Determining Economically Optimal Control Trajectories*, Journal of Economic Entomology **106** (2013) no. 1 366-374.

[11] S. J. Brozak, J. Mohammed-Awel and A. Gumel, *Mathematics of a single-locus model for assessing the impacts of pyrethroid resistance and temperature on population abundance of malaria mosquitoes*, Infectious Disease Modelling, **7** (2022) no. 3, 277-316.

[12] CDC: Center for Disease Control and Prevention. Life Cycle of Anopheles Species Mosquitoes. https://www.cdc.gov/mosquitoes/about/life-cycles/anopheles.html.

[13] Nakul Chitnis, James M. Hyman, and Jim M. Cushing, *Determining important parameters in the spread of malaria through the sensitivity analysis of a mathematical model*, Bull. Math. Biol. **70** (2008), no. 5, 1272–1296, DOI 10.1007/s11538-008-9299-0. MR2421498

[14] CIA World Facebook: Ethiopia, 2017.https://www.cia.gov/library/publications/the-world-factbook/geos/et.html

[15] V. Corbel and R. N'Guessan, *Distribution, Mechanisms, Impact and Management of Insecticide Resistance in Malaria Vectors: A Pragmatic Review*, Intechopen, 2013.

[16] F. Demissie, A. Kebede, T. Shimels, P. Beyene 2009. *Assessment of public health implication of malaria-geohelminth co-infection with an emphasis on hookworm-malaria anemia among suspected malaria patients in asendabo, southwest Ethiopia*, Ethiop Med. J. 47 (2) 153–158.

[17] O. Diekmann, J. A. P. Heesterbeek, and J. A. J. Metz, *On the definition and the computation of the basic reproduction ratio R_0 in models for infectious diseases in heterogeneous populations*, J. Math. Biol. **28** (1990), no. 4, 365–382, DOI 10.1007/BF00178324. MR1057044

[18] P. van den Driessche and James Watmough, *Reproduction numbers and sub-threshold endemic equilibria for compartmental models of disease transmission*, Math. Biosci. **180** (2002), 29–48, DOI 10.1016/S0025-5564(02)00108-6. MR1950747

[19] A. W. F Edwards, *G. H. Hardy (1908) and Hardy-Weinberg equilibrium*, Genetics 2008, 179(3):1143-50.

[20] Facchinelli L., North A.R., Collins C.M., Menichelli M., Persampieri T., Bucci A., Spaccapelo R., Crisanti A., Benedict M.Q, *Large-cage assessment of a transgenic sex-ratio distortion strain on populations of an African malaria vector*, Parasites & Vectors 12:70, 2019.

[21] Richard H. ffrench-Constant, *The Molecular Genetics of Insecticide Resistance*, Genetics, 2013 Aug; 194(4): 807–815.

[22] Gething PW, Casey DC, Weiss DJ et al., ...& Kutz MJ. 2016. *Mapping Plasmodium falciparum Mortality in Africa between 1990 and 2015.* New England Journal of Medicine 375(25):2435-2445.

[23] Hastings A. 1997. *Population Biology: concepts and models*, 1st ed. New York: Springer.

[24] Huijben S, Paaijmans KP. 2017. *Putting evolution in elimination: Winning our ongoing battle with evolving malaria mosquitoes and parasites.* Evolutionary Applications, Special Issue Review and Synthesis 1–16.

[25] S. M. Garba, A. B. Gumel, and M. R. Abu Bakar, *Backward bifurcations in dengue transmission dynamics*, Math. Biosci. **215** (2008), no. 1, 11–25, DOI 10.1016/j.mbs.2008.05.002. MR2459525

[26] Kabula B, Kisinza W, Tungu P, Ndege C, Batengana B, Kollo D, Malima R, Kafuko J, Mohamed M, Magesa S. 2014. Co?occurrence and distribution of East (L1014S) and West (L1014F) African knock?down resistance in Anopheles gambiae sensu lato population of Tanzania. Trop Med Int Health 19(3):331-41.

[27] M. Kuniyoshi, F. Santos 2017. *Mathematical modelling of vector-borne diseases and insecticide resistance evolution*, Journal of Venomous Animals and Toxins including Tropical Diseases 23:34.

[28] B. Levick, A. South, I. A. Hastings, *A two-locus model of the evolution of insecticide resistance to inform and optimise public health insecticide deployment strategies*, PLoS Comput Biol 13(1): e1005327. doi:10.1371/journal.pcbi.1005327

[29] J. Mohammed-Awel, F. Agusto, R. E. Mickens, A. B. Gumel, 2018. *Mathematical assessment of the role of vector insecticide resistance and feeding/resting behavior on malaria transmission dynamics: Optimal control analysis*, Infectious Disease Modeling. 3: 301-321.

[30] Jemal Mohammed-Awel and Abba B. Gumel, *Mathematics of an epidemiology-genetics model for assessing the role of insecticides resistance on malaria transmission dynamics*, Math. Biosci. **312** (2019), 33–49, DOI 10.1016/j.mbs.2019.02.008. MR3943805

[31] Jemal Mohammed-Awel, Enahoro A. Iboi, and Abba B. Gumel, *Insecticide resistance and malaria control: a genetics-epidemiology modeling approach*, Math. Biosci. **325** (2020), 108368, 23, DOI 10.1016/j.mbs.2020.108368. MR4107834

[32] Jemal Mohammed-Awel, Ruijun Zhao, Eric Numfor, and Suzanne Lenhart, *Management strategies in a Malaria model combining human and transmission-blocking vaccines*, Discrete Contin. Dyn. Syst. Ser. B **22** (2017), no. 3, 977–1000, DOI 10.3934/dcdsb.2017049. MR3639152

[33] A. Namias, N.B. Jobe, K.P. Paaijmans, S. Huijben, *The need for practical insecticide-resistance guidelines to effectively inform mosquito-borne disease control programs*, Epidemiology and Global Health Evolutionary Biology. Review article, Aug 6, 2021.

[34] Calistus N. Ngonghala, Sara Y. Del Valle, Ruijun Zhao, and Jemal Mohammed-Awel, *Quantifying the impact of decay in bed-net efficacy on malaria transmission*, J. Theoret. Biol. **363** (2014), 247–261, DOI 10.1016/j.jtbi.2014.08.018. MR3278716

[35] Calistus N. Ngonghala, Jemal Mohammed-Awel, Ruijun Zhao, and Olivia Prosper, *Interplay between insecticide-treated bed-nets and mosquito demography: implications for malaria control*, J. Theoret. Biol. **397** (2016), 179–192, DOI 10.1016/j.jtbi.2016.03.003. MR3482967

[36] Kamaldeen Okuneye, Steffen E. Eikenberry, and Abba B. Gumel, *Weather-driven malaria transmission model with gonotrophic and sporogonic cycles*, J. Biol. Dyn. **13** (2019), no. suppl. 1, 288–324, DOI 10.1080/17513758.2019.1570363. MR3988850

[37] Philip Lounibos, R. L. Escher, 2008. *Sex Ratios of mosquitoes from long-term census of Florida tree holes.* J Am Mosq Control Assoc. 24(1): 11–15.

[38] Hal L. Smith, *Monotone dynamical systems: An introduction to the theory of competitive and cooperative systems*, Mathematical Surveys and Monographs, vol. 41, American Mathematical Society, Providence, RI, 1995. MR1319817

[39] A. South, I. A. Hastings. 2018. *Insecticide resistance evolution with mixtures and sequences: a model-based explanation.* Malaria Journal. 17: 80. https://doi.org/10.1186/s12936-018-2203-y.

[40] World Health Organization. Malaria fact sheets, 2021. `https://www.who.int/news-room/fact-sheets/detail/malaria`.

[41] D.W. Willis and N. Hamon. *Eliminating malaria by 2040 among agricultural households in Africa: potential impact on health, labor productivity, education and gender equality* [version 2; peer review: 2 approved]. Gates Open Research, Jul 2018, 2:33.

Department of Mathematics, Morgan State University, Baltimore, Maryland 21251

Department of Mathematics, University of Maryland, College Park, Maryland 20742; and Department of Mathematics and Applied Mathematics, University of Pretoria, Pretoria 0002, South Africa

Contemporary Mathematics
Volume **793**, 2024
https://doi.org/10.1090/conm/793/15878

A retrospective analysis of COVID-19 dynamics in Mexico and Peru: Studying hypothetical changes in the contact rate

M. Adrian Acuña-Zegarra, Mario Santana-Cibrian, Carlos E. Rodriguez Hernandez-Vela, Ramsés H. Mena, and Jorge X. Velasco-Hernández

ABSTRACT. The COVID-19 pandemic has left many open questions for decision-makers. One of them is knowing the counter-factual effect that the implementation of different mitigation strategies to the ones actually used could have caused. This knowledge would allow to contemplate alternative strategies to deal with future infectious disease emergency. Motivated by the above, we model the observed COVID-19 dynamics for Mexico and Peru and explore the impact of implementing hypothetical stricter mitigation strategies on key days of civic, religious, or political character, that could alter contact and transmission intensities. We explore this impact through hypothetical contact rate reductions during the first year of the pandemic but at different times: i) near its beginning, ii) at the beginning of the second outbreak, and iii) end of the year. The effects of interventions are different for Mexico and Peru, but, in general, they do reduce the cumulative number of cases and deaths. Nonetheless, this study underlines the fact that the time and duration of interventions are important factors to obtain better reductions in prevalence. We also provide evidence that key calendar days are good change points for contact rates and, therefore, may be useful to define strategic time periods for effective interventions, particularly in countries with low testing and lack of contact tracing. We recreate different outbreaks of the COVID-19 disease dynamics in Mexico and Peru and explore the impact of hypothetical interventions that further reduce the contact rate.

1. Introduction

Emerging infectious diseases can be defined as infections that have newly appeared in a population or have existed, but are rapidly increasing in incidence or geographic range [**29**]. Recent examples are H1N1 influenza (2009), Chikungunya (2014), Zika (2015), and COVID-19 (2019 to the present), the latter being the cause, so far, of more than 6.8 million deaths around the world [**27**].

Mathematical modeling has been widely used to study the COVID-19 epidemic. The transmission dynamics has been described with many different methodologies,

Key words and phrases. Mathematical model, superspreading events, key dates, contact rate.

We acknowledge partial support from grants PAPIIT IN115720 and IV100220. JXVH developed part of this work during his sabbatical leave at the Simons Institute, University of Berkeley with partial support from a PASPA-UNAM fellowship. MAAZ thanks ACARUS-UNISON for the use of their computing facilities. CERHV and RHM acknowledge support from PAPIIT grant IG100221.

several of them centered on estimating the effective reproduction number (the average number of secondary cases per infectious case in a population made up of both susceptible and non-susceptible hosts) with some version of the well-known Kermack-McKendrick model [**7**, **22**, **37**, **42**]. During 2020 and 2021, much effort was centered on projecting the COVID-19 pandemic and evaluating the efficacy of the mitigation strategies adopted to contain it [**6**, **19**]. Around the world, the implementation of these measures has varied in strength ranging from strict and mandatory governmental enforcement to a voluntary personal decision. Regardless of the particular version of the mitigation strategy followed, these strategies were based on local factors that combine public health status, economic impact and political conditions [**4**].

The central phenomena of human behavior that we explore are superspreading events occurring in particular calendar dates associated with religious, commercial, or civic holidays specific to each country. During or after these events (depending on their length), the contact rate changes. We, therefore, use these events as change points. We argue that in Mexico and Peru (both middle-income countries with a very stressed economic activity due to the pandemic), these change points (key calendar dates) reflect on the contact rate more clearly than changes imposed by the government (non-pharmaceutical interventions or NPIs) [**13**, **24**]. Using key calendar dates has another advantage from the perspective of forecasting: these dates are known in advance and therefore mobility can be anticipated (short vacations, family visits, buying sprees, etc). Examples of these key dates in Mexico are Mother's day on May 10, Independence day on September 16, and Day of the dead on November 2, among others. For Peru, we have Labor day on May 1, Independence day on July 28, and Santa Rosa de Lima day on August 30, among others.

Here, we present a methodology to estimate a time-dependent contact rate that serves as a baseline to recreate the observed disease dynamics and to generate counterfactual intervention scenarios. In [**13**], the authors used a similar idea but, in our case, to determine changes in the contact rate, we look at particular events occurring on dates related to school, civic or religious periods (vacations, civic holidays, commercial events) that are known in advance each year for a given country. Our results are illustrated considering as examples a subset of Mexican states Queretaro, Quintana Roo, Sonora and the State of Mexico) and some departments of Peru (Arequipa, Cusco, Lima, and Piura). The main objective is to analyze, in retrospective, the impact of hypothetical non-pharmaceutical interventions in several key dates. In particular, we analyze the effect of these interventions in the cumulative number of confirmed cases and deaths.

The COVID-19 epidemic impact on the regional and global economy has influenced decisions on how and when businesses, public centers, tourism, schools and universities can safely reopen [**28**]. For decision-makers, it is important to know the possible effects that the implementation of stricter mitigation strategies could have caused and to compare these effects with the results of the actual strategies during the COVID-19 pandemic. This would allow to envision alternative strategies in the face of a future infectious disease emergency. Such knowledge is pressing in countries that lack the full infrastructure to acquire a more precise or, perhaps we should say, a less uncertain idea of the behavior of the pandemic.

The Chapter is organized as follows. In Section 2, we present the formulation of the mathematical model and the methodology for parameter estimation. In Section 3, we show the results of the model fitting for Mexico and Peru. Section 4 makes a retrospective analysis considering hypothetical interventions that occurred on predefined key dates. Finally, discussion and conclusions can be found in Section 5.

2. Methods

2.1. Mathematical model. A compartmental model is used to describe the evolution of the COVID-19 pandemic. The model considers three classes of infected individuals: Asymptomatic (I), Symptomatic (Y), and Reported (T). Symptomatic individuals recover or report to a health center at a rate ν with ϵ is the proportion of those symptomatic individuals that recover. Similarly, for the dynamics of the reported population (T), individuals recover or die at a rate κ, with $\alpha(t)$ the proportion of those reported individuals that recover. Note that $\alpha(t)$ is a proportion that changes over time. Once reported, infected individuals are effectively isolated and no longer participants in the transmission process. The model allows Susceptible (S) individuals to be Vaccinated (V) with a vaccination rate ψ. It is assumed that the vaccine is not perfect which implies that vaccinated people can be infected. Likewise, it is assumed that vaccinated individuals become susceptible after a certain period. Vital dynamics are also included since this work models the first year of the pandemic. Figure 1 shows the corresponding model diagram.

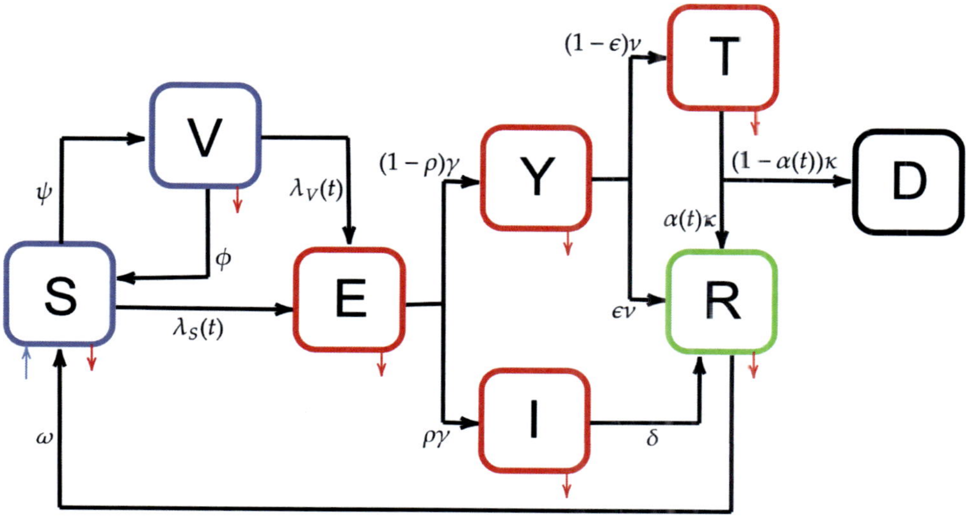

FIGURE 1. **Mathematical model diagram.** There are three types of infectious individuals: asymptomatic, symptomatic and reported. Reported cases do not play a role in transmission. Here, $\lambda_S(t)$ and $\lambda_V(t)$ represent the infection force related to susceptible and vaccinated people, respectively.

The mathematical model is given by the system of ordinary differential equations shown in the system (2.1).

$$
\begin{aligned}
S' &= \omega R + \phi V + \chi M - \beta(t)(Y + qI)\frac{S}{N} - [\chi + \psi]S, \\
V' &= \psi S - (1-\sigma)\beta(t)(Y + qI)\frac{V}{N} - [\phi + \chi]V, \\
E' &= \beta(t)(Y + qI)\frac{S}{N} + (1-\sigma)\beta(t)(Y + qI)\frac{V}{N} - [\gamma + \chi]E, \\
I' &= \rho\gamma E - [\delta + \chi]I, \\
Y' &= (1-\rho)\gamma E - \nu Y - \chi Y, \\
T' &= (1-\epsilon)\nu Y - \kappa T - \chi T, \\
R' &= \delta I + \epsilon\nu Y + \alpha(t)\kappa T - [\omega + \chi]R, \\
D' &= (1 - \alpha(t))\kappa T,
\end{aligned}
\tag{2.1}
$$

Note that, in (2.1), $N = S + V + E + I + Y + R$ and $M = N + T$. Thus, the definition of the population that actively participates in the disease transmission dynamics, N, does not include the reported class T. Table 1 shows a description of all model parameters and the values of those that are fixed and common to every region that is analyzed along the Chapter. Furthermore, Tables 3 and 6 present the values of parameters that change from one region to another.

TABLE 1. **Description of the model's parameters for system 2.1.**

Parameter	Description	Value	Units	Source
ϕ	Vaccine immunity waning rate	0.006	days^{-1}	Assumed
ω	Natural immunity waning rate	0.006	days^{-1}	Assumed
ϵ	Proportion of symptomatic recovered	0.94	%	Assumed
q	Asymptomatic infectiousness reduction	0.45	%	Assumed
γ	Incubation rate	0.1587	days^{-1}	**[14]**
δ	Asymptomatic recovery rate	0.1429	days^{-1}	**[15]**
χ	Natural mortality rate	3.629×10^{-5}	days^{-1}	**[21, 20]**
ρ	Proportion of asymptomatic	0.35	%	**[30]**
ν	Symptomatic output rate by recovery or screening	*	days^{-1}	From data
κ	Reported output rate by recovery or disease mortality	*	days^{-1}	From data
ψ	Vaccination rate	*†	days^{-1}	From data
σ	Vaccine efficacy	*†	%	From data
$\alpha(t)$	Proportion of reported recovered people	*§	%	Estimated
$\beta(t)$	Effective transmission contact rate	*§	days^{-1}	Estimated

* Parameters vary depending on the area of study.
† Values obtained from partial information of available vaccines.
§ Estimates are obtained using a Bayesian approach.

2.1.1. $\beta(t)$ *and* $\alpha(t)$. Although the model depicted in Figure 1 is an extension of the Kermack-McKendrick model, its main feature is how the contact rate (β) and the proportion of reported recovered people (α) are characterized. To account for behavioral change associated to mitigation measures and other changes in the disease dynamics, it is assumed that both parameters, α and β, are time-dependent continuous functions parameterized by $(t_0, t_1, \ldots, t_k, a_1, a_2, \ldots, a_k)$ and $(t_0, t_1, \ldots, t_k, b_1, b_2, \ldots, b_k)$, respectively. Times $t_0, t_1, \ldots, t_k$ are fixed and represent key dates that are associated with civic or religious holidays or perhaps official interventions, where transmission might change. The vectors $(a_1, a_2, \ldots, a_k, b_1, \ldots, b_k)$

are unknown and must be estimated using the observed time series of reported cases and deaths. Details on the estimation process are provided in section 2.2. Once appropriate values for these parameters are found, we employ an interpolation process through a monotone piece-wise cubic Hermite polynomial [**16**] to obtain a smooth continuous function. This is numerically done using function `pchip` of `scipy.interpolate` library in Python 3. The obtained polynomial has two relevant characteristics: i) preserves the shape, which means that, between two key dates, the parameter values will be always increasing or decreasing; and ii) the interpolated function does not overshoot data values, which guarantees that both parameter values remain positive at any point in time,particularly where the number of reported cases is low. It is important to note that both parameters change over time, β is a contact rate with units $1/t$, and α is a dimensionless proportion.

2.1.2. *Vaccination parameters.* Many countries worldwide are using more than one vaccine. However, our model does not incorporate a detailed vaccination dynamic with different vaccines, doses, and efficacy since there is limited information regarding this process in Mexico and Peru.

The lack of information on the vaccination process affects the knowledge of important parameter values. Thus, as a first approach, it is assumed that susceptible individuals are vaccinated at a rate ψ to estimate the vaccination rate. Since the probability of having been vaccinated at time t is $1 - \exp(-\psi t)$ then, if a proportion p of the susceptible population is already vaccinated at time T_V, then the vaccination rate that achieves this coverage is given by

$$\psi = -\frac{\log(1-p)}{T_V}. \tag{2.2}$$

Parameters p, T_V, and therefore, ψ, will vary between regions depending on the initial date of vaccination. For this study, we consider that T_V is the number of days lapsed from the start of vaccination, in each Mexican state or Peruvian department, up to June 16, 2021. The proportion of vaccinated people (p) during that period is approximated based on official communications of the Mexican government [**17**] or available COVID-19 data from Peru [**11**].

In Mexico, up to June 16, 2021, were applied six different vaccines brands (Pfizer-BioNTech, AstraZeneca, SinoVac, Sputnik V, Cansino, and Janssen) [**17**], while in Peruvian departments, vaccines from four different laboratories were administrated (Pfizer-BioNTech, AstraZeneca, Moderna, and Sinopharm) [**11**]. To calculate the vaccine efficacy, we used available information on COVID-19 vaccines provided by each country such as the efficacy of the applied vaccines [**8**]; and the amount of each available vaccine batch (Mexico) [**17**] or the number of applied vaccines (Peru) [**11**]. The estimate of σ is given by a weighted average that includes the aforementioned information. Detailed formulas can be found as supplementary material. Section 3 provides the values of ψ and σ, that vary for each Mexican state or Peruvian department, depending on the available information.

2.2. Parameter estimation.

2.2.1. *Data.* We use the available COVID-19 data of Mexico and Peru [**9, 10, 11, 18**]. Data consists of daily records of reported cases for Mexico, confirmed cases for Peru, and deaths for both countries. We considered data from the start of the pandemic (late February to early March, depending on the specific location) until June 16, 2021.

2.2.2. *Statistical inference.* A Bayesian approach is used to estimate the vector $(a_1, \ldots, a_k, b_1, \ldots, b_k)$ that determines the contact rate $\beta(t)$ and the proportion of recovered reported individuals $\alpha(t)$.

Let $\boldsymbol{\theta} = (a_1, a_2, \ldots, a_k, b_1, b_2, \ldots, b_k)$ the vector of parameters that will be estimated, where $b_i > 0$ and $a_i \in (0, 1)$, for all $i \in \{1, 2, \ldots, k\}$. All the other parameters needed to solve system 2.1 are fixed.

Let W_j and X_j be the random variables that count the number of daily COVID-19 reported infected individuals and deaths at time t_j, respectively, for $j = 1, 2, \ldots n$. Here, t_j represents the number of days since the start of the pandemic in each region. It is assumed that the probability distribution of W_j, conditional on the vector of parameters $\boldsymbol{\theta}$, is a Poisson distribution such that $E[W_j] = C(t_j|\boldsymbol{\theta}) - C(t_{j-1}|\boldsymbol{\theta})$, with $C(t|\boldsymbol{\theta})$ being the cumulative number of reported cases according to System 2.1. Similarly, the probability distribution of X_j, conditional on the vector of parameters $\boldsymbol{\theta}$, is a Poisson distribution such that $E[X_j] = D(t_j|\boldsymbol{\theta}) - D(t_{j-1}|\boldsymbol{\theta})$, with $D(t|\boldsymbol{\theta})$ being the cumulative number of deaths according to the compartmental model. Assuming that variables $W_1, W_2, \ldots, W_n$ and $X_1, X_2, \ldots, X_n$ are conditionally independent, then the likelihood function is given by

$$\begin{aligned}\pi(w_1, \ldots, w_n, x_1, \ldots, x_n|\boldsymbol{\theta}) = &\prod_{j=1}^{n} \frac{[C(t_j|\boldsymbol{\theta}) - C(t_{j-1}|\boldsymbol{\theta})]^{w_j} \exp\{-[C(t_j|\boldsymbol{\theta}) - C(t_{j-1}|\boldsymbol{\theta})]\}}{w_j!} \\ &\times \prod_{j=1}^{n} \frac{[D(t_j|\boldsymbol{\theta}) - D(t_{j-1}|\boldsymbol{\theta})]^{x_j} \exp\{-[D(t_j|\boldsymbol{\theta}) - D(t_{j-1}|\boldsymbol{\theta})]\}}{x_j!} \end{aligned} \tag{2.3}$$

The joint prior distribution for vector $\boldsymbol{\theta}$ is a product of independent uniform distributions and independent log-normal distributions. For parameters $a_1, \ldots, a_k$, the prior is Uniform(0,1), and for $b_1, \ldots, b_k$ the prior is log-normal(0.3, 2). Then

$$\pi(\boldsymbol{\theta}) = \prod_{i=1}^{k} \pi(b_i)\pi(a_i) \propto \prod_{i=1}^{k} \frac{1}{b_i} \exp\left\{ -\frac{[\ln(b_i) - \mu]^2}{2\sigma^2} \right\}.$$

The posterior distribution of the parameters of interest is

$$\pi(\boldsymbol{\theta}|w_1, \ldots, w_n, x_1, \ldots, x_n) \propto \pi(w_1, \ldots, w_n, x_1, \ldots, x_n|\boldsymbol{\theta})\pi(\boldsymbol{\theta}),$$

and it does not have an analytical form since the likelihood function depends on the numerical solution of the ODE system 2.1. We analyze the posterior distribution using an MCMC algorithm called *t-walk* [**12**]. For each region, 3 chains of 8 million iterations are run, from which 200,000 are discarded as burn in. At the end, only 1000 iterations evenly spaced are retained to create the estimations presented in this work. The number of samples that can be used are limited due to the extremely high correlation between samples, which is a common problem in the estimation of parameters in non-linear ODE systems; nonetheless, 1000 samples are enough to provide a good approximation of the posterior distribution.

Table 2 shows the maximum a posteriori probability (MAP) estimate of parameters $(a_1, \ldots, a_k, b_1, \ldots, b_k)$ for the State of Mexico and Lima. Figure 2 shows the resulting $\alpha(t)$ and $\beta(t)$ functions from these MAP estimates.

It is important to mention that confirmed COVID-19 cases can be grouped in different forms in order to represent the epidemic curve. In the case of Mexico, the number of cases can be grouped by date of symptoms onset, by the date when individuals seek medical attention (or tests), and by date of test results. For Mexico, the number reported cases $X_1, \ldots, X_n$ for the estimation process is the number of infected individuals by date when they seek medical attention. In this case,

TABLE 2. Maximum a posteriori estimate (MAP) of the parameters that define functions $\alpha(t)$ and $\beta(t)$.

Key dates of the State of Mexico	b_i	a_i	Key dates of Lima	b_i	a_i
First case	0.828973	0.789712	First case	2.303287	0.005046
NPIs start	0.828973	0.789712	NPIs start	2.303287	0.005046
Childrens day	0.198444	0.718261	Easter	0.265669	0.845001
Mothers day	0.210087	0.708832	Labor day	0.289026	0.843621
Fathers day	0.144456	0.819741	Mothers day	0.240280	0.875993
Independence day	0.161813	0.879569	Fathers day	0.147836	0.879879
Day of the dead	0.171163	0.869444	St Peter and St Paul day	0.258877	0.851755
Buen Fin ends	0.277118	0.845341	Independence day	0.419919	0.873841
Day of the virgin	0.257825	0.868141	Santa Rosa de Lima day	0.366638	0.932971
Christmas	0.023711	0.874383	Battle of Angamos	0.390530	0.954227
New Year	0.473427	0.845068	Saints' day	0.329028	0.955060
Wise men day	0.164426	0.841191	Immaculate Conception day	0.356197	0.938364
Valentines day	0.131805	0.858950	Christmas	0.447588	0.918832
Easter	0.177696	0.884394	New Year	0.771399	0.890371
Childrens day	0.090611	0.910299	Wise men day	0.709509	0.889655
Mothers day	0.192934	0.929212	Valentines day	0.445069	0.851196
Last date*	0.180519	0.953574	Easter	0.685881	0.796018
			Labor day	0.450853	0.817262
			Mothers day	0.310796	0.837250
			Last date*	0.296567	0.847973

* The last date considered for the analysis is June 9, 2021.

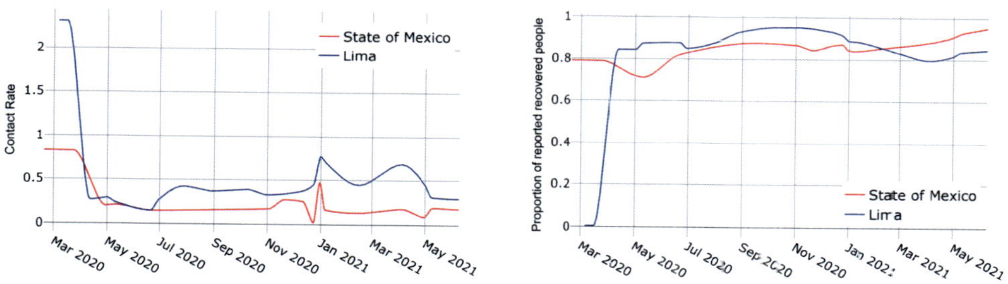

(A) **Baseline time-dependent contact rate.**

(B) **Baseline time-dependent proportion of reported recovered people.**

FIGURE 2. **MAP-based interpolating polynomials of the estimate of parameters for the State of Mexico and Lima.**

parameter ν is referred as the screening rate, the time from symptoms onset to testing. In the case of Peru, reported cases are those that got positive test results at time t. In that case, ν^{-1} represents the time from symptoms onset to test results. Notice that there is no need to modify the compartmental model to handle the difference between these two types of data, it is enough to change the parameter ν, although the interpretation of the model is slightly different.

3. Results

The estimation of the time-varying contact rate $\beta(t)$ is central to our work. We first fit the observed cases for both Mexico and Peru. The estimated function $\beta(t)$ constitutes the baseline on which comparisons are made when evaluating hypothetical interventions. Figure 2A illustrates the baseline time-dependent contact rate for the State of Mexico and Lima, estimated from the open repositories of incidence data available in each country. The perturbations that will be described in the following subsections are local, i.e., only change the contact rate on an specific time interval. For each perturbation, outside this interval the rest of the contact rate remains equal to the baseline $\beta(t)$ plotted in Figure 2A.

3.1. COVID-19 dynamics in Mexico. To exemplify our results, we chose states located at the north, center, and south of Mexico: Sonora, the State of Mexico, Queretaro, and Quintana Roo, respectively. We use the available COVID-19 data of each Mexican state [**18**]. With exception of $\beta(t)$ and $\alpha(t)$, the other parameter values are given in Tables 1 and 3.

TABLE 3. **Parameters of system 2.1 that change from one region to another, for the case of Mexico.**

Parameters	State of Mexico	Queretaro	Sonora	Quintana Roo
ν	0.117647	0.111111	0.111111	0.117647
κ	0.117647	0.111111	0.125	0.111111
ψ	0.001812	0.001693	0.002003	0.002204
σ		0.672245		

The key dates when parameters $\beta(t)$ and $\alpha(t)$ change are listed in Table 4. These dates are the same for all Mexican states, except for the start of the pandemic.

TABLE 4. **Key dates used to estimate the values of the contact rate and the proportion of reported recovered people for some Mexican States.**

Description	Dates	Description	Dates
First case	Varying	Day of the virgin	2020-12-12
NPIs start	2020-03-23	Christmas	2020-12-24
Childrens day	2020-04-30	New Year	2020-12-31
Mothers day	2020-05-10	Wise men day	2021-01-06
Fathers day	2020-06-21	Valentines day	2021-02-14
Independence day	2020-09-16	Easter	2021-04-04
Day of the dead	2020-11-02	Childrens day	2021-04-30
Buen Fin ends	2020-11-21	Mothers day	2021-05-10

Figure 3 shows COVID-19 dynamics of reported cases in comparison with the observed data since the beginning of the pandemic until June 9, 2021. Figures 3 A-D illustrate COVID-19 dynamics of the State of Mexico, Queretaro, Sonora and Quintana Roo, respectively. Note that each epidemic curve has a different behavior. Nevertheless, the proposed scheme based on key dates provides a good approximation of the data.

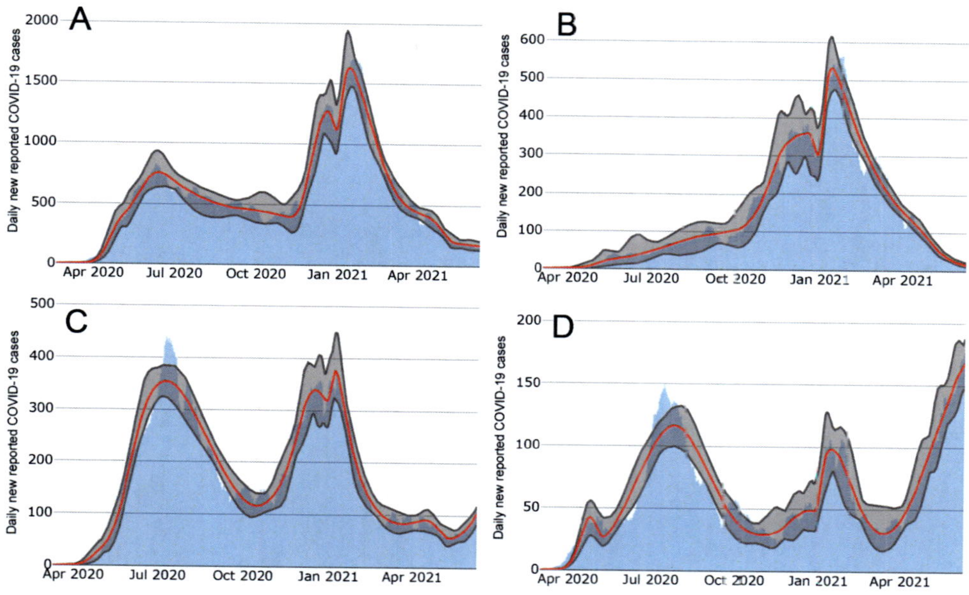

FIGURE 3. **New reported COVID-19 cases in four Mexican states.** (A) State of Mexico, (B) Queretaro, (C) Sonora, and (D) Quintana Roo. Blue bars show reported cases data. Red lines show the median posterior estimates. The gray shadow represents 95% pointwise probability regions.

Also, Figure 4 shows daily mortality given by the model in comparison with the observed deaths until June 9, 2021 for the same Mexican sates. The model also satisfactorily recreates the trajectory of the observed data even when each state has different specific patterns. Observe that the dynamics of reported cases and deaths are typical to each state.

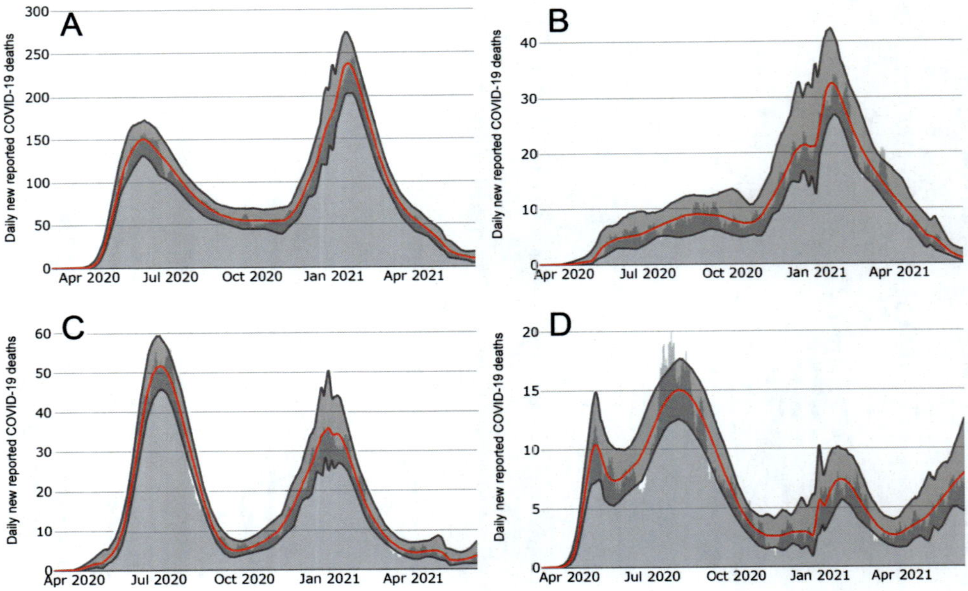

FIGURE 4. **New reported COVID-19 deaths in four Mexican States.** (A) State of Mexico, (B) Queretaro, (C) Sonora, and (D) Quintana Roo. Black bars show reported deaths. Red lines show median posterior estimates. The gray shadow represents 95% pointwise probability regions.

3.2. COVID-19 dynamics in Peru. COVID-19 data for Peru was obtained from [**9**, **10**]. The corresponding key dates used to estimate $\beta(t)$ and $\alpha(t)$ are shown in Table 5. With exception of $\beta(t)$ and $\alpha(t)$, other parameter values are given in Tables 1 and 6.

TABLE 5. **Key dates used to estimate the contact rate and the proportion of reported recovered people by department (political unit) of Peru.**

Description	Dates	Description	Dates
First case	Varying		
NPIs start	2020-03-16	Saints' day	2020-11-01
Easter	2020-04-12	Immaculate Conception day	2020-12-08
Labor day	2020-05-01	Christmas	2020-12-24
Mothers day	2020-05-10	New Year	2021-01-01
Fathers day	2020-06-21	Wise men day	2021-01-06
St Peter and St Paul day	2020-06-29	Valentines day	2021-02-14
Independence day	2020-07-28	Easter	2021-04-04
Santa Rosa de Lima day	2020-08-30	Labor day	2021-05-01
Battle of Angamos	2020-10-08	Mothers day	2021-05-09

TABLE 6. **Parameters of system 2.1 that change from one region to another, for Peru. A row with a single number indicates that it is used for all regions.**

Parameters	Arequipa	Cusco	Lima	Piura
ν		0.111111		
κ	0.117647	0.117647	0.105263	0.125
ψ	0.001299	0.000786	0.001428	0.000755
σ	0.661926	0.67096	0.697056	0.676159

Figures 5 and 6 show observed and fitted COVID-19 confirmed cases and mortality, respectively, in four departments of Peru: Arequipa, Cusco, Lima and Piura. We can see that, despite obvious differences in the shape of the epidemic curve between both countries, the model provides a good description of the observed data. Particularly, notice that for certain regions, such as Piura, the pattern of reported cases differs considerably from the pattern of deaths. The model can describe these two contrasting behaviors for each region thanks to the inclusion of two time-dependent parameters ($\alpha(t)$ and $\beta(t)$); allowing the contact rate to change is not be sufficient to obtain this results.

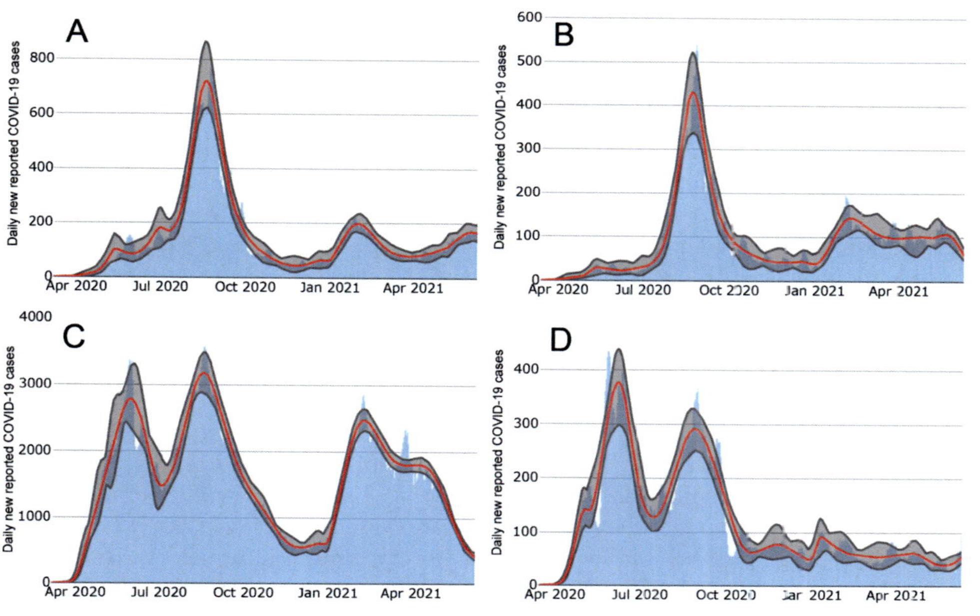

FIGURE 5. **New confirmed COVID-19 cases in four departments of Peru.** (A) Arequipa, (B) Cusco, (C) Lima, and (D) Piura. Blue bars shows confirmed cases. Red lines show the median posterior estimates. The gray shadow represents 95% pointwise probability regions.

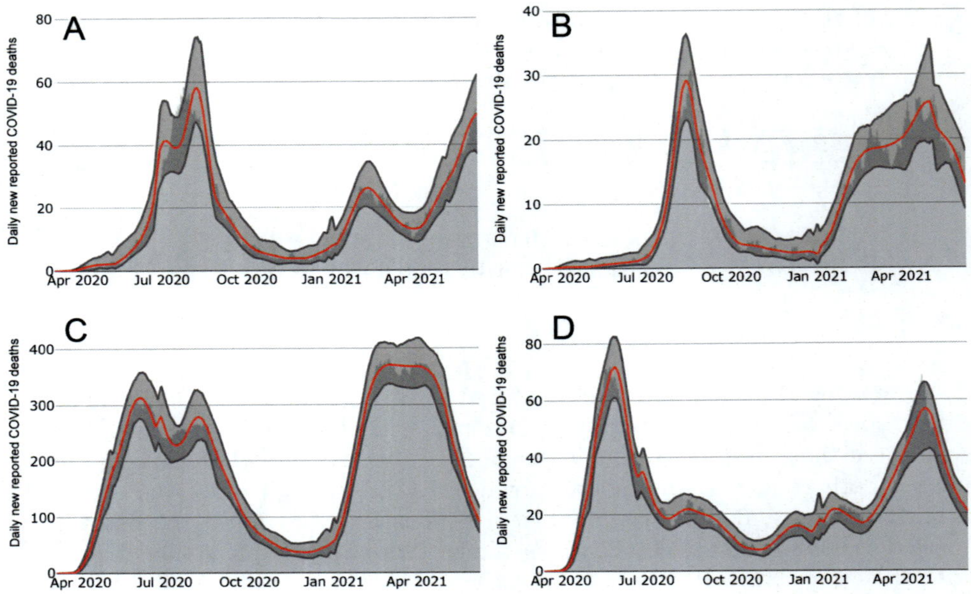

FIGURE 6. **New reported COVID-19 deaths in some departments of Peru.** (A) Arequipa, (B) Cusco, (C) Lima, and (D) Piura. Beige bars shows reported cases data. Red lines show the median posterior estimates. The gray shadow represents 95% pointwise probability regions.

4. Looking back: hypothetical trends under different interventions

As we know, at the beginning of the COVID-19 pandemic, countries around the world implemented different mitigation measures to control COVID-19 disease transmission. However, these measures were largely insufficient to control disease contagion and mortality in many places. A high mortality rate and the occurrence of more than one outbreak in the first year of the pandemic are evidence of this fact.

As we have argued before, key calendar dates are civic, religious or commercial days when a change in disease transmission dynamics might be observed. We use our mathematical model to explore the counterfactual impact of stricter NPIs executed or enforced during some key calendar dates. It is important to mention that, even when NPIs can be of different nature (use of face-mask, curfews, mobility restrictions, etc), we assume that all NPIs have the same aggregated effect in terms of the mathematical model: they all reduce the contact rate. We do not consider an specific type of intervention because there is very limited information regarding the efficacy of any type of NPIs in Mexico and Peru.

We evaluate the hypothetical epidemic trends obtained from the reduction of the contact rate in a neighborhood of a specific key date. For example, suppose that we want to see the effect of reducing in half the contact rate at time t_j (j-th key date). Then, the corresponding estimated contact rate at that time, b_j, will change to $b_j^* = 0.5b_j$. The new time dependent contact rate, $\beta^*(t)$, now is parameterized by $(b_1, \ldots, b_{j-1}, b_j^*, b_{j+1}, \ldots, b_k)$. Due to the interpolation, this means that the contact rate will continuously change in the interval (t_{j-1}, t_{j+1}). The reasoning behind this idea is that changes in the real contact rate are not instantaneous, for it takes time for NPIs to produce the desired effects on the majority of the population. The scenarios presented here consider hypothetical interventions in three periods during the first year of the pandemic: (i) near the beginning of the pandemic, (ii) at the beginning of the second outbreak, and (iii) at the end of the year. Specific intervention dates are presented in Table 7. For each period, we consider scenarios where the contact rate is reduced 25%, 50%, and 75%. We point out that, for each scenario, the interval (t_{j-1}, t_{j+1}) spans from the key date before to the key date after the chosen key date. A different length of this interval is explored in Appendix A.

TABLE 7. **Periods for which transmission contact rate is modified to analyze the effect of stricter NPIs.**

	State of Mexico		Lima	
	From	To	From	To
Near the beginning of the pandemic	2020-04-30	2020-06-21	2020-05-01	2020-06-21
At the beginning of the second outbreak	2020-11-02	2020-12-12	2020-06-29	2020-08-30
At the end of the first year	2020-12-24	2021-01-06	2020-12-24	2021-01-06

4.1. State of Mexico. Figure 7 shows State of Mexico scenarios regarding newly reported COVID-19 cases. Table 2 summarizes the maximum a posteriori estimate (MAP) of the parameters that define functions $\alpha(t)$ and $\beta(t)$. Other parameter values are given in Tables 1 and 3. Dashed black line, dotted black line and solid black line show COVID-19 dynamics when reducing the value of transmission contact rate to 25%, 50%, and 75%, respectively. Red line represents our baseline COVID-19 dynamics.

Figure 7A shows that all reductions in case (i) lead to a period of slow growth longer than observed in the baseline. We have argued before [**35**] that this can be a consequence of the proximity of the intervention day to the first peak of the epidemic. With respect to outbreaks, all contact rate reductions generate outbreaks with lower intensity than the baseline. Moreover, taking June 9, 2021 as the final date of our analysis, the most extreme intervention (75% reduction) produces a reduction close to 69% of the baseline cumulative incidence (see Figure 7D).

Figure 7B illustrates that reductions in case (ii) occur at a period of epidemic growth far from the observed peak. For this case the reductions on incidence are significant; in particular, the most extreme intervention (75% reduction), the incidence changes direction and takes a downward trend. However, if we consider June 9, 2021, as the final date of our analysis, the most extreme intervention in case (ii) is less effective than in case (i). Cumulative incidence presents a reduction close to 46% regarding the baseline dynamics (see Figure 7E). This last may be a consequence of the length of the period perturbation in case (i) being greater than in case (ii).

Finally, Figure 7F shows that acting on a period very close to the peak of the curve (case (iii)) achieves comparatively poor reductions in the cumulative incidence even in the case of the most extreme intervention (where the reduction is close to 20% of the baseline cumulative incidence). However, Figure 7C shows that both medium and extreme interventions notably reduce second-peak levels of incidence which is very useful to avoid the health centers' saturation.

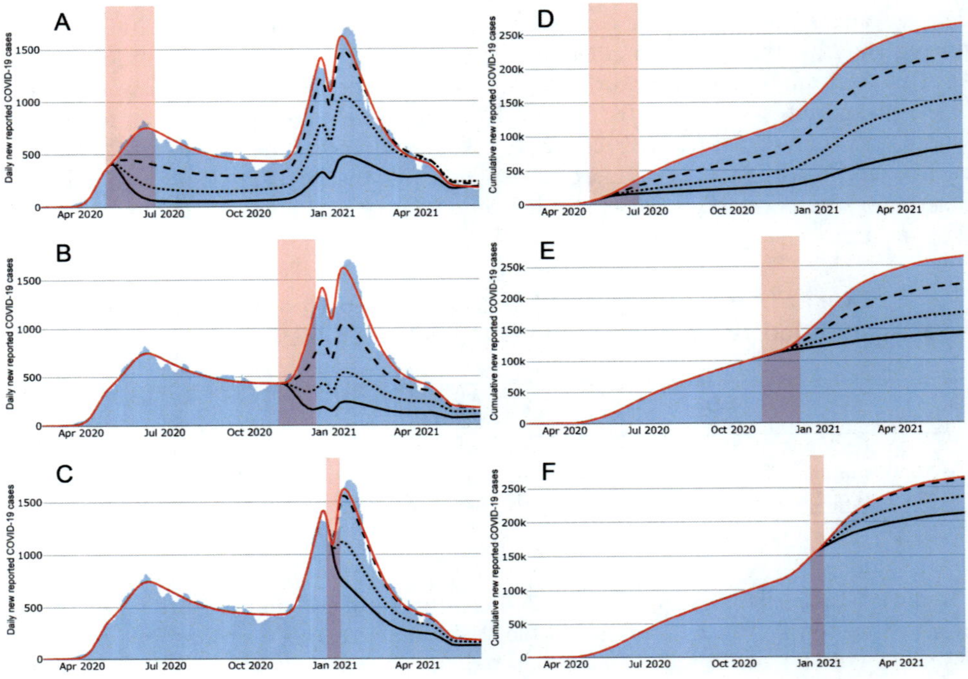

FIGURE 7. **State of Mexico Scenarios - New reported COVID-19 cases.** Left and right columns represent the daily and cumulative reported COVID-19 cases, respectively. Panels A and D illustrate COVID-19 dynamics when perturbing the transmission contact rate around Mother's day. Panels B and E show COVID-19 dynamics when perturbing the transmission contact rate around the end of the *Buen Fin*. Panels C and F illustrate COVID-19 dynamics when perturbing the transmission contact rate near New Year's day. The red shadow represents periods in which the transmission contact rate is modified with respect to the baseline.

4.2. Lima. Figure 8 illustrates Lima scenarios regarding new confirmed COVID-19 cases. Table 2 summarizes the maximum a posteriori estimate (MAP) of the parameters that define functions $\alpha(t)$ and $\beta(t)$. Other parameter values are given in Tables 1 and 6. Dashed black line, dotted black line and solid black line show COVID-19 dynamics when reducing the value of transmission contact rate to 25%, 50%, and 75%, respectively. Red line represents our baseline COVID-19 dynamics.

Figure 8 shows the COVID-19 epidemic curve for Lima. It certainly differs from that observed in the State of Mexico. Even though a curfew was imposed, there was more than one outbreak at the beginning of the pandemic. This suggests that the curfew in Lima was not strict as it could have been considering that in China this measure maintained a low number of cases.

Figure 8A shows case (i) for Lima, showing differences with respect of the trends for the State of Mexico. In this case, this intervention reduces the first outbreak but the level of the second outbreak is higher than that of the baseline scenario. Nevertheless, despite this increase, we can observe a slight reduction in the cumulative incidence overall (see Figure 8D).

Figure 8B illustrates that, for Lima, in case (ii) the intervention occurs when the epidemic curve is in a downward trend and reaches a local minimum. The consequence of the intervention at this time is an effective reduction of the size of the second peak with respect to the baseline scenario. Moreover, as the reduction in the contact rate increases, the appearance of the second outbreak is delayed. In contrast to case (i), the maximum values of the peaks do not exceed those in the baseline scenario. However, similar to case (i), at the final date of our analysis (June 9, 2021), all interventions produce a slight reduction in the cumulative incidence (see Figure 8E).

Figures 8C and F case (iii) shows a poor reduction in prevalence even when achieving a 75% reduction in the contact rate. We observe that a strictier reduction only delays the appearance of the third outbreak, while a medium reduction (50%) generates a plateau. However, on the final date of our analysis (June 9, 2021), case (iii) does not show a reduction with respect to the baseline cumulative incidence (see Figure 8F).

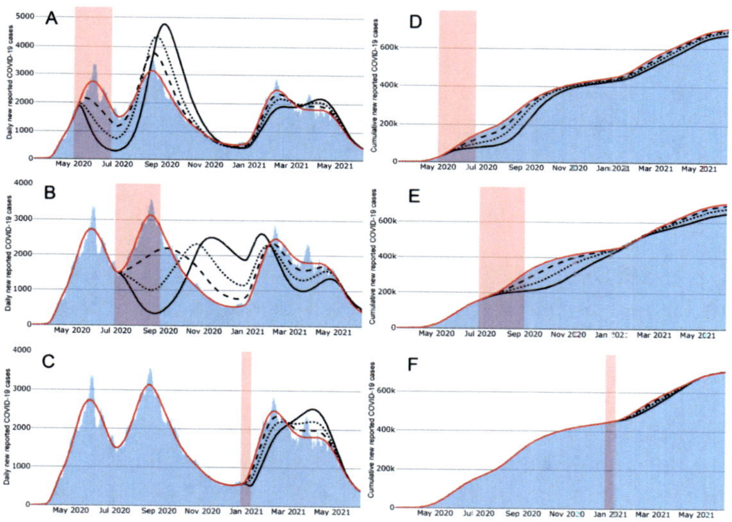

FIGURE 8. **Lima Scenarios - New confirmed COVID-19 cases.** Left and right columns represent the daily and cumulative confirmed COVID-19 cases, respectively. Panels A and D illustrate COVID-19 dynamics when perturbing the transmission contact rate around Mother's day. Panels B and E show COVID-19 dynamics when perturbing the transmission contact rate around Independence day. Panels C and F illustrate COVID-19 dynamics when perturbing the transmission contact rate in the vicinity of New Year's day. The red shadows represents periods in which the transmission contact rate is modified.

5. Discussion and conclusions

Emerging infectious diseases are an important concern for public health. COVID-19 disease is the more recent example that has caused more than 6.6 million deaths and 650 million confirmed cases around the world as of middle December, 2022 [**27**]. This disease has once again shown the role of mobility in the spread of acute respiratory diseases [**35**]. This is consistent with research where superspreading events are found to be [**26**].

The epidemic dynamics observed in all of Mexico have been driven by events associated with heightened mobility and increased social activity [**2, 5, 25, 31, 32, 38, 41, 43**] during holidays and other important calendar key dates. The older population as well as that with comorbidities, such as obesity, diabetes, and hypetension, has been most affected [**34**]. Early in the pandemic, key calendar days as described here, were associated with large gatherings violating the social-distancing policies in effect at the time. These dates provide good support to putative changes in the contact rates. As an example, in Mexico City the first case occurred by the end of February 2020, and the first set of mitigation measures was applied on March 23, 2020. Later that year, there were superspreading events in Mexico City during Easter celebrations (April 6-12) and early May (April 30 to May 10) that shifted the day of maximum incidence to the end of May, and pushed the epidemic into a quasi-stationary state characterized by values of $R_t \approx 1$ [**1, 35**]. This behavior, observed around the world, has been explored in [**23, 36, 39**]. In particular, [**39**] claim that this quasi-linear growth and the maintenance of the effective reproduction number around $R_t \approx 1$ for sustained periods of time, involves critical changes in the structure of the underlying contact network of individuals. In the case of Mexico City, we have argued [**35**] that these changes relate to superspreading events on key calendar dates (Easter holidays and children's, Holy Cross' and Mother's days).

This study extends the ideas presented in [**35**]. We develop a methodology for the description of the epidemic curve based on key calendar days that allows us to recreate the COVID-19 dynamics in different cities. Our main hypothesis assumes that key calendar days are associated to events where the contact rates change and, therefore, are central points to estimate parameters and explain the observed epidemic dynamics. We employed the same methodology for Mexico and Peru obtaining good results both for incidence and death dynamics (see Figures 3 to 6) showing that our approach can recreate different types of COVID-19 dynamics and likely can be extended to other acute respiratory infections.

With our model and the associated methodology, we have explored the impact of hypothetical interventions that reduce the contact rate within a given period. Although our scenarios change the behavior of the epidemic curve, the cumulative incidence never exceeds the baseline case. Both the State of Mexico and Lima show that reducing the contact rate is not enough to obtain low levels of incidence throughout the year. The length of the reduction period and its timing are also important. In Lima, different shapes of the epidemic curve are observed. For example, we observe that an early intervention causes a reduction in the intensity of the first outbreak, but an increase in the peak magnitude of the second outbreak. This phenomenon can be explained by the existence of a large susceptible pool resulting from the early reduction on the contact rate that impacted the implementation of non-pharmaceutical measures. In Peru, on July 1, 2020, mandatory isolation was shifted to a targeted quarantine [**33**] which coincides with the first relevant increase in the contact rate (see Figure 2A). However, a reduction in the intensity of the first outbreak would have been positive because it would have helped to reduce hospitalizations and the saturation of the health centers that were almost reached both in Mexico and Peru in early 2021 [**40**, **3**].

This study has shown the role of key dates in the implementation of non-pharmaceutical interventions (NPIs). Key dates are known in advance, and thus can help to design an intervention strategy ahead of time. Likewise, we have shown that although the enforcement of NPIs will not always generate a reduction in the cumulative incidence, it can greatly impact epidemic outbreaks, reducing the intensity of their peak or delaying the moment of their appearance. Finally, there are still relevant questions to answer mainly centered on planning strategies in advance. We have shown that prior knowledge of key dates (civil, religious, political) can be useful to provide an accurate description of epidemic trends when testing and contact tracing are not widely available.

A. Appendix: looking back

In addition to the scenarios presented above, we consider the case were contact rate reductions occur in shorter periods, lets say, between one week before the perturbation and one week after (see Table 8).

For the following figures, dashed black line, dotted black line and solid black line show COVID-19 dynamics when reducing the value of transmission contact rate to 25%, 50%, and 75%, respectively. Red line represents our baseline COVID-19 dynamics, and the red shadows represent periods in which the transmission contact rate is perturbed.

Figures 9 and 10 show scenarios for the State of Mexico and Lima, respectively. Figures 9A and D illustrate COVID-19 dynamics when perturbing the transmission contact rate around Mother's day. Figures 9B and E show COVID-19 dynamics when perturbing the transmission contact rate around the end of the *Buen Fin.* Figures 9C and F illustrate COVID-19 dynamics when perturbing the transmission contact rate around New Year's day.

Figures 10A and D illustrate COVID-19 dynamics when perturbing the transmission contact rate around Mother's day. Figures 10B and E show COVID-19 dynamics when perturbing the transmission contact rate around Independence day. Figures 10C and F illustrate COVID-19 dynamics when perturbing the transmission contact rate around New Year's day.

TABLE 8. **Periods in which transmission contact rate is perturbed for Figures 9 and 10.**

	State of Mexico		Lima	
	From	To	From	To
Near the beginning of the pandemic	2020-05-03	2020-05-17	2020-05-03	2020-05-17
At the beginning of the second outbreak	2020-11-14	2020-11-28	2020-07-21	2020-08-04
At the end of the first year	2020-12-24	2021-01-07	2020-12-25	2021-01-08

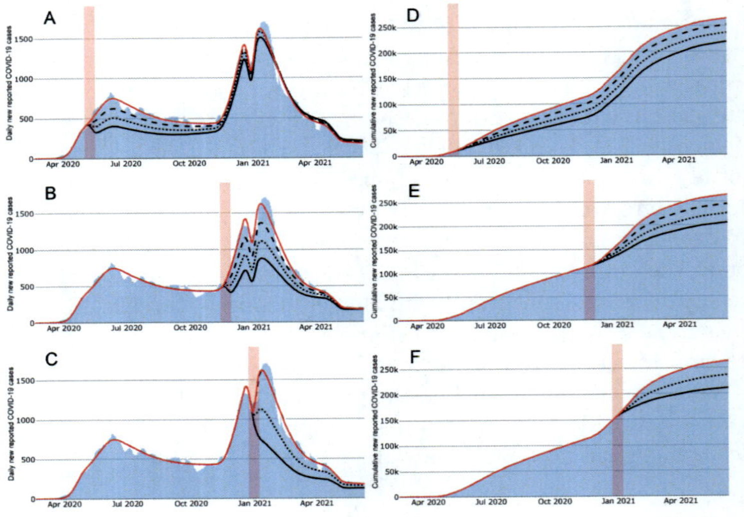

FIGURE 9. **State of Mexico Scenarios - reduction periods of two weeks.** Left and right columns represent the daily and cumulative confirmed COVID-19 cases, respectively. The red shadow represents periods in which the transmission contact rate is perturbing.

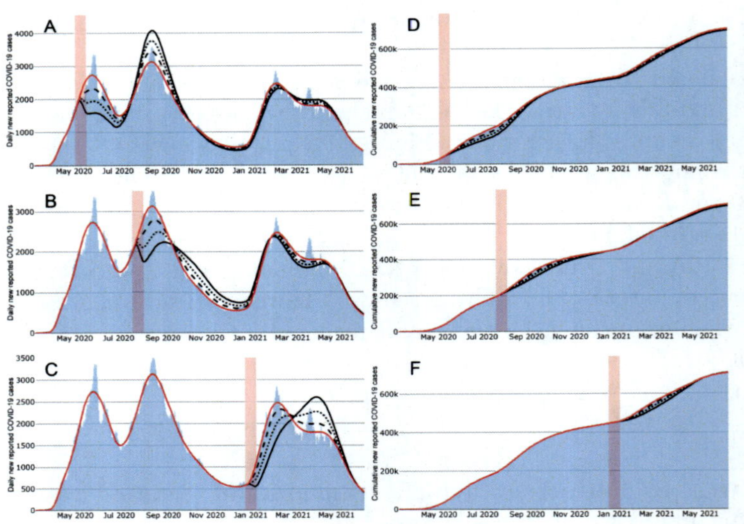

FIGURE 10. **Lima Scenarios - reduction periods of two weeks.** Left and right columns represent the daily and cumulative confirmed COVID-19 cases, respectively. The red shadow represents periods in which the transmission contact rate is perturbing.

Author contributions

Conceptualization: M. Adrian Acuña-Zegarra, Mario Santana-Cibrian, Jorge X. Velasco-Hernández.

Discussion and draft preparation: M. Adrian Acuña-Zegarra, Mario Santana-Cibrian, Carlos E. Rodriguez Hernandez-Vela, Ramsés H. Mena, Jorge X. Velasco-Hernández.

References

[1] Manuel Adrian Acuña-Zegarra, Mario Santana-Cibrian, and Jorge X. Velasco-Hernandez, *Modeling behavioral change and COVID-19 containment in Mexico: a trade-off between lockdown and compliance*, Math. Biosci. **325** (2020), 108370, 9, DOI 10.1016/j.mbs.2020.108370. MR4100865

[2] D. C. Adam, P. Wu, J. Y. Wong, E. H. Lau, T. K. Tsang, S. Cauchemez, G. M. Leung, and B. J. Cowling, *Clustering and superspreading potential of SARS-CoV-2 infections in Hong Kong*, Nature Medicine **26** (2020), no. 11, S. 1714–1719. DOI 10.1038/s41591–020–1092–0.

[3] Agencia EFE, *Perú vuelve a tener sus UCI tan llenas como en el peor momento de la pandemia.* `https://gestion.pe/peru/peru-vuelve-a-tener-sus-uci-tan-llenas-como-en-el-peor-momento-de-la-pandemia-noticia/`. January 2021, Abruf: January 15, 2021. – Gestion

[4] D. Allain-Dupre, I. Chatry, A. Kornprobst, and M.-V. Michalun, *The territorial impact of COVID-19: Managing the crisis across levels of government / OECD*, November 2020. `https://www.oecd.org/coronavirus/policy-responses/the-territorial-impact-of-covid-19-managing-the-crisis-across-levels-of-government-d3e314e1/`. – Forschungsbericht

[5] B. M. Althouse, E. A. Wenger, J. C. Miller, S. V. Scarpino, A. Allard, L. Hebert-Dufresne, and H. Hu, *Superspreading events in the transmission dynamics of SARS-CoV-2: Opportunities for interventions and control*, PLoS Biology **18** (2020), no. 11, S. 1–13. `http://dx.doi.org/10.1371/journal.pbio.3000897`. – DOI 10.1371/journal.pbio.3000897.

[6] N. Askitas, K. Tatsiramos, and B. Verheyden, *Estimating worldwide effects of non-pharmaceutical interventions on COVID-19 incidence and population mobility patterns using a multiple-event study*, Scientific Reports **11** (2021), no. 1, S. 1–13. DOI 10.1038/s41598–021–81442–x.

[7] K. M. Bubar, S. M. Kissler, M. Lipsitch, S. Cobey, Y. H. Grad, D. B. Larremore, and K. Reinholt, *Model-informed COVID-19 vaccine prioritization strategies by age and serostatus*, Science **371** (2020), S. 916–921. `http://dx.doi.org/10.1101/2020.09.08.20190629`. – DOI 10.1101/2020.09.08.20190629

[8] V. M. Camarillo Nava, D. Pérez López, and E. López Rojas, *Eficacia de las vacunas contra el SARS-CoV-2 en México y el mundo*, Atención Familiar **28** (2021), sep., no. 4, 291–295, DOI 10.22201/fm.14058871p.2021.4.80601

[9] Centro Nacional de Epidemiología, prevención y Control de Enfermedades – MINSA, *Casos positivos por COVID-19*, `https://www.datosabiertos.gob.pe/dataset/casos-positivos-por-covid-19-ministerio-de-salud-minsa`, 2020.

[10] Centro Nacional de Epidemiología, prevención y Control de Enfermedades – MINSA, *Fallecidos por COVID-19*, `https://www.datosabiertos.gob.pe/dataset/fallecidos-por-covid-19-ministerio-de-salud-minsa`, 2020.

[11] Centro Nacional de Epidemiología, prevención y Control de Enfermedades – MINSA, *Vacunación contra COVID-19*, `https://www.datosabiertos.gob.pe/dataset/vacunaci%C3%B3n-contra-covid-19-ministerio-de-salud-minsa`, 2020.

[12] J. Andrés Christen and Colin Fox, *A general purpose sampling algorithm for continuous distributions (the t-walk)*, Bayesian Anal. **5** (2010), no. 2, 263–281, DOI 10.1214/10-BA60. MR2719653

[13] J. Dehning, J. Zierenberg, F. Spitner, M. Wibral, J. Neto, M. Wilczek, and V. Priesemann, *Inferring change points in the spread of COVID-19 reveals the effectiveness of interventions*, Science (New York, N.Y.) 9789 (2020), no. May, S. 1–15, DOI 10.1126/science.abb9789

[14] Z. Du, C. Liu, L. Wang, Y. Bai, E. H. Lau, P. Wu, and B. J. Cowling, *Shorter serial intervals and incubation periods in SARS-CoV-2 variants than the SARS-CoV-2 ancestral strain*, Journal of travel medicine **29** (2022), no. 6, S. DOI 10.1093/jtm/taac052

[15] S. E. Eikenberry, M. Mancuso, E. Iboi, T. Phan, K. Eikenberry, Y. Kuang, E. Kostelich, and A. B. Gumel, *To mask or not to mask: Modeling the potential for face mask use by the general public to curtail the COVID-19 pandemic*, Infectious disease modelling **5** (2020), S. 293–308. DOI 10.1016/j.idm.2020.04.001

[16] F. N. Fritsch and J. Butland, *A method for constructing local monotone piecewise cubic interpolants*, SIAM J. Sci. Statist. Comput. **5** (1984), no. 2, 300–304, DOI 10.1137/0905021. MR740848

[17] Gobierno de México, *Conferencias de prensa*, `https://coronavirus.gob.mx/wp-content/uploads/2021/06/2021.06.16_CP_Salud_CTD_COVID-19.pdf`, 2020.

[18] Gobierno de México, *Datos Abiertos*, `https://www.gob.mx/salud/documentos/datos-abiertos-152127`, 2020.

[19] INSP, *Reflexiones sobre la respuesta de Mexico ante la pandemia de COVID19 y sugerencias para enfrentar los próximos retos*, Mexico City : Instituto Nacional de Salud Pública, Mexico, 2020

[20] Instituto Nacional de Estadística e Informática - INEI, *Perú tiene una población de 32 millones 131 mil 400 habitantes al 30 de junio del presente año*, `https://m.inei.gob.pe/media/MenuRecursivo/noticias/notadeprensa123_1.pdf`, 2019.

[21] Instituto Nacional de Estadística y Geografía (INEGI), *Esperanza de vida al nacimiento por entidad federativa según sexo, serie anual de 2010 a 2022*, `https://www.inegi.org.mx/app/tabulados/interactivos/?pxq=Mortalidad_Mortalidad_09_61312f04-e039-4659-8095-0ce2cd284415`, 2022.

[22] H. Petard, *A Contribution to the Mathematical Theory of Big Game Hunting*, Amer. Math. Monthly **45** (1938), no. 7, 446–447, DOI 10.2307/2304150. MR1524340

[23] N. L. Komarova, L. M. Schang, and D. Wodarz, *Patterns of the COVID-19 pandemic spread around the world: Exponential versus power laws: Patterns of the COVID-19 pandemic spread around the world: Exponential versus power laws*, Journal of the Royal Society Interface **17** (2020), no. 170, DOI 10.1098/rsif.2020.0518.

[24] Y. Li, H. Campbell, D. Kulkarni, A. Harpur, M. Nundy, X. Wang, H. Nair, and U. Group, *The temporal association of introducing and lifting non-pharmaceutical interventions with the time-varying reproduction number (R) of SARS-CoV-2: a modelling study across 131 countries*, The Lancet Infectious Diseases **21** (2021), no. 0, S. 193–202. DOI 10.1016/S1473–3099(20)30785–4.

[25] Y. Liu, R. M. Eggo, and A. J. Kucharski, *Secondary attack rate and superspreading events for SARS-CoV-2*, The Lancet **395** (2020), no. 10227, S. e47, DOI 10.1016/S0140–6736(20)30462–1.

[26] J. O. Lloyd-Smith, S. J. Schreiber, P. E. Kopp, and W. M. Getz, *Superspreading and the effect of individual variation on disease emergence*, Nature **438** (2005), no. 7066, S. 355–359, DOI 10.1038/nature04153.

[27] E. Mathieu, H. Ritchie, L. Rodés-Guirao, C. Appel, C. Giattino, J. Hasell, B. Macdonald, S. Dattani, D. Beltekian, E. Ortiz-Ospina, and M. Roser, *Coronavirus Pandemic (COVID-19)*, Our World in Data (2020). – `https://ourworldindata.org/coronavirus`.

[28] B. May, *Research Briefing — Global World: GDP revisions / Oxford Economics*, Oxford, March 2021.

[29] S. Morse, *Factors in the emergence of infectious diseases*, Emerging Infectious Diseases **1** (1995), no. 1, S. 7. DOI 10.3201/eid0101.950102

[30] H. Nishiura, T. Kobayashi, T. Miyama, A. Suzuki, S.-m. Jung, K. Hayashi, R. Kinoshita, Y. Yang, B. Yuan, A. R. Akhmetzhanov, u. a., *Estimation of the asymptomatic ratio of novel coronavirus infections (COVID-19)*, International journal of infectious diseases **94** (2020), S. 154–155, DOI 10.1016/j.ijid.2020.03.020

[31] A. Popa, J. W. Genger, M. D. Nicholson, T. Penz, D. Schmid, S. W. Aberle, B. Agerer, A. Lercher, L. Endler, H. Colaço, M. Smyth, M. Schuster, M. L. Grau, F. Martínez-Jiménez, O. Pich, W. Borena, E. Pawelka, Z. Keszei, M. Senekowitsch, J. Laine, J. H. Aberle, M. Redlberger-Fritz, M. Karolyi, A. Zoufaly, S. Maritschnik, M. Borkovec, P. Hufnagl, M. Nairz, G. Weiss, M. T. Wolfinger, D. von Laer, G. Superti-Furga, N. Lopez-Bigas, E. Puchhammer-Stöckl, F. Allerberger, F. Michor, C. Bock, and A. Bergthaler, *Genomic epidemiology of*

superspreading events in Austria reveals mutational dynamics and transmission properties of SARS-CoV-2, Science Translational Medicine **12** (2020), no. 573, DOI 10.1126/scitranslmed.abe2555.

[32] B. Rader, S. V. Scarpino, A. Nande, A. L. Hill, B. Adlam, R. C. Reiner, D. M. Pigott, B. Gutierrez, A. E. Zarebski, M. Shrestha, J. S. Brownstein, M. C. Castro, C. Dye, H. Tian, O. G. Pybus, M. U. G. Kraemer, *Crowding and the shape of COVID-19 epidemics*, Nature Medicine 0.1038/s41 (2020). DOI 10.1038/s41591–020–1104–0.

[33] C. B. Ramírez, *Gobierno dispone cuarentena focalizada hasta el 31 de julio por Coronavirus*, `https://www.gob.pe/institucion/munijepelacio/noticias/189135-gobierno-dispone-cuarentena-focalizada-hasta-el-31-de-julio-por-coronavirus`. Version: June 2020, Abruf: April 3, 2023. – Plataforma digital única del Estado Peruano

[34] Carlos E. Rodríguez and Ramsés H. Mena, *COVID-19 clinical footprint to infer about mortality*, J. Roy. Statist. Soc. Ser. A **185** (2022), no. suppl. S2, S547–S572. DOI 10.1111/rssa.12947. MR4547522

[35] Mario Santana-Cibrian, Manuel A. Acuña-Zegarra, and Jorge X. Velasco-Hernandez, *Lifting mobility restrictions and the effect of superspreading events on the short-term dynamics of COVID-19*, Math. Biosci. Eng. **17** (2020), no. 5, 6240–6258, DOI 10.3934/mbe.2020330. MR4160271

[36] H. M. Singer, *The COVID-19 pandemic: growth patterns, power law scaling, and saturation*, Physical biology **17** (2020), no. 5, S. 055001. `http://dx.doi.org/10.1088/1478-3975/ab9bf5`. – DOI 10.1088/1478–3975/ab9bf5. – ISSN 14783975

[37] R. Subramanian, Q. He, and M. Pascual, *Quantifying asymptomatic infection and transmission of COVID-19 in New York City using observed cases, serology and testing capacity*, PNAS **118** (2020), no. 9, S. e2019716118. DOI 10.1073/pnas.2019716118.

[38] K. Sun, W. Wang, L. Gao, Y. Wang, K. Luo, L. Ren, Z. Zhan, X. Chen, S. Zhao, Y. Huang, Q. Sun, Z. Liu, M. Litvinova, A. Vespignani, M. Ajelli, C. Viboud, and H. Yu, *Transmission heterogeneities, kinetics, and controllability of SARS-CoV-2*, Science **371** (2021), no. 6526. DOI 10.1126/science.abe2424.

[39] S. Thurner, P. Klimek, and R. Hanel, *A network-based explanation of why most COVID-19 infection curves are linear*, Proceedings of the National Academy of Sciences of the United States of America **1** (2020), no. 37, S. 22684–22689, DOI 10.1073/pnas.2010398117.

[40] M. Varela, *Los hospitales de Ciudad de México bordean el colapso por la segunda ola*, `https://elpais.com/mexico/2021-01-10/los-hospitales-de-ciudad-de-mexico-bordean-el-colapso-por-la-segunda-ola.html`. Version: January 2021.

[41] F. Wong and J. J. Collins, *Evidence that coronavirus superspreading is fat-tailed*, Proceedings of the National Academy of Sciences (2020), S. 202018490. DOI 10.1073/pnas.2018490117.

[42] J. T. Wu, K. Leung, and G. M. Leung, *Nowcasting and forecasting the potential domestic and international spread of the 2019-nCoV outbreak originating in Wuhan, China : a modelling study*, The Lancet S0140-6736 (2020), no. 20, S. 30260–9. DOI 10.1016/S0140–6736(20)30260–9.

[43] Y. Zhang, Y. Li, L. Wang, M. Li, and X. Zhou, *Evaluating transmission heterogeneity and super-spreading event of COVID-19 in a metropolis of China*, International Journal of Environmental Research and Public Health **17** (2020). DOI 10.3390/ijerph17103705.

Departamento de Matematicas, Universidad de Sonora, Sonora, 83000, Mexico
Email address: adrian.acuna@unison.mx

Escuela Nacional de Estudios Superiores Unidad Juriquilla, Universidad Nacional Autónoma de México, Queretaro, 76230, Mexico
Email address: msantana@unam.mx

Instituto de Investigaciones en Matematicas Aplicadas y en Sistemas UNAM, Ciudad de México, 01000, Mexico
Email address: carloserwin@sigma.iimas.unam.mx

Instituto de Investigaciones en Matematicas Aplicadas y en Sistemas UNAM, Ciudad de México, 01000, Mexico
Email address: ramses@sigma.iimas.unam.mx

Instituto de Matematicas UNAM, Juriquilla, 76230, Mexico
Email address: jx.velasco@im.unam.mx

Contemporary Mathematics
Volume **793**, 2024
https://doi.org/10.1090/conm/793/15881

Mathematical analysis and a nonstandard scheme for a model of the immune response against COVID-19

Gustavo Morais Rodrigues Costa, Marcelo Lobosco, Matthias Ehrhardt, and Ruy Freitas Reis

This paper is dedicated in honor of Ronald E. Mickens' 80th birthday.

ABSTRACT. In this work, we consider a compartmental model to describe the immune response to SARS-CoV-2. The model considers the primary cells involved in the body's immune response, antigen-presenting cells, CD4+ and CD8+ T cells, B cells, IgM and IgG antibodies, proinflammatory cytokines, and infected cells of the immune system. The resulting system consists of 15 ordinary differential equations (ODEs) with 38 parameters.

For the numerical solution of this rather large ODE system, we develop a special non-standard finite difference (NSFD) scheme that preserves the positivity of the solutions.

1. Introduction

The first case of the coronavirus disease (COVID-19) was detected in Wuhan, China, in December 2019. The virus rapidly spread, and by February 2020, there were cases of this new disease in 8 countries. Deeply concerned about the alarming spread of the disease, its severity, and the lack of action, the World Health Organization (WHO) declared COVID-19 a global pandemic in March 2020 [**43**].

By December 22, 2022, COVID-19 had already infected more than 655 million people worldwide, 244 million in Europe, 100 million only in the United States of America, and 36 million in Brazil, as shown in Figure 2A. On the other hand, the number of worldwide deaths at the same time was more than 6.67 million, 2 million in Europe, 1.09 million in the United States of America, and 692,652 in Brazil, as shown in Figure 2B. Even after the first dose of vaccine was administered in December 2020 [**31**], more than 3 million people died from COVID-19 infection.

2020 *Mathematics Subject Classification.* Primary 65L05, 92C60.

Key words and phrases. Nonstandard finite difference scheme, COVID-19, compartmental models, computational immunology, differential evolution.

This work has been supported by Universidade Federal de Juiz de Fora (UFJF) through the scholarship BIC/UFJF; by Conselho Nacional de Desenvolvimento Científico e Tecnológico (CNPq) - Brazil grant number 308745/2021-3 and 423278/ 2021-5; by Fundação de Amparo à Pesquisa do Estado de Minas Gerais (FAPEMIG) - Brazil CEX APQ 02830/17, APQ-01226-21, and APQ-02513-22 APQ-02513-22; FINEP - Brazil, grant CLUSTERSGI - FINEP 0062/22 - SOS EQUIPAMENTOS..

Figure 1 illustrates the relationship between the population size of each country, the number of infected people, and the number of deaths.

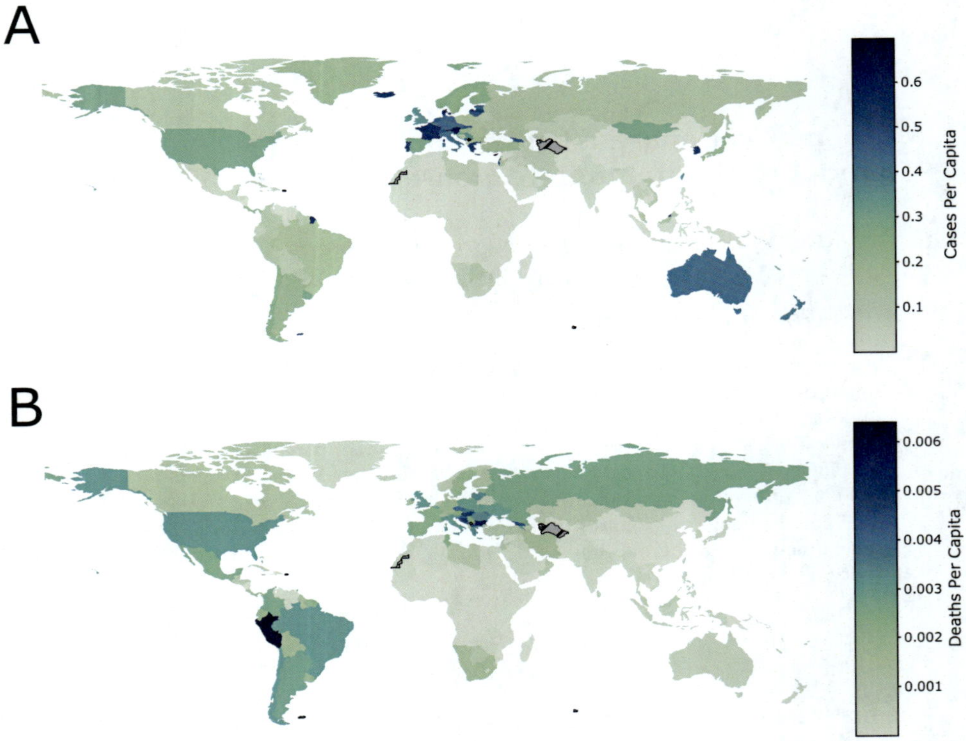

FIGURE 1. Choropleth map showing the incidence of COVID-19 around the globe. Panel A presents the number of COVID-19 cases compared with the total population, while Panel B presents the number of deaths as complications of COVID-19 compared with the total population. Crosshatched black and grey areas identify locations where no data is available.

Even though the pandemic started more than three years ago, there are still open questions regarding COVID-19. Some of these questions can be answered with the help of mathematical models and *in silico* trials. The numerical results obtained from these trials might reveal information that may lead the scientific community to understand this disease. A considerable amount of work makes use of mathematical and computational modeling to understand how the *human immune system* (HIS) works by different means, such as ordinary differential equations (ODEs) [**1**, **2**, **7**, **18**, **24**, **30**, **42**], partial differential equations (PDEs) [**10**, **13**, **25**, **27**, **40**], and stochastic methods [**8**, **44**].

This work uses a mathematical model to represent the immune response against SARS-CoV-2 [**30**]. This model considers the primary cells involved in the body's immune response, antigen-presenting cells, CD4+ and CD8+ T cells, B cells, IgM and IgG antibodies, pro-inflammatory cytokines, and infected immune system cells. The model is composed of 15 ODEs and has 38 parameters. In this work, we

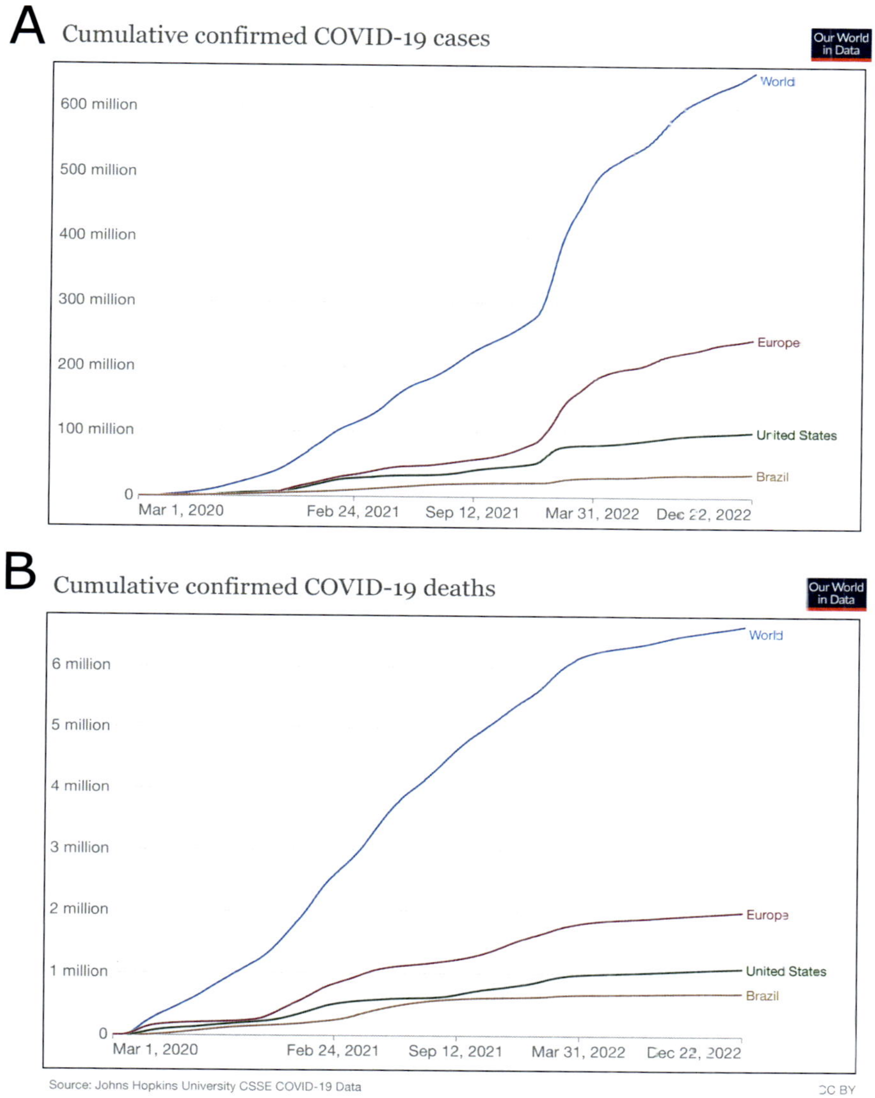

FIGURE 2. Evolution of cumulative confirmed COVID-19 cases (A) and confirmed COVID-19 deaths (B). A) Due to limited testing, the number of confirmed cases is lower than the true number of COVID-19 infections [**31**]. B) Due to varying protocols and challenges in defining the cause of death, the number of confirmed deaths may not accurately represent the true death toll of COVID-19.

adjusted and validated the model using a new set of cohort data [**46**] referring to CD4+ (Effector T Helper Cells), CD8+ (Effector T Killer Cells), and the viremia obtained from COVID-19 patients. *Differential Evolution* (DE) [**39**] was used to adjust the model parameters.

Additionally, sensitivity analysis via Sobol main indexes [**37**] made it possible to gather further information on the immune response to SARS-CoV-2. Furthermore, we employ the *Non-Standard Finite Difference* (NSFD) approach to designing a numerical method specially designed for our system. To the best of our knowledge, this type of numerical scheme has not been used to solve large ODEs systems such as this one. So, this study is interested in demonstrating whether this numerical scheme is able to solve large systems of ODEs.

This work is organized as follows. Section 2 presents a brief introduction to the Immunological System to help achieve a better understanding of the matter at hand. Section 3 presents the mathematical model (3.1), which was solved using the NSFD Scheme presented in Section 3.2, adjusted to the data extracted as in Section 3.3 with the use of DE (Section 3.4), understood with the help of Sensitivity Analysis presented in Section 3.5. Section 4 presents the results obtained from our study, which we delve into further in Section 5. Finally, the last section presents our conclusions and plans for future work.

2. Immunological background

The *human immune system* (HIS) consists of several mechanisms that work together to ensure the health of the human body. Some of its elements act directly on the immunological response to antigens, such as T-killer cells (CD8+). With the help of antigen-presenting cells (APCs) and cytokines, these cells can recognize and eliminate various threats that enter our bodies, such as viruses and bacteria.

The HIS defense against antigens is made up of two main parts: the innate and the adaptive immune systems. The innate system acts as the first layer of defense against any type of pathogen, while the adaptive system is tailored to deal with specific types of antigens, including viruses. The adaptive defense system can evolve throughout our lives to protect us against almost any invader.

When APCs and other stimuli activate T helper cells, a cascade of events is triggered, leading to the activation of B and T-killer cells. The functions of T-killer cells include identifying and destroying virus-infected cells. A distinct strategy to kill viruses involves the production of antibodies. Antibodies opzonize and prepare viruses to be eliminated by phagocytes. Mature B cells are responsible for specific antibody production. Thus, the receptors on the various B cells are so diverse that our B cells can probably recognize any organic molecule that exists and produce specific antibodies for it. After the infection is under control, some B cells differentiate into memory B cells to maintain information about how to repel future invasions of the same antigen. The necessary amount of memory B cells is stored for this purpose. These memory B cells can be reactivated faster to produce specific antibodies against the same invader.

In this work, we focus on the adaptive immune system, working on a model that describes antigen-presenting cells, the T cells, and some types of antibodies (IgG and IgM) besides the virus and the cytokines. The original model [**3**] was adapted to describe the specific response to coronavirus infection [**30**]. A difference is that this paper includes CD4+ and CD8+ data [**46**] to adjust and analyze the results of the *in silico* experiments. The focus on this population can be explained by the

T cells' role in understanding the severity of COVID-19 cases since lymphopenia is an indicator of SARS-CoV-2 infection and a predictive of disease progression.

3. Material and methods

3.1. Mathematical model. In this work, we use a model composed of 15 ODE [**30**]s, where each one represents an element involved in the immune system response to SARS-CoV-2 infection: virus (I); naive (T_{hn}) and effector (T_{he}) T helper (CD4+) cells; naive (T_{kn}) and effector (T_{ke}) T Killer (CD8+) cells; B cells (B); short- (P_s) and long-lived (P_l) plasma cells; B memory cells (B_m); IgM (Ig_M) and IgG (Ig_G) antibodies, and cytokines (C).

The first equation represents SARS-CoV-2 (V) usage of the cell machinery to replicate itself and spread to other cells, as well as its interaction with the immune system [**6**, **16**, **35**]. The first term of (3.16a), $\pi_v V$, expresses the replication of the virus since pi_v represents the growth rate of the SARS-CoV-2. The other terms of (3.16a) indicate the elimination rate of the virus by the immune system. The antibodies can opsonize the virus to facilitate the binding of the virus to receptor molecules present in phagocytes [**23**]. This phenomenon is denoted by the second and the third terms of (3.16a): the term $k_{v1}VI_{gG}$ illustrates the death of the virus due to its opsonization by I_{gG}, and the term $k_{v1}VI_{gM}$ due to its opsonization by I_{gM}, respectively, being k_{v1} the rate at which an antibody opsonizes a virus. The term $k_{v2}VT_{ke}$, where k_{v2} is the clearance rate, represents viral clearance due to the induction of apoptosis of cells infected by the SARS-CoV-2 virus. Finally, $k_{v3}VA_{pm}$depicts the viral clearance by mature APCs, such as macrophages, where the term k_{v3} is the clearance rate.

$$\frac{d}{dt}V = \pi_v V - k_{v1}VI_{gG} - k_{v1}VI_{gM} - k_{v2}VT_{ke} - k_{v3}VA_{pm}. \tag{3.1}$$

Antigen-presenting cells (APC) are found in two stages, naive and mature [**22**]. The second and third equations represent these two stages of the APCs, naive (A_p) and mature (A_{pm}). APCs are those cells specialized in presenting an antigen to a T-cell. Distinct immune cells can act as APC. In this work, we consider that the main APCs are macrophages. In (3.2), the naive APCs' homeostasis and activation are described by the first and second terms, respectively. Pro-inflammatory cytokines influence the homeostatic balance of the APCs [**22**], and, for this reason, in the first term, $\alpha_{ap}(C+1)(A_{p0} - A_p)$, α_{ap} represents the homeostasis rate and $(C+1)$ represents the influence of pro-inflammatory cytokines in the homeostasis. The term $\beta_{ap}A_p\frac{c_{ap1}V}{c_{ap2}+V}$ denotes the conversion of immature APCs into mature ones which explains why the same term appears in (3.3) with a positive sign. Moreover, this last term uses a function to model growth combined with the saturation phenomenon [**14**].

$$\frac{d}{dt}A_p = \alpha_{ap}(C+1)(A_{p0} - A_p) - \beta_{ap}A_p\frac{c_{ap1}V}{c_{ap2}+V}. \tag{3.2}$$

In (3.3), which represents mature APCs, $\beta_{apm}A_{pm}V$ denotes A_{pm} infection by the SARS-CoV-2 virus, where β_{apm} is the infection rate. In other words, this term considers the hypothesis that SARS-CoV-2 also infects immune cells [**9**, **15**], thus causing the production of several pro-inflammatory cytokines (mainly IL-6), which may cause a Cytokine Release Syndrome (CRS) or cytokine storm. The third term,

$\delta_{apm}A_{pm}$, means the natural decay of the mature APCs, where δ_{apm} is the decay rate.

$$\frac{d}{dt}A_{pm} = \beta_{ap}A_p\frac{c_{ap1}V}{c_{ap2}+V} - \beta_{apm}A_{pm}V - \delta_{apm}A_{pm}. \tag{3.3}$$

The dynamics of the infected immune system cells are represented by (3.4). The first term, $\beta_{apm}A_{pm}V$, represents A_{pm} infection, and the second term, $\beta_{tke}T_{ke}V$, represents CD8+ T cell infection. The infection rates are, respectively, β_{apm} and β_{tk}. Infected cells die with a rate δ_{apm}.

$$\frac{d}{dt}I = \beta_{apm}A_{pm}V + \beta_{tke}T_{ke}V - \delta_{apm}I. \tag{3.4}$$

Equation (3.5) represents the population of naive CD4+ T cells (T_{hn}). The term $\alpha_{th}(T_{hn0} - T_{hn})$ represents the homeostasis of CD4+ T cells, where α_{th} is the homeostasis rate. APCs are responsible for activating naive CD4+ T cells [**22**]. The term $\beta_{th}A_{pm}T_{hn}$ denotes the activation of naive CD4+ T cells, where β_{th} is the activation rate.

$$\frac{d}{dt}T_{hn} = \alpha_{th}(T_{hn0} - T_{hn}) - \beta_{th}A_{pm}T_{hn}. \tag{3.5}$$

Equation (3.6) represents the effector CD4+ T cell population (T_{he}). The term $\pi_{th}A_{pm}T_{he}$ represents the proliferation of effector CD4+ T cells, where π_{th} is the proliferation rate. The term $\delta_{th}T_{he}$ represents the natural death of these cells, with δ_{th} representing its death rate.

$$\frac{d}{dt}T_{he} = \beta_{th}A_{pm}T_{hn} + \pi_{th}A_{pm}T_{he} - \delta_{th}T_{he}. \tag{3.6}$$

Equations (3.7) and (3.8) represent the population of naive (T_{kn}) and effector (T_{ke}) CD8+ T cells, respectively. In (3.7), the naive CD8+ T cell's homeostasis and activation are described by the first and second terms, respectively. In the first term, $\alpha_{tk}(C+1)(T_{kn0} - T_{kn})$, α_{tk} represents the homeostasis rate. The term $\beta_{tk}(C+1)A_{pm}T_{kn}$ denotes the activation of naive CD8+ T cells, where β_{tk} is the activation rate. As one can see, pro-inflammatory cytokines (C) influence the homeostatic balance and activation of naive CD8+ T cells.

$$\frac{d}{dt}T_{kn} = \alpha_{tk}(C+1)(T_{kn0} - T_{kn}) - \beta_{tk}(C+1)A_{pm}T_{kn}. \tag{3.7}$$

In (3.8), the term $\pi_{tk}A_{pm}T_{ke}$ represents the proliferation of effector CD8+ T cells. The terms $\beta_{tke}T_{ke}V$ and $\delta_{tk}T_{ke}$ represent the infection and death of effector CD8+ T cells, respectively.

$$\frac{d}{dt}T_{ke} = \beta_{tk}(C+1)A_{pm}T_{kn} + \pi_{tk}A_{pm}T_{ke} - \beta_{tke}T_{ke}V - \delta_{tk}T_{ke}. \tag{3.8}$$

Equation (3.9) represents both naive and effector B cells (B). These populations were considered together in order to simplify the model. The term $\alpha_b(B_0 - B)$ represents the B cell's homeostasis, where α_b is the homeostasis rate. The terms $\pi_{b1}VB$ and $\pi_{b2}T_{he}B$ represent the proliferation of B cells activated by the T-cell independent and T-cell dependent mechanisms [**38**], respectively. The terms $\beta_{ps}A_{pm}B$, $\beta_{pl}T_{he}B$ and $\beta_{bm}T_{he}B$ denote the differentiation of active B cells into short-lived plasma cells, long-lived plasma cells, and memory B cells, respectively. The activation rates are respectively given by β_{ps}, β_{pl} and β_{bm}.

$$\frac{d}{dt}B = \alpha_b(B_0 - B) + \pi_{b1}VB + \pi_{b2}T_{he}B - \beta_{ps}A_{pm}B \\ -\beta_{pl}T_{he}B - \beta_{bm}T_{he}B. \tag{3.9}$$

The dynamics of memory B Cells (B_m) is depicted by (3.10). The logistic growth of memory B cells is represented by the term $\pi_{bm1}B_m\left(1 - \frac{B_m}{\pi_{bm2}}\right)$, *i.e.*, there is a limit to this growth [**4**], being π_{bm1} the limits of the growth rate, and pi_{bm2} of the growth per se.

$$\frac{d}{dt}B_m = \beta_{bm}T_{he}B + \pi_{bm1}B_m\left(1 - \frac{B_m}{\pi_{bm2}}\right) - \gamma_{bm}B_m. \tag{3.10}$$

Equation (3.11) represents the short-lived plasma cells (P_s) [**38**]. The term $\delta_{ps}P_s$ denotes the natural decay of short-lived plasma cells, where δ_{ps} is the decay rate.

$$\frac{d}{dt}P_s = \beta_{ps}A_{pm}B - \delta_{ps}P_s. \tag{3.11}$$

Long-lived plasma cells (P_l) are represented by (3.12). The second term, $\delta_{pl}P_l$, describes the natural decay of long-lived plasma cells, where δ_{pl} is the decay rate. The term $\gamma_{bm}B_m$ depicts the replenishing of P_l by memory B cells, with γ_{bm} representing the production rate.

$$\frac{d}{dt}P_l = \beta_{pl}T_{he}B - \delta_{pl}P_l + \gamma_{bm}B_m. \tag{3.12}$$

Equations (3.13) and (3.14) describe the generation of antibodies. The first term of each equation, $\pi_{ps}P_s$ and $\pi_{pl}P_l$, represent the production of antibodies by their correspondent plasma cells, short-lived and long-lived, respectively. Both π_{ps} and π_{pl} denote the production rates, respectively. The second terms, $\delta_{am}I_{gM}$ and $\delta_{ag}I_{gG}$, describe the natural decay of IgM and IgG antibodies, respectively, with δ_{am} and δ_{ag} representing the decay rate.

$$\frac{d}{dt}I_{gM} = \pi_{ps}P_s - \delta_{am}I_{gM}. \tag{3.13}$$

$$\frac{d}{dt}I_{gG} = \pi_{pl}P_l - \delta_{ag}I_{gG}. \tag{3.14}$$

Finally, the pro-inflammatory cytokine dynamics are described by (3.15). The first term of this equation, $\pi_{c_{apm}}A_{pm}$, depicts the production of cytokines by A_{pm}, being $\pi_{c_{apm}}$ the rate of production. The next term describes the cytokine production by infected immune cells, with π_{ci} representing the production rate. The term $\pi_{c_{tke}}T_{ke}$ represents the cytokine production by T_{ke} cells, where $\pi_{c_{tke}}$ denotes the production rate. The last term depicts the natural decay of cytokines, with δ_c describing the decay rate.

$$\frac{d}{dt}C = \pi_{c_{apm}}A_{pm} + \pi_{ci}I + \pi_{c_{tke}}T_{ke} - \delta_c C. \tag{3.15}$$

For simplification purposes, the model used in this work [**30**] did not include infected and non-infected epithelial cells, as it would have more constants to adjust and without the availability of data to validate these cell populations over time. The quality of the results, especially those related to the virus population, is unaffected by the use of implicit antigen replication, as previous works have demonstrated [**3**, **4**, **26**, **29**]. It is also assumed that the virus is located in the tissue, where it can infect matured immune cells. The naive cells are activated in the bloodstream (APCs), lymph nodes (CD4 and CD8), or after leaving both [**22**, **23**, **38**]. It is also assumed that virus is mainly produced by the epithelial tissue, despite some pieces of evidence showing that infected alveolar macrophages can also assist in virus replication [**15**]. The phagocytic activity by infected cells is assumed to be null in this model. The model used in this work considers that infected cells continue the production of pro-inflammatory cytokines. Finally, the model implicitly accounts for the effects of different pro-inflammatory cytokines.

3.2. The nonstandard scheme. In addition to properties such as stability and consistency, qualitative properties, such as the preservation of positivity, are also important for biological models like the one described in this work. This is exactly where nonstandard finite difference (NSFD) methods come into play to meet these requirements.

NSFD schemes for the numerical integration of ODEs were pioneered by Mickens [**20**]. This section presents an NSFD method to solve the resulting ODE system presented in (3.16). These specialized schemes can be regarded as discrete models of the ODE systems. In this direction, they are designed in order to preserve certain properties such as the positivity of the analytic solution of the ODE system. These schemes can be reformulated in an efficient explicit way, and as a byproduct of the positivity preserving property, the NSFD methods are stable.

For ease of presentation, we rewrite the equations described so far in one ODE system, as shown below.

$$\frac{d}{dt}V = \pi_v V - k_{v1}VI_{gG} - k_{v1}VI_{gM} - k_{v2}VT_{ke} - k_{v3}VA_{pm}, \tag{3.16a}$$

$$\frac{d}{dt}A_p = \alpha_{ap}(C+1)(A_{p0} - A_p) - \beta_{ap}A_p\frac{c_{ap1}V}{c_{ap2}+V}, \tag{3.16b}$$

$$\frac{d}{dt}A_{pm} = \beta_{ap}A_p\frac{c_{ap1}V}{c_{ap2}+V} - \beta_{apm}A_{pm}V - \delta_{apm}A_{pm}, \tag{3.16c}$$

$$\frac{d}{dt}I = \beta_{apm}A_{pm}V + \beta_{tke}VT_{ke} - \delta_{apm}I, \tag{3.16d}$$

$$\frac{d}{dt}T_{hn} = \alpha_{th}(T_{hn0} - T_{hn}) - \beta_{th}A_{pm}T_{hn}, \tag{3.16e}$$

$$\frac{d}{dt}T_{he} = \beta_{th}A_{pm}T_{hn} + \pi_{th}A_{pm}T_{he} - \delta_{th}T_{he}, \tag{3.16f}$$

$$\frac{d}{dt}T_{kn} = \alpha_{tk}(C+1)(T_{kn0} - T_{kn}) - \beta_{tk}(C+1)A_{pm}T_{kn}, \tag{3.16g}$$

$$\frac{d}{dt}T_{ke} = \beta_{tk}(C+1)A_{pm}T_{kn} + \pi_{tk}A_{pm}T_{ke} - \beta_{tke}VT_{ke} - \delta_{tk}T_{ke}, \tag{3.16h}$$

$$\begin{aligned}\frac{d}{dt}B = {}& \alpha_b(B_0 - B) + \pi_{b1}VB + \pi_{b2}T_{he}B - \beta_{ps}A_{pm}B \\ & - \beta_{pl}T_{he}B - \beta_{bm}T_{he}B,\end{aligned} \tag{3.16i}$$

$$\frac{d}{dt}P_s = \beta_{ps}A_{pm}B - \delta_{ps}P_s, \tag{3.16j}$$

$$\frac{d}{dt}B_m = \beta_{bm}T_{he}B + \pi_{bm1}B_m\Big(1 - \frac{B_m}{\pi_{bm2}}\Big) - \gamma_{bm}B_m, \tag{3.16k}$$

$$\frac{d}{dt}P_l = \beta_{pl}T_{he}B - \delta_{pl}P_l + \gamma_{bm}B_m, \tag{3.16l}$$

$$\frac{d}{dt}I_{gM} = \pi_{ps}P_s - \delta_{am}I_{gM}, \tag{3.16m}$$

$$\frac{d}{dt}I_{gG} = \pi_{pl}P_l - \delta_{ag}I_{gG}, \tag{3.16n}$$

$$\frac{d}{dt}C = \pi_{c_{apm}}A_{pm} + \pi_{ci}I + \pi_{c_{tke}}T_{ke} - \delta_c C. \tag{3.16o}$$

One can easily observe that, for positive parameters and positive initial data, the solution of (3.16) remains positive all the time. We will call this the "positivity property".

NSFD schemes go back to a paper by Mickens published in 1989 [**21**]. Their structural properties come from investigations of special groups of ODEs for which no exact finite difference schemes are available. In contrast to conventional finite difference methods, NSFD schemes consider not only stability and consistency order, but also pay attention to qualitative properties, *i.e.*, how well does the discrete model (the NSFD scheme) model the most important properties of the underlying continuous model?

We introduce a time step $h > 0$ and consider temporal grid points $t_n := nh$, $n = 0, 1, \ldots$. In the sequel, we denote by V^n the approximation of V at the grid point t_n, and similarly for the other components in (3.16). In NSFD schemes, derivatives have to be modeled by proper discrete analogs, *i.e.* nonstandard difference quotients of the form, cf. [**20**]

$$\frac{\mathrm{d}}{\mathrm{d}t}V(t_n) \to \frac{V^{n+1} - \psi(h)V^n}{\phi(h)}, \tag{3.17}$$

where $\psi(h) = 1 + \mathcal{O}(h)$ and the denominator function $\phi(h) = h + \mathcal{O}(h^2)$. Using this rather general time discretization in NSFD schemes, our aim is to preserve the positivity of the solution, which is the most important structural property. We choose here $\psi(h) = 1$ and $\phi(h) = h$ and focus on the discretization of the right-hand side of the ODE system given by (3.16).

In an NSFD scheme, the nonlinear terms are approximated in a non-local way, *e.g.* by a suitable function of several grid points, like $V^2(t_n) \approx V^n V^{n+1}$ or $V^3(t_n) \approx (V^n)^2 V^{n+1}$. First, we write down the nonstandard discretization of the 15 components of the ODE system:

$$\begin{aligned}\frac{V^{n+1} - V^n}{\Phi(h)} &= \Pi_v V^{n+1} - k_{v1}I_{gG}^n V^{n+1} - k_{v1}I_{gM}^n V^{n+1} - k_{v2}T_{ke}^n V^{n+1}\\ &\quad - k_{v3}A_{pm}^n V^{n+1},\end{aligned} \tag{3.18a}$$

$$\frac{A_p^{n+1} - A_p^n}{\Phi(h)} = \alpha_{ap}(C+1)\big(A_{p0} - A_p^{n+1}\big) - \beta_{ap}\frac{c_{ap1}V^{n+1}}{c_{ap2} + V^{n+1}}A_p^{n+1}, \tag{3.18b}$$

$$\frac{A_{pm}^{n+1} - A_{pm}^n}{\Phi(h)} = \beta_{ap}\frac{c_{ap1}V^{n+1}}{c_{ap2} + V^{n+1}}A_p^{n+1} - \beta_{apm}V^{n+1}A_{pm}^{n+1} - \delta_{apm}A_{pm}^{n+1}, \tag{3.18c}$$

$$\frac{I^{n+1} - I^n}{\Phi(h)} = \beta_{apm}A_{pm}^{n+1}V^{n+1} + \beta_{tke}T_{ke}^n V^{n+1} - \delta_{apm}I^{n+1}, \tag{3.18d}$$

$$\frac{T_{hn}^{n+1} - T_{hn}^{n}}{\Phi(h)} = \alpha_{th}(T_{hn0} - T_{hn}^{n+1}) - \beta_{th} A_{pm}^{n+1} T_{hn}^{n+1}, \tag{3.18e}$$

$$\frac{T_{he}^{n+1} - T_{he}^{n}}{\Phi(h)} = \beta_{th} A_{pm}^{n+1} T_{hn}^{n+1} + \Pi_{th} A_{pm}^{n+1} T_{he}^{n+1} - \delta_{th} T_{he}^{n+1}, \tag{3.18f}$$

$$\frac{T_{kn}^{n+1} - T_{kn}^{n}}{\Phi(h)} = \alpha_{tk}(C+1)\big(T_{kn0} - T_{kn}^{n+1}\big) - \beta_{tk}(C+1) A_{pm}^{n+1} T_{kn}^{n+1}, \tag{3.18g}$$

$$\begin{aligned}\frac{T_{ke}^{n+1} - T_{ke}^{n}}{\Phi(h)} &= \beta_{tk} A_{pm}^{n+1} T_{kn}^{n+1} + \Pi_{tk} A_{pm}^{n+1} T_{ke}^{n+1} - \beta_{tke} V^{n+1} T_{ke}^{n+1} \\ &\quad - \delta_{tk} T_{ke}^{n+1},\end{aligned} \tag{3.18h}$$

$$\begin{aligned}\frac{B^{n+1} - B^{n}}{\Phi(h)} &= \alpha_b(B_0 - B^{n+1}) + \Pi_{b1} V^{n+1} B^{n+1} + \Pi_{b2} T_{he}^{n+1} B^{n+1} \\ &\quad - \beta_{ps} A_{pm}^{n+1} B^{n+1} - \beta_{pl} T_{he}^{n+1} B^{n+1} - \beta_{pm} T_{he}^{n+1} B^{n+1},\end{aligned} \tag{3.18i}$$

$$\frac{P_{s}^{n+1} - P_{s}^{n}}{\Phi(h)} = \beta_{ps} A_{pm}^{n+1} B^{n+1} - \delta_{ps} P_{s}^{n+1}, \tag{3.18j}$$

$$\frac{B_{m}^{n+1} - B_{m}^{n}}{\Phi(h)} = \beta_{bm} T_{he}^{n+1} B^{n+1} + \Pi_{bm1} B_{m}^{n+1}\left(1 - \frac{B_m^n}{\Pi_{bm2}}\right) - \gamma_{bm} B_{m}^{n+1}, \tag{3.18k}$$

$$\frac{P_{l}^{n+1} - P_{l}^{n}}{\Phi(h)} = \beta_{pe} T_{he}^{n+1} B^{n+1} - \delta_{pl} P_{l}^{n+1} + \gamma_{bm} B_{m}^{n+1}, \tag{3.18l}$$

$$\frac{I_{gM}^{n+1} - I_{gM}^{n}}{\Phi(h)} = \Pi_{ps} P_{s}^{n+1} - \delta_{am} I_{gM}^{n+1}, \tag{3.18m}$$

$$\frac{I_{gG}^{n+1} - I_{gG}^{n}}{\Phi(h)} = \Pi_{pl} P_{l}^{n+1} - \delta_{ag} I_{gG}^{n+1}, \tag{3.18n}$$

$$\frac{C^{n+1} - C^{n}}{\Phi(h)} = \Pi_{c_{apm}} A_{pm}^{n+1} + \Pi_{ci} I^{n+1} + \Pi_{c_{tke}} T_{ke}^{n+1} - \delta_c C^{n+1}, \tag{3.18o}$$

for $n = 0, 1, \ldots$ and a denominator function $\phi(h) = h$. We briefly comment on the chosen discretizations of the nonlinear terms. For example, in the first line of (3.18), we have discretized the quadratic term $A_{pm}(t)V(t)$ by $A_{pm}^n V^{n+1}$ rather than, say, $A_{pm}^n V^n$ or $A_{pm}^{n+1} V^{n+1}$. The rule is that exactly one factor of the variable appearing in the time derivative (here V) must be taken at the new time level $n+1$. This is needed to obtain the positivity-preserving property. In order to not destroy the explicit sequential evaluation, all other variables are taken from the previous time level, unless they are already known from a previous step, like $V^{n+1}A_{pm}^{n+1}$ in the third line of (3.18).

While the NSFD discretization presented in (3.18) of the ODE system defined in (3.16) is formally implicit, it can be easily rearranged to obtain a scheme that can be evaluated sequentially in an explicit and thus efficient way:

$$V^{n+1} = \frac{V^n}{1 + \Phi(h)\big[k_{v1} I_{gG}^n + k_{v1} I_{gM}^n + k_{v2} T_{ke}^n + k_{v3} A_{pm}^n - \Pi_v\big]}, \tag{3.19a}$$

$$A_p^{n+1} = \frac{A_p^n + \Phi(h)\alpha_{ap}(C+1)A_{p0}}{1 + \Phi(h)\big[\alpha_{ap}(C+1) + \beta_{ap}\frac{c_{ap1}V^{n+1}}{c_{ap2}+V^{n+1}}\big]}, \tag{3.19b}$$

$$(3.19c)\quad A_{pm}^{n+1} = \frac{A_{pm}^n + \Phi(h)\beta_{ap}\frac{c_{ap1}V^{n+1}}{c_{ap2}+V^{n+1}}A_p^{n+1}}{1+\Phi(h)\left[\beta_{apm}V^{n+1}+\delta_{apm}\right]},$$

$$(3.19d)\quad I^{n+1} = \frac{I^n + \Phi(h)\left[\beta_{apm}A_{pm}^{n+1}+\beta_{tke}T_{ke}^n\right]V^{n+1}}{1+\Phi(h)\delta_{apm}},$$

$$(3.19e)\quad T_{hn}^{n+1} = \frac{T_{hn}^n + \Phi(h)\alpha_{th}T_{hn0}}{1+\Phi(h)\left[\alpha_{th}+\beta_{th}A_{pm}^{n+1}\right]},$$

$$(3.19f)\quad T_{he}^{n+1} = \frac{T_{he}^n + \Phi(h)\beta_{th}A_{pm}^{n+1}T_{hn}^{n+1}}{1+\Phi(h)\left[\delta_{th}-\Pi_{th}A_{pm}^{n+1}\right]},$$

$$(3.19g)\quad T_{kn}^{n+1} = \frac{T_{kn}^n + \Phi(h)\alpha_{tk}(C+1)T_{kn0}}{1+\Phi(h)(C+1)\left[\alpha_{tk}+\beta_{tk}A_{pm}^{n+1}\right]},$$

$$(3.19h)\quad T_{ke}^{n+1} = \frac{T_{ke}^n + \Phi(h)\beta_{tk}A_{pm}^{n+1}T_{kn}^{n+1}}{1+\Phi(h)\left[\delta_{tk}+\beta_{tke}V^{n+1}-\Pi_{tk}A_{pm}^{n+1}\right]},$$

$$(3.19i)\quad B^{n+1} = \frac{B^n + \Phi(h)\alpha_b B_0}{1+\Phi(h)\left[\alpha_b - \Pi_{b1}V^{n+1}+\beta_{ps}A_{pm}^{n+1}+(\beta_{pl}+\beta_{pm}-\Pi_{b2})T_{he}^{n+1}\right]},$$

$$(3.19j)\quad P_s^{n+1} = \frac{P_s^n + \Phi(h)\beta_{ps}A_{pm}^{n+1}B^{n+1}}{1+\Phi(h)\delta_{ps}},$$

$$(3.19k)\quad B_m^{n+1} = \frac{B_m^n + \Phi(h)\beta_{bm}T_{he}^{n+1}B^{n+1}}{1+\Phi(h)\left[\gamma_{bm}-\Pi_{bm1}\left(1-\frac{B_m^n}{\Pi_{bm2}}\right)\right]},$$

$$(3.19l)\quad P_l^{n+1} = \frac{P_l^n + \Phi(h)\left[\beta_{pe}T_{he}^{n+1}B^{n+1}+\gamma_{bm}B_m^{n+1}\right]}{1+\Phi(h)\delta_{pl}},$$

$$(3.19m)\quad I_{gM}^{n+1} = \frac{I_{gM}^n + \Phi(h)\Pi_{ps}P_s^{n+1}}{1+\Phi(h)\delta_{am}},$$

$$(3.19n)\quad I_{gG}^{n+1} = \frac{I_{gG}^n + \Phi(h)\Pi_{pl}P_l^{n+1}}{1+\Phi(h)\delta_{ag}},$$

$$(3.19o)\quad C^{n+1} = \frac{C^n + \Phi(h)\left[\Pi_{c_{apm}}A_{pm}^{n+1}+\Pi_{ci}I^{n+1}+\Pi_{c_{tke}}T_{ke}^{n+1}\right]}{1+\Phi(h)\delta_c},$$

for $n = 0, 1, \ldots$

The NSFD scheme presented in (3.19) is *positivity-preserving*, *i.e.*, it always produces non-negative solutions for positive data if the step size h is sufficiently small since all parameters and the denominator function are non-negative. Thus, negative values for the solution are avoided, and as a byproduct, stability with respect to the maximum norm is guaranteed, cf. [**11**].

For an application example where exactly this problem occurs and the numerical solution of the standard solver becomes negative, we refer the reader to the literature [**19**].

3.3. Data extraction. Studies available in the scientific literature on individuals infected with SARS-CoV-2 were used to evaluate the mathematical model and its numerical implementation with the NSFD scheme. In our case, viremia and CD4+ and CD8+ levels of patients who survived or died due to COVID-19 were manually extracted from the graphs on CD4+ (T_{he}) and CD8+ (T_{ke}) in Figures 2-b and 2-c of Zhang *et al.* [**46**] and the viral load of Sars-CoV-2 from Figure 2 of To *et al.* [**41**]. This manual process was supported by a tool hosted on `WebPlotDigitizer` website [**32**]. After uploading a target figure to the website, the tool requires some settings regarding axis positions and scale. Then, the desired point data on the figure can be selected to convert it into raw data. The result is a CSV file with comma-separated values containing the x- and y-axis values of each assigned point.

Once we have a coarse grid of extracted point data, we use a first-degree Lagrange interpolation polynomial to approximate the intermediate values [**5**].

3.4. Differential evolution. *Differential evolution* (DE) is a stochastic heuristic algorithm that is considered a robust strategy for global optimization [**28**]. The algorithm is inspired by natural evolution in that it features generations, selections, and mutations to better represent an individual's ability to survive in an environment. DE is thus an evolutionary algorithm and a parallel direct search method. It works in generations and represents a population with a given number of individuals, using vectors of parameters to represent each generation. In addition, each vector has a corresponding mutation rate that takes into account the synonym phenomenon occurring in nature, where some positions of the vector are generated randomly. In this way, the offspring of the population is always composed of individuals resulting from the cross between two other individuals of the previous population, also taking into account any mutation that might occur in one of the new individuals. The generation of the offspring continues until convergence, or the maximum number of iterations is reached.

In this work, we attempted to minimize the error between the CD4+, CD8+, and viremia model solution and the corresponding values from the extracted cohort data. To measure the difference between the cohort (λ) and the approximated data ($\hat{\lambda}$), we use the relative error (R_E), as follows:

$$R_E = \frac{\|\lambda - \hat{\lambda}\|_2}{\|\hat{\lambda}\|_2}. \tag{3.20}$$

To achieve the parameter estimation with DE, we use an objective function. Thus, in this work, the difference between the model results and the cohort data was minimized using the equation that takes into account not only the relative error between the approximated values of virus (V), CD4+(T_{he}) and CD8+(T_{ke}) but also the weight attributed to w_1, w_2 and w_3, respectively. DE thus performs a search for the values of the p parameters that better fit the curves considering the following objective function:

$$\min_p \big(w_1 R_E(V, \hat{V}) + w_2 R_E(T_{he}, \hat{T}_{he}) + w_3 R_E(T_{ke}, \hat{T}_{ke})\big). \tag{3.21}$$

During our study, the algorithm was improved with the results of each run, and adjusting the weight of each relative error in the objective function was a crucial factor in obtaining the best results. In the end, we found that $w_1 = 1$, $w_2 = 1$, and $w_3 = 2$ were the proper values for matching the model results with the cohort data.

3.5. Sensitivity analysis. One method of quantifying the influence of parameters in the model is to perform what is called a *sensitivity analysis* [**17**]. Once applying DE to 38 parameters in a set of 15 ordinary differential equations is no longer practical, we use Sobol main indices to understand which parameters have a greater impact on the equation we are trying to fit and include them in the optimization process performed via DE [**36**, **37**].

The method of Sobol main indices requires as input a statistical distribution of each of the parameters to be analyzed. The result of the method indicates the influence of each parameter in the selected time point, ranging from 0 (lowest value) to 1 (highest value). Moreover, once the influence of each parameter has been evaluated relatively, the sum of all parameter indices in the selected time point is equal to 1. Thus, this type of analysis can be used to quantify the influence of an uncertain input p_i on the model.

Considering a general model and its parameters:

$$G = f(p_1, \ldots, p_n), \tag{3.22}$$

where the output G is a scalar and the input or parameters are independent variables that are also randomized and described by *probability density function* (PDF) distributions since they represent the uncertainty of the system. The main idea is to decompose the variation to relate it to the contribution of each parameter.

To perform the sensitivity analysis, we use the parameters as uniform PDF distributions in the range of $[0.6p_i, 1.4p_i]$ for the CD4+ population and $[0.01p_i, 100p_i]$ for CD8+. Thus, we evaluate the influence of a particular p_i parameter at a particular moment of the model solution compared to the others.

4. Results

This section presents the results of the COVID-19 model used, which was fitted to viremia data of SARS-CoV-2, CD4+ (effector T helper cells), and CD8+ (effector T killer cells) in patients with COVID-19.

4.1. Software specifications. The fitting was done by minimizing the error between the cohort data and numerical results using the DE method. The method `differential_evolution` [**33**] from the `scipy` [**34**] package was used to perform the DE method. In addition, sensitivity analysis using the Sobol index method provides the quantitative influence of the parameters on the model results, which is valuable information for model fitting. The Sobol analysis was performed using the Chaospy tool [**12**]. All simulations were performed using Python 3.9.1. The initial conditions of each population are presented in Tab. 1, while the values if the parameters used to solve the system of ODEs are presented in Tab. 2.

4.2. Influence of parameters in the model. Figure 3 presents the results of the Sobol main indices for both effector T helper and T killer cells. Figure 3A shows that for the population of effector helper T cells, the parameters that had the greatest impact on the simulation result of CD4+ cells were c_{ap1}, β_{tk}, β_{tke}, δ_{apm}, α_{th} and π_{tk}. Figure 3B shows that c_{ap2}, δ_{apm}, β_{ap}, π_v, β_{tk} have a smaller influence on the simulation result of CD8+ cells. Considering this information, most of these parameters are included in the DE variables, where c_{ap1} and β_{tke} belong to the parameters that enter a wider range of the search space (bounds) input for DE.

4.3. Model adjustment. As shown in Figure 4A, the response to COVID-19 infection in patients with severe cases causes a decline in CD4+ around day 30. After a few days of growth, the CD4+ population begins to decline and halves within approximately 10 days. The model results were able to qualitatively capture

TABLE 1. The initial values of each population.

Pop	Value	Unit
A_p	10^6	(cells/mL)
A_{pm}	0	(cells/mL)
I	0	(cells/mL)
T_{hn}	10^6	(cells/mL)
T_{he}	0	(cells/mL)
T_{ke}	500.0×10^3	(cells/mL)
B	0	(cells/mL)
P_s	250.0×10^3	(cells/mL)
P_l	0	(cells/mL)
B_m	0	(cells/mL)
I_{gG}	0	S/CO
I_{gM}	0	S/CO
C	0	(pg/mL)

this behavior. Furthermore, when the standard deviation of the cohort data is considered, the model results fit within the range of the overall simulation.

Figure 4B shows the model results together with cohort data from the CD8+ population in severe cases of COVID-19. The data show a stable state after 40 days of COVID-19 infection. Similar to the cohort data, the fitted model results show qualitatively the same behavior. We can highlight that the CD8+ emerges after decay, but when we consider the standard deviation of the cohort data, the model results fit within the range of the whole simulation.

5. Discussion

As can be seen in Figure 4, the obtained results show that the curves are qualitatively similar and correspond to the standard deviation of the cohort data.

The insights derived from the sensitivity analysis results guide the fitting of the model in the right direction once the application of DE to so many variables (38 parameters) is ineffective in a large system of ODEs. Performing the Sobol index method followed by DE could provide us with valuable information on whether the parameters should be included in DE. Moreover, each population (CD4+ and CD8+) has its own set of parameters affecting its curves, which explains the different sets of parameters used in each of the Sobol index results shown in Figure 3.

The behavior of CD4+ and CD8+ cells is similar to the cohort data (see [**41**]), whereas their population decreases after the first symptoms of COVID-19. This behavior is justified in the scientific literature when one of the features of SARS-CoV-2 infection is a decreased number of white cells in the blood or lymphocytopenia, which is seen in both the simulation and cohort data.

Moreover, the model results presented in this study were obtained by solving the COVID-19 model using a newly developed NSFD scheme, which favors our hypothesis regarding the convergence of the method, apart from the considerable number of equations in the ODE system.

TABLE 2. The values of the parameters used to solve the system.

Parameter	Value	Unit
π_v	6.3482×10^{-1}	(day^{-1})
k_{v1}	0.0098	$(\text{day}^{-1}(\text{mlU/ml})^{-1})$
k_{v2}	2.8469×10^{-5}	$(\text{day}^{-1}(\text{cells/mL})^{-1})$
k_{v3}	0.06452	$(\text{day}^{-1}(\text{cells/mL})^{-1})$
α_{ap}	1	$(\text{day}^{-1}(\text{pg/mL})^{-1})$
β_{ap}	$3.8486x10^{-1}$	$(\text{day}^{-1}(\text{copies/mL})^{-1})$
c_{ap1}	2.4393×10^{5}	(copies/mL)
c_{ap2}	1.1730×10^{12}	(copies/mL)
δ_{apm}	7.0824×10^{-02}	(day^{-1})
β_{apm}	1.9534×10^{-02}	$(\text{day}^{-1}(\text{copies/mL})^{-1})$
β_{tke}	2.2884×10^{-05}	$(\text{day}^{-1}(\text{copies/mL})^{-1})$
α_{th}	1.629510^{-04}	(day^{-1})
β_{th}	1.7124×10^{-05}	$(\text{day}^{-1}(\text{cells/mL})^{-1})$
π_{th}	10^{-8}	$(\text{day}^{-1}(\text{cells/mL})^{-1})$
δ_{th}	1.0518×10^{-01}	(day^{-1})
α_{tk}	1	$(\text{day}^{-1}(\text{pg/mL})^{-1})$
β_{tk}	1.3354×10^{-05}	$(\text{day}^{-1}(\text{pg/mL})^{-1}(\text{cell/mL})^{-1})$
π_{tk}	10^{-8}	$(\text{day}^{-1}(\text{cells/mL})^{-1})$
δ_{tk}	5.6691×10^{-02}	(day^{-1})
α_b	3.5782	(day^{-1})
π_{b1}	8.98×10^{-5}	$(\text{day}^{-1}(\text{copies/mL})^{-1})$
π_{b2}	1.27×10^{-8}	$(\text{day}^{-1}(\text{cells/mL})^{-1})$
β_{ps}	6.00×10^{-6}	$(\text{day}^{-1}(\text{cells/mL})^{-1})$
β_{pl}	5.00×10^{-6}	$(\text{day}^{-1}(\text{cells/mL})^{-1})$
β_{bm}	1.00×10^{-6}	$(\text{day}^{-1}(\text{cells/mL})^{-1})$
δ_{ps}	2.5	(day^{-1})
δ_{pl}	0.35	(day^{-1})
γ_{bm}	0.0009	(day^{-1})
π_{bm1}	1.00×10^{5}	(day^{-1})
π_{bm2}	2500.00	(cells/mL)
π_{ps}	0.087	$(\text{day}^{-1}(\text{cells/mL})^{-1}(\text{S/CO}))$
π_{pl}	0.001	$(\text{day}^{-1}(\text{cells/mL})^{-1}(\text{S/CO}))$
δ_{am}	0.07	(day^{-1})
δ_{ag}	0.07	(day^{-1})
π_{capm}	328.0626434	$(\text{day}^{-1}(\text{pg/mL})(\text{cell/mL})^{-1})$
π_{ci}	0.0064	$(\text{day}^{-1}(\text{pg/mL})(\text{cell/mL})^{-1})$
π_{ctke}	0.01783	$(\text{day}^{-1}(\text{pg/mL})(\text{cell/mL})^{-1})$
δ_c	704.2259	(cells/mL)
V	1.7662×10^{2}	(day^{-1})

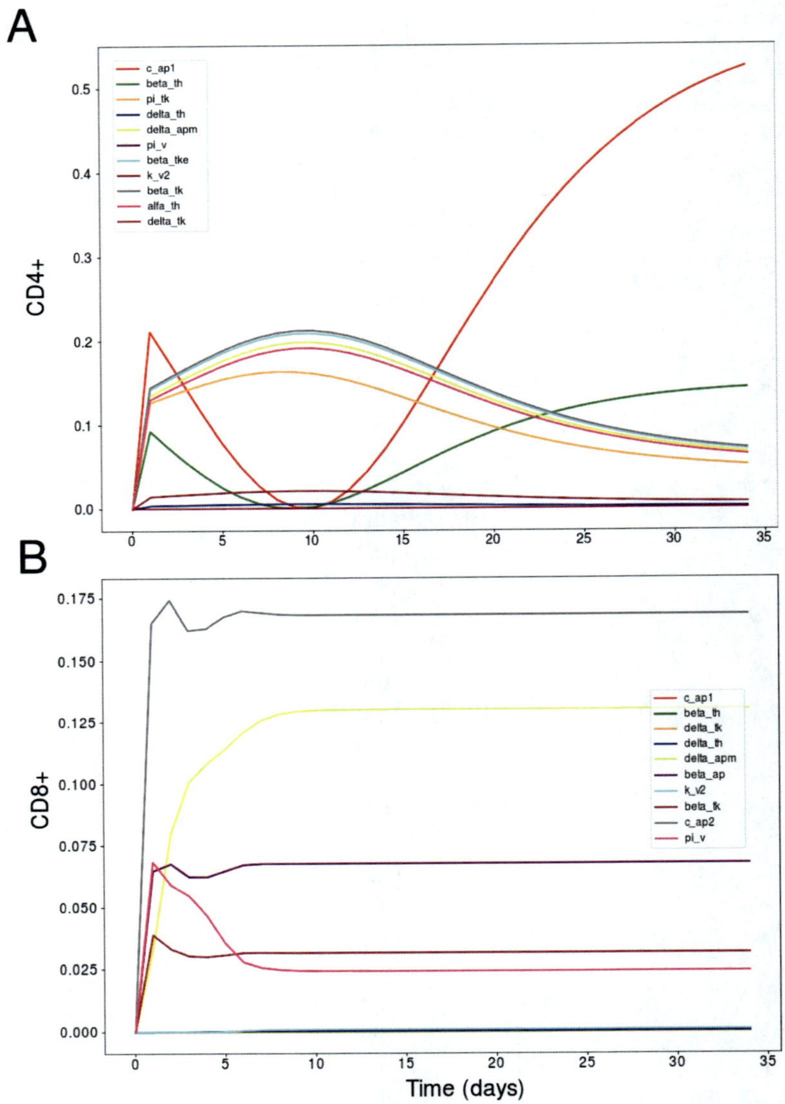

FIGURE 3. The curves in this figure illustrate the influence of each parameter included in the sensitivity analysis on the CD4+ and CD8+ populations. The vertical axis shows the relative influence of each variable on the analysis versus the influence of all included parameters on the Sobol index method. The horizontal axis represents the simulation time such that the overall picture shows the relative influence of each parameter at each time step over 35 days of SARS-CoV-2 infection.

6. Conclusions and future work

The results obtained in this study show that the mathematical model used in the numerical experiments is suitable to capture cohort data on CD4+, CD8+,

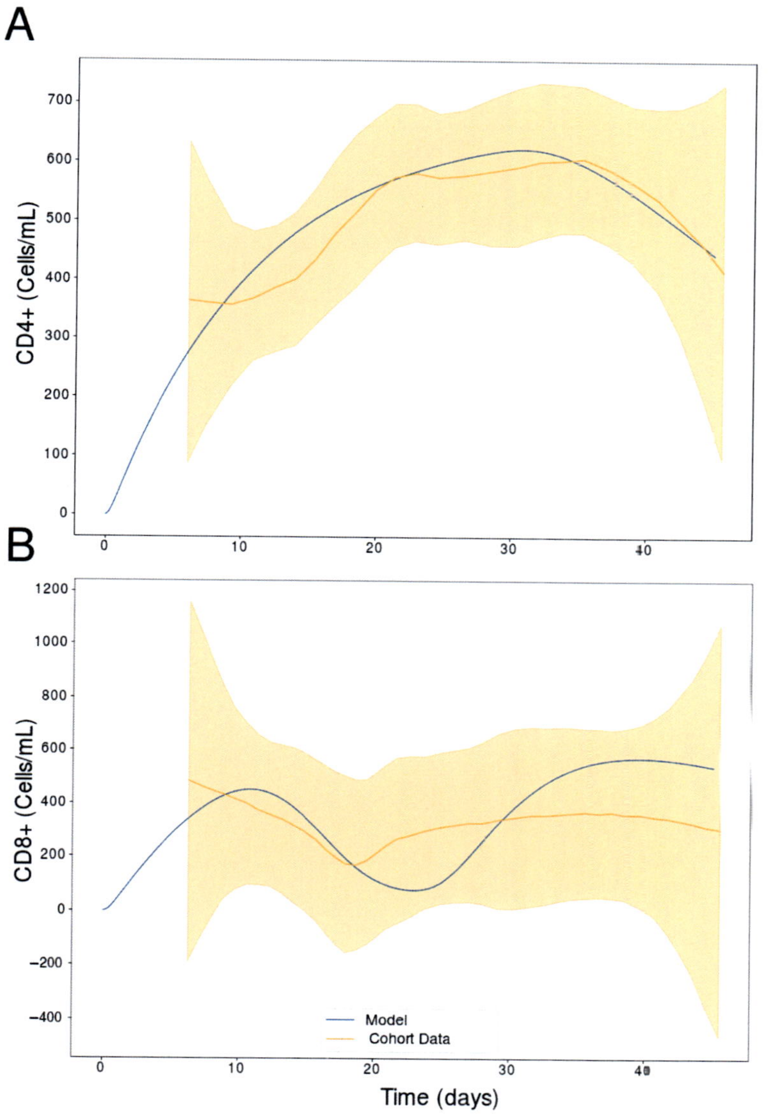

FIGURE 4. The numerical result presented in this figure shows the time variations of the concentrations of CD4+ and CD8+ cells, A and B, respectively. The solution curve, referred to as the "Model" shows the result of solving the ODE system, while the "Cohort Data" curve represents the values obtained from the literature [**45**]. The values of the parameters used to solve the system are given in Table 2, and the initial values can be found in Table 1.

and viremia behavior in severe cases of COVID-19. Moreover, to our knowledge, no NSFD scheme has ever been proposed for such a large ODE system with 15 components and 38 parameters. Thus, this work presents evidence for using NFSD schemes for a large system of equations.

Future work will investigate the robustness of this model for various HIS cells that play key roles in SARS-CoV-2 infections.

References

[1] C. T. H. Baker, G. A. Bocharov, and C. A. H. Paul, *Mathematical modelling of the interleukin-2 T-cell system: a comparative study of approaches based on ordinary and delay differential equation*, Comput. Math. Meth. Med., **1** (1997) no. 2, 117–128.

[2] C. R. B. Bonin, R. W. dos Santos, G. C. Fernandes, and M. Lobosco, *Computational modeling of the immune response to yellow fever*, J. Comput. Appl. Math. **295** (2016), 127–138, DOI 10.1016/j.cam.2015.01.020. MR3418175

[3] C. R. B. Bonin, G. C. Fernandes, R. W. dos Santos, and M. Lobosco, *A qualitatively validated mathematical-computational model of the immune response to the yellow fever vaccine*, BMC Immunol.,**19** (2018), no. 1, 15.

[4] C. R. B. Bonin, G. C. Fernandes, R. M. Menezes, L. A. B. Camacho, L. M. H. da Mota, S. M. B. de Lima, A. C. Campi-Azevedo, O. A. Martins-Filho, R. W. dos Santos, and M. Lobosco, *Quantitative validation of a yellow fever vaccine model*, 2019 IEEE International Conference on Bioinformatics and Biomedicine (BIBM), (2019), 2113–2120.

[5] J. Douglas Faires and Richard Burden, *Numerical methods*, 2nd ed., Brooks/Cole Publishing Co., Pacific Grove, CA, 1998. MR1639937

[6] M. Catanzaro, F. Fagiani, M. Racchi, E. Corsini, S. Govoni, and C. Lanni, *Immune response in COVID-19: addressing a pharmacological challenge by targeting pathways triggered by SARS-CoV-2*, Sign. Transduct. Target. Ther., **5** (2020) no. 1, 1–10.

[7] S. T. Chang, J. J. Linderman, and D. E. Kirschner, *Multiple mechanisms allow mycobacterium tuberculosis to continuously inhibit mhc class ii-mediated antigen presentation by macrophages*, Proc. Ntl. Acad. Sci. United States Amer., **102** (2005) no. 12, 4530–4535.

[8] Dennis L. Chao, Miles P. Davenport, Stephanie Forrest, and Alan S. Perelson, *A stochastic model of cytotoxic T cell responses*, J. Theoret. Biol. **228** (2004), no. 2, 227–240, DOI 10.1016/j.jtbi.2003.12.011. MR2080267

[9] G. G. Davanzo, A. C. Codo, N. S. Brunetti, et al, *SARS-CoV-2 uses cd4 to infect t helper lymphocytes*, medRxiv, 2020.

[10] B. de M. Quintela, R. W. dos Santos, and M. Lobosco, *On the Coupling of Two Models of the Human Immune Response to an Antigen*, BioMed Res. Int., (2014) 2014:e410457.

[11] M. Ehrhardt and R. E. Mickens, *A Nonstandard Finite Difference Scheme for Solving a Zika Virus Model*, Unpublished manuscript, 2017.

[12] Jonathan Feinberg and Hans Petter Langtangen, *Chaospy: an open source tool for designing methods of uncertainty quantification*, J. Comput. Sci. **11** (2015), 46–57, DOI 10.1016/j.jocs.2015.08.008. MR3435044

[13] Jennifer A. Flegg, Helen M. Byrne, Mark B. Flegg, and D. L. Sean McElwain, *Wound healing angiogenesis: the clinical implications of a simple mathematical model*, J. Theoret. Biol. **300** (2012), 309–316, DOI 10.1016/j.jtbi.2012.01.043. MR2902918

[14] S. Goutelle, M. Maurin, F. Rougier, X. Barbaut, L. Bourguignon, M. Ducher, and P. Maire, *The Hill equation: a review of its capabilities in pharmacological modelling*, Fundam. Clin. Pharmacol., **22** (2008) no. 6, 633–648.

[15] R. A. Grant, L. Morales-Nebreda, N. S. Markov, et al, *Circuits between infected macrophages and T cells in SARS-CoV-2 pneumonia*, Nature, **590** (2021) no. 7847, 635–641 .

[16] M. Hoffmann, H. Kleine-Weber, S. Schroeder, et al, *SARS-CoV-2 Cell Entry Depends on ACE2 and TMPRSS2 and Is Blocked by a Clinically Proven Protease Inhibitor*, Cell, **181** (2020) no. 2,271–280.e8.

[17] T. Homma and A. Saltelli, *Importance measures in global sensitivity analysis of nonlinear models*, Reliability Engineering & System Safety, **52** (1996) no. 1, 1–17.

[18] A. M. Jarrett, N. G. Cogan, and M. E. Shirtliff, *Modelling the interaction between the host immune response, bacterial dynamics and inflammatory damage in comparison with immunomodulation and vaccination experiments*, Math. Med. Biol., **32** (2015) no. 3, 285–306.

[19] Maghnia Hamou Maamar, Matthias Ehrhardt, and Louiza Tabharit, *A nonstandard finite difference scheme for a time-fractional model of Zika virus transmission*, Math. Biosci. Eng. **21** (2024), no. 1, 924–962, DOI 10.3934/mbe.2024039. MR4681671

[20] *Applications of nonstandard finite difference schemes*, posted on 2000, xiv+250, DOI 10.1142/9789812813251. MR1831401

[21] Ronald E. Mickens, *Exact solutions to a finite-difference model of a nonlinear reaction-advection equation: implications for numerical analysis*, J. Difference Equ. Appl. **9** (2003), no. 11, 995–1006, DOI 10.1080/1023619031000146959. Dedicated to Professor Ronald E. Mickens on the occasion of his 60th birthday. MR2027162

[22] K. Murphy and C. Weaver, *Janeway's immunobiology*, Garland Science, 2008.

[23] W. E. Paul, *Fundamental Immunology*, Fundamental Immunology. Wolters Kluwer/Lippincott Williams & Wilkins, 2008.

[24] Alan S. Perelson, *Modeling the interaction of the immune system with HIV*, Mathematical and statistical approaches to AIDS epidemiology, Lecture Notes in Biomath., vol. 83, Springer, Berlin, 1989, pp. 350–370, DOI 10.1007/978-3-642-93454-4_17. MR1040598

[25] G. J. Pettet, H. M. Byrne, D. L. S McElwain, and J. Norbury, *A model of wound-healing angiogenesis in soft tissue*, Math. Biosci., **136** (1996) no. 1, 35–63.

[26] A. B. Pigozzo, D. Missiakas, S. Alonso, R. W. Dos Santos, and M. Lobosco, *Development of a Computational Model of Abscess Formation*, Front Microbiol, **9** (2018), 1355.

[27] A. B. Pigozzo, G. C. Macedo, R. W. dos Santos, and M. Lobosco, *On the computational modeling of the innate immune system*, BMC Bioinformatics, **14** (2013) no. 6, S7.

[28] Kenneth V. Price, Rainer M. Storn, and Jouni A. Lampinen, *Differential evolution: A practical approach to global optimization*, Natural Computing Series, Springer-Verlag, Berlin, 2005. MR2191377

[29] R. F. Reis, J. L. Fernandes, T. R. Schmal, B. M. Rocha, R. W. Dos Santos, and M. Lobosco, *A personalized computational model of edema formation in myocarditis based on long-axis biventricular MRI images*, BMC Bioinformatics, **20** (2019) no. 6, 1–11.

[30] R. F. Reis, A. B. Pigozzo, C. R. B. Bonin, B. de M. Quintela, L. T. Pompei, A. C. Vieira, L. de L. Silva, M. P. Xavier, R. Weber dos Santos, and M. Lobosco, *A validated mathematical model of the cytokine release syndrome in severe COVID-19*, Front. Molecular Biosci., **8**, 2021.

[31] H. Ritchie, E. Mathieu, L. Rodés-Guirao, C. Appel, C. Giattino, E. Ortiz-Ospina, J. Hasell, B. Macdonald, D. Beltekian, and M. Roser, *Coronavirus pandemic (COVID-19)*, Our World in Data, 2020. `https://ourworldindata.org/coronavirus`.

[32] A. Rohatgi, Webplotdigitizer: Version 4.3, 2020.

[33] Scipy, scipy.optimize.differential_evolution, accessed on December, 2022.

[34] Scipy, Scipy's homepage, accessed on December, 2022.

[35] J. Shang, Y. Wan, C. Luo, et al, *Cell entry mechanisms of SARS-CoV-2*, Proc. Nat. Acad. Sci., **117** (2020) no. 21, 11727–11734.

[36] I. M. Sobol, *On sensitivity estimation for nonlinear mathematical models*, Matematicheskoe Modelirovanie, **2** (1990), no. 1, 112–118.

[37] I. M. Sobol′, *Global sensitivity indices for nonlinear mathematical models and their Monte Carlo estimates*, Math. Comput. Simulation **55** (2001), no. 1-3, 271–280, DOI 10.1016/S0378-4754(00)00270-6. MR1823119

[38] L. Sompayrac, *How the Immune System Works*, Wiley-Blackwell, Oxford, 2012.

[39] Rainer Storn and Kenneth Price, *Differential evolution—a simple and efficient heuristic for global optimization over continuous spaces*, J. Global Optim. **11** (1997), no. 4, 341–359, DOI 10.1023/A:1008202821328. MR1479553

[40] B. Su, W. Zhou, K. S. Dorman, and D. E. Jones, *Mathematical modelling of immune response in tissues*, Comput. Math. Methods Med. **10** (2009), no. 1, 9–38, DOI 10.1080/17486700801982713. MR2526974

[41] K. K.-W. To, O. T.-Y. Tsang, W.-S. Leung, A. R. Tam, T.-C. Wu, D. C. Lung, C. C.-Y. Yip, J.-P. Cai J. M.-C. Chan, T. S.-H. Chik, et al, *Temporal profiles of viral load in posterior oropharyngeal saliva samples and serum antibody responses during infection by SARS-CoV-2: an observational cohort study*, The Lancet Infectious Diseases, **20** (2020) no. 5, 565–574.

[42] Y. Vodovotz, C. C. Chow, J. Bartels, et al, *In silico models of acute inflammation in animals*, Shock, **26** (2006), no. 3, 235–244.

[43] World Health Organization, *WHO Timeline - COVID-19 - 27 April 2020*, `https://www.who.int/news/item/27-04-2020-who-timeline---covid-19`, 2020. Last accesses on October 22 of 2021.

[44] M. P. Xavier, C. R. Bonin, R. W. dos Santos, and M. Lobosco, *On the use of Gillespie stochastic simulation algorithm in a model of the human immune system response to the Yellow Fever vaccine*, 2017 IEEE Int. Conf. Bioinform. Biomed. (BIBM), (2017), 1476–1482.

[45] C. Zhang, Z. Wu, J.-W. Li, H. Zhao, and G.-Q. Wang, *Cytokine release syndrome in severe COVID-19: interleukin-6 receptor antagonist tocilizumab may be the key to reduce mortality*, Int. J. Antimicr. Agents, **55** (2020) no. 5, 105954.

[46] X. Zhang, Y. Tan, Y. Ling, et al, *Viral and host factors related to the clinical outcome of COVID-19*, Nature, **583** (2020) no. 7816, 437–440.

Institute of Exact Sciences, Department of Computing, Federal University of Juiz de Fora, Juiz de Fora, Brazil; and FISIOCOMP - Laboratory of Computational Fisiology and High-Performance Computing

Email address: `gustavo.morais@estudante.ufjf.br`

Department of Computer Science and Graduate Program in Computational Modeling, Federal University of Juiz de Fora, Juiz de Fora, Brazil; and FISIOCOMP - Laboratory of Computational Fisiology and High-Performance Computing

Email address: `marcelo.lobosco@ice.ufjf.br`

Chair of Applied and Computational Mathematics, University of Wuppertal, Wuppertal, Germany

Email address: `ehrhardt@uni-wuppertal.de`

Department of Computer Science and Graduate Program in Computational Modeling, Federal University of Juiz de Fora, Juiz de Fora, Brazil; and FISIOCOMP - Laboratory of Computational Fisiology and High-Performance Computing

Email address: `ruy.reis@ufjf.br`

Contemporary Mathematics
Volume **793**, 2024
https://doi.org/10.1090/conm/793/15901

Effects of rising temperature and prescribed fire on *Amblyomma Americanum* with ehrlichiosis

Alexander Fulk and Folashade B. Agusto

ABSTRACT. Ehrlichiosis is a nationally notifiable disease in the United States and the prevalence of this disease, as with other tick-borne diseases, has been increasing since at least the year 2000. One aspect that has likely contributed to the increase in the prevalence of this disease is rising temperatures due to climate change. A promising method for control of tick populations is prescribed burning. In this study, we develop a detailed compartmental mathematical model for assessing the effect of rising temperatures on *Amblyomma americanum* populations in endemic and invasion scenarios. The model, which takes the form of an impulsive system of ordinary differential equations, is used to investigate the effects of prescribed burning on infectious ticks to determine if prescribed fire remains effective as temperatures increase under endemic and invasion scenarios. This study shows that, in the absence of prescribed fires, (a) as temperature increases, there are significant increases in the number of infectious questing nymphs and adults. (b) Ehrlichiosis becomes established in the questing nymph and adult tick populations quicker in an invasion scenario as temperature increases. In the presence of prescribed burning, however, our results showed a reduction in the prevalence of infectious questing nymphs and adults. These results with prescribed burning hold regardless of increases in temperature. Thus, this study shows that prescribed burning is an effective control method for *Amblyomma americanum* even in the presence of increasing temperatures (in the range 2 °C to 4 °C).

1. Introduction

Tick-borne diseases have been increasing in incidence over the past several decades and one disease that has become of interest to researchers is human ehrlichiosis. Ehrlichiosis, formally known as human monocytic ehrlichiosis, was recognized as a human disease in the late 1980s and official reporting by the CDC on the disease began in 2000 [**2**, **4**]. *Ehrlichia chaffeensis*, *Ehrlichia ewingii*, and *Ehrlichia muris eauclairensis*, are all bacterium that contribute to the diseases grouped into the term ehrlichiosis, however most cases found in the United States are caused by *Ehrlichia chaffeensis* [**4**]. The bacterium are transmitted to humans via their primary vector, which is *Amblyomma americanum* in the United States. This tick

The first author was suppported by National Science Foundation under EPSCOR Track 2 grant number 192094. The second author was supported by National Science Foundation under EPSCOR Track 2 grant number 192094.

This research was supported by National Science Foundation under EPSCOR Track 2 grant number 192094.

vector is a significant concern mainly due to its extremely aggressive and somewhat indiscriminate biting behavior in all three life stages as well as its potential for transmission of several bacterium that are known to cause disease in humans [**15**].

Although the number of cases of ehrlichiosis remains relatively low, they have been steadily increasing since reporting on the disease began and along with this, there has been an increase in the incidence of the disease as well [**8, 18, 27**]. The annual number of cases of *Ehrlichia chafeensis* ehrlichiosis in the United States when reporting began in 2000 was approximately 200 cases. As of 2019, the most recent year of publicly available annual data on the disease had over 2000 cases reported. These cases have been concentrated in several states throughout the Midwestern, Northeastern, and Southwestern United States. Some states that have consistently seen a relatively high incidence of the disease include Arkansas, Kansas, Kentucky, Missouri, North Carolina, and Tennessee [**4**]. This increasing incidence can be attributed to several factors including climate change, human land modification, and increased prevalence of reservoir hosts, such as white-tailed deer [**6, 32**]. Despite the clear threat to public health that these ticks present, their dynamics and the diseases they transmit remain relatively understudied [**5, 36**].

Prescribed burning is a crucial land management practice that is used to maintain certain ecosystems and reduce the risk of destructive wildfires [**34, 35**]. It may also be used as a possible control measure for tick populations. Overall, prescribed burning has been shown to initially reduce the prevalence of ticks in an area. The effectiveness of this control measure over longer periods of time has been debated in various field studies; however, many of the studies that have been performed to unravel the effect that prescribed burning has on ticks were flawed. Several studies focused on small plots of land that may or may not have been consistently burned prior to the study [**1, 7, 9–11, 19, 20, 25**]. These situations are in contrast with typical prescribed burning practices [**21, 34, 35**]. Often, prescribed burning is performed consistently over a long period of time on large swaths of land [**21**]. Gleim, et al [**13**] performed a study in Northwestern Florida and Southwestern Georgia that addressed the shortcomings given above and found that prescribed burning was effective at reducing the prevalence of ticks in pine and mixed pine forests. Upon further investigation, the author found that prescribed burning can even reduce the prevalence of certain tick-borne diseases [**14**], though this was not the case for ehrlichiosis. Further investigation is needed to determine how effective prescribed burning is at reducing the prevalence of disease in ticks and which tick-borne diseases have the potential for mitigation using this control measure.

Construction and analysis of the dynamics of mathematical models via numerical simulations allow for insights into complex situations that are difficult or impossible to simulate experimentally. Recent modeling efforts have provided further support to the effectiveness of prescribed burning as a control measure for *Ixodes scapularis* ticks [**12, 17**]. In addition, Wallace et. al. [**37**] conducted a modeling study that evaluated the effect of rising temperatures on these ticks and the primary disease that they transmit, Lyme disease. They found that rising temperature has the potential to increase transmission risk to humans due to higher tick numbers and a longer season of transmission. Thus, one of our goals is to explore the effect of rising temperature on *Amblyomma americanum* populations. Further, we aim to determine whether or not prescribed burning remains an effective control measure as temperature increases and if prescribed burning is able to reduce the percentage of infectious ticks.

The outline of the paper is as follows. Section 2 covers the formulation of the mathematical model, the impulsive system used to simulate the effect of prescribed burning, and the calculation of the parameters used to simulate the effect of prescribed burning. In Section 3 we present the results of numerical simulations covering a range of scenarios. Discussion of the numerical simulations is covered in Section 4.1. Finally, conclusions and future directions are given in Section 4.2.

2. Methods and modeling

2.1. Model formulation. To formulate the model for the transmission dynamics of ehrlichiosis among ticks, we adapt some of the approaches used in the literature for modeling ticks dynamics, such as the studies in [**23, 37**]. These include a complete modeling of all intermediate stages of each life-stage of the tick along with temperature-dependent development rates for certain stages and density dependent death rates for feeding ticks. We made a simplifying assumption that the number of susceptible and infectious small and large hosts is constant. We now describe each aspect in detail.

First, the number of eggs (E) are determined, in part, by the number of egg-laying adults. The fecundity of egg-laying adults is reduced in a density-dependent manner by the factor f^{A_F}, with a female having the capacity to lay a maximum of $p = 6000$ eggs if $f^{A_F} = 1$. The function is given as:

$$f^{A_F} = 1 - \left\{0.25 + \left[0.04\ Ln\left(\frac{1.01 + A_{FS} + A_{FI}}{H_{ls} + H_{li}}\right)\right]\right\},$$

where A_{FS} and A_{FI} are susceptible and infectious feeding adults, while H_{ls} and H_{li} are susceptible and infectious large hosts.

The temperature-dependent egg developmental rate is denoted as $d_e(T)$ and they die at a fixed daily per-capita mortality rate μ_e. Transovarial transmission of ehrlichiosis has long been assumed to be unlikely, if not impossible [**22**], and a more recent study seems to support this assumption [**28**], so we do not incorporate it into our model. The equation for the dynamics of the number of eggs laid (E) is given by:

$$\frac{dE}{dt} = pf^{A_F}A_{EL} - [d_e(T) + \mu_e]E.$$

The hardening larvae population (L_H) increases depending on the temperature-dependent egg developmental rate, $d_e(T)$. The population decreases via maturation by a fixed rate d_{hl} as well as via death by the fixed rate μ_{hl}.

$$\frac{dL_H}{dt} = d_e(T)E - (d_{hl} + \mu_{hl})L_H.$$

There are several components describing how the questing larvae population changes through time. First, questing larvae have a certain probability for finding a host, given by:

$$\lambda_{ql} = (H_{ss} + H_{si})^{0.5}.$$

This probability depends on the total number of susceptible and infectious small hosts (H_{ss} and H_{si}, respectively) and is thus fixed. We also considered the average effect that both temperature and day-length have on questing tick activity via the fixed rates θ_l and δ_l, respectively. Finally, questing larvae die at the daily rate μ_{ql}. The equation for the questing larvae (L_Q) population is given by:

$$\frac{dL_Q}{dt} = d_{hl}L_H - \lambda_{ql}(\theta_l + \delta_l)L_Q - \mu_{ql}L_Q.$$

Once questing larvae find a host to feed on, they move onto the susceptible feeding larvae stage. This is the first stage where infection of a larval tick is possible if they are feeding on an infectious host, which occurs at the rate defined by the force of infection for larvae as:

$$\beta_l = \frac{\tau_l H_{si}}{H_{ss} + H_{si}},$$

where τ_l is the probability of infection, H_{si} is the number of small infectious hosts, and H_{ss} is the number of small susceptible hosts. Feeding larvae also die in a density dependent manner described as:

$$\mu_{fl}(L_{FS}, L_{FI}) = 0.25 + \left[0.015Ln\left(\frac{1.01 + L_{FS} + L_{FI}}{H_{ss} + H_{si}}\right)\right].$$

Susceptible feeding larvae (L_{FS}) become infectious feeding larvae (L_{FI}) (at the rate β_l), mature to susceptible engorged larvae (at the rate d_{fl}), or die in a density dependent manner (at the rate $\mu_{fl}(L_{FS}, L_{FI})$) given below as shown in the following equations:

$$\begin{aligned}\frac{dL_{FS}}{dt} &= \lambda_{ql}(\theta_l + \delta_l)L_Q - \beta_l L_{FS} - [d_{fl} + \mu_{fl}(L_{FS}, L_{FI})]L_{FS},\\ \frac{dL_{FI}}{dt} &= \beta_l L_{FS} - [d_{fl} + \mu_{fl}(L_{FS}, L_{FI})]L_{FI}.\end{aligned}$$

Once engorged, regardless of whether or not larvae are susceptible or infectious (L_{ES} and L_{EI}, respectively), they proceed to molt into nymphs in a temperature-dependent manner, designated by $d_{el}(T)$, or die at the fixed rate μ_{el} as shown in the following equation:

$$\frac{dL_{ES}}{dt} = d_{fl}L_{FS} - [d_{el}(T) + \mu_{el}]L_{ES},$$

$$\frac{dL_{EI}}{dt} = d_{fl}L_{FI} - [d_{el}(T) + \mu_{el}]L_{EI}.$$

At this stage, a nymph will harden before questing to look for a host to feed on. Note that once a larvae (or nymph or adult tick) becomes infectious, it remains so for the rest of its life. Consequently, we included infected compartments for hardening nymphs. Susceptible and infected hardening nymphs (N_{HS} and N_{HI}, respectively) mature into susceptible and infectious questing nymphs at the fixed rate d_{hn} or die at the rate μ_{hn}. The equations for the dynamics of the hardening nymphs are given by:

$$\begin{aligned}\frac{dN_{HS}}{dt} &= d_{el}(T)L_{ES} - (d_{hn} + \mu_{hn})N_{HS},\\ \frac{dN_{HI}}{dt} &= d_{el}(T)L_{EI} - (d_{hn} + \mu_{hn})N_{HI}.\end{aligned}$$

Once in the questing nymph stage, *Amblyomma americanum* ticks either die at the rate μ_{qn} or begin feeding on susceptible or infectious small or large hosts. The nymph life-stage is the least discriminant when it comes to host preference and as such, we incorporate the potential for nymphs to feed on both small and large hosts. The host finding probability for questing nymphs is thus given as:

$$\lambda_{qn} = (H_{ss} + H_{si} + H_{ls} + H_{li})^{0.1}.$$

Questing activity of nymphs is affected by both temperature and day-length and we incorporate the average effects of each via the fixed parameters θ_n and δ_n, respectively. Finally, the equations for susceptible and infectious questing nymphs (N_{QS} and N_{QI}, respectively) is given by:

$$\begin{aligned}\frac{dN_{QS}}{dt} &= d_{hn}N_{HS} - \lambda_{qn}(\theta_n + \delta_n)N_{QS} - \mu_{qn}N_{QS},\\ \frac{dN_{QI}}{dt} &= d_{hn}N_{HI} - \lambda_{qn}(\theta_n + \delta_n)N_{QI} - \mu_{qn}N_{QI}.\end{aligned}$$

Once those questing ticks that have survived find a host, they attach and begin feeding, progressing into the feeding nymph stage. We again have at this stage that susceptible nymphs may become infected if feeding on an infectious host. The force of infection, β_n, is defined as:

$$\beta_n = \frac{\tau_n(H_{si} + H_{li})}{H_{ss} + H_{si} + H_{ls} + H_{li}},$$

where τ_n is the probability that a susceptible nymph becomes infected after feeding on an infectious host. If susceptible feeding nymphs do not become infected, they, as well as the infectious feeding nymphs, either progress into engorged nymphs at the rate d_{fn}, or die in a density-dependent manner described as:

$$\mu_{fn}(N_{FS}, N_{FI}) = 0.45 + \left[0.02Ln\left(\frac{1.01 + N_{FS} + N_{FI}}{H_{ss} + H_{si} + H_{ls} + H_{li}}\right)\right].$$

The equations describing the rates of change for susceptible and infectious feeding nymph populations (N_{FS} and N_{FI}, respectively) are:

$$\begin{aligned}\frac{dN_{FS}}{dt} &= \lambda_{qn}(\theta_n + \delta_n)N_{QS} - \beta_n N_{FS} - [d_{fn} + \mu_{fn}(N_{FS}, N_{FI})]N_{FS},\\ \frac{dN_{FI}}{dt} &= \beta_n N_{FS} + \lambda_{qn}(\theta_n + \delta_n)N_{QI} - [d_{fn} + \mu_{fn}(N_{FS}, N_{FI})]N_{FI}.\end{aligned}$$

Those susceptible and infectious nymphs that survived feeding on a host then become engorged. Once engorged, they either die at a fixed rate μ_{en}, or molt into hardening adult ticks based on temperature dependent rate $d_{en}(T)$ as shown in the following equations:

$$\begin{aligned}\frac{dN_{ES}}{dt} &= d_{fn}N_{FS} - [d_{en}(T) + \mu_{en}]N_{ES},\\ \frac{dN_{EI}}{dt} &= d_{fn}N_{FI} - [d_{en}(T) + \mu_{en}]N_{EI},\end{aligned}$$

where N_{ES} denotes the susceptible engorged nymphs and N_{EI} denotes the infected engorged nymphs. Hardening adult ticks follow a similar form to hardening nymphs in that there are susceptible and infected populations of these ticks and they progress at a fixed rate, d_{ha} into susceptible or infectious questing adults. Alternatively, they die at the fixed rate μ_{ha}. The two equations for hardening adults

(A_{HS} and A_{HI}, respectively) are:

$$
\begin{aligned}
\frac{dA_{HS}}{dt} &= d_{en}(T)N_{ES} - (d_{ha} + \mu_{ha})A_{HS}, \\
\frac{dA_{HI}}{dt} &= d_{en}(T)N_{EI} - (d_{ha} + \mu_{ha})A_{HI}.
\end{aligned} \tag{2.1}
$$

Once at the questing stage, susceptible and infectious adult ticks seek out large hosts with the following probability of success:

$$\lambda_{qa} = (H_{ls} + H_{li})^{0.51}.$$

Questing activity of adult ticks is influenced by the average effects of temperature and day-length via θ_a and δ_a, respectively. Susceptible and infectious questing adults may fail to find a host and die at the rate μ_{qa}.The equations for adult ticks (A_{QS} and A_{QI}, respectively) are given as:

$$
\begin{aligned}
\frac{dA_{QS}}{dt} &= d_{ha}A_{HS} - \lambda_{qa}(\theta_a + \delta_a)A_{QS} - \mu_{qa}A_{QS}, \\
\frac{dA_{QI}}{dt} &= d_{ha}A_{HI} - \lambda_{qa}(\theta_a + \delta_a)A_{QI} - \mu_{qa}A_{QI}.
\end{aligned}
$$

Those susceptible and infectious questing adult ticks that do find hosts progress to the feeding stage. At this point, as with the other tick life stages, we have the possibility of a tick becoming infected if feeding on an infectious host. The force of infection for feeding adult ticks is:

$$\beta_a = \frac{\tau_a H_{li}}{H_{ls} + H_{li}},$$

where τ_a is the probability of infection in adult ticks. Once attached to a host, the adult ticks can take up to several days to finish feeding and progress into susceptible and infected engorged adults at the rate d_{fa}. They may also die in a density dependent manner:

$$\mu_{fa}(A_{FS}, A_{FI}) = 0.35 + \left[0.03\ Ln\left(\frac{1.01 + A_{FS} + A_{FI}}{H_{ls} + H_{li}}\right)\right].$$

With these, we have the following equations for susceptible and infectious feeding adults, given as A_{FS} and A_{FI}:

$$
\begin{aligned}
\frac{dA_{FS}}{dt} &= \lambda_{qa}(\theta_a + \delta_a)A_{QS} - \beta_a A_{FS} - [d_{fa} + \mu_{fa}(A_{FS}, A_{FI})]A_{FS}, \\
\frac{dA_{FI}}{dt} &= \beta_a A_{FS} + \lambda_{qa}(\theta_a + \delta_a)A_{QI} - [d_{fa} + \mu_{fa}(A_{FS}, A_{FI})]A_{FI}.
\end{aligned}
$$

Once engorged, the female adults (A_{ES} and A_{EI}) will mature into egg-laying adults at the temperature dependent rate $d_{ea}(T)$, or die at the rate μ_{ea} as given below:

$$
\begin{aligned}
\frac{dA_{ES}}{dt} &= d_{fa}A_{FS} - (d_{ea}(T) + \mu_{ea})A_{ES}, \\
\frac{dA_{EI}}{dt} &= d_{fa}A_{FI} - (d_{ea}(T) + \mu_{ea})A_{EI}.
\end{aligned}
$$

As stated previously, since transovarial transmission seems very unlikely for ehrlichiosis, we collapse egg-laying adults into a single population (A_{EL}). We assume that once the eggs have been laid, all of these adults die:

$$\frac{dA_{EL}}{dt} = d_{ea}(T)(A_{ES} + A_{EI}) - \mu_{ela}A_{EL}.$$

The temperature dependent development rates for eggs ($d_e(T)$), engorged larvae ($d_{el}(T)$), engorged nymphs ($d_{en}(T)$), and engorged adults ($d_{ea}(T)$) are given by:

$$\begin{aligned} d_e(T) &= \frac{1}{7775T^{-1.634}}, \; d_{el}(T) &= \frac{1}{16859T^{-2.139}}, \\ d_{en}(T) &= \frac{1}{4219T^{-1.638}}, \; d_{ea}(T) &= \frac{1}{5451.6T^{-2.023}}. \end{aligned}$$

These rates were adapted from [**23**] and transformed into the reciprocal based on the recommendation in [**38**]. These rates depend on the following 5-term Fourier series function of temperature.

$$\begin{aligned} T(t) = 13.55 + w &- 10.7\,cos(\phi t) - 6.238\,sin(\phi t) \\ &- 0.08032\,cos(2\phi t) + 0.2741\,sin(2\phi t) - 0.5544\,cos(3\phi t) \\ &- 0.7403\,sin(3\phi t) + 0.9966\,cos(4\phi t) - 0.6644\,sin(4\phi t) \\ &- 0.0189\,cos(5\phi t) + 0.01968\,sin(5\phi t), \end{aligned}$$

where $w = 0, 2, 4,^\circ C$, and $\phi = 0.01826$. The functional form of this temperature-dependent function was obtaibed using the curve fitting toolbox in Matlab [**26**]. To generate the data used for deriving the functional form of the temperature-dependent function, we used the mean daily temperature data for each county in Kansas from January 1st, 2020, to December 31st, 2020 and averaged over all of those counties to get the average temperature for Kansas for that year. We removed a day before using the curve fitting toolbox since 2020 was a leap year to get a periodic function of temperature with a period equal to 365 days. The mean temperature data for each county in Kansas was downloaded from the Parameter-elevation Regressions on Independent Slopes Model (PRISM) website [**16**].

We model the dynamics for the ticks population using the assumptions and system of ordinary differential equations described above. Thus, we have the following equations for the eggs and the different stages for the larvae as well as their host seeking behavior:

$$(2.2) \qquad \begin{cases} \frac{dE}{dt} &= pf^{A_F}A_{EL} - [d_e(T) + \mu_e]E \\ \frac{dL_H}{dt} &= d_e(T)E - (d_{hl} + \mu_{hl})L_H \\ \frac{dL_Q}{dt} &= d_{hl}L_H - \lambda_{ql}(\theta_l + \delta_l)L_Q - \mu_{ql}L_Q \\ \frac{dL_{FS}}{dt} &= \lambda_{ql}(\theta_l + \delta_l)L_Q - \beta_l L_{FS} - [d_{fl} + \mu_{fl}(L_{FS}, L_{FI})]L_{FS} \\ \frac{dL_{FI}}{dt} &= \beta_l L_{FS} - [d_{fl} + \mu_{fl}(L_{FS}, L_{FI})]L_{FI} \\ \frac{dL_{ES}}{dt} &= d_{fl}L_{FS} - [d_{el}(T) + \mu_{el}]L_{ES} \\ \frac{dL_{EI}}{dt} &= d_{fl}L_{FI} - [d_{el}(T) + \mu_{el}]L_{EI}. \end{cases}$$

To incorporate the effect of fire on the tick populations we use a system of impulsive differential equations. For time $t \neq nP$, $n = 1, 2, \ldots,$ and P is a predetermined impulsive period, i.e., the time when prescribed fire is implemented. At the predetermined time $t = nP$, we initiate a burn that reduces each tick population by a particular proportion using the following equation:

$$\begin{cases} E(nP^+) &= (1-\nu_e)E(nP^-) \\ L_H(nP^+) &= (1-\nu_l)L_H(nP^-) \\ L_Q(nP^+) &= (1-\nu_l)L_Q(nP^-) \\ L_{FS}(nP^+) &= (1-\nu_l)L_{FS}(nP^-) \\ L_{FI}(nP^+) &= (1-\nu_l)L_{FI}(nP^-) \\ L_{ES}(nP^+) &= (1-\nu_l)L_{ES}(nP^-) \\ L_{EI}(nP^+) &= (1-\nu_l)L_{EI}(nP^-). \end{cases} \tag{2.3}$$

The parameters $\nu_i, i = e, l$, indicate a reduction in each population with values ranging from zero to one hundred percent reduction following a burn. These reduced populations are then used as the initial conditions in the model simulation until the next burn is implemented.

Below are the equations for the nymphs with their different life stages and host seeking behavior in the time between burns ($t \neq nP$):

$$\begin{cases} \dfrac{dN_{HS}}{dt} &= d_{el}(T)L_{ES} - (d_{hn} + \mu_{hn})N_{HS} \\ \dfrac{dN_{HI}}{dt} &= d_{el}(T)L_{EI} - (d_{hn} + \mu_{hn})N_{HI} \\ \dfrac{dN_{QS}}{dt} &= d_{hn}N_{HS} - \lambda_{qn}(\theta_n + \delta_n)N_{QS} - \mu_{qn}N_{QS} \\ \dfrac{dN_{QI}}{dt} &= d_{hn}N_{HI} - \lambda_{qn}(\theta_n + \delta_n)N_{QI} - \mu_{qn}N_{QI} \\ \dfrac{dN_{FS}}{dt} &= \lambda_{qn}(\theta_n + \delta_n)N_{QS} - \beta_n N_{FS} - [d_{fn} + \mu_{fn}(N_{FS}, N_{FI})]N_{FS} \\ \dfrac{dN_{FI}}{dt} &= \beta_n N_{FS} + \lambda_{qn}(\theta_n + \delta_n)N_{QI} - [d_{fn} + \mu_{fn}(N_{FS}, N_{FI})]N_{FI} \\ \dfrac{dN_{ES}}{dt} &= d_{fn}N_{FS} - [d_{en}(T) + \mu_{en}]N_{ES} \\ \dfrac{dN_{EI}}{dt} &= d_{fn}N_{FI} - [d_{en}(T) + \mu_{en}]N_{EI}. \end{cases} \tag{2.4}$$

The nymph populations are reduced at the end of each interval by the proportion $(1 - \nu_n)$ in the following manner at time $t = nP$:

$$\begin{cases} N_{HS}(nP^+) &= (1-\nu_n)N_{HS}(nP^-) \\ N_{HI}(nP^+) &= (1-\nu_n)N_{HI}(nP^-) \\ N_{QS}(nP^+) &= (1-\nu_n)N_{QS}(nP^-) \\ N_{QI}(nP^+) &= (1-\nu_n)N_{QI}(nP^-) \\ N_{FS}(nP^+) &= (1-\nu_n)N_{FS}(nP^-) \\ N_{FI}(nP^+) &= (1-\nu_n)N_{FI}(nP^-) \\ N_{ES}(nP^+) &= (1-\nu_n)N_{ES}(nP^-) \\ N_{EI}(nP^+) &= (1-\nu_n)N_{EI}(nP^-). \end{cases} \tag{2.5}$$

Lastly, the adult tick populations are governed by the following equations in the time between burns ($t \neq nP$):

$$(2.6)\quad \begin{cases} \dfrac{dA_{HS}}{dt} &= d_{en}(T)N_{ES} - (d_{ha} + \mu_{ha})A_{HS} \\ \dfrac{dA_{HI}}{dt} &= d_{en}(T)N_{EI} - (d_{ha} + \mu_{ha})A_{HI} \\ \dfrac{dA_{QS}}{dt} &= d_{ha}A_{HS} - \lambda_{qa}(\theta_a + \delta_a)A_{QS} - \mu_{qa}A_{QS} \\ \dfrac{dA_{QI}}{dt} &= d_{ha}A_{HI} - \lambda_{qa}(\theta_a + \delta_a)A_{QI} - \mu_{qa}A_{QI} \\ \dfrac{dA_{FS}}{dt} &= \lambda_{qa}(\theta_a + \delta_a)A_{QS} - \beta_a A_{FS} - [d_{fa} + \mu_{fa}(A_{FS}, A_{FI})]A_{FS} \\ \dfrac{dA_{FI}}{dt} &= \beta_a A_{FS}\lambda_{qa}(\theta_a + \delta_a)A_{QI} - [d_{fa} + \mu_{fa}(A_{FS}, A_{FI})]A_{FI} \\ \dfrac{dA_{ES}}{dt} &= d_{fa}A_{FS} - [d_{ea}(T) + \mu_{ea}]A_{ES} \\ \dfrac{dA_{EI}}{dt} &= d_{fa}A_{FI} - [d_{ea}(T) + \mu_{ea}]A_{EI} \\ \dfrac{dA_{EL}}{dt} &= d_{ea}(T)(A_{ES} + A_{EI}) - \mu_{ela}A_{EL}. \end{cases}$$

And the adult tick populations are subsequently reduced by the proportion $(1 - \nu_a)$ when a burn occurs at time $t = nP$:

$$(2.7)\quad \begin{cases} A_{HS}(nP^+) &= (1 - \nu_a)A_{HS}(nP^-) \\ A_{HI}(nP^+) &= (1 - \nu_a)A_{HI}(nP^-) \\ A_{QS}(nP^+) &= (1 - \nu_a)A_{QS}(nP^-) \\ A_{QI}(nP^+) &= (1 - \nu_a)A_{QI}(nP^-) \\ A_{FS}(nP^+) &= (1 - \nu_a)A_{FS}(nP^-) \\ A_{FI}(nP^+) &= (1 - \nu_a)A_{FI}(nP^-) \\ A_{ES}(nP^+) &= (1 - \nu_a)A_{ES}(nP^-) \\ A_{EI}(nP^+) &= (1 - \nu_a)A_{EI}(nP^-). \end{cases}$$

The flow diagram for model (2.2)–(2.7) at time $t \neq nP$ between burns is given in Figure 1 and the description of the parameter and the values used in simulations are given in Table 1. Note the arrows from the purple box in Figure 1 signify the temperature-dependent developmental rates.

2.2. The non-impulsive system: special case of the impulsive model (2.2)–(2.7). In this section, we consider a special case of the impulsive transmission model (2.2)–(2.6) in the absence of fire. This model is non-impulsive and a continuous system of time. Following the approach in Guo and Agusto [**17**] we can prove the following lemmas for model positivity and boundedness of solutions.

LEMMA 2.1. *Let the initial data* $F(0) \geq 0$, *where* $F(t) = (E(t), L_H(t), L_Q(t), L_{FS}(t), L_{FI}(t), L_{ES}(t), L_{EI}(t), N_{HS}(t), N_{HI}(t), N_{QS}(t), N_{QI}(t), N_{FS}(t), N_{FI}(t), N_{ES}(t), N_{EI}(t), A_{HS}(t), A_{HI}(t), A_{QS}(t), A_{QI}(t), A_{FS}(t), A_{FI}(t), A_{ES}(t), A_{EI}(t))$. *Then the solutions* $F(t)$ *of the non-impulsive system are non-negative for all* $t > 0$.

LEMMA 2.2. *The region* $\Omega_T \subset \mathbb{R}^{23}_+$ *is positively-invariant for the non-impulsive system with non-negative initial conditions in* $\mathbb{R}^{23}_+$.

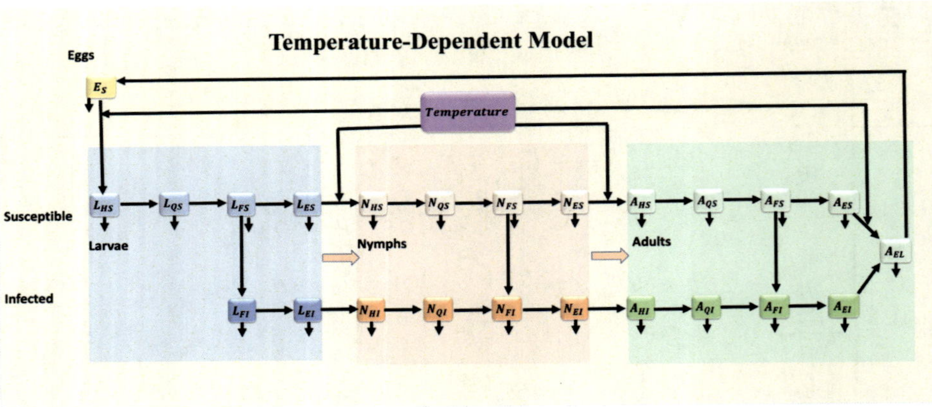

FIGURE 1. Flow diagram of ehrlichiosis transmission model (2.2)–(2.7) in *Amblyomma americanum* with temperature in time between burns ($t \neq nP$).

where,

$$\Omega_T = \Big\{ (E(t), L_H(t), L_Q(t), L_{FS}(t), L_{FI}(t), L_{ES}(t), L_{EI}(t), N_{HS}(t), N_{HI}(t), N_{QS}(t), N_{QI}(t), N_{FS}(t), N_{FI}(t), N_{ES}(t), N_{EI}(t), A_{HS}(t), A_{HI}(t), A_{QS}(t), A_{QI}(t), A_{FS}(t), A_{FI}(t), A_{ES}(t), A_{EI}(t)) \in \mathbb{R}^{23}_{+} \colon N_T(t) \leq \frac{\sigma_T A(t)}{\mu_T} \Big\},$$

$$\mu_T = min(\mu_e, \mu_L, \mu_N, \mu_A), \ \mu_L = min(\mu_{hl}, \mu_{ql}, \mu_{fl}(L_{FS}, L_{FI}), \mu_{el}),$$
$$\mu_N = min(\mu_{hn}, \mu qn, \mu_{fn}(N_{FS}, N_{FI}), \mu_{en}),$$
$$\mu_A = min(\mu_{ha}, \mu_{qa}, \mu_{fa}(A_{FS}, A_{FI}), \mu_{ela}),$$

$N_T = E + L_H + L_Q + L_{FS} + L_{FI} + L_{ES} + L_{EI} + N_{HS} + N_{HI} + N_{QS} + N_{QI} + N_{FS} + N_{FI} + N_{ES} + N_{EI} + A_{HS} + A_{HI} + A_{QS} + A_{QI} + A_{FS} + A_{FI} + A_{ES} + A_{EI}$, *and* $A(t) \leq constant$.

2.3. Estimating the proportion of ticks reduced due to a prescribed burn. Limited data exists on sampling efforts made before and after prescribed burns that might allow us to accurately determine the proportion of each tick life stage reduced following a prescribed burn. So, we used the data from Gleim et. al. [**13**] to calculate the proportions ν_l, ν_n, and ν_a. We made the simplifying assumption that the ν_e parameter was zero. This in part because typical tick sampling methods are not effective at capturing tick eggs; also because tick eggs are typically laid in damp areas, such as deep in a leaf litter layer where prescribed burns may not significantly affect. For the other $\nu_i, i = l, n, a$ parameters, we used the counts in 2011 from burned surrounded by unburned sites for *Amblyomma americanum* in Gleim et. al. [**13**]. We divide the number of ticks in each life stage by the total number of ticks and subtract that proportion from one to get our estimate for the proportion of ticks reduced following a prescribed burn. For example, there were 21 nymphs and 319 *A. americanum* ticks collected in 2011 at the aforementioned sites, so we have ν_n = 1 - (21/319) $\approx$ 0.9341. The value for ν_l was 1 using this

TABLE 1. Description of the variables and parameters for ehrlichiosis transmission model (2.2)–(2.7) in *Amblyomma americanum*. Parameter values and estimates are taken from [**23**].

Parameter	Description	Value
p	Maximum number of eggs that can be laid	6000
μ_e	Death/inviability rate of eggs	0.008
d_{hl}	Progression rate of hardening larva	1/5
μ_{hl}	Death rate of hardening larva	0.00648
θ_l	Average effect of temperature on activity of questing larva	0.002
δ_l	Average effect of day-length on activity of questing larva	0
μ_{ql}	Death rate of questing larva	0.0029
τ_l	Probability of infection in larva	0.1
d_{fl}	Progression rate of feeding larva	1/5
μ_{el}	Death rate of engorged larva	0.005
d_{hn}	Progression rate of hardening nymphs	1/7
μ_{hn}	Death rate of hardening nymphs	0.003
θ_n	Average effect of temperature on activity of questing nymphs	0.01
δ_n	Average effect of day-length on activity of questing nymphs	0.005
μ_{qn}	Death rate of questing nymphs	0.001
τ_n	Probability of infection in feeding nymphs	0.1
d_{fn}	Progression rate of feeding nymphs	1/5
μ_{en}	Death rate of engorged nymphs	0.004
d_{ha}	Progression rate of hardening adults	1/14
μ_{ha}	Death rate of hardening adults	0.0035
θ_a	Average effect of temperature on activity of questing adults	0.007
δ_a	Average effect of day-length on activity of questing adults	0.004
μ_{qa}	Death rate of questing adults	0.00015
τ_a	Probability of infection in feeding adults	0.1
d_{fa}	Progression rate of feeding adults	1/12
μ_{ea}	Death rate of engorged adults	0.003
μ_{ela}	Death rate of egg-laying adults	1

method, however we reduced this slightly to 0.99 since it is rather unlikely that all larvae are killed with a prescribed burn. This also was part of the motivation for us to explore how varying these parameters might affect the tick populations and those results are given in Section 3.2.3.

3. Numerical simulations

The purpose of this study is to explore the effects of rising temperature (due to global warming) on *A. americanum*, prevalence of the disease, and how these dynamics change when prescribed fires are implemented. These goals are achieved by simulating the impulsive model (2.2), (2.4), and (2.6). Before addressing each of these goals, we perform a global sensitivity analysis to determine the impact the model parameters and prescribed fires have on both the number of infectious questing nymph and adult ticks and the prevalence of disease in the tick populations.

3.1. Global sensitivity analysis. The global sensitivity analysis allows us to determine which parameters have the strongest effects on the model outcome as well as the type of relationship (positive or negative) each parameter has with the model outcome. To create representative distributions for the model parameters, we use the Latin Hypercube Sampling method. For details on this method, see [**24**]. Essentially, we generate 1000 samples for each parameter, each generated from a particular distribution, in order to generate the Latin Hypercube Sampling matrix. This creates a multi-dimensional parameter space, from which, we pull various sets of parameters and run our model with each set to gauge the impact of each parameter using their partial rank correlation coefficients. We chose uniform distributions for most model parameters due to the lack of biological data pertaining to parameters governing tick dynamics and the diseases that they transmit. Parameters related to prescribed fire (i.e. the number of burns and the time between burns) are each taken from a Poisson distribution. The Poisson distribution for the number of burns is centered on four, while the Poisson distribution for the time between burns is centered on eighty. Any zeros generated from either of these distributions were subsequently changed to ones to avoid numerical issues implementing the burns. The outcome measure used is either the sum of the number of infectious questing nymphs and adults present at the end of the simulation or the sum of the prevalence of infectious questing nymphs and adults present at the end of the simulation.

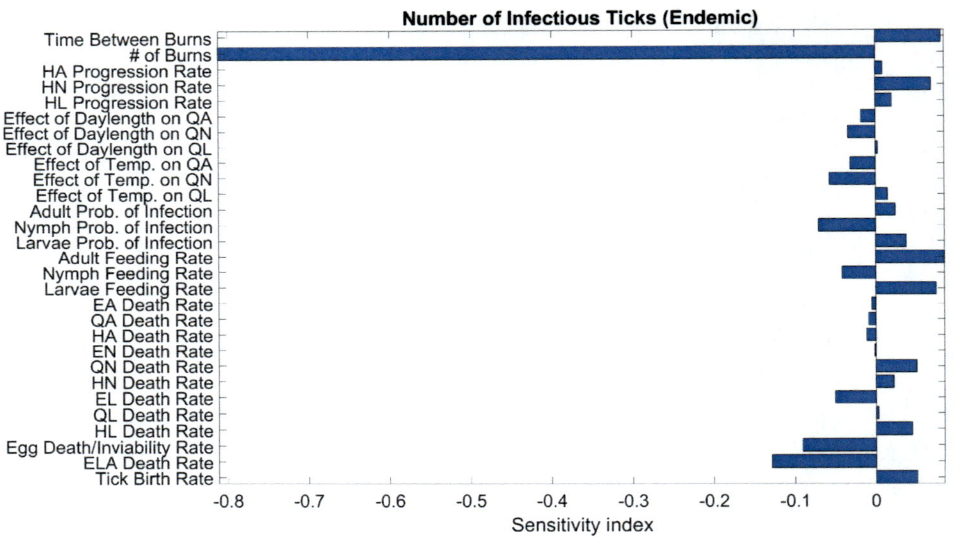

FIGURE 2. Global sensitivity analysis for ehrlichiosis transmission model (2.2), (2.4), and (2.6) with temperature. In this analysis, we considered the sum of the number of infectious nymphs and adults present at the end of a simulation as our outcome measure. The simulations was carried out for an endemic scenario described in Section 3.2. Parameter values are taken from Table 1.

The outcome of the global sensitivity analysis depicted in Figure 2, used the sum of the number of infectious questing nymphs and adults present at the end of the simulation as the outcome measure. In this case, we began each simulation at model equilibrium (i.e. an endemic scenario). The model dynamics had become periodically stable by $t = 7300$ days (20 years) and used the values for each population at that time as our starting point for each simulation. Significant parameters in this analysis included time between burns, number of burns, hardening nymph progression rate (d_{hn}), probability of infection in nymphs (τ_n), adult feeding rate (d_{fa}), larvae feeding rate (d_{fl}), egg death/inviability rate (μ_e), and egg-laying adult death rate (μ_{ela}). From those significant parameters, we find that the number of burns had by far the greatest impact on the number of infectious nymphs and adults present at the end of our simulations with a partial rank correlation coefficient less than -0.8. This negative coefficient intuitively implies that as the number of burns increases, the number of infectious nymphs and adults tends to decrease by the end of the simulation. Other significant parameters that followed this trend were μ_e and μ_{ela}, though their impact on the model outcome was significantly less than the impact of the number of burns. This makes sense due to the fact that an increase of only one in the number of burns kills 93% of nymphs when the burn is initiated. Parameters that had a positive impact on the model outcome were time between burns, d_{hn}, τ_n, d_{fa}, and d_{fl}. These results imply that for nymphs, the most important means of generating new infectious questing nymphs are via progression following infection in the previous life stage as well as increasing the likelihood of becoming infected while feeding on an infectious host. These infected nymphs then move on to become infectious questing adults. However, since the probability of infection (τ_a) is not significant for adult ticks, but the adult feeding rate (d_{fa}) and egg-laying adult death rate (μ_{ela}) are, it is likely that the most important role that the adult ticks play is creating the next generation of ticks to be infected rather than contributing directly to the outcome measure. Since infectious questing adults are maintained at an order of magnitude less than the infectious questing nymphs, it makes sense that it is better to repopulate those nymphs as quickly as possible following the dramatic decrease that occurs when a prescribed fire is implemented.

The model outcome of a global sensitivity analysis is an important consideration as it can have dramatic effects on which parameters are significant in the analysis. Usually, the basic reproduction number, R_0, is used since the goal of many modeling studies is reducing this to a value less than one which has the potential to eliminate the disease from the populations of interest. However, we assume a fixed host population and as such, we cannot calculate R_0 in the usual way (i.e. using the next generation matrix method). We consider this an acceptable assumption since prescribed fire alone is unlikely to successfully reduce R_0 below the necessary threshold due to the fact that it does little to affect the main reservoir host of ehrlichiosis, the white-tailed deer, other than potentially driving them from the area being burned for a short time. It is because of these reasons that we instead evaluate how the sum of the prevalence of infectious questing nymphs and adults is affected by each of the model parameters. We again use an endemic scenario as the starting place for each of our simulations. We observed in Figure 3 that the significant parameters were time between burns, number of burns, the average effect of temperature on questing nymphs (θ_n), probability of infection in nymphs (τ_n), probability of infection in larvae (τ_l), and nymph feeding rate (d_{fn}). From this, we again see that the number of burns has the greatest impact on the sum of the prevalence of infectious nymphs and adults, with a partial rank correlation coefficient value of -0.222. In this case, we see that the impact that this parameter has is significantly less compared to the previous outcome measure. The average effect of temperature on questing nymphs had a significant, though small, negative impact on the sum of the prevalence of infectious questing nymphs and adults.

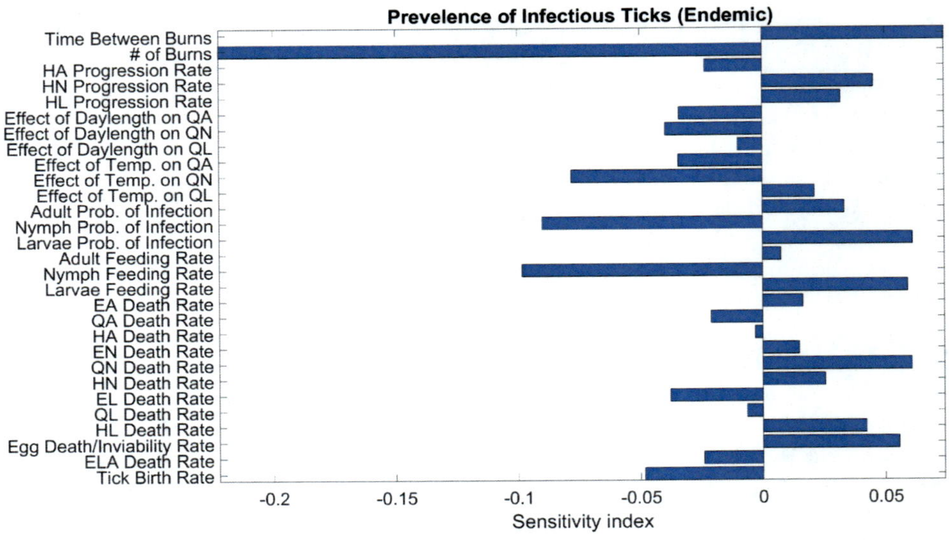

FIGURE 3. Global sensitivity analysis for ehrlichiosis transmission model (2.2), (2.4), and (2.6) with temperature. In this analysis, we considered the sum of the prevalence of infectious nymphs and adults present at the end of a simulation as our outcome measure. The simulations were carried out for an endemic scenario described in Section 3.2.Parameter values are taken from Table 1.

This is somewhat intuitive as the larger this value becomes, the faster infectious questing nymphs will find a host and move to the feeding nymph stage. Both τ_n and d_{fn} had small negative impacts on the outcome measure. The time between burns again had a significant, but small positive impact on the model outcome. The probability of infection in larvae (τ_l) had a small positive impact on the model outcome which is unsurprising since those larvae that become infected progress to become infectious questing nymphs and adults if they survive, directly increasing the numerator of the model outcome.

Next, we evaluate the effect that model parameters have on each of these outcomes when starting at an invasion scenario in terms of the initial conditions.

Starting at the initial conditions stated for the invasion scenario described in Section 3.2.2, we ran the global sensitivity analysis for each of the outcome measures discussed previously in order to determine how the effect of each parameter changes when population and disease dynamics have not yet reached endemic dynamics. We first consider the sum of infectious questing nymphs and adults present at the end of the simulation. We found in Figure 4 that the significant parameters in this case included time between burns, number of burns, progression rate of hardening larvae (d_{hl}), the average effect of temperature on questing nymphs (θ_n), the average effect of temperature on questing larvae (θ_l), adult feeding rate (d_{fa}), larvae feeding rate (d_{fl}), egg death/inviability rate (μ_e), egg-laying adult death rate (μ_{ela}), and tick birth rate (p). Unsurprisingly, the number of burns had the greatest negative impact on the model outcome, though the effect was not as great as it was in

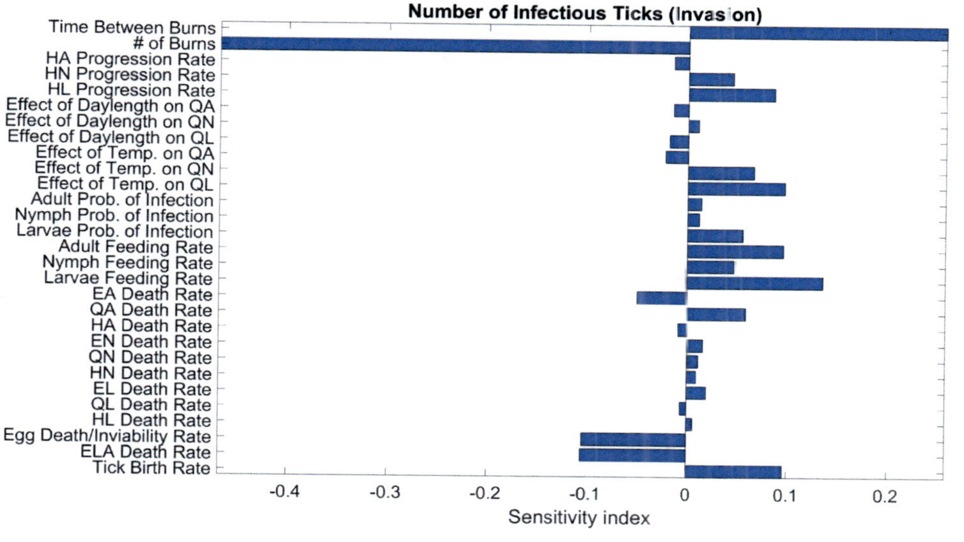

FIGURE 4. Global sensitivity analysis for ehrlichiosis transmission model (2.2), (2.4), and (2.6) with temperature. In this analysis, we considered the sum of the number of infectious nymphs and adults present at the end of a simulation as our outcome measure. The simulations were carried out for an invasion scenario described in Section 3.2. Parameter values are taken from Table 1.

the endemic case. This is due to the fact that prescribed fires alone are not able to prevent the invasion of ticks and will be discussed further in Section 4.1. The time between burns had a much greater impact on the model outcome in this case because the growth in the number of infectious questing nymphs and adults is greater compared to the starting point near zero when compared to the endemic case. The egg death/inviability rate (μ_e) and the egg-laying adult death rate (μ_{ela}) were again significant and negative. Several parameters that increase the number of infectious questing nymphs and adults (d_{hl}, θ_n, θ_l, d_{fa}, and d_{fl}) were important which implies that these rates are much more important in an invasion scenario. This makes sense as it would be easier for this tick species to become established if the environment being invaded is better suited to these ticks as would be the case if some of these parameters are increased (d_{hl}, θ_n, θ_l). These suitable conditions also have the potential to affect the feeding rate of ticks, which generally has a positive effect on the number of infectious nymphs and adults [**33**].

Finally, we explore how the model parameters affect the sum of the prevalence of infectious questing nymphs and adults, see Figure 5. In this case, the significant parameters were number of burns, progression rate of hardening adults (d_{ha}), average effect of temperature on questing adults (θ_a), average effect of temperature on questing larvae (θ_l), probability of infection in larvae (τ_l), and engorged adult death rate (μ_{ea}). In contrast to the earlier global sensitivity analyses results, the parameter that had the greatest impact on the model outcome was the probability of infection in larvae (τ_l). If this probability increases, we tend to see an increase in

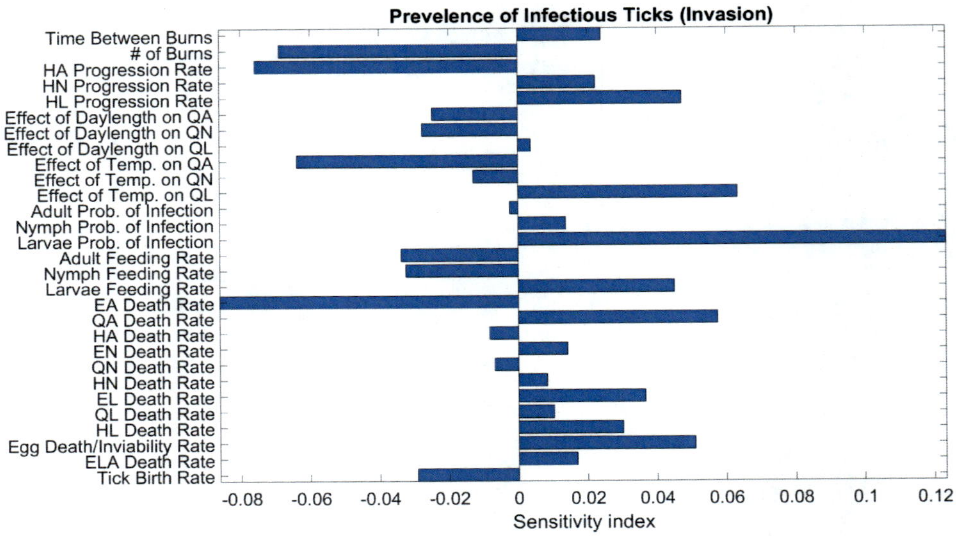

FIGURE 5. Global sensitivity analysis for ehrlichiosis transmission model (2.2), (2.4), and (2.6) with temperature. In this analysis, we considered the sum of the prevalence of infectious nymphs and adults present at the end of a simulation as our outcome measure. The simulations were carried out for an invasion scenario described in Section 3.2. Parameter values are taken from Table 1.

the prevalence of infectious questing nymphs and adults at the end of the simulation. The number of burns did have a small negative impact on the outcome, but it mattered less than the progression rate of hardening adults (d_{ha}) and the engorged adult death rate (μ_{ea}). Larvae and adults seem to play an important role in the disease dynamics in an invasion scenario as they have more significant parameters in this case and will be discussed further in Section 4.1.

3.2. Tick and disease dynamics in an environment with rising temperature. Next, we explore how rising temperature affects tick and disease dynamics in two different contexts that occur in many places throughout the United States, but particularly in the Midwest and throughout the Southeastern United States and also implement prescribed burns to investigate how tick populations change following a burn as temperature increases.

3.2.1. *Endemic dynamics. Amblyomma americanum* are endemic throughout the North- and Southeastern United States as well as much of the Midwest. As such, we investigate the effect of rising temperature on the dynamics of questing nymph and adult tick susceptible and infectious populations at equilibrium. As stated in Section 3.1, we let the model run for 20 years, collected the values of each population at the end of this simulation and used those values as a starting point for our equilibrium simulations. The simulation presented in Figure 6 shows the model results under endemic scenario for current temperature ($w = 0$), and for increase in temperature with $w = 2$, and 4 °C. Note, the worst-case emission scenario projection by the the Intergovernmental Panel on Climate Change (IPCC) suggest warming of no more that 1.5 °C by 2050 or 2080 [**31**].

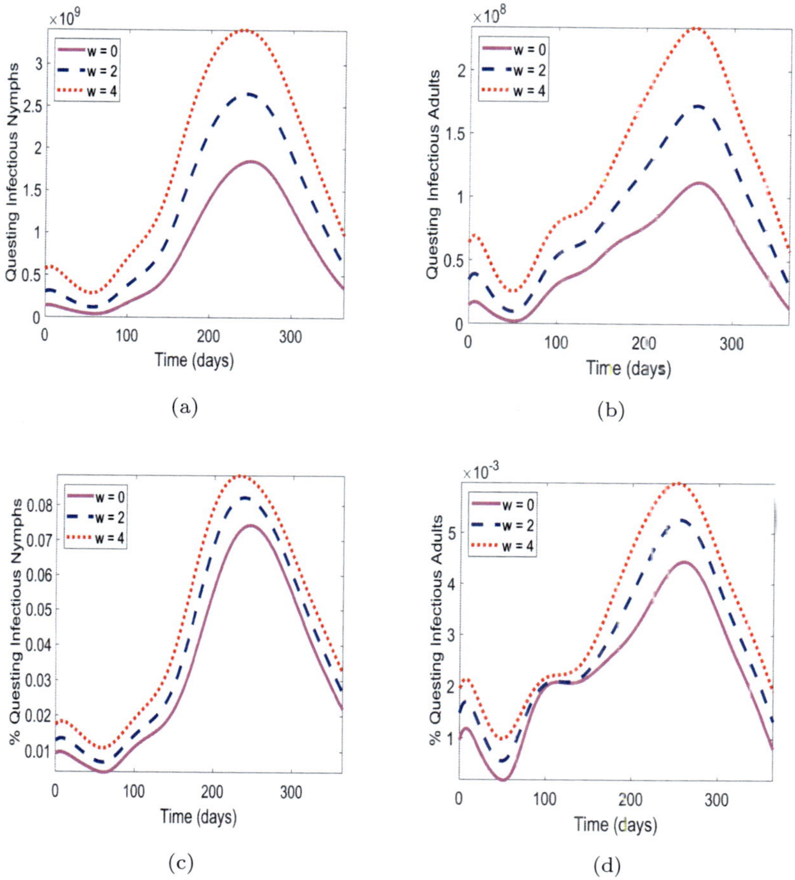

FIGURE 6. Nymph and adult tick population and disease dynamics under endemic scenario for current temperature $w = 0$ and increased temperature $w = 2, 4$ °C. (a) and (b) Questing infectious nymphs and adults; (c) and (d) Prevalence of questing nymphs and adults. Parameter values are taken from Table 1.

From Figure 6, we can see that as temperature increases, we have a roughly two-fold increase in both the number of questing infectious nymph and adult ticks when moving from w = 0 to w = 4. We do see increases in the prevalence of ticks infected with the disease in each tick population as temperature increases, however, this still leaves a relatively low percentage in infectious questing nymphs and adults. This does not however diminish the threat of infectious ticks. Since ticks will be prevalent in much larger numbers as temperature increases and the prevalence of disease in ticks will increase slightly, the more infectious questing ticks humans are going to encounter when in areas that ticks frequent. This is also exacerbated by the fact that these increases are greatest during times when humans are likely to be outdoors in tick infested areas (July-September, days 180-280). Overall, ehrlichiosis poses an increased threat to public health in areas where the disease is prevalent in ticks as temperature increases.

Next, we present the results of implementing four prescribed burns throughout the year in an endemic situation (Figure 7).

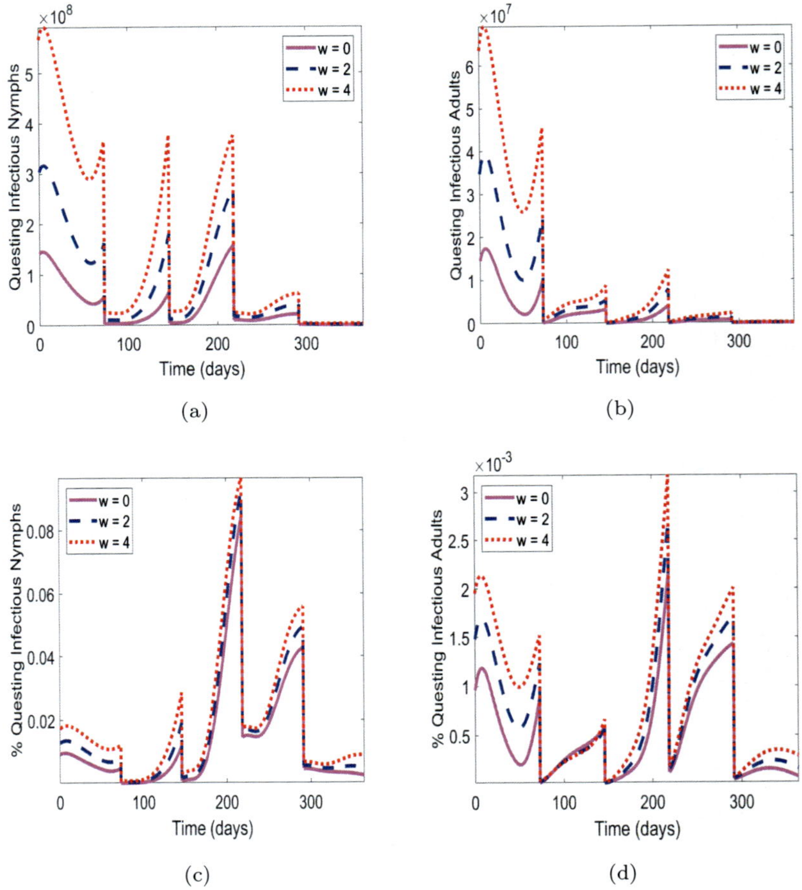

FIGURE 7. Nymph and adult tick population and disease dynamics under endemic scenario with four prescribed burns for current temperature $w = 0$ and increased temperature $w = 2, 4$ °C. (a) and (b) Questing infectious nymphs and adults; (c) and (d) Prevalence of questing nymphs and adults. Parameter values are taken from Table 1.

Because of the increased threat that ehrlichiosis poses, we wanted to investigate whether or not prescribed fire might be a viable solution to mitigating the increased prevalence of both ticks and disease. To evaluate this, we implemented four prescribed burns implemented evenly throughout the year. This is typically unnecessary in an environment that is typically managed using prescribed burns, but it may be done if the goal is to use prescribed burns as a means of tick population control. In this case, we see differences in how each tick-life stage responds following a burn depending on the time of year. We observed in in Figure 7(a), that the infectious questing nymphs have a similar profile of increase following the first and second burns and then very little increase as we approach the end of the year when less ideal temperatures occur. The nymph populations are reduced by two orders of magnitude by day 300 compared to Figure 6(a). We do still see an increased number of ticks as temperature increases, though the difference between each temperature increase is not as severe as the endemic case without fire and this difference diminishes as we reach the end of the year when colder temperatures occur.

Infectious questing adult ticks do not recover as readily as infectious questing nymphs following a prescribed burn. In Figure 7(b), we see an increase in the number of adults following the second burn, but even at the end of this increase, just before the next burn is implemented, the number of adults is roughly an order of magnitude less than in Figure 6(b) at around 200 days. We also see in Figure 7(b) that there is even less of a difference between the results for each increase in temperature compared to the nymphs in Figure 7(a). In Figure 7(c), we see that the percentage of questing infectious adults changes dramatically following each burn after an initial decrease. Following the first burn, the prevalence of disease slowly increases and there is almost no difference in prevalence as temperature increases. Following the second burn, there is a sharp increase in prevalence with a slight distinction as temperature increases. Following the third burn, the prevalence increases similarly regardless of temperature until day 250, after which, we see a greater increase in prevalence of disease as temperature increases. Following the final burn, there we see that the prevalence of disease increases as temperature increases, but from about day 330, it begins to decrease moving into the next year. We observe similar dynamics in Figure 7(d) for prevalence of adult ticks. These results indicate that prescribed fire remains an effective control method for nymph and adult ticks even with temperature increases of up to 4 °C and it also has the potential to reduce the prevalence of disease in ticks in an endemic situation.

3.2.2. *Invasion dynamics.* Invasion of ticks into a new area is a situation that can occur when a female adult tick has finished feeding on a host that has traveled a significant distance while the tick was attached to a new area with little to no ticks. Examples of tick hosts that can travel significant distances before a tick has finished feeding are deer and migratory birds [**3, 29, 30**]. After the female has completed her blood meal, she detaches from the host and lays several thousand eggs. This can lead to new populations of ticks becoming endemic in new areas as well as the possibility of increased spread of tick-borne diseases since the ticks and/or their hosts may be infected [**3, 30**]. To explore these dynamics in the context of rising temperature, we present a simulation with disease and with only an initial number of eggs and a small number of feeding susceptible and infectious ticks and run it to determine the implications that invasion can have for the area.

From Figures 8(a) and 8(b), we see that the number of infectious questing nymphs and adults increases dramatically as temperature increases. They remain much lower in number than the endemic scenario, which indicates that ticks can take several years for these ticks to become fully established after being introduced to a new area. There is a five-fold increase in the peak number of infectious questing nymphs and adults, meaning that ticks will become more dangerous more quickly as temperature increases. The prevalence of disease in these ticks does not follow

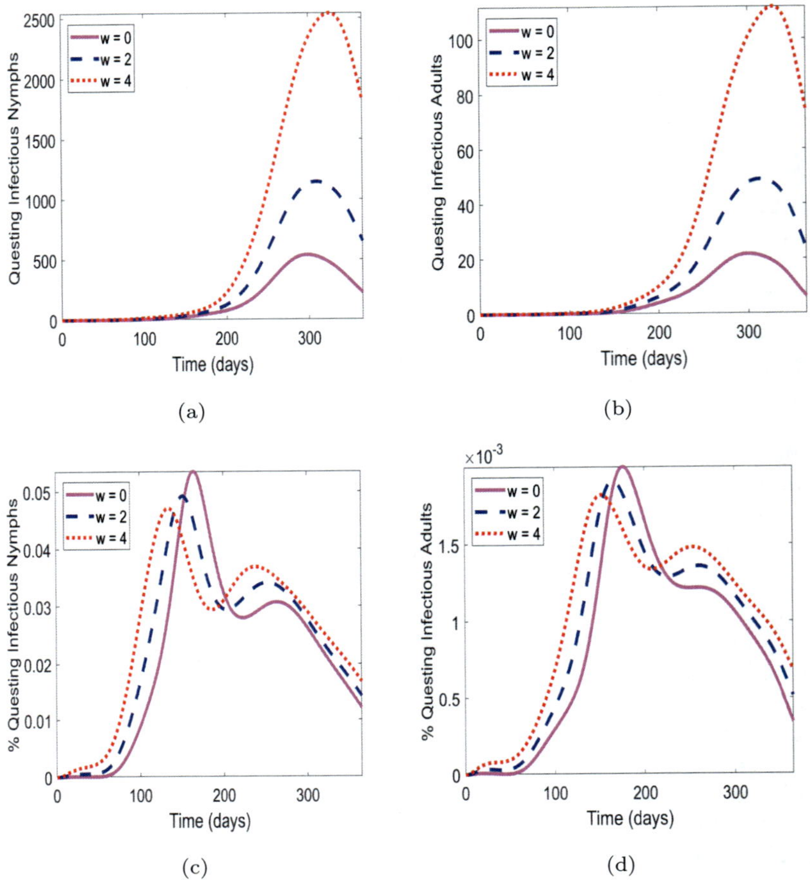

FIGURE 8. Nymph and adult tick population and disease dynamics in an invasion scenario for current temperature $w = 0$ and increased temperature $w = 2, 4$ °C. (a) and (b) Questing infectious nymphs and adults; (c) and (d) Prevalence of questing nymphs and adults. Parameter values are taken from Table 1.

the same trend. We see in Figures 8(c) and 8(d) that the prevalence of disease reaches similar, though slightly lower levels compared to Figure 6(c) and Figure 6(d) within the first year. This is partially due to the fact that we assume the same level of prevalence in tick hosts as the endemic scenario. Since we are in an invasion scenario and ticks are only able to enter a new area via a host moving a significant distance, we assume those hosts have some level of infection a priori. One thing to note about this case is that the peak prevalence of disease is slightly lower as temperature increases. Since every tick population is growing significantly during this first year, and even faster as temperature increases, the rates of infection lag behind slightly and we don't see them flip until roughly day 200 in the simulation. We do see that the prevalence of disease peaks faster and earlier on in the year as temperature increases, which is consistent with previous results [**37**]. This is especially dangerous in an invasion scenario where individuals are likely not aware of the ticks now present in the area and may not be checking themselves for ticks. The lower numbers of ticks in this scenario mitigate that danger temporarily, but ticks will become more and more prevalent each year, increasing the risk that they pose.

For the following scenario in Figure 9, we present the results of implementing four prescribed burns in the area being invaded to explore the effect that burning has on an invasion and to determine if burning can be used to reduce the prevalence of disease in invading ticks.

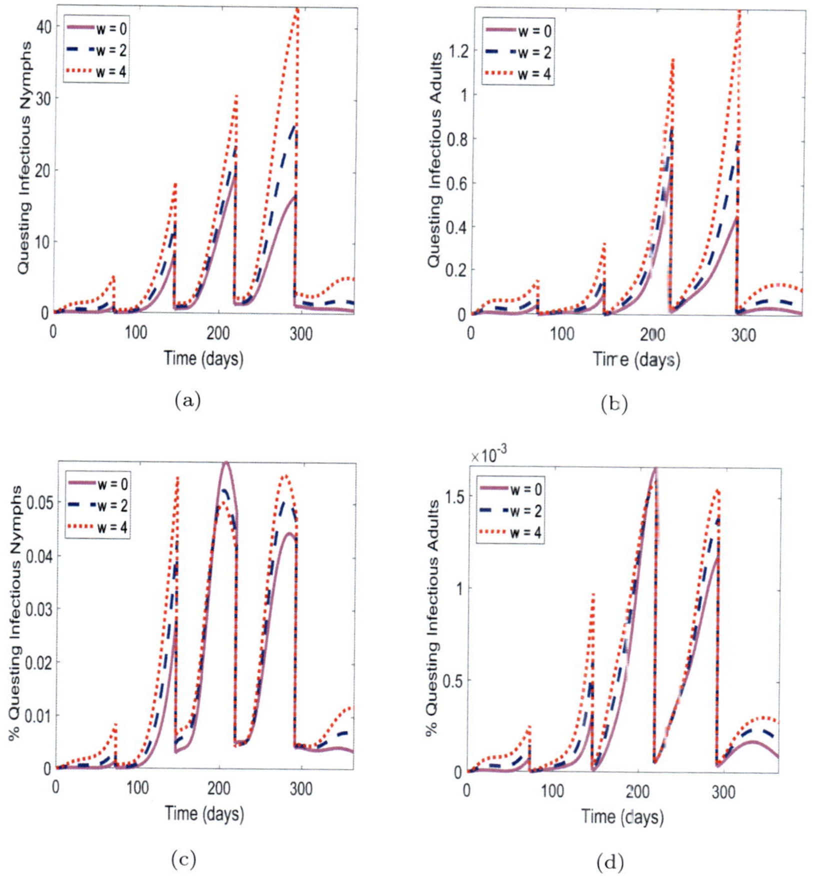

FIGURE 9. Nymph and adult tick population and disease dynamics in an invasion scenario with four prescribed burns for current temperature $w = 0$ and increased temperature $w = 2, 4$ °C. (a) and (b) Questing infectious nymphs and adults; (c) and (d) Prevalence of questing nymphs and adults. Parameter values are taken from Table 1.

Results indicate that prescribed burning is highly effective at slowing establishment of ticks (see Figure 9). Compared to Figure 8, the number of infectious questing nymphs has been reduced from ~2500 to ~50 at their peak in the case of

$w = 4$ (see Figure 9(a)). Increases in the number of nymphs occur faster as temperature increases, but remain relatively low even at the largest temperature increase. The infectious questing adults have been reduced to near zero (Figure 9(b)) even at an increase of temperature of 4 °C, though this does not mean that they pose no threat as the survival of even a single female adult tick means that thousands of eggs may be laid to perpetuate the invasion of these ticks. In addition, we see that the prevalence of disease in both the nymph and adult ticks is again reduced immediately following a burn. From Figure 8(c), the prevalence of ehrlichiosis never surpasses ~0.06% for questing nymphs infected, whereas when prescribed burning is implemented, we see in Figure 9(c) that the infection prevalence surpass 0.06%, though it is important to keep in mind the low numbers of ticks present in this scenario; even if the prevalence of disease is slightly higher, the likelihood of a human encountering an infectious tick is drastically lower since almost none remain in the burned area. An interesting difference in this case is that the prevalence of disease in nymphs and adults is higher for $w = 4$ in most cases, except following the second burn (see Figures 9(c) and 9(d)). In other cases, the trends are similar to the endemic case with fire (see Figures 7(c) and 7(d)).

3.2.3. *Probabilistic implementation of prescribed fire.* To provide a more realistic implementation of the initial effect that prescribed burning has on tick populations, we update $\nu_i, i = e, l, n, a$ values before each burn is performed and set the value to a random value between the baseline and 30% below the baseline value (or above, as with the eggs). The variation in the effectiveness of prescribed burning can occur due to several reasons. Some of which include the escape of ticks on mobile hosts or deep into leaf litter that is not fully burned due to dampness. For a detailed description on how we chose the values of each $\nu_i, i = e, l, n, a$ value for each burn, see Appendix A.

Figure 10 shows the results of a single simulation when the $\nu_i, i = e, l, n, a$ parameters are determined in a probabilistic manner. From this, we see that following a prescribed burn, there is significant variation in how the number of infectious questing nymphs and adults recover following a burn compared to Figures 7(a) and 7(b). There is also significant variation in how the prevalence of disease changes following a prescribed burn compared to Figures 7(c) and 7(d). From Figures 10(a) and 10(b), we see that the trend of a higher number of ticks as temperature increases generally holds, except following the second and third burns in the questing infectious adult population for $w = 4$ and $w = 2$ temperatures, respectively.

We also determined if these trends persist through multiple runs of the same scenario. Figure 11 below shows the results of the average of several simulations.

We ran 100 simulations using this probabilistic implementation of prescribed fire and subsequently averaged the results to see if the trends observed in Figure 10 held generally. The results of each run are displayed in Figure A1 in appendix A. From Figures 11(a) and 11(b), we see that nymphs and adults respond very differently following a prescribed burn depending on the time of year. Early on in the year (i.e. before 220 days), we see the infectious questing nymph population increase above pre-burn levels following the first and second burns. After the third burn, there is a relatively small increase in the number of infectious questing nymphs and once the final burn occurs, we see the population decrease until the end of the year. Infectious questing adult ticks remain in much larger numbers compared to Figure 7(b), and even exceed pre-burn levels following the second burn.

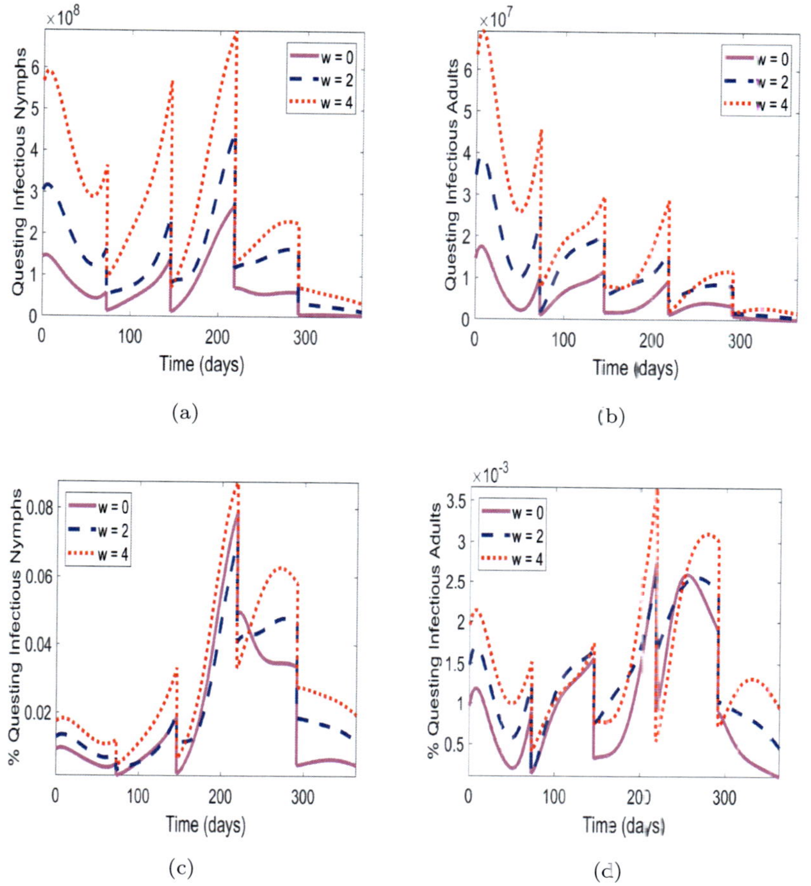

FIGURE 10. The effect of probabilistic prescribed fire under endemic scenario with four prescribed burns for current temperature $w = 0$ and increased temperature $w = 2, 4$ °C. (a) and (b) Questing infectious nymphs and adults; (c) and (d) Prevalence of questing nymphs and adults. Parameter values are taken from Table 1.

The third burn leads to slight increases in each temperature case and there is little difference between the number of tick as temperature increases following the final burn. There are small, but clear increases in the prevalence of questing infectious nymphs in Figure 11(c) as temperature increases and these differences persist even immediately following a burn, unlike what we saw in Figure 7(c). Note that following each burn except the final, the prevalence of infectious questing nymphs increases consistently until the next burn. In Figure 11(d), we observed that the prevalence of infectious questing adult ticks remains surprisingly similar to the case with a static implementation of prescribed fire (see Figure 7(d)). Following the second burn, there is very little difference in the prevalence regardless of an increase in temperature, but this trend does not persist following the second and third burns.

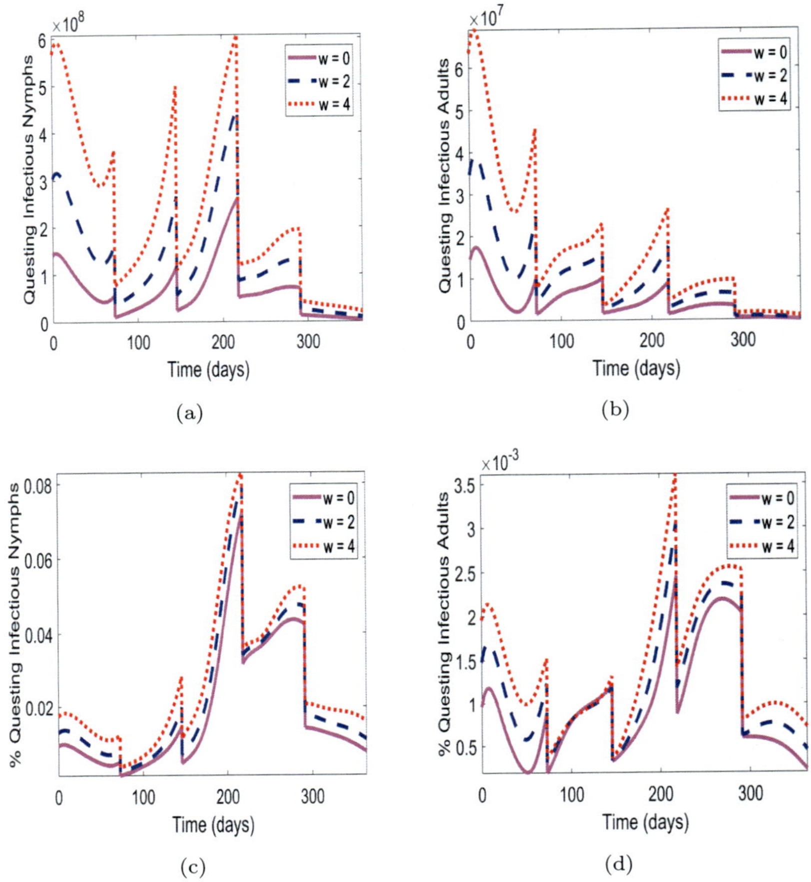

FIGURE 11. The effect of probabilistic prescribed fire under endemic scenario averaged over 100 run with four prescribed burns for current temperature $w = 0$ and increased temperature $w = 2, 4$ °C. (a) and (b) Questing infectious nymphs and adults; (c) and (d) Prevalence of questing nymphs and adults. Parameter values are taken from Table 1.

Following the final burn, we again see an increase in the prevalence until around day 330, after which the prevalence slowly decreases.

Figure 12 displays the results of a single simulation under an invasion scenario with probabilistic prescribed fire. From Figures 12(a) and 12(b), we see that prescribed fire is slightly less effective compared to the static case. We still observe that the higher the temperature, the more infectious questing nymphs and adults, but still at very low numbers compared to the invasion scenario without prescribed fire (Figure 8). The prevalence of infectious questing nymphs and adults shown in Figures 12(c) and 12(d) show significant variation. The prevalence of infectious questing nymphs is similar to the static case following the first burn, but following the second burn we see that an increase in temperature does not imply an increase in the prevalence of questing infectious nymphs as was seen in Figure 8(c) and

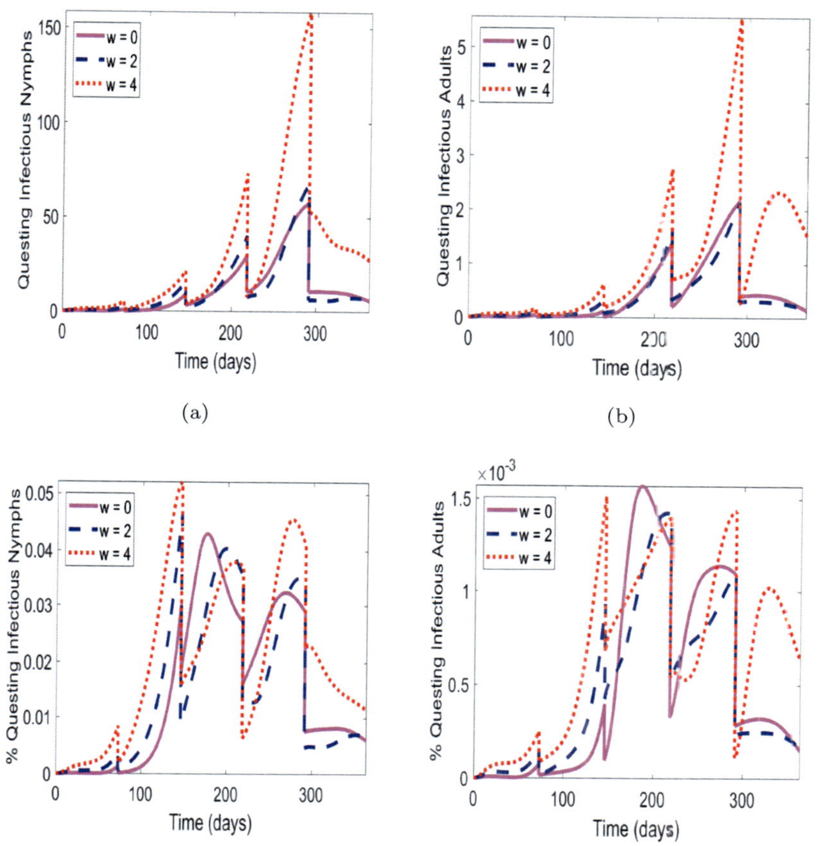

FIGURE 12. The effect of probabilistic prescribed fire in an invasion scenario with four prescribed burns for current temperature $w = 0$ and increased temperature $w = 2, 4$ °C. (a) and (b) Questing infectious nymphs and adults; (c) and (d) Prevalence of questing nymphs and adults. Parameter values are taken from Table 1.

this is also the case following the third and fourth burns as well. Interestingly, the prevalence of questing infectious adult ticks also responds similarly following the first burn, in that it increase consistently and there is a clear increase when going from $w = 0$, 2, and 4 °C. Following the second burn, we see that with current temperatures ($w = 0$) and temperature increase of 2 °C, a higher prevalence of infectious questing adults is achieved for a portion of the time before the next burn. The third burn leaves the base case ($w = 0$) at a higher initial prevalence, but this is quickly overtaken by the $w = 2$ and $w = 4$ °C cases, with the highest prevalence being achieved when $w = 4$ °C. After the final burn, the prevalence decreases for $w = 0$, and increases significantly for $w = 4$ before decreasing in the questing infectious adult population. For the intermediate temperature case ($w = 2$), there is little overall change in prevalence following the final burn.

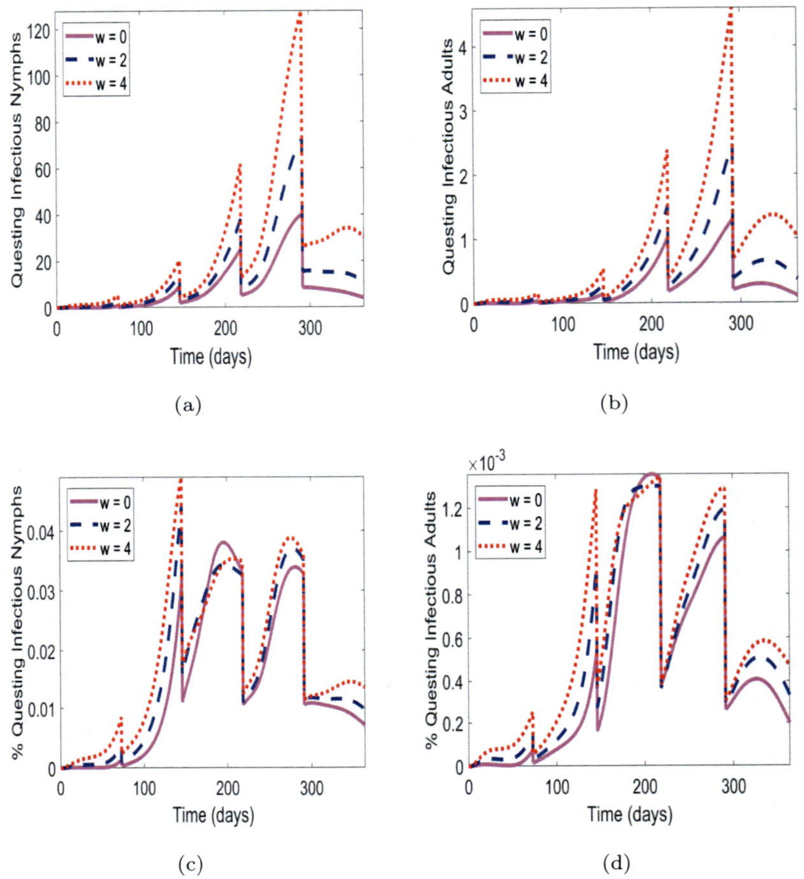

FIGURE 13. The effect of probabilistic prescribed fire in an invasion scenario averaged over 100 runs current temperature $w = 0$ and increased temperature $w = 2, 4$ °C. (a) and (b) Questing infectious nymphs and adults; (c) and (d) Prevalence of questing nymphs and adults. Parameter values are taken from Table 1.

The results of four probabilistic prescribed burns averaged over 100 runs in an invasion scenario is displayed in Figure 13. The results of each individual run are shown together in Figure A2 in Appendix A. As occurred in the endemic case, we see that much of the variation that was seen in a single run of probabilistic prescribed fire is not maintained when averaging over several runs. We again see in Figures 13(a) and 13(b) slightly higher numbers of infectious questing nymphs and adults compared to the static case, though it is still clear that invasion has been delayed thanks to the prescribed burns. In Figures 13(c) and 13(d) we observed that the prevalence of infectious questing nymphs and adults again have commonalities following the first burn in that they increase and there is a greater prevalence of infectious questing nymphs and adults as temperature increases. Following the second burn, there is much less of a difference in prevalence as temperature increases, though the peak reach for the prevalence of questing infectious nymphs is slightly higher for the $w = 0$ case, which was also observed in Figure 9(c). After the third and fourth burns, we see that the trend of increasing prevalence of infectious questing nymphs and adults as temperature increases holds true.

4. Discussion and conclusion

4.1. Discussion. As far as we know, there have been two approaches used thus far to model tick populations and the effects of prescribed burning on those ticks [**12, 17**]. The first by Guo and Agusto [**17**] evaluated the effectiveness of high- and low-intensity prescribed burning as the duration between burns increases via an impulsive system of ordinary differential equations. They found that there was little difference in the number of *I. scapularis* ticks calculated in their simulations with a high- and low-intensity burn interval of six years (see Figure 4 in [**17**]). They also concluded that burn intensity matters more than frequency, which is important since high-intensity burns are likely not able to be performed as often as low intensity burns and thus one may opt for performing a high-intensity burn every few years rather than low-intensity burns more often. The second by Fulk et. al. [**12**] extended the model from [**17**] to a system of impulsive partial differential equations in order to determine how the effects of a spatial setting alter the effectiveness of prescribed burns. There, they found that a crucial determinant of the effectiveness of a prescribed burn was the size of the burn being performed. Since ticks and the hosts that they attach to tend to disperse across an environment, if the size of the burn being performed is smaller than the rate of movement of the ticks/hosts, the effectiveness of the burn may be diminished as they repopulate the burned area by diffusing from the surrounding area. Those models focused on *Ixodes scapularis*, the main vector for Lyme disease in much of the United States.

While it is likely that some of the conclusions hold for *Amblyomma americanum* and other tick species (e.g. high intensity burns being more effective than low intensity burns and smaller prescribed fires being less effective at reducing tick populations) since the ranges of some of these ticks overlap, the physiological differences between *Ixodes scapularis*, *Amblyomma americanum* and other tick species that are prominent in the United States (e.g. *Dermacentor variabilis*) warrant a thorough investigation into the differing effects of prescribed burning on each of these ticks. Gleim et. al. [**13**] provided support to this idea when they found that the distribution of various tick species in southwestern Georgia and northeastern Florida differed by site (e.g. unburned sites surrounded by unburned were dominated by *Amblyomma americanum*, while unburned sites surround by commonly burned areas were dominated by *Ixodes scapularis*).

Due to these findings, it thus important to include more aspects of this tick's biology to properly determine the effect of rising temperature on populations of *Amblyomma americanum* ticks as well as the effects of prescribed burning in both endemic and invasion scenarios. Our results indicated that, depending on the life stage of the tick, populations of *Amblyomma americanum* may be able to recover very quickly following a prescribed burn in an endemic scenario, especially if the burn occurs before the ticks reach their highest peak numbers (see the increases in the nymph population in Figure 7(a) following the first and second burns), but frequent burning over the course of a year significantly reduces the number of ticks.

Adult *Amblyomma americanum* ticks increased at lower rates following a burn (see Figure 7(b)), indicating that timing a prescribed burn when this life stage is most prevalent may lead to increased effectiveness in reducing the number of ticks, especially in the following year since egg-laying adults are the main source of questing larvae.

The results of the invasion scenarios seem to indicate that as temperature increases, *Amblyomma americanum* will be able to establish itself in large numbers much more quickly. When prescribed fire is introduced into an area where these ticks have recently invaded, we see their populations grow at much lower rates. We also see that few infectious questing adult ticks are able to survive the burns, further indicating that timing a burn during times of the year when this life stage is most active may be effective at further slowing the establishment of these ticks. Eradication of ticks from an area using prescribed burning alone likely is not possible [**12**], thus additional control measures should be implemented in conjunction with prescribed burning if the goal is eradication of ticks. Four prescribed burns is an unsustainable number to perform on an annual basis, but it is extremely effective for an initial reduction in tick populations in both endemic and invasion scenarios. If the goal of prescribed burns is tick population management, then as many burns as possible should be performed in a year as possible, followed by an additional burn at regular intervals (annually, or more frequently if possible).

The prevalence of infectious questing nymphs and adults was also evaluated in each of our scenarios. The percentage of infectious questing nymphs and adults increases slightly in an endemic scenario as temperature increases, but the greater threat is the increase in the number of infectious questing nymphs and adults (see Figure 6). The invasion scenario highlighted another likely effect of rising temperatures. Prescribed burning seems to be initially effective at reducing the prevalence of infectious questing nymphs and adults, however the prevalence quickly returns to and exceed pre-burn levels if the burn is performed before approximately day 220 for nymphs. The prevalence of infectious questing adults does not increase in the same way that the nymphs do, meaning that an individual is less likely to encounter an infectious questing adult in both endemic and invasion scenarios and with or without prescribed burning. These factors indicate that prescribed burning may not be the most effective way to reduce the prevalence of disease in tick populations. This is because prescribed burning likely does not kill many of the reservoir hosts of ehrlichiosis. One of the key reservoir hosts is white-tailed deer, which are able to move quickly relative to a prescribed burn and may return to recently burned areas quickly to feed on newly growing vegetation. If eradication of disease is the main goal in an area, then other control methods that directly target these reservoir hosts may lead to a more effective long-term reduction in the prevalence of disease in *Amblyomma americanum*.

4.2. Conclusion. In this study we investigated the effect of rising temperature on *Amblyomma americanum* populations and the prevalence of infectious questing nymphs and adults under endemic and invasion scenarios. We also evaluated the effectiveness of prescribed burning tick populations and the percentage of infectious questing nymphs and adults. We have three main conclusions on the effects of rising temperature in the absence of prescribed burning: (1) as temperature increases, there are significant increases in the number of infectious questing nymphs and adults. (2) The prevalence of infectious questing nymphs and adults increases slightly, but the greater threat to public health is the increase in the number of infectious ticks. (3) Ehrlichiosis becomes established in the questing nymph and adult tick populations quicker under the invasion scenario as temperature increases.

We note that these conclusions with prescribed burning hold regardless of increases in temperature, indicating that prescribed burning remains an effective control method for *Amblyomma americanum* even in the case of a significant increase in temperature of 2 and 4 °C.

There are some limitations to our model that should be mentioned. We assumed a constant number of susceptible and infectious small and large hosts. We do not expect that dynamic populations of these hosts would lead to significantly different conclusions since one of the main hosts for *Amblyomma americanum* (and ehrlichiosis) in many areas is the white-tailed deer, which remains active to varying degrees during much of the year. We hypothesized that timing a prescribed burn to when the adult tick population is near its peak may lead to increased effectiveness of the burn in reducing tick populations in the following year. In Section 4.1, though we are not able to address that question directly with our current model since our goal was not to accurately model tick phenology. This is a very interesting question that likely has a different answer depending on which tick species is being considered and we plan to address it in future works.

Competing interests

The authors have declared no competing interest.

Data availability statement

All data and code produced in this study are available upon reasonable request to the corresponding author.

Appendix A. Probabilistic prescribed fire

To calculate the proportion of ticks reduced following each of the prescribed burns implemented in Section 3.2.3, we used the following scheme. For the egg population, we first defined the range of a uniform distribution on the interval [0, 0.3] for a maximum reduction of 30% and extracted a value from this range at each burn throughout the year. For the other tick populations, we used ranges for a uniform distribution from 30% below to the baseline values described in Section 2.3. So, for the larvae populations, the range of possible values for the uniform distribution was [0.693, 0.99]. For the nymph populations, the range of possible values for the uniform distribution was [0.653919, 0.93417]. For the adult populations, the range of possible values for the uniform distribution was [0.6846392, 0.978056]. These values are used in the impulsive equations described in Section 2.1.

Figure A1 shows the results of each run used to generate the average effect of probabilistic prescribed burning provided in Figure 11. Burns initiated earlier in the year lead to similar increases in tick nymphs as temperature increases. Following the third burn, there is more variation, particularly in the prevalence of disease in tick nymphs. There is much more variation seen in the numbers of adult ticks following the first burn, but not the second. This also holds for the prevalence of disease in the tick populations. Following the third and fourth burns, we again

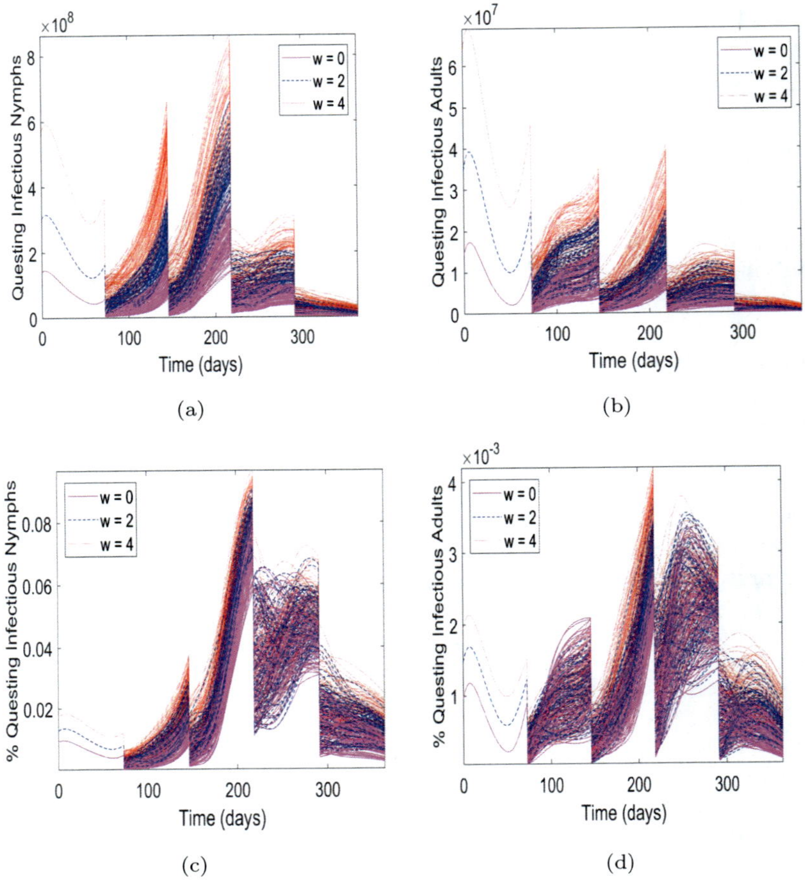

FIGURE A1. The effect of probabilistic prescribed fire in an endemic scenario over 100 runs with four prescribed burns for current temperature $w = 0$ and increased temperature $w = 2, 4\ ^\circ\mathrm{C}$. (a) and (b) Questing infectious nymphs and adults; (c) and (d) Prevalence of questing nymphs and adults. Parameter values are taken from Table 1.

see significant variation. When averaging over all 100 runs, this variation is not maintained.

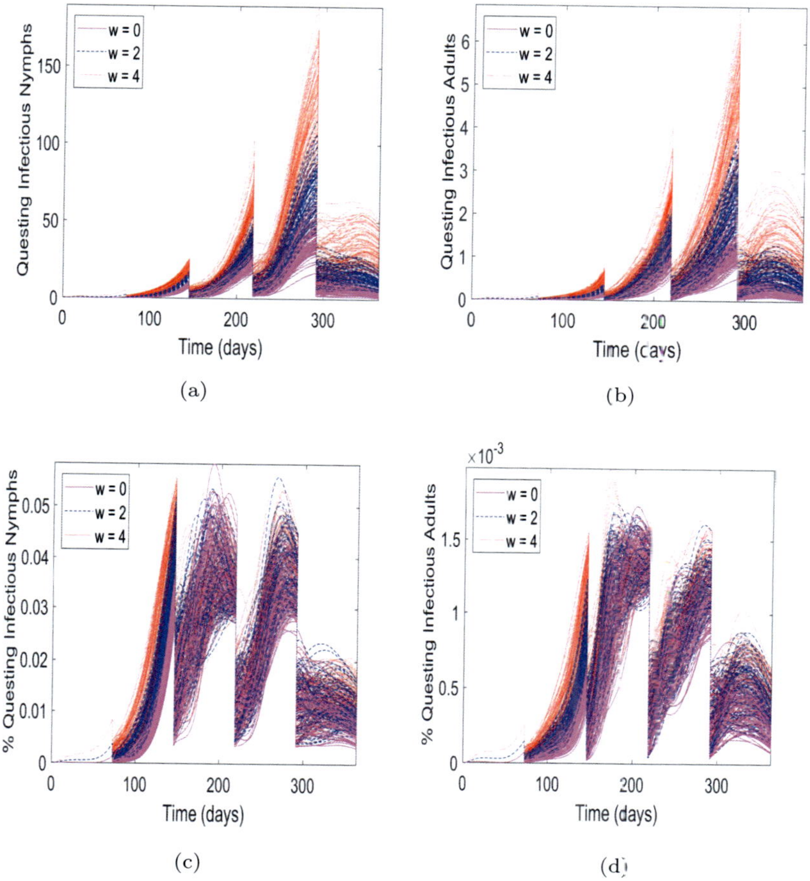

FIGURE A2. The effect of probabilistic prescribed fire in an invasion scenario over 100 runs with four prescribed burns for current temperature $w = 0$ and increased temperature $w = 2, 4\,^\circ$C. (a) and (b) Questing infectious nymphs and adults; (c) and (d) Prevalence of questing nymphs and adults. Parameter values are taken from Table 1.

Figure A2 shows the results of each run used to generate the average effect of probabilistic prescribed burning provided in Figure 13. There is a clear distinction between the numbers of infectious questing nymphs and adults as temperature increases throughout the year, which was not observed in the endemic scenario. Prevalence of infectious questing nymphs and adults varies significantly following the second third and fourth burns, but not following the first burn.

References

[1] Joseph K. Bailey and Thomas G. Whitham, *Interactions among fire, aspen, and elk affect insect diversity: reversal of a community response*, Ecology **83** (2002), no. 6, 1701–1712.

[2] Johan S. Bakken, *The discovery of human granulocytotropic ehrlichiosis*, Journal of Laboratory and Clinical Medicine **132** (1998), no. 3, 175–180.

[3] Alicja M. Buczek, Weronika Buczek, Alicja Buczek, and Katarzyna Bartosik, *The potential role of migratory birds in the rapid spread of ticks and tick-borne pathogens in the changing climatic and environmental conditions in europe*, International Journal of Environmental Research and Public Health **17** (2020), no. 6, 2117.

[4] Centers for Disease Control and Prevention (CDC), *Ehrlichiosis*, Centers for Disease Control and Prevention (2022).

[5] Ijeuru Chikeka and J. Stephen Dumler, *Neglected bacterial zoonoses*, Clinical Microbiology and Infection **21** (2015), no. 5, 404–415.

[6] James E. Childs and Christopher D. Paddock, *The ascendancy of amblyomma americanum as a vector of pathogens affecting humans in the united states*, Annual Review of Entomology **48** (2003), 307–337.

[7] Jack F. Cully, *Lone star tick abundance, fire, and bison grazing in tallgrass prairie.*, Rangeland Ecology & Management/Journal of Range Management Archives **52** (1999), no. 2, 139–144.

[8] F. Scott Dahlgren, Eric J. Mandel, John W. Krebs, Robert F. Massung, and Jennifer H. McQuiston, *Increasing incidence of ehrlichia chaffeensis and anaplasma phagocytophilum in the united states, 2000–2007*, The American Journal of Tropical Medicine and Hygiene **85** (2011), no. 1, 124.

[9] William R. Davidson, Debra A. Siefken, and Lynn H. Creekmore, *Influence of annual and biennial prescribed burning during march on the abundance of amblyomma americanum (acari: Ixodidae) in central georgia*, Journal of Medical Entomology **31** (1994), no. 1, 72–81.

[10] Mark L. Drew and W.M. Samuel, *An evaluation of burning for control of winter ticks, dermacentor albipictus, in central alberta*, Journal of Wildlife Diseases **21** (1985), no. 3, 313–315.

[11] Haley Elizabeth Frater, *Impact of prescribed burning for oak regeneration on forest vegetation, white-footed mouse populations, and lyme disease*, Iowa State University, 2011.

[12] Alexander Fulk, Weizhang Huang, and Folashade B. Agusto, *Exploring the effects of prescribed fire on tick spread and propagation in a spatial setting*, Computational and Mathematical Methods in Medicine **2022** (2022).

[13] Elizabeth R. Gleim, L. Mike Conner, Roy D. Berghaus, Michael L. Levin, Galina E. Zemtsova, and Michael J. Yabsley, *The phenology of ticks and the effects of long-term prescribed burning on tick population dynamics in southwestern georgia and northwestern florida*, PLoS One **9** (2014), no. 11, e112174.

[14] Elizabeth R. Gleim, Galina E. Zemtsova, Roy D. Berghaus, Michael L. Levin, Mike Conner, and Michael J. Yabsley, *Frequent prescribed fires can reduce risk of tick-borne diseases*, Scientific Reports **9** (2019), no. 1, 1–10.

[15] Jerome Goddard and Andrea S. Varela-Stokes, *Role of the lone star tick, amblyomma americanum (l.), in human and animal diseases*, Veterinary Parasitology **160** (2009), no. 1-2, 1–12.

[16] PRISM Climate Group, *Data explorer: Download time series values in bulk*, `https://prism.oregonstate.edu`, 2020, Generated: 2022-09-10.

[17] Emily Guo and Folashade B. Agusto, *Baptism of fire: Modeling the effects of prescribed fire on lyme disease*, Canadian Journal of Infectious Diseases and Medical Microbiology **2022** (2022).

[18] Kristen Nichols Heitman, F. Scott Dahlgren, Naomi A. Drexler, Robert F. Massung, and Casey Barton Behravesh, *Increasing incidence of ehrlichiosis in the united states: a summary of national surveillance of ehrlichia chaffeensis and ehrlichia ewingii infections in the united states, 2008–2012*, The American Journal of Tropical Medicine and Hygiene **94** (2016), no. 1, 52.

[19] A.L. Hoch, Paul J. Semtner, Robert W. Barker, and Jakie A. Hair, *Preliminary observations on controlled burning for lone star tick (acarina: Ixodidae) control in woodlots*, Journal of Medical Entomology **9** (1972), no. 5, 446–451.

[20] H.A. Jacobson and G.A. Hurst, *Prevalence of parasitism by amblyomma americanum on wild turkey poults as influenced by prescribed burning*, Journal of Wildlife Diseases **15** (1979), no. 1, 43–47.

[21] Jeffrey Kane, *prescribed fire*, `https://www.britannica.com/science/prescribed-fire`, 2020, Accessed: 2022-09-09.

[22] S. Wesley Long, Xiaofeng Zhang, Jianzhi Zhang, Randall P. Ruble, Pete Teel, and Xue-Jie Yu, *Evaluation of transovarial transmission and transmissibility of ehrlichia chaffeensis (rickettsiales: Anaplasmataceae) in amblyomma americanum (acari: Ixodidae)*, Journal of Medical Entomology **40** (2003), no. 6, 1000–1004.

[23] Antoinette Ludwig, Howard S. Ginsberg, Graham J. Hickling, and Nicholas H. Ogden, *A dynamic population model to investigate effects of climate and climate-independent factors on the lifecycle of amblyomma americanum (acari: Ixodidae)*, Journal of Medical Entomology **53** (2016), no. 1, 99–115.

[24] Simeone Marino, Ian B. Hogue, Christian J. Ray, and Denise E. Kirschner, *A methodology for performing global uncertainty and sensitivity analysis in systems biology*, J. Theoret. Biol. **254** (2008), no. 1, 178–196, DOI 10.1016/j.jtbi.2008.04.011. MR2971156

[25] Thomas N. Mather, David C. Duffy, and Scott R. Campbell, *An unexpected result from burning vegetation to reduce lyme disease transmission risks*, Journal of Medical Entomology **30** (1993), no. 3, 642–645.

[26] MATLAB, *version 9.10.0 (r2021a)*, The MathWorks Inc., Natick, Massachusetts, 2021.

[27] Michael Mogg, Hsiao-Hsuan Wang, Adam Baker, Zakary Derouen, Jennifer Borski, and William E. Grant, *Increased incidence of ehrlichia chaffeensis infections in the united states, 2012 through 2016*, Vector-Borne and Zoonotic Diseases **20** (2020), no. 7, 547–550.

[28] Thomas C. Moore, Laura A. Pulscher, Luke Caddell, Michael E. von Fricken, Benjamin D. Anderson, Battsetseg Gonchigoo, and Gregory C. Gray, *Evidence for transovarial transmission of tick-borne rickettsiae circulating in northern mongolia*, PLoS Neglected Tropical Diseases **12** (2018), no. 8, e0006696.

[29] Michael E. Nelson, L. David Mech, and Paul F. Frame, *Tracking of white-tailed deer migration by global positioning system*, Journal of Mammalogy **85** (2004), no. 3, 505–510.

[30] Nicholas H. Ogden, L. Robbin Lindsay, K. Hanincová, Ian K. Barker, Michel Bigras-Poulin, Dominique F. Charron, Audrey Heagy, Charles M. Francis, Christopher J. O'Callaghan, I. Schwartz, et al., *Role of migratory birds in introduction and range expansion of ixodes scapularis ticks and of borrelia burgdorferi and anaplasma phagocytophilum in canada*, Applied and Environmental Microbiology **74** (2008), no. 6, 1780–1790.

[31] Intergovernmental Panel on Climate Change, *Global warming of 1.5° c: An ipcc special report on the impacts of global warming of 1.5° c above pre-industrial levels and related global greenhouse gas emission pathways, in the context of strengthening the global response to the threat of climate change, sustainable development, and efforts to eradicate poverty*, Intergovernmental Panel on Climate Change, 2018.

[32] C.D. Paddock and M.J. Yabsley, *Ecological havoc, the rise of white-tailed deer, and the emergence of amblyomma americanum-associated zoonoses in the united states*, Wildlife and Emerging Zoonotic Diseases: the Biology, Circumstances and Consequences of Cross-Species Transmission (2007), 289–324.

[33] Eric L. Rulison, Roger A. Lebrun, and Howard S. Ginsberg, *Effect of temperature on feeding period of larval blacklegged ticks (acari: Ixodidae) on eastern fence lizards*, Journal of Medical Entomology **51** (2014), no. 6, 1308–1311.

[34] Kevin C. Ryan, Eric E. Knapp, and J. Morgan Varner, *Prescribed fire in north american forests and woodlands: history, current practice, and challenges*, Frontiers in Ecology and the Environment **11** (2013), no. s1, e15–e24.

[35] US Forest Service, *US Forest Service — prescribed fire*, `https://www.fs.usda.gov/managing-land/prescribed-fire`, Accessed: 2022-09-09.

[36] David H. Walker, Christopher D. Paddock, and J. Stephen Dumler, *Emerging and re-emerging tick-transmitted rickettsial and ehrlichial infections*, Medical Clinics of North America **92** (2008), no. 6, 1345–1361.

[37] Dorothy Wallace, Vardayani Ratti, Anita Kodali, Jonathan M. Winter, Matthew P. Ayres, Jonathan W. Chipman, Carissa F Aoki, Erich C. Osterberg, Clara Silvanic, Trevor F. Partridge, et al., *Effect of rising temperature on lyme disease: Ixodes scapularis population dynamics and borrelia burgdorferi transmission and prevalence*, Canadian Journal of Infectious Diseases and Medical Microbiology **2019** (2019).

[38] Jianhong Wu and Xue Zhang, *Transmission dynamics of tick-borne diseases with co-feeding, developmental and behavioural diapause*, Springer Nature, 2020.

Department of Ecology and Evolutionary Biology, University of Kansas, Lawrence, Kansas

Email address: alxjfulk@gmail.com

Department of Ecology and Evolutionary Biology, University of Kansas, Lawrence, Kansas

Email address: fbagusto@gmail.com

Contemporary Mathematics
Volume **793**, 2024
https://doi.org/10.1090/conm/793/15905

Modeling T-cell repertoire response to a viral infection with short immunity

Elena N. Naumova, Maryam B. Yassai, Jack Gorski, and Yuri N. Naumov

ABSTRACT. At the molecular level, we observed that immune response to the influenza virus becomes diverse upon repeated viral exposures and can be presented as a fractal self-similar system. We proposed an adaptation of the Markov chain process to model the dynamics of memory T-cell repertoire considering typical experimental conditions. We derived model-inspired characteristics to quantify repertoires' temporal stability and detected significant age-specific differences. These differences indicate that transitional probabilities are subject to change as the immune system evolves and could potentially serve as a marker of immunosenescence. We concluded that for infection with short immune memory, a boost via periodic vaccinations or direct exposures is critical for sustaining diverse immune responses.

Introduction

Modeling the spread of influenza in humans has been explored in many domains—from deterministic models to detailed examination of global viral evolution—and in many scales—from molecular to population levels. Despite global efforts to control the spread of influenza with systematic approaches to vaccine development and tracking via national surveillance systems, this seasonal infection continues to pose sizable large-scale outbreaks and global pandemic threats. The high pandemic potential is driven by rapid and frequent viral mutations and short immune memory. Immune memory to influenza must be much shorter than the human average life expectancy, which is 70 years, worldwide [**1**]. The duration of immune memory to a specific influenza strain is difficult to measure. Based on recent studies, it was reported that re-exposure to new influenza strains occurs every 5 years [**2**], whereas protective immunity, mediated mostly by B-cells, lasts about 4 to 8 months [**3–6**]. It means that a person could experience at minimum a dozen of re-infections over a lifetime.

The immune system represents the major system with a large cellular component dedicated to the generation of adaptive memory to pathogens. The immune

All authors were supported in part by NIH Grants U19 AI062627 and NO1 AI50032. The first author was also supported by IARPA Grant 2017-17072100002.

This research is based upon work supported in part by the Office of the Director of National Intelligence (ODNI), Intelligence Advanced Research Projects Activity (IARPA), via 2017-17072100002 (Naumova–PI). This work was supported by NO1 AI50032 and U19 AI062627 (Gorski -PI) from the National Institutes of Health.

system is tasked with protecting against encountered pathogens and it does so remarkably well although incompletely, partially because of short immune memory. In our early work, we created a framework to formalize, measure, quantify, and model the dynamics of selection, sustainment, and re-engagement of virus-specific immune T-cells that form immune memory to a previously encountered pathogen [**6**]. We focused on key mechanisms that govern the formation, maturation, and senescence of clonal repertoires of CD8 T-cells which are responsible for the detection and recognition of a viral peptide in a host. We presented this framework from a system point of view and defined the immune system as a collection of specialized cells in a multi-cell organism with its evolving structure to ensure organism survival. We explored the dynamics of the immune system using CD8 T-cells expressing T-cell receptors [$\alpha\beta$-TCR heterodimer], that have the 'best structural fitness' for pathogen recognition and clearance. As clonotypes are the clonal offspring of a fully recombined unmutated common ancestor and are distinguished by the nucleotide sequence of the rearranged T-cell receptor, their gene recombination are unique and can be used for tracking clonal cells within a pool of cells of different clonalities. The collections of clonotypes, finely tuned for pathogen recognition, form T-cell repertoire.

To aid the quantitation of T-cell repertoires in experimental conditions, we proposed a naming convention to keep track of the clonotypes [**7**]. The clonotype naming strategy incorporates labeling of essential elements that characterize genetic sequences (V-, J-, C-genes, encoding TCR α-chain, and V-, D-, J-, C-genes, encoding β-chain) that allows extraction and analysis of biologically relevant information from a unique clonotype identifier (when clonal identities are based on TCR β-chain usage). This approach applies to all fields of current cellular immunobiology [**8**, **9**]. Clonotyping, the original process of designing an experiment to define the repertoire in response to pathogen viral peptides, has been now widely replaced by full gene sequence and identification of peptide or epitope-specific clonotypes [**10**]. Current high-throughput sequencing methods use unique molecular identifiers (UMIs) introduced during cDNA synthesis to distinguish between single RNA molecules and minimize the impact of PCR amplification and sequencing errors, and specific algorithms for correction of this particular data type [**11**, **12**].

While imperfect, the early experiments allowed the detection of common patterns, distributional features, and ways in which TCR repertoire could change due to high viral load [**13**] or due to immunosenescence [**14**]. Most importantly, the experimental base and naming convention allowed us to analyze repertoires from many donors and at multiple time points. With these first experiments of studying the CD8 T-cell response to an influenza-derived epitope, we established that the response was polyclonal and represented by multiple clonotypes or clones with amino acid motifs within CDR3 regions of TCR α- and β-chains [**14**–**17**]. These findings indicated the potential for the repertoires to be diverse and likely complex.

In the mid-late 90s, the common view was that a mid-aged individual possesses a few clones expressing T-cell receptors with the 'best fit' for epitope recognition, likely due to several exposures to flu over the life course. We found the opposite – many clones responded to a viral epitope. Furthermore, the clones responded with different intensities or frequencies. We introduced several key repertoires characteristics to quantify and formally compare clonotypic distributions based on clonal ranks and rank frequencies [**15**, **16**]. We proposed a measure of clonotypic diversity based on the average number of copies per clonotype and the highest observable rank [**18**]. By combining and aligning repertoires from several donors and at multiple time points, the repertoires' self-similar fractal structure emerges.

TCR system's scale-free behavior

The accurate repertoire mapping, analysis, and interpretation of its evolution over the lifetime and altered by illness require novel tools for data acquisition, database formation, and statistical analysis. We found that if we count unique clonotypes and rank them according to their frequency, the emerging patterns follow a power-like law distribution, a phenomenon commonly observed in nature. Further analysis allowed us to suggest that repertoires represent an example of a complex scale-free fractal system. This led us to examine repertoire dynamics *in silico* [**6**] and verify their behaviors *in vitro* [**16**]. This framework: the nomenclature, repertoire characteristics, modeling agreements, and normalization of rank-frequency allowed us to formulate and test new hypotheses related to repertoire stability and immunosenescence.

A scale-free behavior has been observed in many living systems at micro- and macro-scales. These systems continue to stimulate interest in theoretical studies, including the understanding of immune system maturation and aging. It is this component of immunity that is the most instructive in understanding the life stages of humans. In the experimental studies of the adaptive immune system, we observed a scale-free network governing the repertoire of memory T-cells [**16**]. Theoretical explanation of experimental findings has been described by the small-world construction [**19**] as a special case of the scale-free network [**20**]. We simulated the fractal behavior mimicking immune memory - its generation, maintenance, and senescence [**6**] and experimentally illustrated the general stability of the power-law structures and age-related changes [**14**]. Our recent theoretical work confirmed the assumptions that multiple 'expansion-contraction' cycles define the robustness of immune response and correspond to memory formation [**6, 21**]. Our observations agreed with the findings of Saito and Narikiyo, who proposed a dynamical network of the adaptive immune system as a self-organized critical state in which the avalanche feedback reinforcement may reduce immunosenescence [**22**].

At the population level, seasonal outbreaks of influenza infections are common. We proposed that multiple exposures to influenza could serve as a marker of immunological age [**14, 18**]. In the cohort of healthy donors, each encounter with an infectious agent is unique for every person. Yet, the commonality in responses formed immunological kinship among all affected individuals, manifested by a preserved T-cell clonal pool. The differences in responses to seasonal flu and changes in repertoire diversity allowed us to make an inference to immunological kinship and immunological age. We proposed that repertories' periodic 'expansion-contraction' cycles are related to recurrent exposure to the influenza virus and informed the lifetime evolution of immune memory [**21**]. These experimental data indicate that recurrent exposures to influenza alter the diversity of immune response.

These observations lead us to explore conditions governing the 'stable' and 'volatile' components of T-cell repertoires. Understanding clonotypes' temporal stability allows us to learn how immune responses mature and age. Our experimental work indicated that the stability of influenza-specific T-cells in children is low, so not many identical clonotypes were frequently observed in longitudinal samples

[**14**]. In the mid-aged cohort, the stability is more prominent, especially in dominant clonotypes. In the older cohort, the loss of stability varied. For some people, repertoires showed child-like properties of low stability; and for some, repertoires maintained stable high-ranking clonotypes but were losing stability at intermediate ranks. Thus, aging can disturb the distribution patterns of stable clonotypes.

In our earlier work, we measured stability as a fraction of the times a clonotype was observed in longitudinal samples [**14**]. Although this measure is simple and intuitive, it does not capture the pattern of clonotype appearances. In this communication we proposed a framework to define possible pathways of clonotypic stability and provided a refined approach for assessing clonotypic stability and associated properties of T-cell repertoires. We applied a Markov chain, a discrete-time discrete-valued random process that follows a specific property stating that the probability distribution of the next state (future) depends only on the probability distribution of the current state. We derived model parameters based on experimental longitudinal data from children, middle-aged and older donors collected over several years, illustrated age-related acceleration of immunosenescence, and offered potential implications and directions for future research.

Modeling framework

We assumed that the repertoire temporal stability is observable, and the circulating pool is a depot of selected clonotypes. We assumed that stability changes with age, and we could detect meaningful changes at a reasonable time interval. We assumed that exposure to flu can occur over time and we could measure the effect of repeated exposures. We also assumed that the recurrent exposures over a lifetime inform and reinforce responses in a self-similar manner. Overall repertoire temporal stability, measured by frequency of longitudinal occurrence, is governed by clonotype dominance in the pool of continuously circulating clonotypes. We now aim to quantify the repertoire stability in a more refined way using specific experimental settings.

We started with a simple Markov process that consists of two states representing repertoire observable (P) and non-observable (A) components at each timestep. We assumed that some fraction of the repertoire can be permanently lost from circulation, and some can come back. The probabilities for transitioning from state to state are defined by the state space transition matrix, a nxn square matrix that describes the stochastic behavior of a Markov chain, and in the case of a 2-state Markov chain can be mathematically expressed using a 2x2 transition matrix. We defined transitional probabilities for being observed (or present) after being observed (α), for disappearing (or becoming absent) after being observed (β), for staying absent (δ), and reappearing after being absent (γ).

In the simplest case, we restricted clonotypes to return to a circulating pool (Figure 1), so the transitional probabilities α and β fully define the speed of exhausting the circulation pool of unique clonotypes in a repertoire. We introduced a repertoire initial (new) pool (N^o), a sustained (stay) pool (N), and an outflow (gone) pool (N^*). In a simulated experiment, we can define the initial pool, examine the dynamic and estimate both the sustained and outflow pool at any discrete time step. In experimental settings, this scenario imitates a sequence of longitudinal measurements of a repertoire size, specifically the number of unique clonotypes

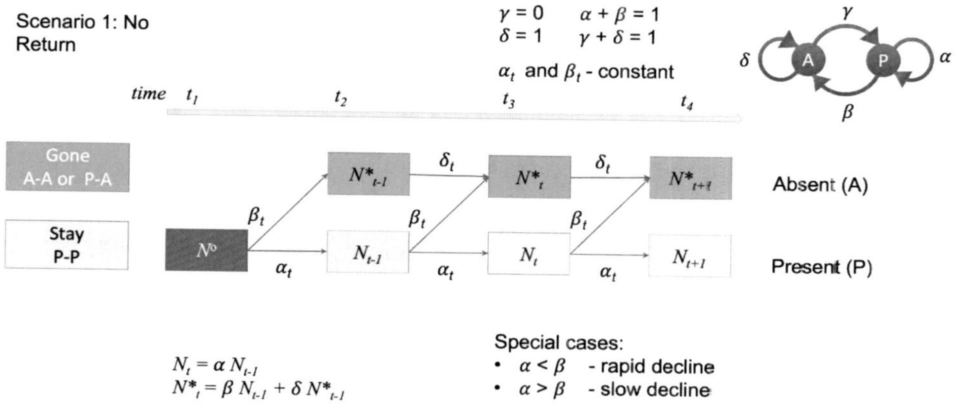

FIGURE 1. A simulated environment for modeling TCR repertoire stability: a closed Markov steady state process with no return (Scenario 1) and a corresponding flow diagram illustrating sequential steps for simulation or estimation of a repertoire initial (new) pool (N^o), a sustained (stay) pool (N), and an outflow (gone) pool (N^*) and the transitional probabilities α and β.

observed at each time point. The transitional probabilities α and β can be estimated, as $\alpha_t = f\ (N_t, N_1)$, say as N_t/N_{t-1}, and $\beta_t = 1$ - α_t, and then α and β can be formally compared to infer whether a repertoire shows a slow $\alpha > \beta$ or fast $\alpha < \beta$ decline.

Next, we considered that clonotypes can come back to a circulation pool (Figure 2), so the transitional probabilities α and β are affected by δ and γ and alter the speed of exhausting the pool. This scenario allowed us to introduce a 'return' pool that we could measure in experimental conditions and thus, provide an estimate for reappearing after being absent (γ) and to have a better grasp on a size of an outflow pool and probabilities for staying absent (δ). A simple summary of how many times a given clonotype was observed in a longitudinal experiment provides a gradient of frequencies that roughly equates to: a) a fraction of the most stable clonotypes that 'stay' over the study period and quantify the sustained pool (N); b) a fraction of clonotypes observed only once – a conservative estimate of an outflow (gone) pool (N^*); and c) a fraction of clonotypes observed more than once - an estimate of a variable component of a repertoire pool (N^{**}).

To make this model better resemble the assumed dynamic, we allowed the pool of clonotypes to be replenished over time, thus imitating a draw of a new sample without and with return options, Scenarios 3 and 4, respectively (Figures 3 and 4). These scenarios allowed us to introduce a pool of clonotypes that disappear and reappear during the observation period and further refine the concept of stability. In the simulated environment, we could further explore conditions that deplete the pool of stable clonotypes and refine the composition of transitional probabilities.

To finalize the framework and consider some finite conditions, we introduced a pool of clonotypes that fully disappeared from circulation over the observation

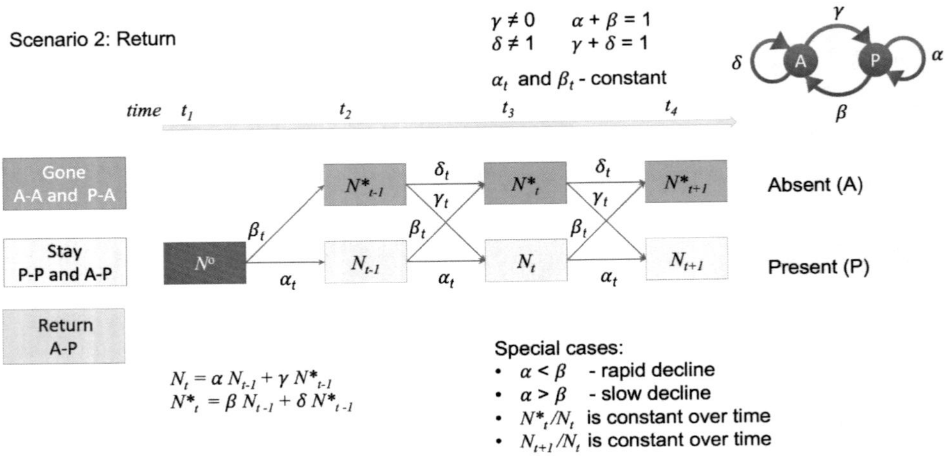

FIGURE 2. A simulated environment for modeling TCR repertoire stability: a closed Markov steady state process with a potential return (Scenario 2) and a corresponding flow diagram illustrating sequential steps for simulation or estimation of a repertoire initial (new) pool (N^o), a sustained (stay) pool (N), an outflow (gone) pool (N^*), and the transitional probabilities α, β, γ, and δ.

period and focus on the transitions to a stable pool (Figure 5). We added a 'Return' pool of clonotypes that are in flux. Using the proposed structure, we can further refine the composition of transitional probabilities for complete disappearance from a circulation pool or 'hard loss' (δ); quantify 'soft loss' (β_1, and β_2 and γ_2), the potential for persistence (α_1 and γ_1), and the most stable pool (α_2). In experimental settings, the counting of clonotypes in each specific pool must be set up in a way that allows to repeat and recategorize clonotypes at each iteration. We proposed to focus on the quantification of four pools taking into consideration all samples from each individual at each timestep: 1 – a pool of 'new' clonotypes or not yet observed in the prior samples (New, or N-pool); 2 – a pool of clonotypes observed in a prior sample (Stay, or S-pool); 3 – a pool of clonotypes observed in a given sample and earlier but was absent in a prior sample (Return, or R-pool); 4 – a pool of clonotypes disappeared from a subsequent sample (Gone, or G-pool, or Absent). We also consider two composite pools: 1 – a pool of clonotypes observed at a given sample that contains clonotypes from N-, S-, and R-pool (so-called, Present); and 2 – a pool of clonotypes observed at to date (Observed).

Estimation from experimental data

At each timestep, we assigned each clonotype to a relevant pool: N (New), R (Return), S (Stay), and G (Gone) and counted all clonotypes that were present and absent. We also counted the number of clonotypes to reflect the status on a previous step to assess the clonotypic potential for persistence (or a path from N, S, and R pools to S-pool, as precursors for estimating α_1, α_2, and γ_1, respectively.

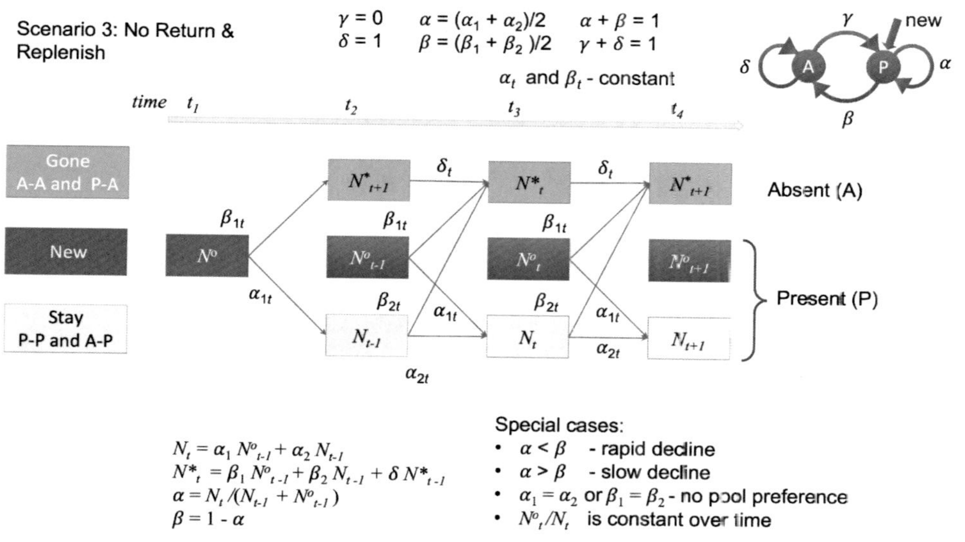

FIGURE 3. A simulated environment for modeling TCR repertoire stability: a Markov steady state process with replenishment but no return (Scenario 3) and a corresponding flow diagram illustrating sequential steps for simulation or estimation of a repertoire initial (new) pool (N^o), a sustained (stay) pool (N), an outflow (gone) pool (N^*), and the transitional probabilities α_1, α_2, β_1, and β_2.

Similarly, to assess the clonotypic potential for hard and soft loss, we counted clonotype arrived in G-pool from G, N, S, and R pools, as precursors for estimating δ, β_1, β_2, and γ_2, respectively. Supplementary Tables SD1-16 contain raw data and summaries for estimated pools and estimates for each person at each timestep.

The summaries for total counts and corresponding fractions are shown in Supplementary Table S2. We then estimated transitional probabilities and their averages and derived several characteristics to quantify stability, expansion, contraction, and recovery in terms of fractions of pools at given and prior timesteps (Table 1). The characterization of various pools of clonotypes that are present, gone from circulation, disappeared at least once and returned to the pool, and stayed for the full study period is useful to better understand the dynamics. Finally, we provided the estimates of monthly rates of expansion, persistence, contraction, and hard loss as the average number of clonotypes in specific pools. These estimates are corrected for duration variability of available measurements to allow the comparison of various groups. Based on known dates of blood draw we estimated the duration for each timestep in days (D_t) and expressed monthly rates using a constant of M=30 days. We provided both individual durations and the total number of days between the first and last blood collections for each person (Supplementary Tables S2 and S3).

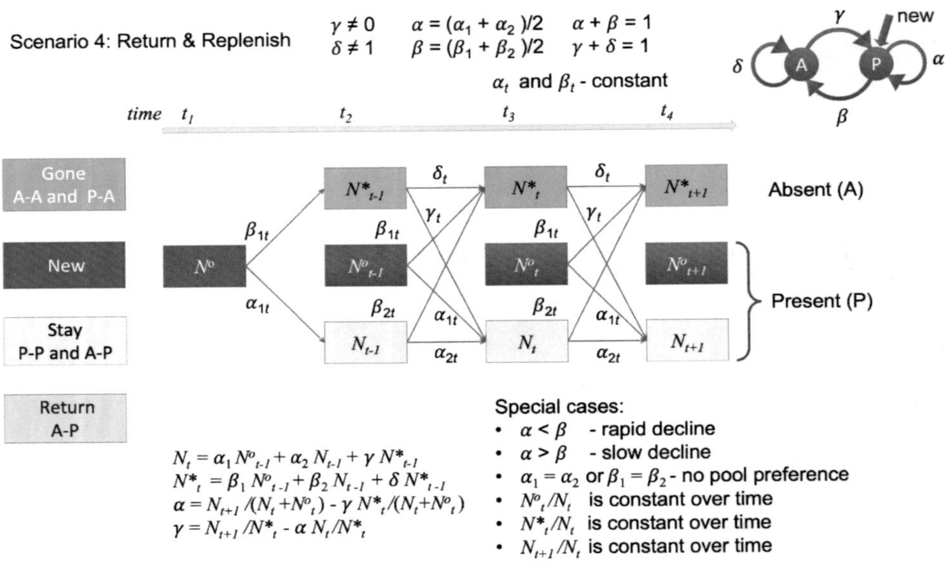

FIGURE 4. A simulated environment for modeling TCR repertoire stability: a Markov steady state process with replenishment and a potential return (Scenario 4) and a corresponding flow diagram illustrating sequential steps for simulation or estimation of a repertoire initial (new) pool (N^o), a sustained (stay) pool (N), a return pool (N^{**}), and an outflow (gone) pool (N^*), and the transitional probabilities α_1, α_2, β_1, β_2, γ, and δ.

Motivating example

We obtained 126 blood draw samples from 16 healthy individuals (5 children – between 7 and 14 y.o., 5 middle-aged adults – between 39 and 48 y.o., and 6 older adults – between 68 and 78 y.o.). Each person contributed about 7 longitudinal measurements (from 5 to 13 measurements) for about 862 days (from 413 to 1389 days) or 2.4 years (Supplementary Table S1). The average duration between consecutive blood draws was 115 days or about 4 months.

From each sample we determined the number of identical clonotypes, reflecting a specific response to influenza as reported in [**14**]. Briefly, the blood samples were used for isolating the CD8 T-cells that were seeded in cell cultures containing an influenza-specific peptide and kept *in vitro* for 7 days with autologous cells. Irradiation of autologous cells precluded autologous cell proliferation yet ensured that CD8 T-cells were able to recognize and lyse autologous cells expressing an influenza-specific peptide. This *in vitro* model mimics the biological events that take place in the human organism during influenza viral infection.

Estimation from experimental data

At each timestep, we assigned each clonotype to a relevant pool: N (New), R (Return), S (Stay), and G (Gone) and counted all clonotypes that were present and absent. We also counted the number of clonotypes to reflect the status on a

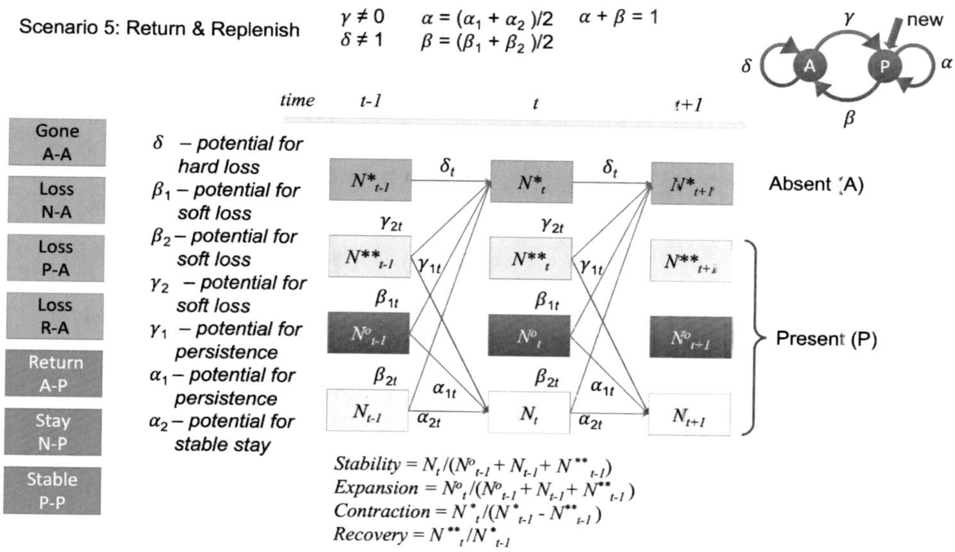

FIGURE 5. A simulated environment for modeling TCR stability: a Markov steady state process with replenishment and a potential return (Scenario 5) and a corresponding flow diagram illustrating sequential steps for simulation or estimation of a repertoire initial (new) pool (N^o), a sustained (stay) pool (N), a return pool (N^{**}), and an outflow (gone) pool (N^*), and the transitional probabilities α_1, α_2, β_1, β_2, γ_1, γ_2, and δ.

previous step to assess the clonotypic potential for persistence (or a path from N, S, and R pools to S-pool, as precursors for estimating α_1, α_2, and γ_1, respectively. Similarly, to assess the clonotypic potential for hard and soft loss, we counted clonotype arrived in G-pool from G, N, S, and R pools, as precursors for estimating δ, β_1, β_2, and γ_2, respectively. Supplementary Tables SD1-16 contain raw data and summaries for estimated pools and estimates for each person at each timestep.

The summaries for total counts and corresponding fractions are shown in Supplementary Table S2. We then estimated transitional probabilities and their averages and derived several characteristics to quantify stability, expansion, contraction, and recovery in terms of fractions of pools at given and prior timesteps (Table 1). The characterization of various pools of clonotypes that are present, gone from circulation, disappeared at least once and returned to the pool, and stayed for the full study period is useful to better understand the dynamics. Finally, we provided the estimates of monthly rates of expansion, persistence, contraction, and hard loss as the average number of clonotypes in specific pools. These estimates are corrected for duration variability of available measurements to allow the comparison of various groups. Based on known dates of blood draw we estimated the duration for each timestep in days (D_t) and expressed monthly rates using a constant of M=30 days. We provided both individual durations and the total number of days between the first and last blood collections for each person (Supplementary Tables S2 and S3).

TABLE 1. Proposed characteristics of TCR clonotypic dynamics based on transitions across pools.

Characteristic	**Equation**	**Comments**
Transitional probabilities		
Most stable, α_2	$= (N_t \mid N_{t-1})/N_{t-1}$	Based on transition to S-pool from S-pool at a prior timestep
Persistence, α_1	$= (N_t \mid N^o_{t-1})/N^o_{t-1}$	Based on transition to S-pool from N-pool at a prior timestep
Persistence, γ_1	$= (N_t \mid N^{**}_{t-1})/N^{**}_{t-1}$	Based on transition to S-pool from R-pool at a prior timestep
Soft loss, β_1	$= (N^*_t \mid N^o_{t-1})/N^o_{t-1}$	Based on transition to G-pool from N-pool at a prior timestep
Soft loss, β_2	$= (N^*_t \mid N_{t-1})/N_{t-1}$	Based on transition to G-pool from S-pool at a prior timestep
Soft loss, γ_2	$= (N^*_t \mid N^{**}_{t-1})/N^{**}_{t-1}$	Based on transition to G-pool from R-pool at a prior timestep
Hard loss, δ	$= (N^*_t \mid N^*_{t-1})/N^*_{t-1}$	Based on transition to G-pool from G-pool at a prior timestep
Temporal characteristics		
Stability	$= N_t/(N^o_{t-1} + N_{t-1} + N^{**}_{t-1})$	Based on a fraction of clonotypes in S-pool relative to all present clonotypes in N-, S-, and R-pools at a prior timestep
Expansion	$= N^o_t/(N^o_{t-1} + N_{t-1} + N^{**}_{t-1})$	Based on a fraction of clonotypes in N-pool relative to all present clonotypes in N-, S-, and R-pools at a prior timestep
Contraction	$= N^*_t/(N^*_{t-1} - N^{**}_{t-1})$	Based on a fraction of clonotypes in G-pool relative to clonotypes in G-pool after accounting for R-pool at a prior timestep
Recovery	$= N^{**}_t/N^*_{t-1}$	Based on fraction of clonotypes in R-pool relative to clonotypes in G-pool at a prior timestep
Monthly rates as number of clonotypes per month		
Monthly rate of expansion	$= \mathrm{M}(N^o_t)/D_t$	Based on a pool of new clonotypes at each timestep
Monthly rate of persistence	$= \mathrm{M}(N^o_t + N_t + N^{**}_t)/D_t$	Based on a pool of all clonotypes present at each timestep
Monthly rate of contraction	$= \mathrm{M}(N^*_t - N^*_{t-1})/D_t$	Based on an increment of G-pool at each timestep
Monthly rate of contraction due to hard loss	$= \mathrm{M}(N^*_t \mid N^*_{t-1} - N^*_{t-1} \mid N^*_{t-2})/D_t$	Based on an increment of G-pool at each timestep

Results

Figure 6 illustrates an example of the pool counts for all available time points for donor C1, indicating that most clonotypes (72%) left the observed circulation pool and newly recruited clonotypes (67%) are replenishing the pool of clonotypes

Duration in days		42	112	228	97	142	142	45	C1 donor Total	(%)
Observed	37	56	80	110	128	142	162	166	881	
Present	37	23	30	37	33	23	36	29	248	28%
Gone		33	50	73	95	119	126	137	533	72%
Return			3	4	9	5	13	8	42	17%
New	37	19	24	30	18	14	20	4	166	67%
Stay		4	3	3	6	4	3	17	40	16%

FIGURE 6. An example of clonotype counting for four individual (New, Stay, Return, and Gone) and two combined (Present, Observed) pools for C1-donor at 8 timepoints spaced about 115 days apart and total number of clonotypes in each pool. The fractions for present and absent clonotypes are given with respect to the observed; the fractions for new, stayed, and returned are given with respect to the present (Supplementary Tables SD1-16)

that are in the circulation. On average almost equal amount of clonotypes returned or remained in the circulation pool.

The summaries for total counts and corresponding fractions for each person are provided in Supplementary Table S2. Table 2 shows averages and standard deviations for a) the number of all clonotypes observed at any timestep (Observed); b) fractions of clonotypes that are present (P-pool) or absent (G-pool) relative to all observed; c) fractions of new, returned, and stayed clonotypes (N-, R, S- pools, respectively) as contributors to the P-pool; d) fractions of clonotypes contributing to the S-pool based on their pool at a prior step, and e) fractions of clonotypes contributing to the G-pool based on their pool at a prior step. Across three age groups, we observed substantial repertoire renewal over time, with new clonotypes contributing 50-70% to all clonotypes present in a blood sample. About 20% of clonotypes that were present in N-, S-, and R-pools at a given timestep are likely to stay. In adults, the pool of stable clonotypes was the most prominent. At the same time, 90% of all clonotypes are leaving the circulation pool, mostly due to hard loss.

Transitional probabilities for hard loss indicate that the circulating pool can be depleted relatively fast. With $\alpha < \beta$, a fast depletion is common in all age groups (0.33 vs 0.72, 0.49 vs 0.59, and 0.39 vs 0.64, in children, adults, and older adults, respectively). As $\alpha_2 < \alpha_1$ and $\beta_2 < \beta_1$, the clonotypes that arrive with the new blood draw are more likely to disappear at the next time step than to stay. As $\gamma_1 < \gamma_2$, in age groups the recovery is weak. In fact, in older adults, there were several instances that no clonotypes returned to the circulation pool. About 3-5 new

TABLE 2. Estimated characteristics of TCR clonotypic dynamics based on transitions across pools for children, adults, and older adults.

	Children		Adults		Older Adults	
	Average	St.dev	Average	St.dev	Average	St.dev
Pool composition						
Observed at any timestep	657	252.9	785	296.4	441	189.5
Present (P-pool)	0.354	0.067	0.383	0.043	0.298	0.043
New (N-pool)	0.683	0.044	0.508	0.088	0.629	0.137
Return (R-pool)	0.148	0.016	0.161	0.033	0.120	0.074
Stay (S-pool)	0.170	0.042	0.332	0.069	0.251	0.111
N-S - persistence	0.475	0.118	0.269	0.135	0.338	0.150
S-S - persistence	0.332	0.133	0.585	0.087	0.506	0.162
R-S - persistence	0.193	0.084	0.146	0.058	0.156	0.141
Absent (G (Gone)	0.646	0.067	0.617	0.043	0.702	0.043
G-G - hard loss	0.633	0.075	0.655	0.055	0.717	0.061
N-G - soft loss	0.290	0.058	0.240	0.061	0.225	0.082
S-G - soft loss	0.041	0.018	0.050	0.018	0.032	0.013
R-G - soft loss	0.037	0.004	0.054	0.016	0.026	0.017
Transitional probabilities						
Most stable, α_2	0.446	0.129	0.720	0.061	0.634	0.113
Persistence, α_1	0.216	0.184	0.258	0.191	0.158	0.107
Persistence, γ_1	0.380	0.137	0.360	0.117	0.376	0.135
Soft loss, β_1	0.891	0.057	0.894	0.054	0.909	0.067
Soft loss, β_2	0.554	0.129	0.280	0.061	0.366	0.113
Soft loss, γ_2	0.620	0.137	0.640	0.117	0.624	0.135
Hard loss, δ	0.883	0.030	0.858	0.040	0.937	0.037
Temporal characteristics						
Stability	0.202	0.045	0.406	0.064	0.290	0.054
Expansion	0.705	0.208	0.537	0.273	0.610	0.226
Recovery	0.117	0.030	0.142	0.040	0.063	0.037
Contraction	1.562	0.131	1.660	0.346	1.514	0.269
Monthly rates as number of clonotypes per month						
Rate of expansion	5.05	1.10	5.38	1.94	3.02	0.79
Rate of persistence	9.04	2.53	13.44	3.28	5.34	1.09
Rate of contraction	4.91	1.46	5.85	3.03	3.49	1.22
Rate of 'hard loss'	5.71	2.93	5.44	2.29	3.48	0.86

clonotypes are recruited each month, 5-13 clonotypes remain in the circulating pool and 3-6 are likely to leave the pool, suggesting the needed continuity in replenishing the repertoires.

Discussion

In this modeling study we introduced a framework to examine the dynamic of renewal in a T-cell repertoire reflecting the response to influenza peptides in human. We demonstrated the flow of circulating clonotypes using the Markov chain steady state process to capture the renewal, sustained, recovered, and outflow pools of clonotypes. We defined several characteristics to explore the dynamics and quantify the maintenance of adaptive immunity, such as stability, expansion, and contraction. We demonstrated the potential of estimating transitional probabilities from experimental data and comparing several characteristics across three age groups. Based on the observed dynamics we concluded that for an infection with the short immune memory, like an influenza, the 'hard loss' is likely to be high and the renewal pool should be sufficient to sustain the circulation.

In our early work, we defined stability based on the frequency of observing a clonotype in longitudinal samples; we explored fractal distributions of repertoires and concluded that clonotypes of high frequency are most stable over time [14, 18]. The most complex and most stable repertoires were observed in mid-aged adults; in children, repertoires were not stable and among the elderly repertoires were losing stability. The system first develops distributional complexity, which incorporates stability, and then shows signs of loss of both.

The newly proposed estimates support the early findings and further refined our understanding of the observed dynamics. We observed that while the overall patterns of temporal stability are similar in three age groups, there are significant differences in specially derived characteristics informed by the Markov chain model. These differences indicate that transitional probabilities are subject to change as the immune system evolves with age. The tendency for hard loss could potentially serve as a marker of immunosenescence. We concluded that for infections with short immune memory, a boost via periodic vaccinations or direct exposures is responsible for sustaining the memory pool.

There are several limitations: the examined repertoires were measured with a somewhat restrictive method since the limited number of plasmid DNA containing TCR α- or β-chain sequences can be collected. The new generation of methods using single-cell mRNA sequencing based on next-generation sequencing technology provides a more complete picture and will be our next step in examining the dynamics [**23**–**26**]. The estimation of transitional probabilities requires at minimum four longitudinal samples, preferably collected at regular intervals, and such data are not commonly available. In this study we did not utilize information on clonotype frequency, influenza exposure history, yet this could be an informative way to learn about potential reasons for high variability in transition probabilities and stability characteristics. We anticipate that clonotypes that are of high frequency or dominant are likely to contribute the most to a stable pool and can formally test this hypothesis in future analysis. In future studies we also could measure how clonotypes reappear with respect to a resent exposure to flu to have a better grasp on the effects of vaccination or infection on repertoire characteristics.

By considering the proposed framework, in future studies we could define scenarios for continuous oscillations, stabilization, potential repertoire collapse, and changes associated with the known exposures. This framework could guide the *in silico* investigations to explore the timeline of aging, development, and senescence of the immune system. As the immune response undergoes a series of expansion-contraction cycles, we could quantify the changes related to vaccination or direct exposure to a pathogen.

The framework could be applied to better understand age-specific trajectories of repertoire evolution and how they relate to chronological and immunological aging. We distinguish three scenarios of aging and senescence: early decline and deceleration, strong effects of compensatory mechanisms, and prolong maintenance and rapid acceleration [**21**]. While there are no experimental conditions to test these scenarios in real life, it would be possible to create *in silico* investigations to form new hypotheses and a sound basis for estimating the immunological age.

At the population level, each person's encounter with an infectious agent is unique. Yet, the commonality in responses formed "immunological kinship" among all affected individuals, manifested by a preserved T-cell clonal pool. The diverse responses to flu and changes in diversity allow us to make an inference to "immunological kinship" by measuring the stable and reoccurring pools of clonotypes. The measure of kinship could be a plausible and more precise alternative to herd immunity assessment. There are no studies yet that connect the circulation pools of clonotypes to circulating influenza strains. The proposed framework creates a platform to initiate such investigations.

Conclusions

This study was inspired by Professor Ronald E. Mickens through several lively conversations during NIMBioS seminars and through his enthusing work on constructing and examining the properties of discrete-time models for a periodic infection with a short-lived immunity [**27, 28**]. These early papers demonstrated that "discrete (difference equation) and continuous (differential equation) models of the same system may lead to vastly divergent conclusions. ... In particular, a discrete model may give rise to oscillatory or chaotic dynamics while the corresponding continuous model predicts long term steady-state motion" [**27**]. This idea forced us to think about how we could explore the dynamics of the T-cell repertoire and consider the intricacies and challenges of experimental conditions that offer data for enabling computer simulations. Another valuable discussion point that enlightened our conversations was the general question of the possibility of preparing a paper that could be comprehensive and useful for both mathematicians and immunologists. In science, disciplinary specialization offered tremendous progress, yet it also created technical shortcuts, jargon, and systems of abbreviations that stifled mutual understanding. Multiple cross-disciplinary efforts aim to bridge the disciplinary communication gaps and enable knowledge co-creation. This paper represents our attempt to reach readers from both fields by avoiding equation intimidation and jargon overload.

Acknowledgments

The views and conclusions contained herein are those of the authors and should not be interpreted as necessarily representing the official policies, either expressed or implied, of ODNI, IARPA, or the U.S. Government. The U.S. Government is authorized to reproduce and distribute reprints for governmental purposes notwithstanding any copyright annotation therein.

Supplementary Material

Supplementary Material contains repertoires summaries and is available by request.

References

[1] World Health Organization, *GHE: Life expectancy and healthy life expectancy 2022* [Available from: `https://www.who.int/data/gho/data/themes/mortality-and-global-health-estimates/ghe-life-expectancy-and-healthy-life-expectancy`].

[2] Kucharski AJ, Lessler J, Read JM, Zhu H, Jiang CQ, Guan Y, et al, *Estimating the life course of influenza A(H3N2) antibody responses from cross-sectional data*, PLoS Biol. **13** (2015) no. 3 e1002082.

[3] Skowronski DM, Tweed SA, De Serres G, *Rapid decline of influenza vaccine-induced antibody in the elderly: Is it real, or is it relevant?*, J Infect Dis. **97** (2008) no. 4 490-502.

[4] Song JY, Cheong HJ, Hwang IS, Choi WS, Jo YM, Park DW, et al, *Long-term immunogenicity of influenza vaccine among the elderly: Risk factors for poor immune response and persistence*, Vaccine. **28** (2010) no. 23 3929-35.

[5] Castilla J, Martinez-Baz I, Martinez-Artola V, Reina G, Pozo F, Garcia Cenoz M, et al, *Decline in influenza vaccine effectiveness with time after vaccination, Navarre, Spain, season 2011/12*, Euro Surveill. **18** (2013) no. 5

[6] Naumova EN, Gorski J, Naumov YN, *Simulation studies for a multistage dynamic process of immune memory response to influenza: experiment in silico*, Ann Zool Fennici. **45** (2008) no. 5 369-84.

[7] Yassai MB, Naumov YN, Naumova EN, Gorski J, *A clonotype nomenclature for T cell receptors*, Immunogenetics. **61** (2009), no. 7 493-502.

[8] International ImMunoGeneTics information system System. [Available from: `https://www.imgt.org`.]

[9] Shugay M, Bagaev DV, Zvyagin IV, Vroomans RM, Crawford JC, Dolton G, et al, *VDJdb: a curated database of T-cell receptor sequences with known antigen specificity*, Nucleic Acids Res. **46** (2018) no. D1 D419-D27

[10] Rosati E, Dowds CM, Liaskou E, Henriksen EKK, Karlsen TH, Franke A, *Overview of methodologies for T-cell receptor repertoire analysis*, BMC Biotechnology. **17** (2017) no. 1 61.

[11] Kivioja T, Vähärautio A, Karlsson K, Bonke M, Enge M, Linnarsson S, et al, *Counting absolute numbers of molecules using unique molecular identifiers*, Nature Methods. **9** (2012) no. 1 72-4.

[12] Fu Y, Wu PH, Beane T, Zamore PD, Weng Z, *Elimination of PCR duplicates in RNA-seq and small RNA-seq using unique molecular identifiers*, BMC Genomics. **19** (2018) no. 1 531.

[13] Naumov YN, Naumova EN, Clute SC, Watkin LB, Kota K, Gorski J, et al, *Complex T cell memory repertoires participate in recall responses at extremes of antigenic load*, J Immunol. **177** (2006) no. 3 2006-14.

[14] Naumova EN, Yassai MB, Demos W, Reed E, Unruh M, Haribhai D, et al, *Age-based dynamics of a stable circulating CD8 T cell repertoire component*, Front Immunol. 2019;10:1717.

[15] Naumov YN, Hogan KT, Naumova EN, Pagel JT, Gorski J, *A class I MHC-restricted recall response to a viral peptide is highly polyclonal despite stringent CDR3 selection: implications for establishing memory T cell repertoires in "real-world" conditions*, J Immunol. **160** (1998) no. 6) 2842-52.

[16] Naumov YN, Naumova EN, Hogan KT, Selin LK, Gorski J, *A fractal clonotype distribution in the CD8+ memory T cell repertoire could optimize potential for immune responses*, J Immunol. **170** (2003) no. 8 3994-4001.

[17] Naumov YN, Naumova EN, Yassai MB, Gorski J, *Selective T cell expansion during aging of CD8 memory repertoires to influenza revealed by modeling*, J Immunol. **186** (2011) no. 11 6617-24.

[18] Naumova EN, Gorski J, Naumov YN, *Two compensatory pathways maintain long-term stability and diversity in CD8 T cell memory repertoires*, J Immunol. **183** (2009) no. 4 2851-8.

[19] Ruskin HJ, Burns J, *Weighted networks in immune system shape space. Physica A: Statistical Mechanics and its Applications*, 2006;365(2):549-55.

[20] Réka Albert and Albert-László Barabási, *Statistical mechanics of complex networks*, Rev. Modern Phys. **74** (2002), no. 1, 47–97, DOI 10.1103/RevModPhys.74.47. MR1895096

[21] Naumova EN, Naumov YN, Gorski J, *Measuring Immunological Age: From T Cell Repertoires to Populations*, Fulop T, Franceschi C, Hirokawa K, Pawelec G, editors. Handbook of

Immunosenescence: Basic Understanding and Clinical Implications. Cham: Springer International Publishing; 2018. p. 1-62.

[22] Saito S, Narikiyo O, *Scale-free dynamics of somatic adaptability in immune system.* Biosystems. **103** (2011) no. 3 420-4.

[23] Song I, Gil A, Mishra R, Ghersi D, Selin LK, Stern LJ, *Broad TCR repertoire and diverse structural solutions for recognition of an immunodominant CD8(+) T cell epitope*, Nat Struct Mol Biol. **24** (2017) no. 4 395-406.

[24] Kamga L, Gil A, Song I, Brody R, Ghersi D, Aslan N, et al, *CDR3a drives selection of the immunodominant Epstein Barr virus (EBV) BRLF1-specific CD8 T cell receptor repertoire in primary infection*, PLoS Pathog. **15** (2019) no. 11 e1008122.

[25] Khatun A, Kasmani MY, Zander R, Schauder DM, Snook JP, Shen J, et al, *Single-cell lineage mapping of a diverse virus-specific naive CD4 T cell repertoire*, Journal of Experimental Medicine. **218** (2020) no. 3

[26] Piper C, Hainstock E, Yin-Yuan C, Chen Y, Khatun A, Kasmani MY, et al, *Single-cell immune profiling reveals a developmentally distinct CD4+ GM-CSF+ T-cell lineage that induces GI tract GVHD*, Blood Adv. **6** (2022) no. 9 2791-804.

[27] Mickens RE, *A discrete-time model for the spread of periodic diseases without immunity*, Biosystems. **26** (1992) no. 3 193-8.

[28] Ronald E. Mickens, *An exact discretization of Michaelis–Menten type population equations*, J. Biol. Dyn. **5** (2011), no. 5, 391–397, DOI 10.1080/17513758.2010.515690. MR2834492

Division of Nutrition Epidemiology and Data Science, Friedman School of Nutrition Science and Policy, Tufts University, Boston, Massachusetts, 02111
Email address: elena.naumova@tufts.edu

The Blood Research Institute, Versiti Wisconsin, Milwaukee, Wisconsin, 53142
Email address: jmyassai@versiti.org

The Blood Research Institute, Versiti Wisconsin, Milwaukee, Wisconsin, 53142
Email address: jgorski@versiti.org

Smart Diagnostics Medica, Boston, Massachusetts, 02141
Email address: yuri.naumov@backbaygroup.com

Contemporary Mathematics
Volume **793**, 2024
https://doi.org/10.1090/conm/793/15885

Geometric approach to the construction of Leah-type periodic functions: Basic and analytic properties

Sandra A. Rucker

ABSTRACT. We again consider functions associated with the cube-root oscillator and present a construction of the associated three basic periodic functions using the geometric properties of its two-dimensional phase-space curve. In addition, to the evaluation of these functions for specific values of the independent variable θ, $0 \leq \theta < 2\pi$, we also derive a number of other relations among the three periodic functions and their derivatives. These new results extend previous work of the author and Mickens.

1. Introduction

The current work extends previous investigations of Mickens and Rucker [**1**], other collaborators [**2**], and colleagues [**3, 4, 5**], who have studied the properties of the solutions to the cube-root oscillator (CRO) [**6**]

$$\frac{d^2x}{dt^2} + x^{\frac{1}{3}} = 0. \tag{1.1}$$

A concise summary of what is known about the mathematical features of the CRO is given in the paper of Rucker [**7**]. However, the main goals here are to introduce solutions of the CRO which are a priori defined in terms of its two-dimensional phase-space employing polar coordinates. We will discover that this representation produces three bounded, periodic functions that we label the Leah-cosine, Leah-sine, and Leah-dine functions. These functions are closely related to our previous work [**1, 2**], but they, in general are not identical to them. Thus, the results presented here may be considered as an independent contribution on the CRO and its solutions.

In the next section, we introduce the background required to define and understand the three periodic Leah-functions from a geometric perspective. For this analysis the independent variable will be taken to be the polar angle, θ, $0 \leq \theta < 2\pi$, rather than the dynamic time variable, t. This section also gives a number of the fundamental geometric features of the Leah-functions. Section 3 provides derivations of various analytical relations among the Leah-functions and discusses the issue of the existence of Taylor series.

2020 *Mathematics Subject Classification.* Primary 33B10, 33B99, 33E30; Secondary 34C05.
Key words and phrases. Periodic functions, special functions, nonlinear oscillations, cube-root oscillator.

In Section 4, we show that exact, closed-formed solutions may be obtained for the Leah-functions, but they require the solution of a cubic equation. Section 5 provides an algorithm which permits the numerical calculation of the three periodic Leah-functions. Finally, in Section 6, we summarize our results and give a short listing of unresolved issues relating to the CRO equation and its solutions.

2. Geometric definitions of the Leah functions

Consider the curve

$$H(x,y) \equiv \frac{y^2}{2} + \left(\frac{3}{4}\right) x^{\frac{4}{3}} = \frac{3}{4}. \tag{2.1}$$

Its plot is given in Figure 1. Note that

$$H(x,y) = constant > 0, \tag{2.2}$$

is the first-integral [**1**, **6**] for the CRO differential equation, see Eq.(1.1), if y is defined to be

$$y \equiv -\frac{dx}{dt}. \tag{2.3}$$

Consequently, the system equations are

$$\frac{dx}{dt} = -y, \qquad \frac{dy}{dt} = x^{\frac{1}{3}}. \tag{2.4}$$

Also, by inspection, we see that $H(x,y)$ is invariant under the coordinate transformations

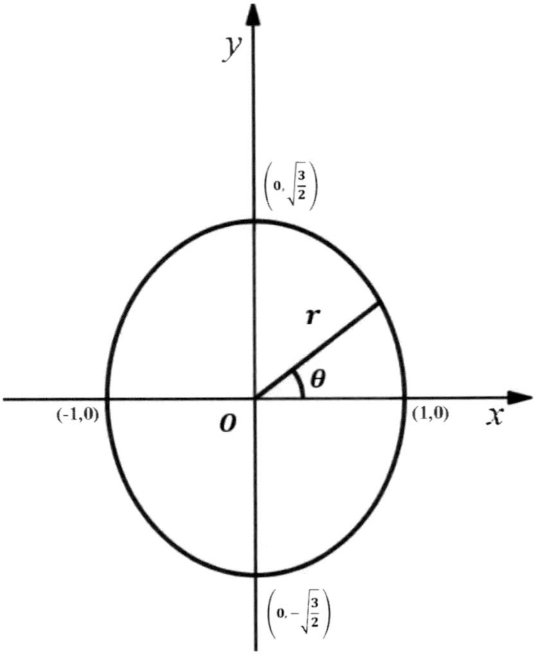

FIGURE 1. Plot of $H(x,y) = \frac{3}{4}$, and definition of the polar coordinates, r and θ.

$$(2.5)\quad \begin{cases} T_1: & x \to -x, \quad y \to y \quad \text{(reflection in the y-axis)}, \\ T_2: & x \to x, \quad y \to -y \quad \text{(reflection in the x-axis)} \\ T_3: & x \to -x, \quad y \to -y \quad \text{(inversion through the origin)}. \end{cases}$$

An important consequence of these invariant transformations is that we need only examine what occurs in the first-quadrant of the $x-y$ plane. Results for the other three quadrants may be determined from applying Eq.(2.5). Let $P(x,y)$ be a point on the closed curve $H(x,y) = \frac{3}{4}$ [**1**, **6**]. Let r be a ray from the origin out to P(x,y) and let the angle between this ray and the positive x-axis be θ. The variables r and θ are the corresponding polar coordinates of $P(x,y)$. We define the three Leah-functions, with respect to the notation in Figure 1, as follows:

$$(2.6)\quad \begin{cases} c(\theta) & = Lc(\theta) \equiv r(\theta)cos(\theta) \\ s(\theta) & = Ls(\theta) \equiv r(\theta)sin(\theta) \\ d(\theta) & = Ld(\theta) \equiv r(\theta) \end{cases}$$

These definitions indicate that the Leah-functions are being considered as functions of the geometric angle, θ, rather than as the dynamical variable time, t. We call them, respectively, the Leah-cosine, Leah-sine, and Leah-dine functions. This nomenclature is consistent with the labelling of the Jacobi elliptic functions [**8**].

From the definitions, given by Eq.(2.6), and from inspection of Figure 1, it follows that the three Leah-functions have the properties:

(i) $c(\theta)$ and $s(\theta)$ are periodic period $2(\pi)$; while $d(\theta)$ has period π, i.e.,

$$(2.7)\quad \begin{cases} c(\theta + 2\pi) & = c(\theta), \\ s(\theta + 2\pi) & = s(\theta), \\ d(\theta + \pi) & = d(\theta). \end{cases}$$

(ii) $c(\theta)$ and $d(\theta)$ are even functions while $s(\theta)$ is odd , i.e.,

$$(2.8)\quad \begin{cases} c(-\theta) & = +c(\theta), \\ s(-\theta) & = -s(\theta), \\ d(-\theta) & = +d(\theta). \end{cases}$$

(iii) The following bounds hold,

$$(2.9)\quad \begin{cases} -1 & \leq c(\theta) \leq 1, \\ -\sqrt{\frac{3}{2}} & \leq s(\theta) \leq \sqrt{\frac{3}{2}}, \\ 1 & \leq d(\theta) \leq \sqrt{\frac{3}{2}}. \end{cases}$$

(iv) Values of $(c,\ s,\ d)$ for special angles

$$(2.10)\quad c(0) = 1,\ c\left(\frac{\pi}{2}\right) = 0,\ c(\pi) = -1,\ c\left(\frac{3\pi}{2}\right) = 0,$$

$$(2.11)\quad s(0) = 0,\ s\left(\frac{\pi}{2}\right) = \sqrt{\frac{3}{2}},\ s(\pi) = 0,\ d\left(\frac{3\pi}{2}\right) = -\sqrt{\frac{3}{2}}.$$

$$(2.12)\quad d(0) = 1,\ d\left(\frac{\pi}{2}\right) = \sqrt{\frac{3}{2}},\ d(\pi) = 1,\ d\left(\frac{3\pi}{2}\right) = \sqrt{\frac{3}{2}}.$$

3. Algebraic and analytic properties

From Figure 1, it is clear that

$$r^2 = x^2 + y^2, \tag{3.1}$$

or using the above Leah-function notation

$$d^2 = c^2 + s^2. \tag{3.2}$$

This is the Pythagorean identity. Likewise, from Eq(2.1), it follows that

$$s^2 + \left(\frac{3}{2}\right) c^{\frac{4}{3}} = \frac{3}{2}. \tag{3.3}$$

Observe that the Pythagorean identity involves all three Leah-functions, while the result of Eq.(3.3) connects only the Leah-sine and Leah-cosine functions.

Expressions for the first-derivatives of $(c,\ s,\ d)$ may be obtained as follows. Taking the derivative of Eq.(3.3), with respect to θ and dividing the result by two, gives

$$ss\,' + c^{\frac{1}{3}} c\,' = 0, \tag{3.4}$$

where $s\,' = \mathrm{d}s/\mathrm{d}\theta$, etc. The simplest expressions for $s\,'$ and $c\,'$, such that they can be separated in Eq. (3.4)

$$c\,' = -s, \qquad s\,' = c^{\frac{1}{3}} \tag{3.5}$$

Carrying out the same procedure for Eq.(3.2) yields the result

$$dd\,' = cc\,' + ss\,'. \tag{3.6}$$

However, we have,

$$\frac{c}{d} = \cos\theta, \ \frac{s}{d} = \sin\theta, \tag{3.7}$$

consequently from Eqs.(3.5) and (3.7), it follows that

$$d\,' = (sin\theta)c^{\frac{1}{3}} - (cos\theta)s. \tag{3.8}$$

Summarizing, Eqs.(3.5) and (3.8) gives the first derivatives of $(c,\ s,\ d)$ in terms of the $(c,\ s,\ d)$ functions. They may also be considered as a system of three, nonlinear, coupled, ordinary first-order differential equations whose solutions are the Leah-functions. For this interpretation, the initial conditions can be taken to be

$$c(0) = 1, \quad s(0) = 0, \quad d(0) = 1. \tag{3.9}$$

From Eqs.(3.5) and (3.8), we can calculate the slopes or derivatives for special values of θ; they are

$$c\,'(0) = 0,\ c\,'\left(\frac{\pi}{2}\right) = -\sqrt{\frac{3}{2}},\ c\,'(\pi) = 0,\ c\,'\left(\frac{3\pi}{2}\right) = +\sqrt{\frac{3}{2}}, \tag{3.10}$$

$$s\,'(0) = 1,\ s\,'\left(\frac{\pi}{2}\right) = 0,\ s\,'(\pi) = -1,\ s\,'\left(\frac{3\pi}{2}\right) = 0, \tag{3.11}$$

$$d\,'(0) = 0,\ d\,'\left(\frac{\pi}{2}\right) = 0,\ d\,'(\pi) = 0,\ d\,'\left(\frac{3\pi}{2}\right) = 0. \tag{3.12}$$

We can also calculate expressions for the second-derivatives of (c, s, d). For example, the derivative of the first expression in Eq.(3.5) is

$$c\,'' = -s\,' = -\left(c^{\frac{1}{3}}\right), \tag{3.13}$$

or

$$c\,'' + c^{\frac{1}{3}} = 0. \tag{3.14}$$

Carrying out the same procedure for the second expression in Eq.(3.5) gives

$$s\,'' = \left(\frac{1}{3}\right) c^{-\frac{2}{3}} c\,', \tag{3.15}$$

but

$$c^{\frac{2}{3}} = \left(c^{\frac{1}{3}}\right)^2 = (-s\,')^2 = (s\,')^2, \tag{3.16}$$

and it follows that

$$s\,'' + \left[\frac{1}{3(s\,')^2}\right] s = 0. \tag{3.17}$$

Finally, in a similar fashion, we can show that $d\,''(\theta)$ satisfies the relation

$$d\,'' + (sin\theta)\left[1 - \frac{1}{3c^{2/3}}\right] s = 0. \tag{3.18}$$

Of related interest is the fact that the Leah-cosine function's Taylor series can be easily calculated using the following initial conditions,

$$Lc(0) = 1, \quad dLc(0)/d\theta = 0. \tag{3.19}$$

(Remember that $Lc(\theta)$ = c(θ) and $dLc(0)/$dθ = c $'(0)$. Some of the early terms of this representation are [**2**, **9**]

$$\begin{aligned} Lc(\theta) &= c(\theta) \\ &= 1 - \left(\frac{1}{2!}\right)\theta^2 + \left(\frac{1}{3}\right)\left(\frac{1}{4!}\right)\theta^4 \\ &\quad + \left(\frac{5}{9}\right)\left(\frac{1}{6!}\right)\theta^6 + \left(\frac{115}{27}\right)\left(\frac{1}{8!}\right)\theta^8 + O\left(\theta^{10}\right) \end{aligned} \tag{3.20}$$

Professor George E. Parker has evaluated this series out to terms $O(\vartheta^{38})$ [**9**], while Mann et.al.[**2**] have provided an argument to support the proposition that the radius of convergence is $\frac{\pi}{2}$, i.e., the series is convergent in the interval,

$$-\frac{\pi}{2} < \theta < \frac{\pi}{2}. \tag{3.21}$$

An interesting feature of the expansion is that all of the coefficients, calculated to date, are positive except for that of the θ^2 term. Further, the coefficient of the n-th term has the structure [**2**],

$$\begin{aligned} Lc(\theta) &= 1 - \left(\frac{1}{2!}\right)\theta^2 + \left(\frac{1}{3}\right)\left(\frac{1}{4!}\right)\theta^4 + \dots \\ &\quad + D_n\left[\frac{\theta^{2n}}{3^{(n-1)}(2n)!}\right] + \dots, \end{aligned} \tag{3.22}$$

where D_n is an increasing positive function of n. Note that the first three positive D_n are

$$D_2 = 1, \quad D_3 = 5, \quad D_4 = 115. \tag{3.23}$$

In fact, the D_n appear to increase very fast and our efforts to find a formula for them have not yet succeeded. Since c $'(\theta)$ = -s(θ), the corresponding Taylor series for Ls(θ) = s(θ) is

$$\begin{aligned} Lc(\theta) = \theta &- \left(\frac{1}{3}\right)\left(\frac{1}{3!}\right)\theta^3 - \left(\frac{5}{9}\right)\left(\frac{1}{5!}\right)\theta^5 \\ &- \left(\frac{115}{27}\right)\left(\frac{1}{7!}\right)\theta^7 + O\left(\theta^9\right), \end{aligned} \tag{3.24}$$

and it converges over the interval given in Eq.(3.21).

4. Exact solutions in polar coordinates

Exact, closed form solutions can be constructed for the three Leah-functions. To see this, start with the first-integral, Eq.(2.1), and substitute into it the definitions of the Leah-functions,

$$\begin{cases} x(\theta) &= r(\theta)cos\theta \equiv Lc(\theta) \\ y(\theta) &= r(\theta)sin\theta \equiv Ls(\theta) \\ r(\theta) &= Ld(\theta). \end{cases} \tag{4.1}$$

Doing this gives after simplification the result,

$$(sin\theta)^2 r^2 + \left[\left(\frac{3}{2}\right)(cos\theta)^{\frac{4}{3}}\right] r^{\frac{4}{3}} - \left(\frac{3}{2}\right) = 0 \tag{4.2}$$

If we define z as

$$z = r^{\frac{2}{3}} \tag{4.3}$$

then in the variable z, Eq.(4.2) takes the form,

$$(sin\theta)^2 z^3 + \left[\left(\frac{3}{2}\right)(cos\theta)^{\frac{4}{3}}\right] z^2 - \left(\frac{3}{2}\right) = 0, \tag{4.4}$$

and this equation is a cubic polynomial equation in z. Using Descartes' rule of signs [**10**], it can be shown that Eq.(4.4) has a single, real, positive root; the other roots are either complex conjugates or negative. Since $r = z^{\frac{3}{2}}$, only the positive root is meaningful. An explicit expression can be written out for this root and it will be denoted by z^*. Note that z^* depends on θ; therefore, we have

$$r(\theta) = [z^*(\theta)]^{\frac{3}{2}}, \tag{4.5}$$

and the three Leah-functions take the form,

$$\begin{cases} Lc(\theta) &= c(\theta) = [z^*(\theta)]^{\frac{3}{2}} cos\theta \\ Ls(\theta) &= s(\theta) = [z^*(\theta)]^{\frac{3}{2}} sin\theta \\ Ld(\theta) &= [z^*(\theta)]^{\frac{3}{2}} \end{cases} \tag{4.6}$$

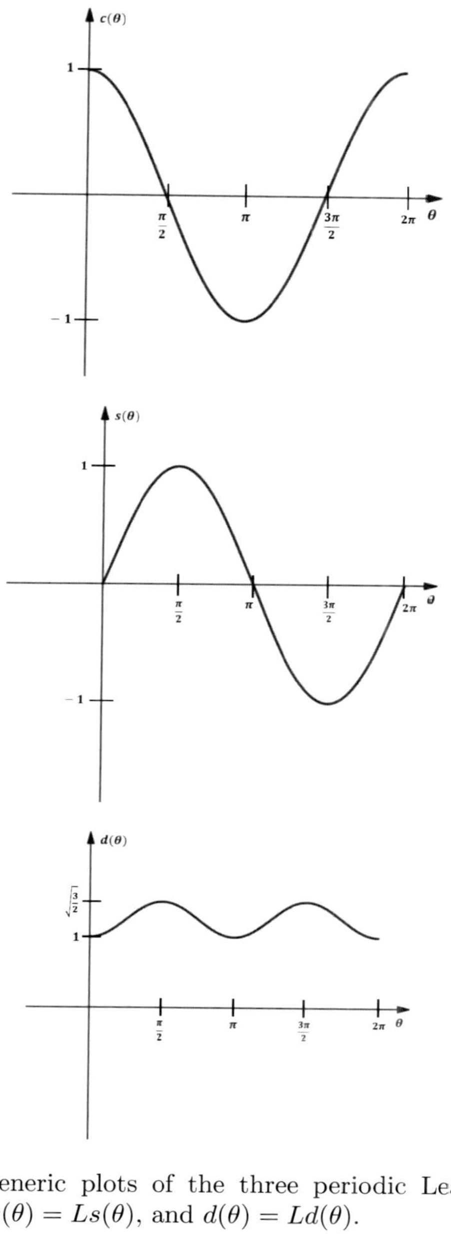

FIGURE 2. Generic plots of the three periodic Leah-functions: $c(\theta) = Lc(\theta), s(\theta) = Ls(\theta)$, and $d(\theta) = Ld(\theta)$.

5. Calculating numerical values of the Leah-functions

Numerical values may be calculated for the three periodic Leah-functions. The procedure is direct and can be accomplished by carrying out the following steps:

(1) Select the angle $\overline{\theta}$, where $0 < \overline{\theta} < \frac{\pi}{2}$.
(2) Solve for the positive root, z^*, in Eq.(4.4) .

(3) Calculate $r(\overline{\theta})$ using Eq.(4.5), i.e.,

$$r(\overline{\theta}) = \left[z^{*}\left(\overline{\theta}\right)\right]^{\frac{3}{2}}. \tag{5.1}$$

(4) Calculate $Lc(\overline{\theta})$, $Ls(\overline{\theta})$ and $Ld(\overline{\theta})$ using Eq.(4.6).

It should be noted that given values of $Lc(\theta)$, $Ls(\theta)$ and $Ld(\theta)$ in the interval, $0 < \theta < \frac{\pi}{2}$, values of these functions outside this interval may be determined from a use of the symmetry transformations listed in Eq.(2.5).

6. Discussion

The importance of the Leah-functions arises from the fact that they provide mathematical solutions to the physical problem where the force is proportional to the cube-root of the displacement [**1**, **3**, **4**, **6**]; see Eq.(1.1). This nonlinear oscillator differential equation has been extensively investigated [**3**, **4**, **5**, **6**] and is the standard model equation for one-dimension oscillating systems having a single term for the force which is raised to an odd power fraction [**6**], i.e.,

$$\frac{d^2x}{dt^2} = -x^p, \tag{6.1}$$

where

$$p = \frac{2n+1}{2m+1}; \quad n = (0, 1, 1, 2, \ldots), \quad m = (0, 1, 2, \ldots), \quad n < m. \tag{6.2}$$

For the CRO, $n = 0$ and $m = 1$. Observe that for this class of differential equations, all the solutions are periodic, but none of the standard pertubation procedures can be applied since they all rely on the presence of a linear term in the representation of the force [**6**, **11**] and such is not the case for the CRO equation. The following is a summary of the major results achieved in this work:

(i) defined geometrically the three periodic Leah-functions in terms of the properties of the curve, $y^2 + \left(\frac{3}{2}\right) x^{\frac{4}{3}} = \frac{3}{2}$;

(ii) explicit exact formulas for the Leah-functions expressed in terms of a polar representation;

(iii) presented the steps required to numerically evaluate the Leah-functions;

(iv) shown the existence of two Pythagorean identities for the Leah-functions; see Eqs.(3.2) and (3.3).

Future work on this topic will include investigations on the following issues:

(a) The Fourier series representations for the three Leah-functions.

(b) The discovery of sum formulas for the Leah-functions, i.e., relationships similar to the trigonometric expressions,

$$\begin{cases} sin(x+y) & = (sinz)(cosy) + (cosx)(siny) \\ cos(x+y) & = (cosx)(cosy) - (sinx)(siny). \end{cases} \tag{6.3}$$

The resolution of these two issues has to date proven to be very difficult.

References

[1] Ronald E. Mickens and Sandra A. Rucker, *Summary of some properties of the Leah-cosine function*, Dynamic systems and applications. Vol. 6, Dynamic, Atlanta, GA, 2012, pp. 267–269. MR3077427

[2] J. Mann, A. Scrouse and R.E. Mickens, *Preliminary investigation on the properties of the Leah-cosine and -sine functions, Abstracts of Papers Presented to the American Mathematical Society*, **32 #1**, (Winter 2012), 138.

[3] H. L. Zhang, *Application of He's frequency amplitude formulation to an $x^{1/3}$ force nonlinear oscillator*, International Journal of Nonlinear Sciences and Numerical Simulation, **9 #3** (2008), 297–300.

[4] Augusto Beléndez, *Homotopy perturbation method for a conservative $x^{1/3}$ force nonlinear oscillator*, Comput. Math. Appl. **58** (2009), no. 11-12, 2267–2273, DOI 10.1016/j.camwa.2009.03.011. MR2557356

[5] B.M. Ikramul Haque and M.M. Ayub Hassain, *An effective solution of the cube-root truly nonlinear oscillator: Extended iteration procedure*, International Journal of Differential Equations, **6** (2021), Article ID 7819209, doi.org/10.1155/2021/7819209.

[6] Ronald E. Mickens, *Truly nonlinear oscillations*, World Scientific Publishing Co. Pte. Ltd., Hackensack, NJ, 2010, DOI 10.1142/9789814291668. MR2583028

[7] Sandra A. Rucker, *Leah-cosine and -sine functions: definitions and elementary properties*, Mathematics of continuous and discrete dynamical systems, Contemp. Math., vol. 618, Amer. Math. Soc., Providence, RI, 2014, pp. 265–280, DOI 10.1090/conm/618/12341. MR3223118

[8] Paul F. Byrd and Morris D. Friedman, *Handbook of elliptic integrals for engineers and scientists*, Die Grundlehren der mathematischen Wissenschaften, Band 67, Springer-Verlag, New York-Heidelberg, 1971. Second edition, revised. MR0277773

[9] Professor G. Edgar, *Private communications to Ronald E. Mickens*, Department of Mathematics, James Madison University, Harrisonburg, VA, (2010-2012).

[10] Frank W. J. Olver, Daniel W. Lozier, Ronald F. Boisvert, and Charles W. Clark (eds.), *NIST handbook of mathematical functions*, U.S. Department of Commerce, National Institute of Standards and Technology, Washington, DC; Cambridge University Press, Cambridge, 2010. With 1 CD-ROM (Windows, Macintosh and UNIX). MR2723248

[11] Ronald E. Mickens, *Generalized trigonometric and hyperbolic functions*, CRC Press, Boca Raton, FL, 2019, DOI 10.1201/9780429446238. MR3931307

Department of Mathematical sciences, Clark Atlanta University, Atlanta, Georgia 30314

Email address: `srucker@cau.edu`

Contemporary Mathematics
Volume **793**, 2024
https://doi.org/10.1090/conm/793/15884

Discursion on a paper of R. E. Mickens and J. E. Wilkins, Jr.

Isom H. Herron

Dedicated to Ronald E. Mickens for his 80th birthday.

ABSTRACT. The heat equation and a nonlinear counterpart predict the temperature at the end of a semi-infinite rod, when the heat flux is specified there. Both equations have reductions to ordinary differential equations which simplify their solutions. For the nonlinear equation, the problem becomes $y''(z) = y(z)^2 - zy(z)y'(z)$, $y'(0) = -\sqrt{3}, y(\infty) = 0$ on $[0, \infty)$. A variational approximation to the initial value $y(0)$ of the nonlinear boundary value problem was found by employing an integral constraint suggested by the equation itself. The approximation is more accurate for the initial value than that of Mickens and Wilkins but misses their pointwise description of the solution.

1. Introduction: the origins of the problem

In a very interesting paper published in 2019, two masters, our honoree, Ronald Mickens, along with his friend and mentor, the late J. Ernest Wilkins, Jr., showed results in which they investigated an approximation to the solution of a nonlinear second order ordinary differential equation (ODE) for a function $y(z)$ [**2**]. The title of that paper, which be will be referred to as MW, contains the differential equation, mentioned in the abstract (see equation 2.33 below). The major task was to determine the value $y(0)$.

The partial differential equation (PDE) which led to the work of MW is a nonlinear diffusion equation. In what follows, a comparison is made with two treatments of the linear heat equation one of which, by similarity methods, is applicable to the nonlinear PDE. The other method, by integral transforms is only applicable to linear equations. However, its value for the heat equation provides further evidence of the behavior of the comparable solutions. Finally, in this work, a variational method is introduced in order to determine $y(0)$ as the stationary value of a functional. The value obtained thereby compares favorably with an approximate value obtained by Mickens and Wilkins and the known numerical value.

2020 *Mathematics Subject Classification.* 34B18, 34A45, 34B40, 35A24, 35Q79.
Key words and phrases. boundary-value, variational, approximation.
This paper is in final form and no version of it will be submitted for publication elsewhere.

2. The governing diffusion equations

The presentation here could be entirely taken from earlier treatments [**11**, §5.3], [**2**], based on work of Dresner [**3**, §4.7]. It has a physical modeling origin. The problems of heat conduction in solids, though classical mathematically, continue to be treated because of modern applications [**12**]. The current nonlinear partial differential equation arose in the problem of the thermal expulsion of a fluid from a long, semi-infinite ($x \geq 0$), slender heated tube. The temperature is $u(x,t)$ to be determined for $t > 0$. The connection of it to the ordinary differential equation is described subsequently. The particular problem of their interest corresponds to heat conduction wherein the specific heat is proportional to the temperature (i.e. u), which is sometimes the case for metals at low temperatures, while the thermal conductivity is constant; here the boundary condition at $x = 0$ is a Neumann condition corresponding to constant heat flux at that end. This is a practical problem in some contexts, where the heat flux may be prescribed more easily than the temperature.

As with other treatments, begin with the nonlinear partial differential equation for $u(x,t)$:

$$u\frac{\partial u}{\partial t} = \frac{\partial^2 u}{\partial x^2}, x > 0, t > 0 \tag{2.1}$$

where

$$u > 0 \text{ for } 0 < x < \infty,\ \ 0 < t < \infty,$$

and the following initial and boundary-value conditions are required to hold

$$u(x,0) = 0,\ \ x > 0, \tag{2.2}$$

$$u(\infty,t) = 0,\ \ t > 0 \tag{2.3}$$

$$\frac{\partial u}{\partial x}(0,t) = -1,\ \ t > 0. \tag{2.4}$$

The nonlinear equation (2.1) bears a resemblance to its familiar linear counterpart, to which we make brief digression.

2.1. A brief look at the heat equation.

The non-dimensional form of the heat equation is

$$\frac{\partial u}{\partial t} = \frac{\partial^2 u}{\partial x^2}. \tag{2.5}$$

Owing to the nature of the side conditions, not all methods of solution appear to be immediately applicable. We will examine two possible types of solutions, separable solutions and similarity solutions.

Separable solutions. As the basis of Fourier's original analysis it was found that separable solutions of the form

$$u(x,t) = a_n \cos(n\pi x)\, e^{-n^2\pi^2 t}, n = 0, 1, 2, \ldots \tag{2.6}$$

formally satisfy (2.5). However, since the $x-$interval is semi-infinite, to complete the solution of (2.5) given (2.2)-(2.4), a summation of (2.6) is not appropriate. Another approach is to use a Laplace transform in t [**1**, §5.4.1],

$$U(x,s) = \mathcal{L}_t\{u(x,t)\} = \int_0^\infty u(x,t)\, e^{-st} dt. \tag{2.7}$$

A generalization of (2.6), where $(n\pi)^2$ is replaced by a continuous variable s and integration over s replaces summation over n is the means of defining the inverse Laplace transform.

Applying $\mathcal{L}_t$ to (2.5) and performing the usual integrations provides the expression (2.7) which turns out to be

$$U(x,s) = \frac{e^{-x\sqrt{s}}}{s^{3/3}}.$$

To find its inverse, a suitable table or computer system may be employed for this evaluation, or of course contour integration could be used. The explicit solution is found to be expressible as

$$u(x,t) = \mathcal{L}_t^{-1}\left\{\frac{e^{-x\sqrt{s}}}{s^{3/3}}\right\} = \frac{1}{\sqrt{\pi}}\int_0^t \left[\frac{e^{-x^2/(4\tau)}}{\sqrt{\tau}}\right] d\tau, \tag{2.8}$$

within a factor an antiderivative of the **heat kernel**

$$K(x,t) = \frac{1}{\sqrt{4\pi t}} e^{-x^2/(4t)}, \tag{2.9}$$

which is called the *fundamental solution* to the heat equation for the infinite rod [**4**, §5-4].

Still another method of obtaining (2.8) would be by a reflection method about $x = 0$, using (2.9). See [**4**, §5-4].

Now expand (2.8) in terms of a special function as

$$u(x,t) = 2\sqrt{\frac{t}{\pi}} e^{-x^2/t} - x \operatorname{erfc}\left(\frac{x}{2\sqrt{t}}\right). \tag{2.10}$$

A plot of this solution is easy to obtain (See Figure 1). Though the flux at the boundary $x = 0$ is constant in this problem, the temperature is not and is thus from (2.10):

$$u(0,t) = 2\sqrt{\frac{t}{\pi}}, t > 0. \tag{2.11}$$

Alternatively, the less used cosine transform, in x, can be employed [**1**]. This may be viewed as a different generalization of the separable solution (2.6), replacing $n\pi$ by a continuous variable ω, and integration over ω now replacing summation over n in the definition of the inverse transform.

Define

$$U_c(\omega,t) = \sqrt{\frac{2}{\pi}} \int_0^\infty u(x,t)\cos(\omega x)\, dx, \tag{2.12}$$

then inversely

$$u(x,t) = \sqrt{\frac{2}{\pi}} \int_0^\infty U_c(\omega,t)\cos(\omega x)\, d\omega. \tag{2.13}$$

This is well suited to solve (2.5) with the side conditions (2.2)-(2.4), because of the occurrence of the derivative in (2.4). The solution turns out to be expressible as

$$u(x,t) = \frac{2}{\pi}\int_0^\infty \left(\frac{1-e^{-\omega^2 t}}{\omega^2}\right)\cos(\omega x)\, d\omega. \tag{2.14}$$

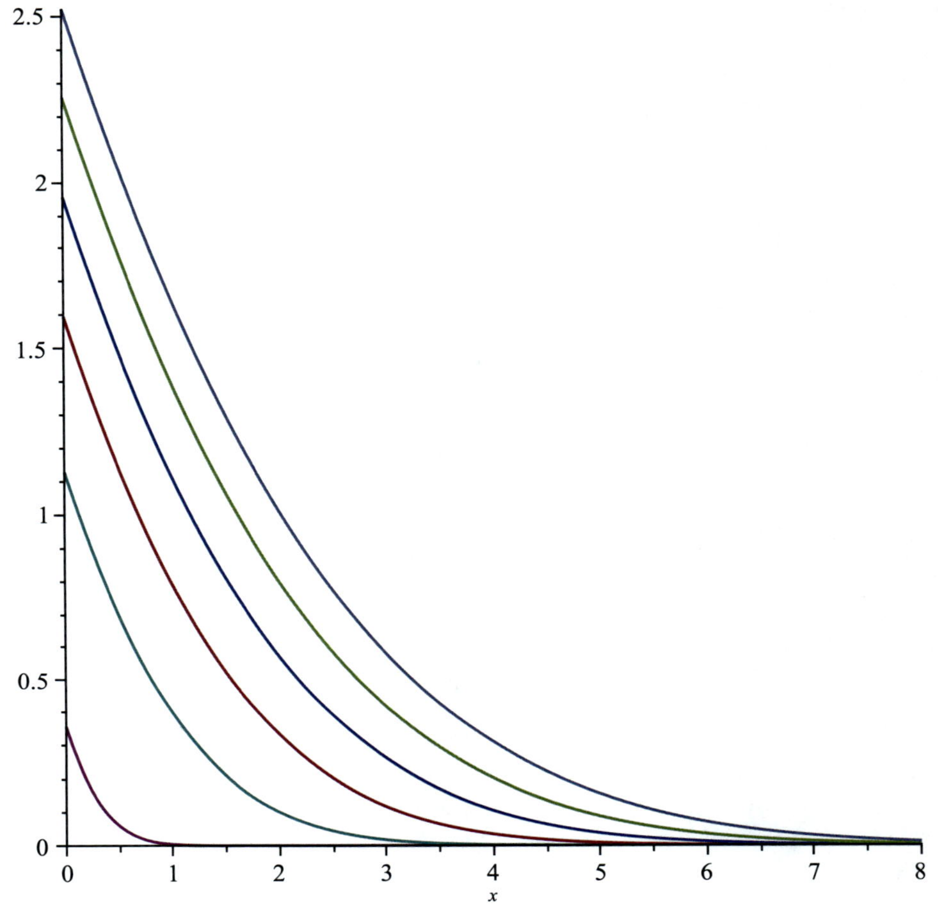

FIGURE 1. The solution $u(x,t)$ for $t = \frac{1}{10}$, 1, 2, 3, 4, 5 from bottom to top.

The result is, as before

$$u(0,t) = \frac{2}{\pi}\int_0^\infty \left(\frac{1-e^{-\omega^2 t}}{\omega^2}\right) d\omega = 2\sqrt{\frac{t}{\pi}}, t > 0. \tag{2.15}$$

The condition (2.4) is confirmed from (2.14):

$$\begin{aligned}\frac{\partial u}{\partial x}(x,t) &= -\frac{2}{\pi}\int_0^\infty \left(\frac{1-e^{-\omega^2 t}}{\omega}\right)\sin(\omega x)\, d\omega \\ &= -\operatorname{erfc}\left(\frac{x}{2\sqrt{t}}\right), t > 0, \\ \frac{\partial u}{\partial x}(0,t) &= -1.\end{aligned}$$

Similarity solutions. Alternatively, because of the special nature of the boundary conditions, similarity solutions might be useful. A more general approach, embraced by MW will be developed later.

For purposes of illustration, suppose that instead of (2.4), the condition at $x = 0$ is given by

$$u(0,t) = 1. \tag{2.16}$$

This particular boundary condition arises in a study of the translating motion of a plate in a viscous fluid [**9**, §5.3.2]. Traditionally, authors have suggested to search for a solution of the form [**8**, §10.4]

$$u(x,t) = f\left(x/t^{1/2}\right),$$

where $\xi = x/t^{1/2}$ is called the **similarity variable,** suggested by a dimensional analysis of the heat equation.

Since

$$\frac{\partial u}{\partial t} = -\frac{1}{2}\frac{x}{t^{1/2}}f'(\xi),$$
$$\frac{\partial u}{\partial x} = \frac{1}{t^{1/2}}f'(\xi),$$

and

$$\frac{\partial^2 u}{\partial x^2} = \frac{1}{t}f''(\xi),$$

a linear ODE follows from (2.5) which is first order for f':

$$f''(\xi) = -\frac{1}{2}f'(\xi),$$

whose general solution is

$$f' = c_1 e^{-\frac{\xi^2}{2}}.$$

Another integration gives the similarity solution of the heat equation

$$u(x,t) = f\left(x/t^{1/2}\right) = c_1 \int_0^{x/t^{1/2}} e^{-\sigma^2/4} d\sigma + c_2. \tag{2.17}$$

The boundary conditions, are applied since $\xi = 0$ for $x = 0$, $\xi \to \infty$ as $x \to \infty$ for $t > 0$, and $\xi \to \infty$ as $t \to 0$ for $x > 0$.

$$f(0) = 1, f \to 0 \text{ as } \xi \to \infty.$$

This gives $c_2 = 1, c_1 = -\frac{1}{\sqrt{\pi}}$.

2.2. The nonlinear diffusion equation. Turning to the nonlinear PDE (2.1), we first search for possible separable solutions. Put

$$u(x,t) = \phi(x)G(t)$$

in (2.1) to obtain

$$(\phi G)(\phi G') = \phi'' G.$$

For $G \neq 0$, this leads to

$$\frac{\phi''}{\phi^2} = G' = k,$$

some separation constant. Thus $G = kt$ is a linear function while

$$\phi'' = k\phi^2 \tag{2.18}$$

is a nonlinear second order ODE. If a solution is sought satisfying (2.2)-(2.4), that is not possible since the last condition is to be independent of time. However, generic solutions of (2.18) can be studied in terms of the Weierstrass p−function with argument x (`https://dlmf.nist.gov/23`).

2.3. Similarity transformations. The nonlinear equation (2.1) may also be amenable to a similarity transformation, especially since it has only two terms. The similarity approach here is more systematic. With a view towards studying more general PDEs an overall change of variables is envisioned.

DEFINITION ([**11**, §4.2]). *A one parameter family of stretching transformations.* Consider the change of variables on xtu−space of the form

$$\bar{x} = \varepsilon^a x, \bar{t} = \varepsilon^b t, \bar{u} = \varepsilon^c u, \tag{2.19}$$

where a, b, c are constants and ε is a real parameter restricted to some open interval I containing $\varepsilon = 1$.

The idea is to write the original equation in terms of the stretched variables.

From (2.19) we observe that

$$u_t = \frac{\partial u}{\partial \bar{t}}\frac{\partial \bar{t}}{\partial t} = \varepsilon^{-c}\varepsilon^b \bar{u}_{\bar{t}} = \varepsilon^{b-c}\bar{u}_{\bar{t}}, \tag{2.20}$$

$$u_{tt} = \varepsilon^{2b-c}\bar{u}_{\bar{t}\bar{t}}, \tag{2.21}$$

$$u_x = \frac{\partial u}{\partial \bar{x}}\frac{\partial \bar{x}}{\partial x} = \varepsilon^a \bar{u}_{\bar{x}}\varepsilon^{-c} = \varepsilon^{a-c}\bar{u}_{\bar{x}}, \tag{2.22}$$

$$u_{xx} = \varepsilon^{2a-c}\bar{u}_{\bar{x}\bar{x}}. \tag{2.23}$$

To abbreviate and streamline the derivations, set

$$p = u_x, q = u_t, r = u_{xx}, v = u_{tt}. \tag{2.24}$$

Then using (2.20)-(2.23) in (2.24) find that

$$\begin{aligned} \bar{p} = \bar{u}_{\bar{x}} &= \varepsilon^{c-a}p, \\ \bar{q} = \bar{u}_{\bar{t}} &= \varepsilon^{c-b}q, \\ \bar{u}_{\bar{x}\bar{x}} = \bar{r} &= \varepsilon^{c-2a}r, \\ \bar{u}_{\bar{t}\bar{t}} = \bar{v} &= \varepsilon^{c-2b}u_{tt} = \varepsilon^{c-2b}v. \end{aligned}$$

There is a theorem of Logan [**11**, §4.2]: If the second order PDE

$$G(x, t, u, u_x, u_t, u_{xx}, u_{xt}, u_{tt}) = 0 \tag{2.25}$$

is *invariant* under the one-parameter family T_ε of stretching transformations (2.19), the transformation

$$u = t^{c/b}y(z), z = \frac{x}{t^{a/b}} \tag{2.26}$$

reduces the PDE (2.25) to a second order ODE of the form

$$g(z, y, y', y'') = 0.$$

The method Logan advocates is to write the PDE in abbreviated form as

$$G(\bar{x}, \bar{t}, \bar{u}, \bar{p}, \bar{q}, \bar{r}, \bar{s}, \bar{v}) = f(\varepsilon)\, G(x, t, u, p, q, r, s, v)$$

to demonstrate the invariance, where $s = u_{xt}$.

We have already seen that the heat equation is amenable to a similarity transformation with special boundary conditions. Logan [**11**, §4.2, Exercise 2] poses the heat equation (2.5) with conditions (2.2)-(2.4) as solvable by the similarity method. We provide that solution here. The abbreviated PDE is simply

$$q - r = 0. \tag{2.27}$$

The stretching transformation is found from

$$\bar{q} - \bar{r} = \varepsilon^{c-b} q - \varepsilon^{c-2a} r = \varepsilon^{c-b} (q - r) = 0,$$

provided that

$$b = 2a.$$

The family of possible transformations are

$$u = t^{c/2a} y(z), z = \frac{x}{\sqrt{t}}. \tag{2.28}$$

Restrictions on the constants a and c can be determined by the initial and boundary conditions.

Substitute into (2.27) to get

$$y'' + \frac{z}{2} y' - \frac{c}{2a} y = 0.$$

Apply the boundary condition (2.4) at the left:

$$\frac{\partial u}{\partial x}(0,t) = -1 \ (t > 0) \Longrightarrow \frac{t^{c/2a}}{t^{1/2}} y'(0) = -1,$$

which is possible only if

$$c = a.$$

The initial condition (2.2):

$$u(x,0) = 0, x > 0,$$

and end condition

$$\lim_{x \to \infty} u(x,t) = 0, t > 0,$$

both lead to

$$\lim_{z \to \infty} y = 0. \tag{2.29}$$

So a solution of the problem

$$y'' + \frac{z}{2} y' - \frac{1}{2} y = 0 \tag{2.30}$$

is desired satisfying $y'(0) = -1$ and (2.29). From the earlier discussion we know the ultimate outcome, but to further illustrate the method, employ a cosine transform (2.12) on (2.30), since a boundary problem is to be solved. This method works even though (2.30) has a variable coefficient. The resulting representation for the solution is

$$y(z) = \frac{2}{\pi} \int_0^\infty \left(\frac{1 - e^{-\omega^2}}{\omega^2} \right) \cos(\omega z) \, d\omega.$$

Again observe that $y(0) = 2/\sqrt{\pi}$, in agreement with (2.11), (2.15) and (2.28).

Next we apply similarity methods to the nonlinear diffusion equation (2.1).

In this case with (2.24), the abbreviated PDE is

$$uq - r = 0. \tag{2.31}$$

Look for a stretching transformation (2.19) under which (2.31) is invariant. Obtain

$$\bar{u}\bar{q} - \bar{r} = (\varepsilon^c u)\left(\varepsilon^{c-b} q\right) - \varepsilon^{c-2a} r = 0.$$

So, require that

$$2c - b = c - 2a \Rightarrow c = b - 2a.$$

According to the theorem, the expected similarity transformation is

$$u = t^{(b-2a)/b} y(z), z = \frac{x}{t^{a/b}}.$$

Restrictions on the constants a and b can be determined by the initial and boundary conditions. Computing u_x leads to

$$u_x = t^{(b-3a)/b} y' \Rightarrow u_x(0,t) = t^{(b-3a)/b} y'(0) = -1.$$

For the boundary condition to be independent of t, take $b - 3a = 0$.

The authors therefore employ the similarity transformations

$$u(x,t) = t^{1/3} y(z), z = \frac{x}{\sqrt{3} t^{1/3}}, \quad t > 0. \tag{2.32}$$

(The factor $\sqrt{3}$ is taken for esthetic reasons.) Putting these assumed forms into (2.1) leads to

$$y''(z) = (y(z))^2 - z y(z) y'(z). \tag{2.33}$$

The resulting boundary conditions on $y(z)$ are

$$y'(0) = -\sqrt{3}, \lim_{z \to \infty} y(z) = 0. \tag{2.34}$$

with the requirements

$$y(z) > 0, y'(z) < 0, \tag{2.35}$$

where $y'(z) = dy(z)/dz$, etc.

One notices from (2.32) that at the end $x = 0$, the temperature increases at a rate proportional to $t^{1/3}$, while the heat equation predicts a rate of increase proportional to $t^{1/2}$ in (2.11).

2.4. The results of Mickens and Wilkins. It has been clear since the original derivation of the equation, that a simple, traditional, analytical solution to the problem (2.33)–(2.34) is not possible. However "features and properties" of the solution were easily found [**2**, §3]. It was several of these that MW made use of in constructing their approximate solution. Without explicitly saying it, the principle employed was that of "dynamic consistency" [**5**],[**6**].

In particular, MW were able to show:

(*i*) Bounds on the solution are

$$0 < y(z) < \frac{2}{z^2}, z > 0. \tag{2.36}$$

(*ii*) Setting $z = 0$ in (2.33) gives

$$y''(0) = (y(0))^2. \tag{2.37}$$

(*iii*) By integrating (2.33) from $z = 0$ to $z = \infty$:

$$\int_0^\infty (y(z))^2 dz = \frac{2}{\sqrt{3}}.$$

Next, MW made a rational function approximation

$$y_a(z) = \frac{A}{1 + Bz + Cz^2} \tag{2.38}$$

which was useful in arriving at a numerical approximation to $y(0)$.

This was accomplished by finding the values of (A, B, C) in terms of $y(0) = A$, and $B = \sqrt{3}/A$ by (2.34), $C = A/2$ by noting that $y_a \sim 2/z^2$ as $z \to \infty$ by (2.35) and (2.36). Putting (2.38) into (2.33) results in a cubic equation for A. The one real root is $A = 1.537656171$.

3. The variational formulation

The derivation of the functionals in section 2.4 is suggestive of another possibility.

For instance, if both sides of (2.33) are multiplied by $y(z)$ and integrated from 0 to ∞, one finds that

$$\begin{aligned}\int_0^\infty y(z)\, y''(z)\, dz &= \int_0^\infty \left[(y(z))^3 - z\,(y(z))^2\, y'(z)\right] dz \\ &= \int_0^\infty \left[(y(z))^3 - z\frac{d}{dz}\frac{(y(z))^3}{3}\right] dz,\end{aligned}$$

and after integration by parts and application of (2.34),

$$y(0)\sqrt{3} - \int_0^\infty (y'(z))^2\, dz = \int_0^\infty \frac{4}{3}(y(z))^3\, dz, \tag{3.1}$$

so

$$y(0) = \frac{1}{\sqrt{3}}\int_0^\infty \left[(y'(z))^2 + \frac{4}{3}(y(z))^3\right] dz. \tag{3.2}$$

MW note that in (2.32), "...the factor, $\sqrt{3}$, is introduced into the definition of z for convenience; any constant will do, but this particular one is that which is used in the research literature [**3**], [**11**]". So we generalize it here, to see what effect it may have to replace it by b. As a result the left boundary condition in (2.34) would read $y'(0) = -b$, and this in (3.1) suggests a more general functional.

Furthermore, we introduce a class of functions related to y, called $w(z)$ which satisfy the same conditions namely, w' belongs to $\mathfrak{L}^2(0, \infty)$, $w(0)$ is finite and an analogous functional to (3.2) may be defined in anticipation of the introduction of variational methods. (See [**7**], for example.)

The proposition that is to be pursued here then, is that a suitable approximation to the solution may be found from the *stationary value* [**10**] of the functional

$$w(0) = \frac{1}{b}\int_0^\infty \left[(w'(z))^2 + \frac{4}{3}(w(z))^3\right] dz, \tag{3.3}$$

where $b > 0$ is a constant.

Perform a *first variation* of (3.3) to obtain:

$$\begin{aligned}\delta w(0) &= \frac{1}{b}\int_0^\infty \left[2w'(z)\,\delta w'(z) + 4(w(z))^2\,\delta w(z)\right] dz \\ &= \frac{2w'(z)\,\delta w(z)}{b}\bigg|_0^\infty - \frac{1}{b}\int_0^\infty \left[2w''(z) - 4(w(z))^2\right]\delta w(z)\, dz. \qquad (3.4)\end{aligned}$$

As $\delta w(0) = 0$, and $\delta w, w' \to 0$ as $z \to \infty$:

$$\frac{1}{b}\int_0^\infty \left[2w''(z) - 4(w(z))^2\right]\delta w(z)\,dz = 0,$$

leading to the Euler-Lagrange equation for the *stationary solution* w:

$$w'' = 2w^2. \tag{3.5}$$

This may be converted to a first order differential equation by multiplying both sides by w' :

$$w'w'' = 2w^2w'$$

to obtain

$$\frac{d}{dz}\left(\frac{w'^2}{2}\right) = \frac{d}{dz}\left(\frac{2w^3}{3}\right)$$

which integrates to

$$\frac{1}{2}w'^2 = \frac{2}{3}w^3 + c_1.$$

Since we must have $w \to 0, w' \to 0$ as $z \to \infty$, then $c_1 = 0$. We conclude that

$$\frac{dw}{dz} = \pm\sqrt{\frac{4}{3}w^3}. \tag{3.6}$$

For $w > 0$ and $dw/dz < 0$, take the negative sign. There is some freedom in finding the the stationary solution since, effectively, no boundary conditions were employed at $z = 0$. In particular, the value of $w'(0)$ does not need to be specified. Then separating variables in (3.6):

$$-w^{-3/2}dw = \frac{2}{\sqrt{3}}dz,$$

integrating

$$2w^{-1/2} = \frac{2}{\sqrt{3}}(z + c_2)$$

and

$$w = \frac{3}{(z + c_2)^2}. \tag{3.7}$$

Now using (3.3) evaluate

$$\begin{aligned} w(0) &= \frac{1}{b}\int_0^\infty \left[(w'(z))^2 + \frac{4}{3}(w(z))^3\right]dz \\ &= \frac{1}{b}\int_0^\infty \left[\left(\frac{-6}{(z + c_2)^3}\right)^2 + \frac{4}{3}\left(\frac{27}{(z + c_2)^6}\right)\right]dz = \frac{72}{5b(c_2)^5} \\ &= \frac{3}{(c_2)^2}, \text{ by (3.7).} \end{aligned}$$

So

$$c_2 = \left(\frac{24}{5b}\right)^{1/3}.$$

The objective is to estimate

$$w(0) = \frac{3}{c_2^2} = 3\left(\frac{5b}{24}\right)^{2/3} = \left(\frac{75b^2}{64}\right)^{1/3}.$$

So, when $b = \sqrt{3}$, as in (3.1), then

$$w(0) = 1.52055 \equiv y(0)_{\text{approx}} \tag{3.8}$$

This compares favorably to the computed numbers $y(0) = 1.5116$ found by Dresner [**3**, §4.7] and $y(0) = 1.5111$ by Logan [**11**, §5.3]; it is within 0.63% of the latter. The approximate value computed by MW was $y_a(0) = 1.5377$, that is a 1.76% difference of Logan's result.

In conclusion, it should be mentioned that the stationary function (3.7) is not a good pointwise approximation to the solution $y(z)$ on $(0, \infty)$ satisfying (2.33)-(2.34), because $w'(0) = -5\sqrt{3}/4$, though it does satisfy (2.35). One notices, interestingly, that each of the approximate functions is a rational function, the one employed here and those of Dresner [**3**, §4.7] and MW; all expressible as the reciprocal of some quadratic. In that sense, each of the obtained approximate values of $y(0)$ are illuminated, in light of the bounds on y stated in (2.36).

Acknowledgment

The author thanks Prof. Abba Gummel for his invitation to submit this chapter.

References

[1] M. Ya. Antimirov, A. A. Kolyshkin, and Rémi Vaillancourt, *Applied integral transforms*, CRM Monograph Series, vol. 2, American Mathematical Society, Providence, RI, 1993, DOI 10.1090/crmm/002. MR1203773

[2] R. E. Mickens and J. Ernest Wilkins, Jr., *Estimation of* $y(0)$ *for the boundary-value problem:* $y''(z) = y(z)^2 - zy(z)y'(z), y'(0) = -\sqrt{3}, y(\infty) = 0$, Comm. in Appl. Anal. 23 (2019), 137-146

[3] Lawrence Dresner, *Similarity solutions of nonlinear partial differential equations*, Research Notes in Mathematics, vol. 88, Pitman (Advanced Publishing Program), Boston, MA, 1983. MR721168

[4] Ronald B. Guenther and John W. Lee, *Partial differential equations of mathematical physics and integral equations*, Dover Publications, Inc., Mineola, NY, 1996. Corrected reprint of the 1988 original. MR1384540

[5] Ronald E. Mickens, *Dynamic consistency: a fundamental principle for constructing nonstandard finite difference schemes for differential equations*, J. Difference Equ. Appl. **11** (2005), no. 7, 645–653, DOI 10.1080/10236190412331334527. American Mathematical Society Special Session on Difference Equations and Discrete Dynamics. MR2173250

[6] Ronald E. Mickens and Isom H. Herron, *Approximate rational solutions to the Thomas-Fermi equation based on dynamic consistency*, Appl. Math. Lett. **116** (2021), Paper No. 106994, 7, DOI 10.1016/j.aml.2020.106994. MR4197233

[7] G. A. Bliss, *An Integral Inequality*, J. London Math. Soc. **5** (1930), no. 1, 40–46, DOI 10.1112/jlms/s1-5.1.40. MR1574997

[8] R. C. Haberman, *Applied Partial Differential Equations*, Fifth edition, Pearson, Boston, 2013

[9] Isom H. Herron and Michael R. Foster, *Partial differential equations in fluid dynamics*, Cambridge University Press, Cambridge, 2014. Paperback edition of the 2008 original [MR2441490]. MR3468027

[10] Francis B. Hildebrand, *Methods of applied mathematics*, 2nd ed., Dover Publications, Inc., New York, 1992. MR1160353

[11] J. David Logan, *An introduction to nonlinear partial differential equations*, 2nd ed., Pure and Applied Mathematics (Hoboken), Wiley-Interscience [John Wiley & Sons], Hoboken, NJ, 2008, DOI 10.1002/9780470287095. MR2397521

[12] M. N. Özişik, *Boundary Value Problems in Heat Conduction*, Dover, New York, 1968

Department of Mathematical Sciences, Rensselaer Polytechnic Institute, Troy, New York 12180

Current address: 1845 Audubon Dr. NE, Atlanta, GA 30329

Email address: herroi@rpi.edu,iherron321@gmail.com

Selected Published Titles in This Series

793 **Abba Gumel, Editor,** Mathematical and Computational Modeling of Phenomena Arising in Population Biology and Nonlinear Oscillations, 2024

792 **Shaoming Guo, Zane Kun Li, and Brian Street, Editors,** Recent Developments in Harmonic Analysis and its Applications, 2024

791 **Mee Seong Im, Bach Nguyen, and Arik Wilbert, Editors,** Algebraic and Topological Aspects of Representation Theory, 2024

790 **Peter Crooks and Alexandru I. Suciu, Editors,** Compactifications, Configurations, and Cohomology, 2023

789 **I. S. Krasil′shchik, A. B. Sossinsky, and A. M. Verbovetsky, Editors,** The Diverse World of PDEs, 2023

788 **I. S. Krasil′shchik, A. B. Sossinsky, and A. M. Verbovetsky, Editors,** The Diverse World of PDEs, 2023

787 **Hemen Dutta, Editor,** Mathematical Modelling, 2023

786 **Hemen Dutta, Editor,** Mathematical Modelling, 2023

785 **A. Leroy and S. K. Jain, Editors,** Algebra and Coding Theory, 2023

784 **Dinh-Liem Nguyen, Loc Hoang Nguyen, and Thi-Phong Nguyen, Editors,** Advances in Inverse Problems for Partial Differential Equations, 2023

783 **Alla Detinko, Michael Kapovich, Alex Kontorovich, Peter Sarnak, and Richard Schwartz, Editors,** Computational Aspects of Discrete Subgroups of Lie Groups, 2023

782 **Galina Filipuk, Alberto Lastra, and Sławomir Michalik, Editors,** Recent Trends in Formal and Analytic Solutions of Diff. Equations, 2023

781 **Emanuel Indrei, Diego Marcon, and Levon Nurbekyan, Editors,** Geometric and Functional Inequalities and Recent Topics in Nonlinear PDEs, 2023

780 **Erik Koelink, Stefan Kolb, Nicolai Reshetikhin, and Bart Vlaar, Editors,** Hypergeometry, Integrability and Lie Theory, 2022

779 **Samuele Anni, Valentijn Karemaker, and Elisa Lorenzo García, Editors,** Arithmetic, Geometry, Cryptography, and Coding Theory 2021, 2022

778 **Carlos Galindo, Alejandro Melle Hernández, Julio José Moyano-Fernández, and Wilson A. Zúñiga-Galindo, Editors,** p-Adic Analysis, Arithmetic and Singularities, 2022

777 **Bang-Yen Chen, Nicholas D. Brubaker, Takashi Sakai, Bogdan D. Suceavă, Makiko Sumi Tanaka, Hiroshi Tamaru, and Mihaela B. Vajiac, Editors,** Differential Geometry and Global Analysis, 2022

776 **Aaron Wootton, S. Allen Broughton, and Jennifer Paulhus, Editors,** Automorphisms of Riemann Surfaces, Subgroups of Mapping Class Groups and Related Topics, 2022

775 **Fernando Galaz-García, Cecilia González-Tokman, and Juan Carlos Pardo Millán, Editors,** Mexican Mathematicians in the World, 2021

774 **Randall J. Swift, Alan Krinik, Jennifer M. Switkes, and Jason H. Park, Editors,** Stochastic Processes and Functional Analysis, 2021

773 **Nicholas R. Baeth, Thiago H. Freitas, Graham J. Leuschke, and Victor H. Jorge Pérez, Editors,** Commutative Algebra, 2021

772 **Anatoly M. Vershik, Victor M. Buchstaber, and Andrey V. Malyutin, Editors,** Topology, Geometry, and Dynamics, 2021

771 **Nicolás Andruskiewitsch, Gongxiang Liu, Susan Montgomery, and Yinhuo Zhang, Editors,** Hopf Algebras, Tensor Categories and Related Topics, 2021

770 **Stéphane Ballet, Gaetan Bisson, and Irene Bouw, Editors,** Arithmetic, Geometry, Cryptography and Coding Theory, 2021

For a complete list of titles in this series, visit the AMS Bookstore at **www.ams.org/bookstore/conmseries/**.